# Quantum Theory of Scattering

Ta-You Wu
Takashi Ohmura

Dover Publications, Inc.
Mineola, New York

*Bibliographical Note*

iThis Dover edition, first published in 2011, is an unabridged republication of the work originally published in 1962 by Prentice-Hall, Inc., Englewood Cliffs, New Jersey. An Errata List has been added to this edition.

*Library of Congress Cataloging-in-Publication Data*

Wu, Ta-you.

Quantum theory of scattering / Ta-you Wu, Takashi Ohmura. — Dover ed.

p. cm.

Originally published: Englewood Cliffs, N.J. : Prentice-Hall, 1962.

Summar: "This volume addresses aspects and applications of the quantum theory of scattering in atomic and nuclear collisions. An encyclopedic source of pioneering work, it serves as a self-contained text and reference for students and professionals in the fields of chemistry, physics, and astrophysics. Numerous graphs, tables, footnotes, appendices, and bibliographies. 1962 edition" — Provided by publisher.

Includes bibliographical references and index.

ISBN-13: 978-0-486-48089-3 (pbk.)

ISBN-10: 0-486-48089-5 (pbk.)

1. Quantum scattering. I. Ohmura, Taskashi. II. II. Title.

QC794.6.S3W79 2011
530.12—dc22

2010052832

Manufactured in the United States by Courier Corporation
48089501
www.doverpublications.com

**Dedicated to our teachers and friends:**

*Professor Emeritus Harrison M. Randall, University of Michigan*

*Professor Yu-Tai Yao, University of Peking*

*Professor Takahiko Yamanouchi, University of Tokyo*

# Preface

The study of collision processes has been a very important field of physics. A knowledge of the properties of atomic and molecular collisions is of basic importance in many fields, such as the properties of gases, the theory of chemical reactions and collision processes in astrophysics. In the case of the atomic nuclei, most of our present knowledge about nuclear interactions has come from the study of nuclear collision processes. During the last ten years or so, there have been considerable developments in the theory of scattering, both in its general and formal aspects and in its applications. There have appeared recently a number of excellent articles, some of a monograph nature, on many aspects of atomic and nuclear collisions.

However, a general text introducing to the less initiated the main ideas and methods of the scattering theory seems to be still lacking. The object of the present volume is to provide such a text.

The general plan of the present work is more fully described in the Introduction, and the choice of the topics in the Table of Contents. The inclusions and omissions have been governed partly by the familiarity and the limitations of the authors and partly by the predetermined size of the volume. An attempt has been made to present the various topics in an elementary way, but with sufficient details to make the accounts a useful introduction. In many cases, the original papers have been followed closely to make easier their further study by the reader. Different approaches to the same result have been included in some cases, for it is believed that, in a text of this nature, "instructiveness" takes precedence over conciseness and elegance.

There are places where the arrangement of material is somewhat awkward (for example, the treatment of the *S* matrix is spread out over a few sections), but it is hoped that with the cross references in the text and with the help of the index, the reader will find the relevant subsections or paragraphs on a given topic.

A list of references is given at the end of each section. From these the reader can find fuller treatments and further references to the literature. In some cases the authors may have inadvertently failed to refer to some important contributions. For this they ask the forgiveness of all authors.

The authors wish to express their appreciation to Dr. Harry E. Moses for reading a few sections of the manuscript and offering many useful comments.

T. Y. WU
*National Research Council of Canada*

T. OHMURA
*Department of Physics*
*Tokyo University*

# Contents

## I

## GENERAL THEORY OF SCATTERING OF A PARTICLE BY A CENTRAL FIELD

## 3

## COLLISION BETWEEN COMPOSITE PARTICLES 181

# 6



# 7

# Introduction

The subject of the scattering, or collision, between particles is a very large one, in both the more theoretical aspects and the applications to many fields of physics, chemistry and astrophysics. A comprehensive account of the status of this large subject up to about 1947 has been given by Mott and Massey in their well-known work *The Theory of Atomic Collisions*. In the last fifteen years or so, there have been great developments in many directions in the subject, namely, the general theories of scattering, many "technical aspects" of the theory, and the experimental studies in nuclear and atomic collisions. Many of these theoretical and experimental advances have been the result of their mutual stimulations. We shall try, in the following, to make a brief survey of these recent developments, thereby making clearer the plan of the present work.

Up to about 1948 when the high energy accelerators opened up the field of elementary particle physics, one of the main interests in experimental and theoretical nuclear physics was the study of nucleon–nucleon scattering and the deduction of some information concerning the law of nucleon–nucleon interaction from the scattering data. These studies have led directly to the following theoretical developments. From the analysis of the low energy proton–proton scattering data, Breit and his associates, and later on Blatt, Jackson, Bethe and others, found the methods of the scattering length and of the effect range (§ E), which were first derived from Schwinger's variational principle in the present form (§ D). The problem of deducing a potential from the observed phase shifts (which are obtained from the scattering cross sections $\sigma(\vartheta)$ on the basis of the theory of scattering) has led to many mathematical investigations, culminating in the Gel'fand-Levitan integral equation (§ G). On the more technical side, the theories of scattering by non-central (tensor and $L \cdot S$) fields and of the polarization of nucleons by scattering have been worked out (§§ H, I, J). These theories have in turn led to the many scattering and polarization experiments. In the absence of a completely satisfactory (meson field theoretic) theory of nuclear forces, the phenomenological two-nucleon potentials deduced from these empirical data constitute in a large measure our present knowledge about the nucleon–nucleon interaction (§ K).

Also on the technical side, considerable progress has been made by Hulthén, Schwinger, Kohn and others, in the developments of various forms of the variational method for the solution of the scattering problem (§ D). Recently some stationary properties of the scattering length and the effective range have been formulated (§ E). Investigations have been made of the convergence of the most often used Born approximations (§§ B, C).

In the general theory of scattering, the scattering, or collision, or simply the $S$ matrix has been introduced by Wheeler, Heisenberg and developed by a large number of authors (§§ A, F, P, Q, R, S, W). In the theory of the time-dependent Schrödinger equation, the unitary operator method (§ O) has been further developed, and various variational principles, the relation with the $S$ operator, transition probabilities and the Schrödinger (time-independent) integral equation for scattering in general, have been established by Lippmann, Schwinger, Gell-Mann, Goldberger and others (§§ P, Q). The time-dependent equation has also been studied by Friedrichs in a more mathematical manner in terms of spectral representations (§ R). The rigorous mathematical proof of the existence and the unitarity of the $S$ matrix has been given for some classes of interactions (§ S). With the Green's function, the theory developed by Feynman has played a basic role in the advances in the field theory of quantum electrodynamics (§ O).

In the last decade, a great deal of experimental work has been done on various types of nuclear reactions. A large category of reactions, most of which characterized by a resonance effect (i.e., the presence of maxima in the cross sections for certain sharply defined energies of the incident particle), is best interpreted on the basis of the compound nucleus theory of Bohr (§ T). There are other reactions which, from the observed angular distribution of the scattered particle, are best interpreted as "direct reactions" in the sense that the incident particle does not form a compound nucleus with the target particle but interacts only with the nucleons near the nuclear surface (§ V). There are other types of nuclear reactions, such as the Coulomb excitation, which we shall not treat in the present work.

For the "direct reactions" such as the stripping and the pickup processes, the theoretical treatment is usually by means of the Born approximation. But for the more general case of the collision between a nucleon and a nucleus, a phenomenological approach has been developed in which the interaction between the incident and the target particle is represented by a complex potential. The imaginary part of this potential is to describe the absorption due to inelastic scattering (§§ N, U). The application of such a potential, with adjustable (energy-dependent) parameters in the shapes and depths in the central and the $(L \cdot S)$ potentials in both the real and the imaginary part, to a large amount of experimental data has been carried out in the last few years.

A much more general method than this for describing scattering in general (including nuclear reactions) is the method of the $S$ matrix (§§ A, P, Q, R, S, W). But for scattering in general and for nuclear reactions involving the compound nucleus in particular, a still more general method is that of the $R$ matrix of Wigner and Eisenbud. The philosophy of this theory is roughly as follows. On account of the lack of knowledge about the detailed processes going on inside a nucleus in a reaction, the theory will not deal with these details inside the nucleus, but will express the properties (wave functions) inside the nucleus by means of the $R$ matrix whose elements are functions of certain energies, their widths and the nuclear radius pertaining to the nucleus for a particle channel (mode) of the reaction. These quantities are taken to be adjustable parameters. The theory expressing the cross sections of the reactions in terms of $R$ then relates the cross sections to these parameters (§§ T, X). The theory does not depend on any detailed models for the reaction mechanisms and is hence very general. But in a way it is also "formal" rather than "physical" (in the classical sense).

A still more general approach to the problem of collision is the recent theory of dispersion relations. These relations were discovered in 1926–27 by Kronig and Kramers and, in the case of the dispersion of light, consist of relations connecting the real and the imaginary parts of the index of refraction. The extension to the case of potential scattering recently (1954) made by Goldberger, Gell-Mann and Thirring has opened up a most active field in the last few years. The dispersion relations are then relations connecting the imaginary part of the scattering amplitude in the forward direction and an integral containing the total cross section for all energies. Relations for the scattering amplitude in other than the forward directions have also been found. The great importance of these dispersion relations lies in their generality; they do not depend on any specific assumptions about the interactions but only on some general principles such as the physically plausible one of causality. They have been of the greatest interest in connection with the scattering involving elementary particles (such as pion–nucleon scattering) for which a detailed knowledge of their interaction is still absent. It is beyond the scope of the work to take up this rapidly developing field, but an account of the historical origin and the derivation of the dispersion relations for the potential scattering of a non-relativistic particle is included in §§ Y, Z.

Now, coming back to the more familiar fields of atomic and molecular collisions, while there have been no really new fundamental developments, we can still count some progress. Thus the low energy electron–hydrogen scattering has been measured with great accuracy with the advanced techniques now available on the one hand, and has been calculated also with great refinements by the variational method and the method of scattering length on the other (§ L). This problem is of interest in that it is, in the theory of scattering, next in simplicity (or complexity) to the two-body (electron–proton) problem, just as the helium atom is to the hydrogen atom in the bound state problem.

Of great interest and importance in the fields of gaseous chemical reactions, the excitation, ionization, dissociation and recombination processes in the upper atmosphere of the earth, and of astrophysics, is a large category of collisions involving an electron and an atom (or a molecule) or two atoms (or molecules). In recent years, a large amount of studies of various approximate methods and of calculations of some simpler processes have been carried out, notably by Massey, Bates and their associates. Unfortunately, the nature of the problem is such that, with the present experience, it is still difficult to know with accuracy the errors of the various approximations in a given specific case. The various approximate methods are given, together with a bibliography of this subject, in Sect. M.

In the problems of the scattering of a particle by a system of particles, the theory of Watson on multiple scattering is important in the introduction of an optical potential and of the scattering operator (§ N) which plays a basic role in the theory of Brueckner of the many-body system. There is another problem of multiple scattering which deals with the scattering of a charged particle in going through a thin foil. The theory is of importance in the analysis of the scattering of cosmic ray particles in slabs of matter. A short account of the recent theoretical work is given in Sect. N.

There are many topics in the field of scattering that have not been covered in the present work. The most conspicuous omission is the scattering involving the various elementary particles. The reasons for this omission are that (i) the processes involve generally the creation or annihilation of these particles and they should be treated by the quantized field theories rather than the (non-relativistic) quantum mechanics used in the present work; (ii) both the experimental and the theoretical literature are very extensive and the authors are not familiar enough with them to write about them.

Another interesting and important topic that has not been treated is the energy dependence of reactions (nuclear or atomic) in the neighborhood of the threshold. Wigner has obtained the energy dependence for various types of processes by the use of the *R* matrix theory, and these results are general, quite independent of any detailed calculations. To this work only a reference has been made in Sect. M, references 75–82 and on pp. 378–79.

There are other omissions. The present work has been prepared, not as a monograph, but as a text introducing the student to some of the recent developments in a very broad field. The emphasis has been on the basic ideas and methods for treating various types of problems, and not on the actual applications, nor on the experimental investigations. For those who wish to go deeper into any particular topics, references have been made, at the end of each section, to the many recent monographs and review articles. In view of the availability of the extensive bibliographies given in these references, it is hoped that the omissions of topics and the lack of judicious judgment in the choice of references will not really harm the serious student.

# ERRATA

| *page* | *line* | *Correction* |
|---|---|---|
| 3 | 20 | Instead of "$\lambda(2\pi d)$," read "$\lambda/(2\pi d)$" |
| 47 | 9 | Delete "$\lambda'(k = 0)$" |
| 58 | (D10) | Replace $k$ by $-k$ |
| 59 | 2, 3, 5, 8, 9 | Change $\lambda$ to $\xi$, and replace $k$ by $-k$ |
| | (D18), (D19) | Delete "$\delta\lambda =$ " |
| 60 | 10 | Read "$\ldots k\lambda - I = \ldots$" |
| | −11 | Read "$\ldots k\lambda - I$" |
| 109 | (G28) | Add "$\varepsilon > 0$" |
| 113 | 3–5 | Delete "If we know ... δ(t)" Insert: The $S(k)$ is analytic in the region $\lvert\mathrm{Im}(k)\rvert < \mu$ except poles, and the zeroes of $S$ (which correspond to the binding energies, $-k_i^2$) can be determined by the phase shift function $\delta(k)$, $k > 0$, if $\mu > k_i > 0$. One example is found in the work of Regge, Nuo. Cim. **14**, 951 (1959). (This, however, does not seem to have been proved rigorously.) Thus if the potential $U(r)$ is known a priori to satisfy (G38) for any finite $\mu$ (as in the case of the Gaussian or the square well potential), all the binding energies and the potential are uniquely determined by $\delta(k)$, $k > 0$. (See Barut, A. O. and Ruei, K. H., *J. Math. Phys.* **2**, 181 (1961).) |
| 118 | 5 | Read "... conditions for (G58) ..." |
| 132 | −9 | Insert after "are": $\sqrt{2\pi}$ times |
| 134 | −14 | Read "... by $M/\hbar$" |
| 143 | above Reference | Read "... refs. 4 and 6" |
| 151 | References | Move refs. 3 and 4 on p. 151 to ref. 7 on p. 143. |

| | | |
|---|---|---|
| 154 | after (J11a) | Add "The formulas are also valid even if $\boldsymbol{k}$, $\boldsymbol{k}'$ are considered as vectors in the laboratory system. |
| 155 | 9 | Delete "All the formulas ... laboratory system" |
| 184 | (L8) | Delete "$\frac{1}{2}$" |
| 194 | (L47) | Read $(t^2 + 9)^3$ in the denominator. |
| 204 | 14 | Delete "$\frac{1}{2}\gamma^2 =$" |
| 343 | −6 | Read "unitarity" |
| 356 | Footnote | Read "The conditions are more precisely specified by<br>1. $V(r)$ is Lebesque integrable in any finite interval not including the origin,<br>2. For some $R > 0$,<br>$\int_0^R r\lvert V(r)\rvert\, dr < \infty, \qquad \int_R^\infty \lvert V(r)\rvert\, dr < \infty$." |
| 357 | 15 | Read "$= \{(k^2/2m)C_{lmn}, \quad (k_{ln}^2/2m)C'_{lm}(k)\}$" |
| 385 | 3 | Read "... $A_{\mathrm{res}}$ in (T41). ..." |
| 392 | −5 | Instead of "$\sigma_{ce}^{(l)}$," read "$\sigma_{se}^{(l)}$" |
| 424 | −10 | Last word, read "at" instead of "is" |
| 438 | (X24) | Read "$\left.\frac{\partial(r_B\phi_s)}{\partial \pi_s} \middle/ (r_B\phi_s)\right\rvert_{rB=aB} =$" |
| 439 | −6 | Replace $\sum_i$ by $\prod_i$ |

# Quantum Theory of Scattering

chapter

1

# General Theory of Scattering of a Particle by a Central Field

The simplest collision processes are the scattering of a particle $A$ by a potential field $V(r)$ and the elastic scattering of a particle of kind $A$ with a given momentum by a target particle $B$. We wish to know the relative probabilities of the particle $A$ being scattered in various directions. It is obvious that these probabilities are determined by the field $V(r)$ or the interaction $V(r_{AB})$ between the two particles. The mathematical theory of scattering enables us to deduce these probabilities from a given $V(r)$ or $V(r_{AB})$, and conversely, to deduce some information concerning $V(r)$, or $V(r_{AB})$, from the experimentally observed scattering cross sections.

In this chapter, we will confine ourselves to (i) the stationary-state method of treating the scattering problem, and (ii) the case where the interaction $V$ between the two particles is central, i.e., $V = V(|\mathbf{r}_A - \mathbf{r}_B|) = V(r_{AB})$.#

Let a particle of mass $m$, energy $E$, and momentum $\mathbf{p} = \hbar\mathbf{k}$,

$$2mE = \hbar^2 k^2, \tag{A1}$$

be scattered by a central field of potential $V(r)$. The Schrödinger equation of the particle is

$$[\Delta + k^2 - U(r)]\,\psi(\mathbf{r}) = 0, \qquad U(r) \equiv \frac{2m}{\hbar^2} V(r). \tag{A2}$$

We seek a solution of (A2) which asymptotically represents an incident plane wave (along the $z$-direction) and a spherical outgoing (scattered) wave

$$\psi(\mathbf{r}) \rightarrow e^{ikz} + \frac{e^{ikr}}{r} f(\vartheta), \tag{A3}$$

where $\vartheta$ is the angle between the direction of the scattered wave ($\mathbf{k}'$) and the original direction $\mathbf{k}$. The angle $\phi$ around $\mathbf{k}$ does not appear because of the symmetry of $V(r)$. The intensity of the scattered wave in the solid angle $d\Omega = \sin\vartheta\, d\vartheta\, d\phi$ is proportional to

$$|f(\vartheta)|^2 \sin\vartheta\, d\vartheta\, d\phi. \tag{A4}$$

As $f(\vartheta)$ defined by (A3) has the dimension of length, we can define (A4) as the differential cross section

$$d\sigma = |f(\vartheta)|^2 \sin\vartheta\, d\vartheta\, d\phi = \sigma(\vartheta) \sin\vartheta\, d\vartheta\, d\phi, \tag{A5}$$

and the total cross section

$$\sigma(E) = \iint |f(\vartheta)|^2 \sin\vartheta\, d\vartheta\, d\phi = 2\pi \int |f(\vartheta)|^2 \sin\vartheta\, d\vartheta. \tag{A6}$$

$\sigma$ gives a measure of the total probability that the particle of energy $E$ is scattered in any way by the field $V(r)$. The problem is to find $f(\vartheta)$ in terms of the potential $V(r)$. The importance of this theory is that on the basis of the theoretical relation between $f(\vartheta)$ and $V(r)$, it is possible to deduce some information about $V(r)$ from the experimentally measured $\sigma(\vartheta)$.

Equation (A5) is usually derived in the following manner. The flux of scattered particle (i.e., the number of particles per unit time) through a large sphere of radius $R$ within the solid angle $d\Omega$ is

$$F\, d\Omega = \frac{\hbar}{2im}\left(\frac{\partial \psi_{sc}}{\partial r}\psi_{sc}^* - \frac{\partial \psi_{sc}^*}{\partial r}\psi_{sc}\right) R^2\, d\Omega \qquad \text{at } r = R. \tag{A5a}$$

---

# It is assumed that $V(r)$ is such that $rV(r) \rightarrow 0$ as $r \rightarrow \infty$ (see Sect. A.6), and, if attractive, $r^2 V(r) \rightarrow 0$ as $r \rightarrow 0$ (for the existence of a solution). See further conditions on $V(r)$ in Sect. A.3.

On substituting,

$$\psi_{sc} \to \frac{1}{r} f(\vartheta)\, e^{ikr}, \qquad R \to \infty,$$

into (A5a) we get

$$F\, d\Omega = (\hbar k/m)\, |f|^2\, d\Omega.$$

The differential cross section is obtained if $F\,d\Omega$ is divided by the flux of incident particle per unit area per unit time, which is just the velocity $v = \hbar k/m$ of the incident particle.

This procedure is, however, only permissible when there is no interference between the incident wave ($e^{ikz}$) and the scattered wave. Under the usual experimental conditions, this interference can be neglected at the point of observation. The incident beam goes through narrow slits and must be represented by a wave packet restricted in the $x$–$y$ plane. Accordingly, the scattering amplitude calculated on the approximation of an incident plane wave is good only when the slit width $d$ (i.e., the width of the incident beam) is larger than the wavelength $\lambda$ of the incident beam, and is also larger than the sum of the range of interaction between the incident particle and the target particle (or particles), and the characteristic length of the size of the target particle (or particles). If $d \lesssim \lambda$, the diffraction of the incident wave (by the edge of slits) will be observed in an angle $\lesssim \lambda/(2\pi d)$, and the incident wave will reach the detector when placed within the angle $\sim \lambda(2\pi d)$. If $d$ is smaller than the size of the target or than the range of interaction, the value of $f(\vartheta)$ thus obtained will be quite different from the value of $f(\vartheta)$ derived from the plane wave assumption.

The angle $\lambda/(2\pi d)$ in usual experimental conditions is so small that we can safely assume that the incident wave does not fall on the detector. For example, let $d$ be equal to 0.01 cm. The angle is about $x^{-1/2} \times 10^{-6}$ radian for $x$ eV electrons, and about $y^{-1/2} \times 10^{-10}$ radian for $y$ MeV nucleons (protons or neutrons).

However, the interference between the incident (approximately plane) wave and the scattered wave at the vicinity of the target (not at the position of detector) is very important in quantum mechanical scattering phenomena. This effect has been included in determining the scattering amplitude $f$; hence it is sufficient to consider $f$ alone. "Shadow scattering" (see Sect. T.8) is caused by interference of this nature near the "black" body.

It is clear that the scattering of one particle by another, both having no internal degrees of freedom (except for spin) and interacting according to a potential $V(\mathbf{r}_{12})$, can be described by (A2) for the relative coordinate $\mathbf{r} = \mathbf{r}_{12}$, where $m$ is now the reduced mass and $E$ the energy of relative motion. It is by means of this theory and the theory discussed in the next chapter that one attempts to arrive at some conclusions concerning the interaction $V(\mathbf{r}_1 - \mathbf{r}_2)$ between two nucleons from the scattering data.

section

# A Partial Wave Analysis

## 1. FAXEN-HOLTZMARK'S THEORY

The following method was first suggested by Faxen and Holtzmark, and is analogous to a method devised by Lord Rayleigh in the classical theory of scattering: $f(\vartheta)$, being a function of $\vartheta$, can be expanded in terms of the Legendre polynomials

$$f(\vartheta) = \sum_l a_l \frac{1}{2k} i(2l + 1) P_l (\cos \vartheta), \tag{A7}$$

where the coefficient $a_l$, in general complex, is to be found in terms of $V(r)$. Let us also expand the incident plane wave in terms of $P_l (\cos \vartheta)$

$$\begin{aligned} e^{ikz} &= \exp (ikr \cos \vartheta) \\ &= \sum_{l=0} (2l + 1) i^l j_l(kr) P_l (\cos \vartheta).^{\#} \end{aligned} \tag{A8}$$

---

\# We shall give the relations between the Bessel functions of the first, second and third kinds and their associated spherical Bessel functions. The notation $N_{l+1/2}(x)$ is that of Jahnke–Emde, Tables of Functions. $N_{l+1/2}(x)$ is denoted by $Y_{l+1/2}(x)$ in Watson's treatise on Bessel Functions.

$$\begin{aligned} j_l(x) &= \sqrt{\frac{\pi}{2x}} J_{l+1/2}(x), \\ N_{l+1/2}(x) &= (-1)^{l+1} J_{-l-1/2}(x), \\ n_l &= (-1)^{l+1} \sqrt{\frac{\pi}{2x}} J_{-l-1/2}, \\ H^{(1)}_{l+1/2}(x) &= J_{l+1/2}(x) - i(-1)^l J_{-l-1/2}(x) = J_{l+1/2}(x) + iN_{l+1/2}(x), \end{aligned}$$

For large distances $r$, the asymptotic behavior of the Bessel function leads to

$$j_l(kr) \to \frac{\sin\left(kr - \frac{l\pi}{2}\right)}{kr}. \tag{A9}$$

On putting (A7), (A8), (A9) into (A3), one obtains the asymptotic form for $\psi(\mathbf{r})$

$$\psi(\mathbf{r}) \to \frac{1}{2ikr}\sum_{l=0}(2l+1)\, i^l\left\{(1-a_l)\exp\left[i\left(kr - \frac{l\pi}{2}\right)\right] - \exp\left[-i\left(kr - \frac{l\pi}{2}\right)\right]\right\} P_l(\cos\vartheta)$$

$$= \frac{1}{2ikr}\sum(2l+1)\left[(1-a_l)\, e^{ikr} - (-1)^l e^{-ikr}\right] P_l(\cos\vartheta). \tag{A10}$$

---

$$H^{(1)}_{-l-\frac{1}{2}}(x) = J_{-l-\frac{1}{2}}(x) + i(-1)^l J_{l-\frac{1}{2}}(x),$$

$$H^{(2)}_{l+\frac{1}{2}}(x) = J_{l+\frac{1}{2}}(x) + i(-1)^l J_{-l-\frac{1}{2}}(x),$$

$$H^{(2)}_{-l-\frac{1}{2}}(x) = J_{-l-\frac{1}{2}}(x) - i(-1)^l J_{l+\frac{1}{2}}(x),$$

$$h_l^{(1)}(x) = j_l(x) + in_l(x), \qquad h_l^{(2)}(x) = j_l(x) - in_l(x), \text{ etc.}$$

All $j_l(x)$, $n_l(x$, $h_l(x)$ satisfy the equation

$$\frac{d^2\phi}{dx^2} + \frac{2}{x}\frac{d\phi}{dx} + \left[1 - \frac{l(l+1)}{x^2}\right]\phi = 0.$$

The asymptotic forms are

$$j_l(x) \to \frac{1}{x}\sin\left(x - \frac{l\pi}{2}\right),$$

$$n_l(x) \to -\frac{1}{x}\cos\left(x - \frac{l\pi}{2}\right),$$

$$h_l^{(1)}(x) \to -\frac{i}{x}\exp\left[i\left(x - \frac{l\pi}{2}\right)\right],$$

$$h_l^{(2)}(x) \to \frac{i}{x}\exp\left[-i\left(x - \frac{l\pi}{2}\right)\right],$$

$$j_0(x) = \frac{\sin x}{x}, \qquad j_1(x) = \frac{\sin x - x\cos x}{x^2},$$

$$j_2(x) = \frac{3}{x}j_1(x) - j_0(x),$$

$$n_0(x) = -\frac{\cos x}{x}, \qquad n_1(x) = -\frac{\cos x + x\sin x}{x^2},$$

$$n_2(x) = \frac{3}{x}n_1(x) - n_0(x), \text{ etc.}$$

Now, since $V(r)$ is spherically symmetric (and hence axially symmetric about $\mathbf{k}$), the wave function $\psi(\mathbf{r})$ in (A2) can be put in the form#

$$\psi(\mathbf{r}) = \sum_l \frac{(2l+1)}{k} i^l \frac{1}{r} u_l(r) P_l(\cos\vartheta), \tag{A12}$$

where $u_l(r)$ is given by the equation

$$\frac{d^2u_l}{dr^2} + \left[k^2 - U(r) - \frac{l(l+1)}{r^2}\right] u_l(r) = 0. \tag{A13}$$

For large $r$, the two independent solutions of (A13) behave as $rj_l(kr)$ and $rn_l(kr)$. We seek a solution $u_l(r)$ of (A13) which vanishes at the origin and which, for large $r$, asymptotically behaves as

$$\begin{aligned} u_l(r) &\rightarrow kr[A_l j_l(kr) - B_l n_l(kr)] \\ &= C_l \sin\left(kr - \frac{l\pi}{2} + \delta_l\right), \end{aligned} \tag{A14}$$

where $A_l$, $B_l$ are constants and $C_l$, $\delta_l$ are simply related to $A_l$, $B_l$ by

$$\tan\delta_l = \frac{B_l}{A_l}, \qquad C_l^2 = A_l^2 + B_l^2.$$

Putting (A14) into (A12), one obtains

$$\psi(\mathbf{r}) \rightarrow \sum_l \frac{(2l+1)}{2ikr} C_l\,[\exp(ikr + i\delta_l) - (-1)^l \exp(-ikr - i\delta_l)]\,P_l(\cos\vartheta) \tag{A15}$$

On identifying (A15) with (A10), one obtains

$$\begin{aligned} a_l &= [1 - \exp(2i\delta_l)], \\ C_l &= \exp(i\delta_l). \end{aligned} \tag{A16}$$

On putting this expression for $a_l$ in (A7), one obtains

$$f(\vartheta) = \frac{1}{2ik} \sum_l (2l+1)[\exp(2i\delta_l) - 1]\,P_l(\cos\vartheta), \tag{A17}$$

and (A6) gives

$$\sigma = \frac{4\pi}{k^2} \sum_l (2l+1) \sin^2\delta_l. \tag{A18}$$

---

# Sometimes (A12) is written in the form

$$\psi(\mathbf{r}) = \frac{1}{kr} \sum (2l+1)\, i^l \exp(i\delta_l)\, u_l(kr)\, P_l(\cos\vartheta), \tag{A12a}$$

where $u_l(kr)$ is normalized by the asymptotic form

$$u_l(kr) \rightarrow \sin(kr - \tfrac{1}{2}l\pi + \delta_l). \tag{A14a}$$

## 2. OPTICAL THEOREM

From (A17), it is seen that $f(\vartheta)$ is essentially complex. We may rewrite $f(\vartheta)$ in the form

$$f(\vartheta) = |f(\vartheta)|\, e^{i\eta(\vartheta)}$$
$$= \sqrt{\sigma(\vartheta)}\, e^{i\eta(\vartheta)}, \quad \text{by (A5).} \tag{A19}$$

From (A17), we obtain

$$\sigma(\vartheta) = \frac{1}{k^2} \sum_{l,m} (2l + 1)(2m + 1) \sin \delta_l \sin \delta_m \times \cos (\delta_l - \delta_m)\, P_l (\cos \vartheta)\, P_m (\cos \vartheta), \tag{A19a}$$

and from (A19), (A17),

$$\sqrt{\sigma(\vartheta)} \sin \eta(\vartheta) = \operatorname{Im} f(\vartheta)$$
$$= \frac{1}{k} \sum_l (2l + 1) \sin^2 \delta_l\, P_l (\cos \vartheta), \tag{A20}$$

where $\operatorname{Im} f(\vartheta)$ denotes the coefficient of $i$ in $f(\vartheta)$. That $f(\vartheta)$ is complex can readily be seen from (A20), since, on putting $\vartheta = 0$, for the imaginary part of the scattered amplitude in the forward direction, we obtain the following important relation

$$\operatorname{Im} f(0) = \frac{1}{k} \sum (2l + 1) \sin^2 \delta_l$$
$$= \frac{k}{4\pi}\, \sigma, \tag{A21}$$

which does not vanish. (A21) is known as the *optical theorem*. [For a generalized theorem covering inelastic scatterings, see Chapter 3, (N19).]

## 3. PHASE SHIFTS

All the relations (A17)–(A21) above are exact relations (i.e., involving no approximations). $\delta_l$ represents the phase shift whose meaning is clear from (A14), namely, $\delta_l$ is the phase difference between the asymptotic solution (A14) of (A13) with suitable normalization and the field-free asymptotic solution $v_l(r)$ of

$$\frac{d^2 v_l}{dr^2} + \left[k^2 - \frac{l(l + 1)}{r^2}\right] v_l(r) = 0, \tag{A22}$$

namely,

$$v_l = r j_l(kr) \to \frac{1}{k} \sin \left(kr - \frac{l}{2}\pi\right). \tag{A23}$$

Obviously, $\delta_l$ is determined by the potential $V(r)$. To obtain $\delta_l$ in terms of $V(r)$, we have, from (A13) and (A22),

$$\int_0^\infty \left( v_l \frac{d^2u_l}{dr^2} - u_l \frac{d^2v_l}{dr^2} \right) dr = \int_0^\infty v_l\, U(r)\, u_l\, dr,$$

or

$$\left[ v_l \frac{du_l}{dr} - u_l \frac{dv_l}{dr} \right]_0^\infty = \int_0^\infty v_l\, U(r)\, u_l\, dr: \tag{A24}$$

By (A14) and (A23), we obtain the following *exact* relation between $\delta_l$ and the potential $V(r)$

$$\sin \delta_l = -\int_0^\infty \sqrt{\frac{\pi kr}{2}}\, J_{l+\frac{1}{2}}(kr)\, U(r)\, u_l(r)\, dr. \tag{A25}$$

In this equation $u_l(r)$ is the solution of (A13) that is bounded at the origin $r = 0$, and that is so normalized that asymptotically for large $r$,

$$u_l(r) \rightarrow \frac{1}{k} \sin\left(kr - \frac{l\pi}{2} + \delta_l\right). \tag{A26}$$

Since $u_l(r)$ in (A25) contains $\delta_l$, it would seem that (A25) is of no practical use as far as the calculation of $\delta_l$ is concerned, since by the time $u_l(r)$ is found from (A13), $\delta_l$ would already be known. We shall see, however, that a useful approximation can be obtained for $\delta_l$ by replacing $u_l(r)$ in (A25) by $v_l(r)$. (See Sect. C on Born approximation.)

The following are explanatory remarks concerning $\delta_l$:

i) From (A13) and (A22), it is seen that for an attractive field $u_l(r)$ is shifted inward relative to $v_l(r)$, and for a repulsive field $u_l(r)$ is shifted outward relative to $v_l(r)$, i.e.,

$$\begin{aligned} \delta_l &> 0 \quad \text{for attractive field,} \\ \delta_l &< 0 \quad \text{for repulsive field.} \end{aligned}$$

ii) Classically, $l\hbar \simeq p\rho$, where $p$ is the momentum of the particle and $\rho$ the impact parameter. The summation of $l$ in (A17) and (A18) for the partial waves $l = 0, 1, \ldots$ in (A7) is equivalent to the integration of all values of the impact parameters in the classical theory.

iii) For large $k$ and $l$, the phase shifts can be calculated by the Born approximation [see (C14–C18)]. From (C18), $\delta_l$ is seen to be of order

$$\delta_l \simeq -\frac{1}{2k}\, U(r_0)\, r_0,$$

where $r_0$ is the classical distance of closest approach. For large $l$, $r_0$ is $\simeq \rho$, the impact parameter. The series for the total cross section $\sigma$ in (A18) behaves, for large $l$ and small $\delta_l$, ($l\hbar = p\rho = \hbar k\rho$)

$$\sigma \propto \sum (2l + 1)\, \delta_l^2 \rightarrow 2 \int^{\infty} l\, d_l\, \delta_l^2 = \tfrac{1}{2} \int^{\infty} d\rho \cdot \rho^3\, U^2(\rho).$$

In order that this may converge, $U(r)$ must decrease with distance at a rate which is greater than $1/r^2$.

**iv)** For the scattering amplitude in the forward direction, $f(0)$, we have from (A17),

$$f(0) \propto \sum (2l + 1)\, \delta_l \rightarrow k \int^{\infty} d\rho \cdot \rho^2\, U(\rho).$$

In order that this may converge, $U(\rho)$ must decrease with distance faster than $1/r^3$.

**v)** For low energy scattering by a potential of the asymptotic form $c/r^n$, the variations of the phase shifts for various $l$ are

$$\begin{aligned} &\delta_l \propto k^{2l+1} && \text{for} \quad 2l < n - 3, \\ &\delta_l \propto k^{2l+1} \ln k && \text{for} \quad 2l = n - 3, \\ &\delta_l \propto k^{n-2} && \text{for} \quad 2l > n - 3, \end{aligned} \tag{A27}$$

(See ref. 8, p. 403.) For very small energies, only $l = 0$ ($s$ waves) contribute to the scattering if $n \geq 3$. In Sect. L, we shall have an example of $n = 4$ (potential due to the polarization of a hydrogen atom in the ground state by a charge). There $\delta_0 \propto k$ and $\delta_l \propto k^2$ for $l \geq 1$. For short range potential ($1/r^n$, $n \rightarrow \infty$), we always have $\delta_l \propto k^{2l+1}$. Thus for small $k$, $\delta_l$ decreases with increasing $l$. The cross section tends to a constant value as $k \rightarrow 0$ only for $n \geq 3$.

**vi)** For low energy scattering by an attractive $V(r)$ having bound states, the variation of $\delta_l$ with the "strength" of $V(r)$ is discussed in Sect. E.3.

**vii)** For scattering by a potential $V(r)P$, where $P$ is a parity operator [inversion of coordinates about the origin of the coordinate system, see Sect. (H10A)],

$$P\mathbf{r} = -\mathbf{r}, \tag{A28}$$

(this, in the case of the collision between two identical particles, corresponds to an exchange of the two particles, i.e., to the Bartlett operator discussed in Sect. H.5), we have

$$PY_{lm}(\vartheta, \varphi) = Y_{lm}(\pi - \vartheta, \pi + \varphi) = (-1)^l\, Y_{lm}(\vartheta, \varphi). \tag{A29}$$

This has the effect that in obtaining (A13) from (A2), the $V(r)$ in (A13) will have a factor $(-1)^l$. This changes the $\delta_l$ from (A13) for all odd $l$.#

### 4. RESONANCE SCATTERING

In many reactions involving the collision of a nucleon with a nucleus, it has been found that the reaction cross section has a number of maxima at certain energies of the incident nucleon. This has been interpreted as a resonance phenomenon when the energy of the incident particle is exactly right to excite the compound nucleus (formed by the nucleus and the nucleon) to one of its excited states. We shall take up this problem of resonance in nuclear reactions in Sects. T and X. But the concept of resonance scattering also

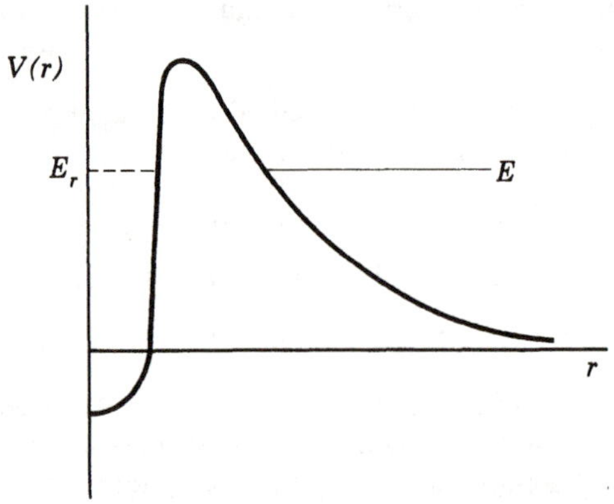


**Figure 1.** Resonance scattering of a particle by a potential having a decaying state at energy $E_r$.

enters in the simple problem of the scattering of a particle by a potential. Consider, for example, the collision between an $\alpha$ particle or a proton and a nucleus, the interaction being approximated by an attractive well, surrounded by a Coulomb barrier. In Fig. 1, let the curve represent the effective potential

---

# If one makes the Born approximation (C8a), then $\delta_l^{(1)}$ (Ex) for $V(r)P$ is simply related to the $\delta_l^{(1)}$ for $V(r)$ by

$$\delta_l^{(1)}\,(\text{Ex}) = (-1)^l \delta_l^{(1)}.$$

Corresponding to this Born approximation and the property

$$P_l\,(\cos\vartheta) = (-1)^l P_l\,[\cos(\pi - \vartheta)],$$

it follows that the scattered amplitude $f_{\text{Ex}}(\vartheta)$ for $V(r)P$ is related to the $f(\vartheta)$ of $V(r)$ by

$$f_{\text{Ex}}(\vartheta) = f(\pi - \vartheta).$$

This relation is of course not valid for the *exact* $f(\vartheta)$.

for angular momentum $l$ states, and let us assume that $E_r$ is one of the quasi-stationary, or decaying, states of this $l$. When the system (the nucleus and the additional particle) is in one of these states, it has a probability of decaying, resulting in the emission of the additional particle. This is essentially the problem of the $\alpha$ decay of radioactive nuclei, and the theories of Gamow and of Gurney and Condon are well known. We shall now outline the theory of scattering and the phenomenon of resonance in such cases.

The problem can be treated with the usual theory dealing with real energy eigenvalues (ref. 7), but, for simplicity, we shall use the method of complex energies first introduced by Gamow. Let us assume that the spacings of these decaying levels are large compared with the "width" of the levels, and let the energy of the system "at $E_r$" be

$$E_r - \frac{i}{2}\Gamma, \qquad E_r, \Gamma \text{ real}, \qquad \Gamma > 0.$$

Then

$$|\exp[-i(E_r - \frac{i}{2}\Gamma)t/\hbar]|^2 = \exp(-\Gamma t/\hbar),$$

so that the probability of decay per unit time is

$$w = \frac{\Gamma}{\hbar}. \tag{A30}$$

This gives $\Gamma$, according to the uncertainty principle, the meaning of the "width" of the state.

To describe the scattering, let the asymptotic form of the radial part of the wave function [(A15), (A34)] be

$$\frac{u_l(r)}{r} \to \frac{1}{r}\left\{\exp\left[-i\left(kr - \frac{l\pi}{2} + \delta_l\right)\right] - \exp\left[i\left(kr - \frac{l\pi}{2} + \delta_l\right)\right]\right\}, \tag{A31}$$

or, in the alternative form,

$$\frac{u_l(r)}{r} \to \frac{1}{r}[A_l(E)\, e^{-ikr} - A_l^*(E)\, e^{ikr}], \tag{A31a}$$

where $A_l(E)$ are complex functions (of complex $E$) $A_l(E) = i^l \exp(-i\delta_l)$, $A_l^*(E) = (-i)^l \exp(i\delta_l)$ such that when $E = E_r - \frac{1}{2}i\Gamma$,#

$$A_l\left(E_r - \frac{i}{2}\Gamma\right) = 0$$

# We denote by $E_r - (i/2)\Gamma$ a resonance or decay level for the angular momentum $l$. We have omitted a subscript $l$ from $\Gamma_l$.

and only the outgoing wave $e^{ikr}$ remains. In the neighborhood of $E_r - (i/2)\Gamma$, we expand $A_l$, obtaining

$$A_l(E) = \left[E - \left((E_r - i\frac{\Gamma}{2}\right)\right] a_l + \cdots$$

and hence

$$\frac{u_l(r)}{r} \simeq \left[E - \left(E_r - \frac{i}{2}\Gamma\right)\right] a_l \frac{e^{-ikr}}{r} - \left[E - \left(E_r + \frac{i}{2}\Gamma\right)\right] a_l^* \frac{e^{ikr}}{r}, \tag{A31b}$$

$$= i\Gamma\, a_l^* \frac{e^{ikr}}{r} \quad \text{at} \quad E = E_r - \frac{i}{2}\Gamma.$$

From the equation of continuity, namely the total probability current, $(\hbar k/\mu)\,|i\Gamma a_l^*|^2$, is equal to the decay probability per unit time, $\Gamma/\hbar$, we get

$$|a_l|^2 = \frac{1}{\hbar v \Gamma}, \tag{A31c}$$

where $v$ is the velocity $\hbar k/\mu$. The $a_l$, or $\Gamma$ are to be determined from the exact potential in Fig. 1.

On comparing (A31) and (A31b), we get for the phase shifts $\delta_l$

$$\exp(2i\delta_l) = e^{i\pi l} \frac{a_l^*}{a_l} \cdot \frac{E - E_r - \frac{i}{2}\Gamma}{E - E_r + \frac{i}{2}\Gamma}.$$

If we denote by

$$\exp[2i\delta_l(0)] \equiv e^{i\pi l} \frac{a_l^*}{a_l},$$

then

$$\exp(2i\delta_l) = \exp[2i\delta_l(0)]\left[1 - \frac{i\Gamma}{E - E_r + \frac{i}{2}\Gamma}\right]. \tag{A32a}$$

Since $\delta_l \to \delta_l(0)$ when $|E - E_r| \gg \Gamma$, $\delta_l(0)$ is thus the phase shift very far from the resonance level. In the neighborhood of the resonance level,

$$\delta_l = \delta_l(0) - \tan^{-1} \frac{\Gamma}{2(E - E_r)}. \tag{A32}$$

It is seen from this expression that when the energy of the incident particle $E$ increases from below $E_r$ to above $E_r$, $\delta_l$ changes by $\pi$.

On putting (A32a) into (A17) (if $E_r$ is the only resonance level for $l$), we have

$$f(\vartheta) = \frac{1}{2ik} \sum (2l + 1)\{\exp[2i\delta_l(0)] - 1\} P_l(\cos\vartheta)$$
$$- \frac{1}{2k}(2l + 1) \frac{\Gamma}{E - E_r + \frac{i}{2}\Gamma} \exp[2i\delta_l(0)]\, P_l(\cos\vartheta). \tag{A32b}$$

The first sum is called the "potential scattering", while the second term is called the "resonance scattering" amplitude. The cross section $|f(\vartheta)|^2$ will contain the contributions from each, and will also contain an interference effect. The resonance cross section is given by

$$d\sigma_l = \frac{1}{4k^2}(2l+1)^2 \frac{\Gamma^2}{(E-E_r)^2 + \frac{1}{4}\Gamma^2} [P_l(\cos\vartheta)]^2 \, d\Omega$$

which, when $E = E_r \pm (\Gamma/2)$, drops to half its maximum value at resonance $E = E_r$,

$$(d\sigma_l)_{\max} = \frac{1}{k^2}(2l+1)^2 [P_l(\cos\vartheta)]^2 \, d\Omega.$$

Hence $\Gamma$ is called the half-width of the resonance level. The maximum value $(d\sigma_l)_{\max}$ above is seen to be determined by the energy alone. At resonance, it is seen from (A32) that

$$\delta_l = (n + \tfrac{1}{2})\pi$$

where $n$ is either a positive or a negative integer.

In the treatment described above, we have assumed that there is only elastic scattering. In the actual case of a particle (including a neutron) colliding with a nucleus, there are inelastic scatterings as well. We shall discuss this further in Sect. T later. For a brief but clear treatment of this topic, see ref. 8.

It should be noted here that there is another type of "resonance" scattering, represented by the resonance scattering of slow neutrons by a proton due to the "virtual" level of the deuteron. This similarity in terminology is rather misleading, since it is quite different from the resonance scattering treated above in that the cross section does not have the energy dependence $[(E - E_r) + (\Gamma^2/4)]^{-1}$, [see (F30)], and secondly, the virtual level has been defined merely for the purpose of convenience, and the virtual state wave function behaves as $e^{ar}$ $(a > 0)$ as $r \to \infty$ and hence does not correspond to a physical state. This type of "resonance" will be further clarified in Sect. F.

## 5. SCATTERING MATRIX

The scattering amplitude and cross section have been expressed in terms of the phase shifts in (A17) and (A18). For the more general theoretical developments (see Chapters 4, 5, 6), it is convenient to introduce the "scattering matrix". For central field scattering, the scattering matrix $S$ can be introduced in the following manner.

The asymptotic form of the incident plane wave is given by (A8) and (A9), which can also be written in the equivalent form

$$e^{ikz} \to \sum_{l=0}^{\infty} \frac{1}{2kr}(2l+1)\, i^{l+1} \{\exp[-i(kr - \tfrac{1}{2}l\pi)] - \exp[i(kr - \tfrac{1}{2}l\pi)]\} P_l(\cos\vartheta). \quad \text{(A33)}$$

The wave function with the outgoing wave condition (A3) is given by (A10) which can also be written in the following form

$$\Psi(\mathbf{r}) \to \sum \frac{1}{2kr} (2l + 1)\, i^{l+1} \{\exp[-i(kr - \tfrac{1}{2}l\pi)] - S_l(k) \exp[i(kr - \tfrac{1}{2}l\pi)]\} P_l(\cos\vartheta). \tag{A34}$$

On comparing (A34) with (A15) and (A16), it is seen that

$$S_l(k) = \exp[2i\delta_l(k)]. \tag{A35}$$

The matrix $S$, which in the present case of central field scattering has $S_l$ for its diagonal elements and zero for its nondiagonal elements, is called the scattering matrix. In terms of $S_l$, $f(\vartheta)$ and $\sigma$ are given by

$$f(\vartheta) = \frac{1}{2ik} \sum (2l + 1)(S_l - 1) P_l(\cos\vartheta), \tag{A36}$$

$$\sigma = \frac{1}{k^2} \sum (2l + 1)\, |1 - S_l(k)|^2. \tag{A37}$$

The meaning of $S(k)$ in (A34) is as follows: The incident particle plane wave is expressed as a superposition of ingoing and outgoing spherical waves in (A34). Because of the scattering potential $V(r)$, the outgoing spherical wave is modified, by the factor $S_l(k)$ for the $l$-partial wave. $S_l(k)$ depends on $l$ and $k$, and functionally on $V(r)$. If $V(r) = 0$, then $S_l(k) = 1$ and $f(\vartheta) = 0$, i.e., there is no scattering—an obvious result.

The $S$ matrix defined in (A35) is seen to be unitary, i.e., $S S^\dagger = 1$. A more general definition of $S$ and a discussion of its properties will be taken up in Chapters 4 and 6. We shall only remark here in passing that the unitary property of $S$ and the optical theorem (A21) are intimately connected with each other. Thus from (A36), one has

$$\operatorname{Im} f(k, 0) = \frac{1}{2i} [f(k, 0) - f^*(k, 0)]$$

$$= \frac{1}{4k} \sum (2l + 1)[2 - S_l - S_l^*]. \tag{A38}$$

On the other hand, the total cross section is given by

$$\sigma = \int f(k, \vartheta) f^*(k, \vartheta)\, d\Omega = \frac{4\pi}{4k^2} \sum (2l + 1)(S_l - 1)(S_l^* - 1) \tag{A39}$$

so that $S_l S_l^\dagger$ $(= S_l S_l^*$ in this case$) = 1$ leads to the relation (A21). The optical theorem (A21) and the unitarity of $S$ are both statements of the conservation of particle flux in the scattering process. (See Chapter 6, Sect. W.2.)

For a more general optical theorem which includes inelastic collisions, see Chapter 3, Sect. N (N19).

NOTE: There are several matrices connected with the $S$-matrix, which are occasionally used to characterize collisions. Among them the reactance matrix $X$, reaction matrix $K$ and the transition matrix $T$ (see Sect. P), are defined by

$$X = i\frac{1 - S}{1 + S}, \qquad K = -2X, \qquad T = S - 1.$$

The matrix $X$ is diagonal and has $\tan \delta_l$ as the diagonal element for scattering by central potentials. Another similar matrix ($R$-matrix) will appear in Sect. X.

The matrix $T$ has matrix elements $T_l = 2i \exp(i\delta_l) \sin \delta_l$. It seems desirable to obtain $T$ from the general scattering equation. We shall use the Schrödinger equation in the integral form (B14)

$$\psi(\mathbf{r}) = e^{i\mathbf{k}\cdot\mathbf{r}} - \frac{1}{4\pi}\int \frac{\exp(ik|\mathbf{r} - \mathbf{r}'|)}{|\mathbf{r} - \mathbf{r}'|} U(r')\,\psi(\mathbf{r}')\,d\mathbf{r}',$$

which is equivalent to (A2) together with the asymptotic condition (A3). Using (A8) for $e^{i\mathbf{k}\cdot\mathbf{r}}$ and (A12) for $\psi(\mathbf{r})$, we can write the above equation in the form

$$\sum (2l + 1)\, i^l \left[\frac{u_l(kr)}{kr} - j_l(kr)\right] P_l(\cos\vartheta)$$
$$= -\frac{1}{4\pi}\sum (2l + 1)\, i^l \int \frac{\exp(ik|\mathbf{r} - \mathbf{r}'|)}{|\mathbf{r} - \mathbf{r}'|} U(r')\,\frac{u_l(kr')}{kr'} P_l(\cos\vartheta')\,d\mathbf{r}', \quad \text{(AA-1)}$$

where $\vartheta'$ = angle between $\mathbf{k}$ and $\mathbf{r}'$, $\vartheta$ = angle between $\mathbf{k}$ and $\mathbf{r}$.

We shall use the expression

$$\frac{\exp(ik|\mathbf{r} - \mathbf{r}'|)}{|\mathbf{r} - \mathbf{r}'|} = ik\sum (2l + 1)\, j_l(kr')\, h_l^{(1)}(kr)\, P_l(\cos\xi), \qquad r > r',$$

where $\xi$ is the angle between $\mathbf{r}$ and $\mathbf{r}'$ and $h_l^{(1)}$ is the Hankel function $h_l^{(1)}$ given in (A11). Expanding $P_l(\cos\xi)$ in terms of

$$P_n^m(\cos\vartheta)\, P_n^m(\cos\vartheta') \exp[im(\varphi - \varphi')]$$

and integrating over $\vartheta'$, $\varphi'$, the right-hand side of (AA-1) becomes

$$-\sum (2l + 1)\, i^l P_l(\cos\vartheta)\left[ih_l^{(1)}(kr)\int_0^r j_l(kr')\, U(r')\, u_l(kr')\, r'\, dr'\right.$$
$$\left. + j_l(kr)\int_r^\infty ih_l^{(1)}(kr')\, U(r')\, u_l(kr')\, r'\, dr'\right].$$

for large values of $r$, the second term approaches zero,

$$ih_l^{(1)}(kr) \to \frac{1}{kr} \exp\left[i\left(kr - \frac{l\pi}{2}\right)\right],$$

and using (A15), we have from (AA-1)

$$\frac{1}{2kr} \sum (2l+1)\, i^{l+1} \left[1 - \exp(2i\delta_l)\right] \exp\left[i\left(kr - \frac{l\pi}{2}\right)\right] P_l(\cos\vartheta)$$

$$= -\frac{1}{2kr} \sum (2l+1)\, i^{l+1}\, T_l \exp\left[i\left(kr - \frac{l\pi}{2}\right)\right] P_l(\cos\vartheta), \qquad \text{(AA-2)}$$

where

$$T_l = \frac{2}{i} \int_0^\infty j_l(kr')\, U(r')\, u_l(kr')\, r'\, dr'. \qquad \text{(AA-3)}$$

On multiplying (AA-2) by $d\cos\vartheta$ and integrating over $\cos\vartheta$, we obtain

$$1 - \exp 2i\delta_l = -T_l, \qquad \text{(AA-4)}$$

or by (A35),

$$T_l = S_l - 1. \qquad \text{(AA-5)}$$

From (AA-4), we obtain $4\sin^2\delta_l = |T_l|^2$, or from (AA-3) and (A25),

$$T_l = 2i \exp(i\delta_l) \sin\delta_l, \qquad \text{(AA-6)}$$

keeping in mind the fact that $u_l(kr)$ in (A25) is normalized according to (A26), whereas $u_l(kr)$ in (AA-2) and (AA-3), is normalized according to (A14) and (A16).

In Sect. P, (P27), we shall come across this operator $T$ for the case of potential scattering.

## 6. SCATTERING BY A COULOMB FIELD

An exact solution of (A13) for $f(\vartheta)$ in analytic form is known only for a very limited number of fields $V(r)$; of these, the most important is the Coulomb field. Consider the collision between two particles of charges $Ze$ and $Z'e$, of reduced mass $m$, and relative velocity $v(mv = \hbar k)$. Equation (A13) is

$$\frac{d^2u_l}{dr^2} + \left[k^2 - U - \frac{l(l+1)}{r^2}\right] u_l = 0, \qquad U(r) = \frac{2k\alpha}{r}, \qquad \text{(A40)}$$

$$\alpha = \frac{mZZ'e^2}{k\hbar^2} = \frac{ZZ'e^2}{\hbar v}. \qquad \text{(A40a)}$$

Firstly, it must be noted that the method described in Sect. A.1 is not applicable for the case of Coulomb field. This can easily be demonstrated in the following manner: Consider, for simplicity, the $S$ wave $l = 0$ of (A40). For large values of $r$, the rate of change of $V(r)$ is small and the *J-W-K-B* method can be applied, giving the asymptotic solution

$$u_l = \frac{1}{k} \sin \int^r (k^2 - U)^{1/2}\, dr.$$

The phase shift defined in (A14) and (A23) is#

$$\delta_0 = \lim_{r \to \infty} \int_{r_0}^{r} [(k^2 - U)^{1/2} - k]\, dr$$

which is not a constant but is a logarithmic function of $r$. This is a consequence of the long-range nature of the Coulomb field which calls for a special treatment. A detailed account of the theory of scattering by a Coulomb field can be found in the work of Mott and Massey (ref. 4). We shall give only a brief summary of the results here.

For the scattering problem, it is convenient to use spherical polar coordinates. The equations (A13), (A40) have been treated in these coordinates by Gordon (ref. 9). But the Schrödinger equation (A2) for Coulomb fields can be solved exactly in a more elegant manner in parabolic coordinates (Temple, ref. 10)

$$\zeta = r - z, \qquad \eta = r + z, \qquad \varphi = \tan^{-1}(y/x)$$

As we shall see, the Coulomb field affects a particle even at large distances in such a way that the wave function is not a plane wave at large distances, and the parabolic coordinates are suitable for describing the "distorted" waves by the Coulomb field. On account of the axial symmetry, we shall have no dependence of the wave function describing the scattering on the angle $\varphi$. On making the Ansatz

$$\Psi(\mathbf{r}) = e^{ikz} F(\zeta), \tag{A41}$$

Equation (A2) becomes

$$\zeta \frac{d^2F}{d\zeta^2} + (1 - ik\zeta)\frac{dF}{d\zeta} - \alpha k F = 0. \tag{A42}$$

There are two independent solutions of (A42). The one which is regular at $r = 0$ is the hypergeometric series

$$F(\zeta) = {}_1F_1(-i\alpha; 1; ik\zeta), \tag{A43}$$

where

$${}_1F_1(a; b; y) = 1 + \frac{a}{b \cdot 1} y + \frac{a(a+1)}{b(b+1)1 \cdot 2} y^2 + \cdots \tag{A43a}$$

---

# It is also seen that for potentials decreasing at large values of $r$ as $r^{-n}$ where $n < 1$, the phase shift is divergent so that the method discussed in Sect. A.1 is not applicable.

The asymptotic expansion at large values of $y$ is given by

$$
{}_1F_1(a; b; y) \to \frac{\Gamma(b)}{\Gamma(b-a)} (-y)^{-a}\, G(a, a-b+1; -y) + \frac{\Gamma(b)}{\Gamma(a)} e^y\, y^{a-b}\, G(1-a, b-a; y), \tag{A44}
$$

where

$$
G(s, t; y) = 1 + \frac{st}{y(1)} + \frac{s(s+1)t(t+1)}{y^2(2!)} + \cdots .
$$

With these, we obtain for the regular solution in (A43), to terms of order $1/\zeta$, the asymptotic expression

$$
F(\zeta) \to \frac{e^{\frac{1}{2}\pi\alpha}}{\Gamma(1+i\alpha)} \left(1 - \frac{\alpha^2}{ik\zeta}\right) \exp(i\alpha \ln k\zeta) - \frac{i\, e^{\frac{1}{2}\pi\alpha}}{\Gamma(-i\alpha)} \frac{e^{ik\zeta}}{k\zeta} \exp(-i\alpha \ln k\zeta).
$$

For the asymptotic $\Psi(\mathbf{r})$ of (A41), with the incident beam normalized to unit flux, we have

$$
\begin{aligned}
\Psi(\mathbf{r}) &\to \text{incident wave} \qquad\qquad + \text{scattered wave} \\
&\to \left[1 - \frac{\alpha^2}{ik(r-z)}\right] \exp[ikz + i\alpha \ln k(r-z)] \\
&\qquad + \frac{1}{r} \exp(ikr - i\alpha \ln 2kr) f(\vartheta),
\end{aligned} \tag{A45}
$$

where

$$
f(\vartheta) = \frac{ZZ'e^2}{2mv^2} \frac{1}{\sin^2(\vartheta/2)} \exp[-i\alpha \ln \tfrac{1}{2}(1 - \cos\vartheta) + i\pi + 2i\delta_0], \tag{A46}
$$

$$
\exp(2i\delta_0) = \frac{\Gamma(1+i\alpha)}{\Gamma(1-i\alpha)}.
$$

All factors containing $\alpha$ in (A45) arise from the $1/r$ dependence of $U(r)$. The deviation of the incident wave from the plane wave is due to the "distortion" of the wave function by the long-range Coulomb field; the same is true for the scattered wave. The scattered intensity is, however, independent of $\alpha$,

$$
|f(\vartheta)|^2 = \left(\frac{ZZ'e^2}{2mv^2}\right)^2 \frac{1}{\sin^4(\vartheta/2)}, \tag{A47}
$$

which is the same as the Rutherford formula derived from classical mechanics. We shall see in Sect. C that the same result (A47) is also obtained in the Born approximation. These coincidences are peculiar to the Coulomb field.

In spherical coordinates, (A13) can be written, with

$$
\Psi(\mathbf{r}) = \sum (2l+1)\, i^l \exp(i\delta_l) \frac{1}{kr} u_l(r) P_l(\cos\vartheta),
$$

$$
u_l = r^{l+1} e^{ikr} f_l(r),
$$

in the form

$$
r\frac{d^2 f_l}{dr^2} + 2(l+1+ikr)\frac{df_l}{dr} + 2k[(l+1)i - \alpha] f_l = 0, \tag{A48}
$$

which is of the form (A42), so that the solution $f_l$ is given by the hypergeometric series (A43). There are two independent solutions of (A13) or (A48). The one which is regular at $r = 0$ is

$$F_l(r) = ku_l(r)$$
$$= kr\, e^{-\frac{1}{2}\pi\alpha} \frac{|\Gamma(l + 1 + i\alpha)|}{(2l + 1)!} (2kr)^l\, e^{ikr}\, {}_1F_1(i\alpha + l + 1; 2l + 2; -2ikr) \tag{A49}$$

which, asymptotically, is

$$F_l(r) = ku_l(r) \to \sin(kr - \tfrac{1}{2}l\pi + \delta_l - \alpha \ln 2kr), \tag{A49a}$$

$$\delta_l = \arg \Gamma(l + 1 + i\alpha). \tag{A49b}$$

The other solution $G_l(r)$, which is irregular at $r = 0$, may be chosen such that asymptotically for large values of $r$, it is

$$G_l(r) = ku_l(r) \to \cos(kr - \tfrac{1}{2}l\pi + \delta_l - \alpha \ln 2kr). \tag{A50}$$

From (A49), one obtains the value of the square of the wave function $v_0(r)/r$ at $r = 0$ for $S$-wave

$$e^{-\pi\alpha}\, |\Gamma(1 + i\alpha)|^2 = \frac{2\pi\alpha}{e^{2\pi\alpha} - 1} \equiv C^2. \tag{A51}$$

For repulsive Coulomb field and low velocities, i.e., $\alpha > 1$, this $C^2$ is $\simeq 2\pi\alpha e^{-2\pi\alpha}$. The factor $e^{-2\pi\alpha}$ measures the penetration of the particle into the Coulomb barrier and is called the penetration factor.

(A49) and (A50) for $F_l(r)$ and $G_l(r)$ involve complex numbers and are not convenient for calculations. Expansions involving only real quantities have been given by Yost, Wheeler and Breit, and for repulsive Coulomb fields, tables of functions for small values of $r$ have been published by these authors and others (refs. 11, 12).

For the $s$ wave, the two independent solutions of (A48) are usually normalized according to the following asymptotic forms:

$$\begin{aligned} F_0(r) &\to \sin(kr - \alpha \ln 2kr + \delta_0), \\ G_0(r) &\to \cos(kr - \alpha \ln 2kr + \delta_0). \end{aligned} \tag{A52}$$

For small values of $r$, on expanding (A49), we have

$$F_0(r) \to Ckr\, e^{ikr}\, (1 + \alpha kr + \cdots). \tag{A53}$$

The solution $G_0(r)$ for small values of $r$ is (refs. 12 and 13)

$$G_0(r) = C^{-1}\{1 + 2\alpha kr\, [\ln(2\alpha kr) + 2\mathscr{C} - 1 - \ln \alpha + \operatorname{Re} \Psi(-i\alpha)] + \cdots\} \tag{A54}$$

where $\mathscr{C}$ = Euler's constant = 0.5772..., and

$$\Psi(x) = \frac{d}{dx} \ln \Gamma(1 + x), \tag{A55}$$

$$\operatorname{Re} \Psi(-i\alpha) = -\mathscr{C} + \alpha^2 \sum_{n=1}^{\infty} \frac{1}{n(n^2 + \alpha^2)}. \tag{A56}$$

These results will be used in Sect. E, Appendix 3.

## 7. DELAY-TIME, OR COLLISION LIFETIME, MATRIX

According to Schrödinger's wave packet theory, the velocity $v$ of a particle is identified with the group velocity which is given by $d\nu/dk$, where $\nu = E/\hbar$. We shall consider the simplest wave packet consisting of two individual waves whose radial part $\varphi$ has the asymptotic form (A34)

$$\varphi(r, t) \to e^{-i\nu t}\left\{\exp\left[-i\left(kr - \frac{l\pi}{2}\right)\right] - \exp(2i\delta_l)\exp\left[i\left(kr - \frac{l\pi}{2}\right)\right]\right\}. \tag{A57}$$

Let $\varphi_+(r)$ and $\varphi_-(r)$ have frequencies $\nu \pm \Delta\nu$, wave numbers $k \pm \Delta k$ and phase shift $\delta_l \pm \Delta\delta_l$. The wave packet $\varphi$ is

$$\begin{aligned}\varphi &\equiv \varphi_+ + \varphi_- \\ &\to 2\left\{\exp\left[-i\left(kr + \nu t - \frac{l\pi}{2}\right)\right]\cos(r\,\Delta k + t\,\Delta\nu)\right. \\ &\qquad \left. - \exp\left[i\left(kr - \nu t - \frac{l\pi}{2} + 2\delta_l\right)\right]\cos(r\,\Delta k - t\,\Delta\nu + 2\,\Delta\delta_l)\right\}.\end{aligned} \tag{A58}$$

The first term represents an ingoing wave which has a maximum amplitude at

$$r = -\frac{d\nu}{dk}t = -vt, \qquad (\Delta k \to 0), \qquad t < 0, \tag{A59}$$

and the second term represents an outgoing wave with a maximum at

$$r = vt - 2\frac{d\delta_l}{dk}, \qquad (\Delta k \to 0), \qquad t > 0. \tag{A60}$$

From (A59) and (A60) we see that the outgoing particles are delayed by a time

$$(\Delta t)_l = \frac{2\,d\delta_l}{v\,dk} = 2\hbar\frac{d\delta_l}{dE}, \tag{A61}$$

relative to the case of no interaction ($\delta_l = 0$). This delay time has been discussed in detail by Eisenbud, Bohm and Wigner (refs. 14–16). It is seen that $\Delta t$ can be expressed in terms of the $S$ matrix (A35) in the form

$$(\Delta t)_l = -i\hbar\frac{dS_l}{dE}S_l^*, \qquad S_l = \exp(2i\delta_l), \tag{A61a}$$

Now for (elastic) collisions, a *collision lifetime* $Q$ can be defined as the difference between the time spent by the particle in the region of the scattering potential and the time spent in the same region in the absence of the scattering potential, i.e.,

$$Q_l = \lim_{R \to \infty} \int^R \left( \psi_l^* \psi_l - \frac{2}{4\pi v r^2} \right) dr, \tag{A62}$$

where $v$ is the velocity of the particle and $\psi_l$ is the wave function normalized to unit incoming flux. It has been shown by Smith (ref. 17) that for elastic collisions, the delay time $\Delta t$ in (A61) is identical with $Q$.

$$Q_l = (\Delta t)_l = -i\hbar \frac{dS_l}{dE} S_l^*. \tag{A61b}$$

For the limit $R \to \infty$ to exist in the above definition of $Q_l$, the scattering potential must decrease with distance faster than $1/r$.

The form (A61b) for the collision lifetime (for the angular momentum $l$) is convenient for the generalization to inelastic collisions. In this case, $S$ will be a non-diagonal matrix. We shall resume this in Sect. W.5.

The following examples will further explain the meaning and the two names (delay time and lifetime) of $\Delta t$. The first example is a rather simple one. Consider an infinitely high repulsive potential (hard core) with the radius $a$. The $s$ phase shift $\delta_0$ is equal to $-ka$, hence by (A61) $\Delta t$ is $-2a/v$, of which the absolute value is just the time the particle would have spent in going in and out of the hard core. $|\Delta t|$ is smaller for higher $l$ waves, since only a small portion of higher $l$ waves reaches the core on account of the centrifugal barrier.

A second example concerns the sharp resonance scattering, whose main feature will be represented by the single-level formula (A32) (see also Sect. U and X),

$$\delta = \text{const} + \tan^{-1} \frac{\Gamma/2}{E_r - E}.$$

From (A61) we have

$$\Delta t = \hbar\Gamma[(E_r - E)^2 + (\Gamma/2)^2]^{-1}, \tag{A63}$$

which at exact resonance has the value $\Delta t_r$

$$\Delta t_r = \frac{4\hbar}{\Gamma}. \tag{A64}$$

The average time delay of the scattered beam over the resonance region is given by

$$\overline{\Delta t} = \int_0^\infty \frac{(\Gamma/2)\, \Delta t\, dE}{\pi[(E_r - E)^2 + (\Gamma/2)^2]} = \frac{2\hbar}{\Gamma}, \tag{A65}$$

since the scattering cross section itself is proportional to

$$[(E_r - E)^2 + (\Gamma/2)^2]^{-1}$$

(if the small potential scattering is omitted). When a long wave train is used# so that the incident beam has sufficiently narrow energy spread at exact resonance, we shall observe that the "intermediate" state has a lifetime twice the decay lifetime. This fact may be understood if we notice that the collision lifetime includes both the time required to form the metastable state and the time for it to break up again. The metastable state will be defined as the eigenfunction of the lifetime matrix, of which the eigenvalue is twice the decay lifetime.

## 8. SPREADING OF A WAVE PACKET IN SCATTERING

We have seen in the preceding subsection that a particle incident upon the scattering centre is delayed by the time given by (A61). This will be further examined by considering another typical wave packet.

The radial part of the $l$ wave of the wave packet can be written asymptotically in the following way,

$$\varphi_l(r, t) \to \int_0^\infty f(k)\left\{\exp\left[-i\left(kr - \frac{l\pi}{2}\right)\right]\right.$$
$$\left. - \exp(2i\delta_l)\exp\left[i\left(kr - \frac{l\pi}{2}\right)\right]\right\} e^{-i\nu t}\, dk. \quad \text{(A66)}$$

If we take a Gaussian distribution for $f(k)$

$$f(k) = A\exp[-a^2(k - k_0)^2], \quad \text{(A67)}$$

(A66) becomes

$$\varphi_l(r, t) \to A\exp\left\{-\left[\frac{(r + vt)^2}{4b} + i\left(k_0 r - \frac{l\pi}{2} + \Omega t\right)\right]\right\}$$
$$\times \int \exp\left\{-b\left[k' + i\frac{(r + vt)}{2b}\right]^2\right\} dk'$$
$$- A\exp\left\{-\left[\frac{(r - vt)^2}{4b} + i\left(-k_0 r + \frac{l\pi}{2} + \Omega t\right)\right]\right\}$$
$$\times \int \exp(2i\delta_l)\exp\left\{-b\left[k' + i\frac{(-r + vt)}{2b}\right]^2\right\} dk', \quad \text{(A68)}$$

where $b = a^2 - \dfrac{it\hbar}{2m}$, $v = \dfrac{\hbar k_0}{m}$, $\Omega = \dfrac{\hbar k_0^2}{2m}$.

The first term represents the incoming beam with the centre at $r = -vt$, ($t \leqslant 0$). The integral in the first term yields $(\pi/b)^{1/2}$ if the contribution from the region $k < 0$ in (A67) is negligible. The second term represents the outgoing wave after scattering. To integrate the second term we shall assume

---

# This is the case here. Note that $\Delta k \to 0$ in (A59).

that the phase shift $\delta_l$ is a smooth function of energy so that it is a good approximation to set

$$\delta_l(k) = \delta_l(k_0) + \frac{d\delta_l(k_0)}{dk}(k - k_0) + \frac{d^2\delta_l(k_0)}{2dk^2}(k - k_0)^2 \tag{A69}$$

for the vicinity of $k = k_0$: $k_0 - (1/a) \lesssim k \lesssim k_0 + (1/a)$.
$\varphi_l(r, t)$ of (A68) now takes the form

$$\begin{aligned}\varphi_l(r, t) \to A\sqrt{\frac{\pi}{b}}\exp\left\{-\left[\frac{(r + vt)^2}{4b} + i\left(k_0 r - \frac{l\pi}{2} + \Omega t\right)\right]\right\} \\ - A\sqrt{\frac{\pi}{b - i\eta}}\exp\left[2i\delta_l(k_0)\right] \\ \times \exp\left\{-\left[\frac{\left(r - vt + 2\dfrac{d\delta_l}{dk}\right)^2}{4(b - i\eta)} + i\left(-k_0 r + \frac{l\pi}{2} + \Omega t\right)\right]\right\}\end{aligned} \tag{A70}$$

where $\eta \equiv [d^2\delta_l(k_0)]/dk^2$. The probability density of the first term is

$$\varphi_l^*\varphi_l \to \frac{|A|^2\pi}{\left|a^2 - \dfrac{it\hbar}{2m}\right|}\exp\left\{-\left[\frac{(r + vt)^2}{2\left|a^2 - \dfrac{it\hbar}{2m}\right|}\right]\right\}, \qquad t \ll 0, \tag{A71}$$

while the probability density of the outgoing particle is given by

$$\varphi_l^*\varphi_l \to \frac{|A|^2\pi}{\left|a^2 - \dfrac{it\hbar}{2m} - i\eta\right|}\exp\left\{-\left[\frac{\left(-r + vt - 2\dfrac{d\delta_l(k_0)}{dk}\right)^2}{2\left|a^2 - \dfrac{it\hbar}{2m} - i\eta\right|}\right]\right\}, \qquad t \gg 0. \tag{A72}$$

A comparison of (A71) and (A72) shows that (i) the outgoing particle is delayed by the time $\Delta t = (2/v)(d\delta_l/dk)$ after the scattering, (ii) the spread (in space) of the beam is given by

$$\sqrt[4]{a^4 + \left(\frac{\hbar t}{2m}\right)^2} \qquad \text{for} \quad t \ll 0,$$

and

$$\sqrt[4]{a^4 + \left(\frac{\hbar t}{2m} + \frac{d^2\delta_l}{dk^2}\right)^2} \qquad \text{for} \quad t \gg 0.$$

Therefore the expansion or spreading of wave packet is hastened by the time $(2m/\hbar)(d^2\delta/dk^2)$ compared with the case of no interaction.

The time delay in scattering is an observable effect. However, the delay in resonance scattering given by (A63) can not be measured accurately if the energy spread of the beam is less than $\Gamma$, because the spread (in space) of the wave packet is necessarily larger than $v\hbar/\Gamma$ by virtue of the uncertainty principle of quantum mechanics. (A63) will be meaningful only statistically. If the packet is so closely confined in space that the energy spread covers the resonance curve (A32) entirely, the time delay is not given by (A65) because of the invalidity of the assumption $\Delta k \to 0$ in (A59). The average time delay becomes just $\hbar/\Gamma$. The delayed wave, however, does not significantly interfere with the "quickly" scattered waves with different $l$, therefore the observed cross section can not be given by the average cross section in energy $|f(\vartheta)|_{\text{Av}}$ which has been derived by the stationary state treatment assuming a plane wave for the incoming particle. One of the examples of this circumstance will appear in nuclear reactions. (See Sect. U.)

### REFERENCES

The theory of partial waves is given in the classic paper

1. Faxen, H. and Holtzmark, J., Zeits. f. Physik **45**, 307 (1927).

The optical theorem is first given by

2. Feenberg, E., Phys. Rev. **40**, 40 (1932);

See also

3. Bohr, N., Peierls, R. E. and Placzek, G., Nature **144**, 200 (1939).

For an extensive treatment of the theory of scattering:

4. Mott, N. F. and Massey, H. S. W., *Theory of Atomic Collisions*, Clarendon Press, Oxford, 2nd ed. (1949);
5. Condon, E. U., Rev. Mod. Phys. **3**, 43 (1931);
6. Condon, E. U. and Morse, P. M., Rev. Mod. Phys. **4**, 577 (1932);
7. Cf. Breit, G., article in Handbuch der Physik Vol. **41/1**, Springer (1959), p. 24;
8. Landau, L. D. and Lifshitz, E. M., *Quantum Mechanics*, Pergamon Press, London, (1958), § 119;
9. Gordon, W., Zeits. f. Physik **48**, 180 (1928);
10. Temple, G., Proc. Roy. Soc. London, **A121**, 673 (1928);
11. Yost, F. L., Wheeler, J. A. and Breit, G., Phys. Rev. **49**, 174 (1936); Terrest, Mag. and Atmos. Elect. **40**, 443 (1935); Wicher, E. R., Terrest. Mag. and Atmos. Elect. **41**, 389 (1936);
12. Hull, M. H. and Breit, G., article in Handbuch der Physik Vol. **41/1**, Springer (1959). This article gives a complete treatment of Coulomb wave functions and bibliography of tables of these functions;

**13.** Sexl, Th., Zeits. f. Physik **56,** 72 (1929);

**14.** Eisenbud, L., Dissertation (unpublished), Princeton University (1948), quoted in refs. 16, 17;

**15.** Bohm, D., *Quantum Theory*. Prentice-Hall, New York (1951), pp. 257–261;

**16.** Wigner, E. P., Phys. Rev. **98,** 145 (1955). Lower bound for $d\delta_l/dE$, by the use of the $R$-matrix;

**17.** Smith, F. T., Phys. Rev. **118,** 349; **119,** 2098 (1960).

section

# B Integral Equation for Scattering

In Sect. A, we have the exact theory of the scattering of a particle by a potential field on the basis of the solution of the Schrödinger (differential) equation (A2), subject to the asymptotic conditions corresponding to the physical situation of an incident plane wave and outgoing spherical waves. We shall now transform the Schrödinger equation into an integral equation in a form that automatically ensures the correct prescribed asymptotic conditions.

## 1. INTRODUCTION: GREEN'S FUNCTION

The Schrödinger equation (A2)

$$[\Delta + k^2 - U(r)]\,\Psi(\mathbf{r}) = 0 \tag{B1}$$

can be put in the form

$$(\mathscr{H}_0 - E)\,\Psi(\mathbf{r}) = \chi(\mathbf{r}) \tag{B2}$$

where

$$\mathscr{H}_0 = -\Delta, \qquad E = k^2, \qquad \chi(\mathbf{r}) = -U(r)\,\Psi(\mathbf{r}). \tag{B3}$$

The asymptotical condition (A3) is

$$\Psi(\mathbf{r}) \to e^{ikz} + \frac{e^{ikr}}{r} f(\vartheta). \tag{B4}$$

From the theory of differential equations, the solution of equation (B2) satisfies the equation

$$\Psi(\mathbf{r}) = \phi(\mathbf{r}) + \int G(\mathbf{r}, \mathbf{r}')\,\chi(\mathbf{r}')\,d\mathbf{r}', \tag{B5}$$

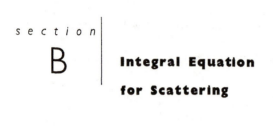

where $\phi(\mathbf{r})$ is a particular solution of the homogeneous equation

$$(\mathscr{H}_0 - E)\,\phi_{\mathbf{k}}(\mathbf{r}) = 0, \qquad \phi_{\mathbf{k}}(\mathbf{r}) = e^{i\mathbf{k}\cdot\mathbf{r}}, \tag{B6}$$

and $G(\mathbf{r}, \mathbf{r}')$ is the Green's function of $\mathscr{H}_0 - E$, i.e.,

$$(\mathscr{H}_0 - E)\,G(\mathbf{r}, \mathbf{r}') = \delta(\mathbf{r} - \mathbf{r}'). \tag{B7}$$

To prove (B5) and to find Green's function, let a particular integral of the inhomogeneous equation (B2) be expanded in terms of the complete, orthogonal set $\phi_{\mathbf{k}}(\mathbf{r})$ in (B6)

$$\Psi(\mathbf{r}) = \int c_{\mathbf{k}'}\,\phi_{\mathbf{k}'}(\mathbf{r})\,d\mathbf{k}'.$$

The coefficients $c_{\mathbf{k}'}$ are determined from (B2) thus:

$$(2\pi)^3 c_{\mathbf{k}'} = \frac{-1}{k^2 - k'^2}\int \phi^*_{\mathbf{k}'}(\mathbf{r}')\,\chi(\mathbf{r}')\,d\mathbf{r}'.$$

We can now write

$$\Psi(\mathbf{r}) = \int G_{\mathbf{k}}(\mathbf{r}, \mathbf{r}')\,\chi(\mathbf{r}')\,d\mathbf{r}', \tag{B8}$$

where

$$(2\pi)^3 G_{\mathbf{k}}(\mathbf{r}, \mathbf{r}') = -\int \frac{\phi_{\mathbf{k}'}(\mathbf{r})\,\phi^*_{\mathbf{k}'}(\mathbf{r}')}{k^2 - k'^2}\,d\mathbf{k}'. \tag{B9}$$

It is seen that the function $G_{\mathbf{k}}(\mathbf{r}, \mathbf{r}')$ satisfies the equation (B7), by virtue of the closure relation

$$\int \phi^*_{\mathbf{k}}(\mathbf{r}')\,\phi_{\mathbf{k}}(\mathbf{r})\,d\mathbf{k} = (2\pi)^3\delta(\mathbf{r}' - \mathbf{r}). \tag{B10}$$

On adding a particular solution of the homogeneous equation (B6) to (B8), we obtain the general solution (B5). Since $\chi(\mathbf{r})$ contains the unknown function $\Psi(\mathbf{r})$ itself, (B5)

$$\Psi(\mathbf{r}) = \phi_{\mathbf{k}}(\mathbf{r}) - \int G(\mathbf{r}, \mathbf{r}')\,U(\mathbf{r}')\,\Psi(\mathbf{r}')\,d\mathbf{r}' \tag{B11}$$

is in fact an integral equation for $\Psi$. Equation (B11) is the Schrödinger equation (B1) in integral form.

The advantage of putting (B1) in the form (B11) is that, by an appropriate choice of $\phi_{\mathbf{k}}$ and $G_{\mathbf{k}}(\mathbf{r}, \mathbf{r}')$ in (B11), it is possible to have the desired condition (B4) incorporated into (B11). To understand this, consider instead of (B9) the following:

$$(2\pi)^3 G^+_{\mathbf{k}}(\mathbf{r}, \mathbf{r}') = -\lim_{\varepsilon\to 0}\int \frac{\phi_{\mathbf{k}'}(\mathbf{r})\,\phi^*_{\mathbf{k}'}(\mathbf{r}')\,d\mathbf{k}'}{k^2 - k'^2 + i\varepsilon}, \qquad \varepsilon > 0. \tag{B12}$$

On using (B6) for $\phi_{\mathbf{k}}$ and carrying out the integrations over the directions of $\mathbf{k}'$, we have, for the integral in (B12),

$$\frac{2\pi}{i\rho}\int_0^\infty \frac{e^{ik'\rho} - e^{-ik'\rho}}{k^2 - k'^2 + i\varepsilon}\,k'\,dk', \qquad \boldsymbol{\rho} = \mathbf{r} - \mathbf{r}'.$$

This integral can be seen to be, up to terms of the order of $\varepsilon$,

$$\cong \frac{\pi}{i\rho} \int_{-\infty}^{\infty} \frac{e^{ik'\rho} - e^{-ik'\rho}}{\left(k + \dfrac{i\varepsilon}{2k} + k'\right)\left(k + \dfrac{i\varepsilon}{2k} - k'\right)} k'\, dk'.$$

The first integral can be replaced by a contour integral along the real axis and a semi-circle in the upper half complex $k$-plane. The pole is at $k' = k + (i\varepsilon/2k)$ and the residue $\frac{1}{2} \exp \{i[k + (i\varepsilon/2k)]\rho\}$. The second integral can be shown to be the same as the first by a transformation $k' = -k''$. Thus by Cauchy's theorem one gets for (B12)

$$G_{\mathbf{k}}^{+}(\mathbf{r}, \mathbf{r}') = \frac{\exp(ik|\mathbf{r} - \mathbf{r}'|)}{4\pi|\mathbf{r} - \mathbf{r}'|}, \tag{B13}$$

and (B11) becomes

$$\Psi^{+}(\mathbf{r}) = e^{ikz} - \frac{1}{4\pi} \int \frac{\exp(ik|\mathbf{r} - \mathbf{r}'|)}{|\mathbf{r} - \mathbf{r}'|} U(\mathbf{r}')\Psi^{+}(\mathbf{r}')\, d\mathbf{r}'. \tag{B14}$$

Asymptotically, for large values of $r$ ($r \gg r'$),

$$|\mathbf{r} - \mathbf{r}'| \simeq r - r' \cos \Theta,$$

where $\Theta$ is the angle between $\mathbf{r}'$ and $\mathbf{r}$, $\mathbf{r}$ being the direction along which we are calculating the scattered wave $\mathbf{k}'$, i.e.,

$$ik|\mathbf{r} - \mathbf{r}'| \simeq ikr - ikr' \cos \Theta = ikr - i\mathbf{k}' \cdot \mathbf{r}'. \tag{B15}$$

Hence, asymptotically.

$$\Psi^{+}(\mathbf{r}) \to e^{ikz} - \frac{e^{ikr}}{r} \cdot \frac{1}{4\pi} \int \exp(-ik' \cdot \mathbf{r}')\, U(r')\Psi^{+}(\mathbf{r}')\, d\mathbf{r}', \tag{B16}$$

which satisfies the condition (B4) for outgoing scattered waves. From this and (B4), we obtain the scattering amplitude in the following expression

$$f(\vartheta) = -\frac{1}{4\pi} \int \exp(-i\mathbf{k}' \cdot \mathbf{r}')\, U(r')\Psi^{+}(\mathbf{r}')\, d\mathbf{r}', \tag{B17}$$

where $\Psi^{+}(\mathbf{r})$ is the solution of the integral equation (B14).

Had we wanted a solution of (B2) with ingoing scattered waves, we have only to use $G_{\mathbf{k}}^{-}(\mathbf{r}, \mathbf{r}')$ defined by (B12) with $i\varepsilon$ replaced by $-i\varepsilon$. (B13) would be replaced by

$$G_{\mathbf{k}}^{-}(\mathbf{r}, \mathbf{r}') = \frac{\exp(-ik|\mathbf{r} - \mathbf{r}'|)}{4\pi|\mathbf{r} - \mathbf{r}'|}, \tag{B13a}$$

with the corresponding change in (B14).

The integral equation (B11) for the case of outgoing and ingoing waves

$$\Psi^{\pm}(\mathbf{r}) = e^{ikz} - \int G_{\mathbf{k}}^{\pm}(\mathbf{r}, \mathbf{r}')\, U(r')\Psi^{\pm}(\mathbf{r}')\, d\mathbf{r}' \tag{B18}$$

with

$$G_{\mathbf{K}}^{\pm}(\mathbf{r}, \mathbf{r}') = \lim_{\varepsilon \to 0} -(2\pi)^{-3} \int \frac{\phi_{\mathbf{K}'}(\mathbf{r})\, \phi_{\mathbf{K}'}(\mathbf{r}')}{E - \mathscr{H}_0 \pm i\varepsilon}\, d\mathbf{k}' = \frac{\exp(\pm ik|\mathbf{r} - \mathbf{r}'|)}{4\pi|\mathbf{r} - \mathbf{r}'|}, \quad \text{(B19)}$$

can be put in the following symbolic form

$$\Psi^{\pm}(\mathbf{r}) = \phi(\mathbf{r}) + \frac{1}{E - \mathscr{H}_0 \pm i\varepsilon} U(r)\, \Psi^{\pm}(\mathbf{r}). \quad \text{(B20)}$$

The integral operator $1/(E - \mathscr{H}_0 \pm i\varepsilon)$ has the meaning

$$\frac{1}{E - \mathscr{H}_0 \pm i\varepsilon} \chi(\mathbf{r}) = (2\pi)^{-3} \lim_{\varepsilon \to 0} \int d\mathbf{k}' \frac{1}{k^2 - k'^2 \pm i\varepsilon} \phi_{\mathbf{K}'}(\mathbf{r}) \int \phi_{\mathbf{K}'}^{*}(\mathbf{r}')\, \chi(\mathbf{r}')\, d\mathbf{r}'. \quad \text{(B21)}$$

In general, for any Hamiltonian $H$ and any function $\chi$ (in the same space in which $H$ operates), the meaning of $[1/(E - H \pm i\varepsilon)]\chi$ is as follows: Let $\chi$ be expanded in terms of the eigenfunctions $\Phi_n$ of $H$

$$\chi = \sum \int \Phi_n(\Phi_n, \chi) = \sum_n \int \Phi_n \int \Phi_n^{*}\, \chi\, d\mathbf{r}.$$

Then

$$\frac{1}{E - H \pm i\varepsilon} \chi(\mathbf{r}) = \lim_{\varepsilon \to 0} \sum_n \int \frac{1}{E - E_n \pm i\varepsilon} \Phi_n(\mathbf{r}) \int \Phi_n^{*}(\mathbf{r}')\, \chi(\mathbf{r}')\, d\mathbf{r}'. \quad \text{(B22)}$$

In particular, if $1/(E - H \pm i\varepsilon)$ operates on an eigenfunction $\Phi_n$ of $H$ itself, then

$$\frac{1}{E - H \pm i\varepsilon} \Phi_n(\mathbf{r}) = \frac{1}{E - E_n} \Phi_n(\mathbf{r}). \quad \text{(B22a)}$$

Apart from the convenience of the form (B16), (B18), or (B20) in the general theory of scattering (Chapters 3, 4 and 6), from the practical point of view the integral form immediately suggests the method of solution by iteration or Fredholm's method as briefly described below.

## 2. SOLUTION OF INTEGRAL EQUATION

The integral Schrödinger equation insuring outgoing wave condition is given in (B14), which, on substituting $\lambda U(r)$ for $U(r)$ in (B3), where $\lambda$ is a constant, is

$$\psi(\mathbf{x}) = e^{i\mathbf{k}\cdot\mathbf{x}} + \lambda \int K(\mathbf{x}, \mathbf{y})\, \psi(\mathbf{y})\, d\mathbf{y}, \quad \text{(B23)}$$

where

$$K(\mathbf{x}, \mathbf{y}) = -\frac{1}{4\pi} \frac{1}{|\mathbf{x} - \mathbf{y}|} \exp(ik|\mathbf{x} - \mathbf{y}|)\, U(y). \quad \text{(B24)}$$

Let $\mathbf{k}'$ be the wave vector of the scattered wave so that the momentum transfer $\boldsymbol{\tau}$ is

$$\boldsymbol{\tau} = \mathbf{k} - \mathbf{k}', \qquad |\boldsymbol{\tau}| = 2k \sin\frac{\vartheta}{2}, \tag{B25}$$

where $\vartheta$ is the angle of scattering. The asymptotic form (B16), now expressed in terms of $\boldsymbol{\tau}$, is

$$\psi(\mathbf{x}) \rightarrow e^{i\mathbf{k}\cdot\mathbf{x}} + \frac{1}{x} e^{ikx} f(\mathbf{k}, \boldsymbol{\tau}). \tag{B26}$$

The scattering problem is now one of solving the integral equation (B23), with the potential $U(\mathbf{r})$ appearing in the kernel $K$ given by (B24). For a given physical problem, $U(r)$ is given and one may consider $\lambda = 1$. But from a more general, mathematical point of view, one may regard $\lambda$ as a parameter of the integral equation (B23), which may be put in the form

$$\psi(\lambda; \mathbf{k}, \mathbf{x}) = e^{i\mathbf{k}\cdot\mathbf{x}} + \lambda \int K(\mathbf{x}, \mathbf{y})\, \psi(\lambda; \mathbf{k}, \mathbf{y})\, d\mathbf{y}. \tag{B23a}$$

(B23), or (B23a), is known as Fredholm's equation.

Equation (B23) may be solved by various methods such as that of successive approximations and the variational method. The latter will be discussed in Sect. D. The method of successive approximations, known as Born approximations, consists in iterating (B23) and is hence a series expansion in powers of $\lambda$. We shall discuss this Born expansion and the question of convergence in Sect. C. Equation (B23a) can also be solved by Fredholm's method (ref. 5), according to which the solution, if it exists, is

$$\psi(\lambda; \mathbf{k}, \mathbf{x}) = e^{i\mathbf{k}\cdot\mathbf{x}} + \lambda \int \frac{\Delta(\lambda; \mathbf{x}, \mathbf{y})}{\Delta(\lambda)}\, e^{i\mathbf{k}\cdot\mathbf{y}}\, d\mathbf{y}, \tag{B27}$$

where

$$\Delta(\lambda; \mathbf{x}, \mathbf{y}) = K(\mathbf{x}, \mathbf{y}) + \sum_{n=1}^{\infty} \frac{1}{n!} (-\lambda)^n \int \cdots \int d\mathbf{x}_1 \ldots d\mathbf{x}_n \times \begin{vmatrix} K(\mathbf{x}, \mathbf{y}) & K(\mathbf{x}, \mathbf{x}_1) & \ldots & K(\mathbf{x}, \mathbf{x}_n) \\ K(\mathbf{x}_1, \mathbf{y}) & K(\mathbf{x}_1, \mathbf{x}_1) & \ldots & K(\mathbf{x}_1, \mathbf{x}_n) \\ \ldots & & & \\ K(\mathbf{x}_n, \mathbf{y}) & K(\mathbf{x}_n, \mathbf{x}_1) & \ldots & K(\mathbf{x}_n, \mathbf{x}_n) \end{vmatrix} \tag{B27a}$$

$$\Delta(\lambda) = 1 + \sum_{n=1}^{\infty} \frac{1}{n!} (-\lambda)^n \int \cdots \int d\mathbf{x}_1 \ldots d\mathbf{x}_n \times \begin{vmatrix} K(\mathbf{x}_1, \mathbf{x}_1) & \ldots & K(\mathbf{x}_1, \mathbf{x}_n) \\ \ldots & & \\ K(\mathbf{x}_n, \mathbf{x}_1) & \ldots & K(\mathbf{x}_n, \mathbf{x}_n) \end{vmatrix} \tag{B27b}$$

Both series $\Delta(\lambda; \mathbf{x}, \mathbf{y})$, $\Delta(\lambda)$ can be shown to be convergent for all values of $\lambda$, provided $K(\mathbf{x}, \mathbf{y})$ is continuous. If $\Delta(\lambda) \neq 0$, then $\psi(\lambda; \mathbf{k}, \mathbf{x})$ above is the solution of (B23a). Jost and Pais, and Khuri have applied Fredholm's method to (B23a), and have investigated the condition on $U(y)$, which enters in $K(\mathbf{x}, \mathbf{y})$ in (B24), for the series $\Delta(\lambda; \mathbf{x}, \mathbf{y})$, $\Delta(\lambda)$ to be convergent, and the conditions on the values of $\lambda$ for the existence of a solution. These studies, in addition to giving a formal series solution (in power of $\lambda$) to (B23a), form the basis of the investigation of the convergence of the Born expansion which is perhaps the most frequently used method of obtaining approximate solutions and which will be discussed in the next section. In the present section, we shall briefly summarize the results of these authors. Further accounts of the bound states in this method and in the theory of the $S$ matrix will be given in Sect. F. For the mathematical proofs of the results given below, refer to the original papers.

In the scattering problem, the kernel $K(\mathbf{x}, \mathbf{y})$ in (B24) is singular at $\mathbf{x} = \mathbf{y}$. One may circumvent this by formally multiplying both the numerator, $\Delta(\lambda; \mathbf{x}, \mathbf{y})$, and the denominator, $\Delta(\lambda)$, in (B27) by $\exp[\lambda \int K(\mathbf{x}, \mathbf{x})\, d\mathbf{x}]$, thereby obtaining expressions for $\Delta(\lambda; \mathbf{x}, \mathbf{y})$, $\Delta(\lambda)$ in which all the $K(\mathbf{x}_i, \mathbf{x}_i)$, $i = 1, \ldots, n$, in (B27a, b) are replaced by 0.# Or, instead of applying

---

# This can be seen by expanding the determinants in (B27a), multiplying $\Delta(\lambda; \mathbf{x}, \mathbf{y})$ by

$$\exp[\lambda \int K(\mathbf{x}, \mathbf{x})\, d\mathbf{x}] = 1 + \lambda \int K(\mathbf{x}, \mathbf{x})\, d\mathbf{x} + \frac{\lambda^2}{2!}\left[\int K(\mathbf{x}, \mathbf{x})\, d\mathbf{x}\right]^2 + \cdots,$$

collecting terms in $(-\lambda)^n$ and showing that the $(-\lambda)^n$ term of $\Delta(\lambda; \mathbf{x}, \mathbf{y}) \exp[\lambda \int K(\mathbf{x}, \mathbf{x})\, d\mathbf{x}]$ is

$$\frac{1}{n!}(-\lambda)^n \int \cdots \int d\mathbf{x}_1 \ldots d\mathbf{x}_n \begin{vmatrix} K(\mathbf{x}, \mathbf{y}) & K(\mathbf{x}, \mathbf{x}_1) & \ldots & K(\mathbf{x}, \mathbf{x}_n) \\ K(\mathbf{x}_1, \mathbf{y}) & 0 & \ldots & K(\mathbf{x}_1, \mathbf{x}_n) \\ \ldots & \ldots & 0 & \ldots \\ \cdot & \cdot & \cdot & \cdot \\ K(\mathbf{x}_n, \mathbf{y}) & K(\mathbf{x}_n, \mathbf{x}_1) & \ldots & 0 \end{vmatrix}$$

by means of the diagonal expansion of a determinant (refer to A. C. Aitken, *Determinants and Matrices*, 8th ed., Oliver and Boyd, Edinburgh (1954), p. 87). Similarly,

$$\Delta(\lambda) \exp[\lambda \int K(\mathbf{x}, \mathbf{x})\, d\mathbf{x}]$$

$$= 1 + \sum_{n=1}^{\infty} \frac{1}{n!}(-\lambda)^n \int \ldots \int d\mathbf{x}_1 \ldots d\mathbf{x}_n \begin{vmatrix} 0 & K(\mathbf{x}_1, \mathbf{x}_2) & \ldots & K(\mathbf{x}_1, \mathbf{x}_n) \\ K(\mathbf{x}_2, \mathbf{x}_1) & 0 & \ldots & \ldots \\ \cdot & \cdot & \cdot & \cdot \\ K(\mathbf{x}_n, \mathbf{x}_1) & K(\mathbf{x}_n, \mathbf{x}_2) & \ldots & 0 \end{vmatrix}$$

Fredholm's method directly to (B23a), the equation is iterated once in the following equation:

$$\psi(\lambda; \mathbf{k}, \mathbf{x}) = F(\mathbf{k}, \mathbf{x}) + \lambda^2 \int K_2(\mathbf{x}, \mathbf{y})\, \psi(\lambda; \mathbf{k}, \mathbf{y})\, d\mathbf{y}, \tag{B28}$$

where

$$F(\mathbf{k}, \mathbf{x}) = e^{i\mathbf{k}\cdot\mathbf{x}} + \lambda \int K(\mathbf{x}, \mathbf{y})\, e^{i\mathbf{k}\cdot\mathbf{y}}\, d\mathbf{y}, \tag{B29}$$

and

$$\begin{aligned} K_2(\mathbf{x}, \mathbf{y}) &= \int K(\mathbf{x}, \mathbf{z})\, K(\mathbf{z}, \mathbf{y})\, d\mathbf{z} \\ &= \frac{1}{(4\pi)^2} \int \frac{1}{|\mathbf{x} - \mathbf{z}|} \exp(ik|\mathbf{x} - \mathbf{z}|)\, U(z) \frac{1}{|\mathbf{z} - \mathbf{y}|} \exp(ik|\mathbf{z} - \mathbf{y}|) \\ &\qquad \times U(y)\, d\mathbf{z}. \end{aligned} \tag{B29a}$$

The solution of (B28) by Fredholm's method is

$$\psi(\mathbf{k}, \mathbf{x}) = F(\mathbf{k}, \mathbf{x}) + \lambda^2 \int \frac{\Delta(\lambda^2, k; \mathbf{x}, \mathbf{y})}{\Delta(\lambda^2, k)} F(\mathbf{k}, \mathbf{y})\, d\mathbf{y}, \tag{B30}$$

where

$$\begin{aligned} \Delta(\lambda^2, k; \mathbf{x}, \mathbf{y}) &= K_2(\mathbf{x}, \mathbf{y}) + \sum_{n=1}^{\infty} \frac{1}{n!} (-\lambda^2)^n \\ &\qquad \times \int \dots \int d\mathbf{x}_1 \dots d\mathbf{x}_n\, B^{(n)}(\mathbf{k}; \mathbf{x}, \mathbf{y}, \mathbf{x}_1, \dots, \mathbf{x}_n) \end{aligned} \tag{B31}$$

and

$$\begin{aligned} \Delta(\lambda^2, k) &= 1 + \sum_{n=1}^{\infty} \frac{1}{n!} (-\lambda^2)^n \\ &\qquad \times \int d\mathbf{x}_1 \dots \int d\mathbf{x}_n\, D^{(n)}(k; \mathbf{x}_1, \dots, \mathbf{x}_n). \end{aligned} \tag{B32}$$

$B^{(n)}$, $D^{(n)}$ are the Fredholm determinants

$$B^{(n)}(\mathbf{k}; \mathbf{x}, \mathbf{y}, \mathbf{x}_1, \dots \mathbf{x}_n) = \begin{vmatrix} K_2(\mathbf{x}, \mathbf{y}) & K_2(\mathbf{x}, \mathbf{x}_1) & \dots & K_2(\mathbf{x}, \mathbf{x}_n) \\ K_2(\mathbf{x}_1, \mathbf{y}) & K_2(\mathbf{x}_1, \mathbf{x}_1) & \dots & K_2(\mathbf{x}_1, \mathbf{x}_n) \\ \cdot & \cdot & \cdot & \cdot \\ K_2(\mathbf{x}_n, \mathbf{y}) & K_2(\mathbf{x}_n, \mathbf{x}_1) & \dots & K_2(\mathbf{x}_n, \mathbf{x}_n) \end{vmatrix}, \tag{B33}$$

$$D^{(n)}(k; \mathbf{x}_1, \dots, \mathbf{x}_n) = \begin{vmatrix} K_2(\mathbf{x}_1, \mathbf{x}_1) & \dots & K_2(\mathbf{x}_1, \mathbf{x}_n) \\ \cdot & \cdot & \cdot \\ \cdot & \cdot & \cdot \\ K_2(\mathbf{x}_n, \mathbf{x}_1) & \dots & K_2(\mathbf{x}_n, \mathbf{x}_n) \end{vmatrix}. \tag{B34}$$

It can be shown that if $U(r)$ satisfies the conditions, for some finite M, M′,

$$|U(r)| \leqslant \frac{M'}{r^2}, \tag{B35}$$

$$\int_0^\infty r|U(r)|\,dr \leqslant M < \infty, \tag{B35a}$$

then the series (B30), (B31), (B32) are uniformly and absolutely convergent for any finite $|\lambda|$ (Jost and Pais, ref. 3, Appendix III), and further, that for *real* $\lambda$, $\psi(\mathbf{k}, \mathbf{x})$ will have no singularities for real $k$, except possibly at $k = 0$.

On writing (B30), with (B29) in the form

$$\psi(\mathbf{k}, \mathbf{x}) = e^{i\mathbf{k}\cdot\mathbf{x}} + \psi_{sc}(\mathbf{k}, \mathbf{x}) \to e^{i\mathbf{k}\cdot\mathbf{x}} + \frac{1}{x} e^{ikx} f(k, \tau), \tag{B36}$$

one gets

$$\psi_{sc}(\mathbf{k}, \mathbf{x}) = \lambda \int K(\mathbf{x}, \mathbf{y})\, e^{i\mathbf{k}\cdot\mathbf{y}}\, d\mathbf{y} + \lambda^2 \int \frac{\Delta(\lambda^2, k; \mathbf{x}, \mathbf{y})}{\Delta(\lambda^2, k)} F(\mathbf{k}, \mathbf{y})\, d\mathbf{y}, \tag{B37}$$

and for the scattered amplitude,

$$f(k, \tau) = \lim_{x\to\infty} x e^{-ikx} \psi_{sc}(\mathbf{k}, \mathbf{x}). \tag{B38}$$

On expanding $\Delta(\lambda^2, k;\ \mathbf{x}, \mathbf{y})$ in (B31), using (B29) for $F(\mathbf{k}, \mathbf{y})$, for $f(k, \tau)$ one obtains

$$\begin{aligned} 4\pi f(k, \tau) = &-\lambda \int e^{i\mathbf{k}'\cdot\mathbf{y}}\, U(y)\, e^{i\mathbf{k}\cdot\mathbf{y}}\, d\mathbf{y} - \lambda^2 \iint e^{-i\mathbf{k}'\cdot\mathbf{z}}\, N_2(\mathbf{z}, \mathbf{y})\, e^{i\mathbf{k}\cdot\mathbf{y}}\, d\mathbf{z}\, d\mathbf{y} \\ &- \lambda^3 \iint e^{i\mathbf{k}'\cdot\mathbf{z}}\, N_3(\mathbf{z}, \mathbf{y})\, e^{i\mathbf{k}\cdot\mathbf{y}}\, d\mathbf{z}\, d\mathbf{y} \\ &- \lambda^4 \iint e^{-i\mathbf{k}'\cdot\mathbf{z}} \frac{N_4(\mathbf{z}, \mathbf{y})}{\Delta(\lambda^2, k)} e^{i\mathbf{k}\cdot\mathbf{y}}\, d\mathbf{z}\, d\mathbf{y} \\ &- \lambda^5 \iint e^{i\mathbf{k}'\cdot\mathbf{z}} \frac{N_5(\mathbf{z}, \mathbf{y})}{\Delta(\lambda^2, k)} e^{i\mathbf{k}\cdot\mathbf{y}}\, d\mathbf{z}\, d\mathbf{y} \end{aligned} \tag{B39}$$

where

$$\begin{aligned} &N_2(\mathbf{z}, \mathbf{y}) = U(z)\, K(\mathbf{x}, \mathbf{y}), \qquad N_3(\mathbf{z}, \mathbf{y}) = U(z) K_2(\mathbf{z}, \mathbf{y}), \\ &N_4(\mathbf{z}, \mathbf{y}) = U(z) \int K(\mathbf{z}, \mathbf{x}_1)\, \Delta(\lambda^2, k; \mathbf{x}_1, \mathbf{y})\, d\mathbf{x}_1, \\ &N_5(\mathbf{z}, \mathbf{y}) = U(z) \iint K(\mathbf{z}, \mathbf{x}_1)\, \Delta(\lambda^2, k; \mathbf{x}_1, \mathbf{x}_2)\, K_2(\mathbf{x}_2, \mathbf{y})\, d\mathbf{x}_1\, d\mathbf{x}_2. \end{aligned} \tag{B40}$$

On introducing $\boldsymbol{\tau} = \mathbf{k} - \mathbf{k}'$ as in (B25) and

$$\mathbf{r} = \mathbf{y} - \mathbf{z}, \qquad \boldsymbol{\pi} = \tfrac{1}{2}(\mathbf{k} + \mathbf{k}'), \qquad \mathbf{R} = \tfrac{1}{2}(\mathbf{y} + \mathbf{z}), \tag{B41}$$

where $|\mathbf{k}'| = |\mathbf{k}| = k$, one gets

$$(\boldsymbol{\pi}\cdot\boldsymbol{\tau}) = 0, \qquad \pi^2 = k^2 - \tfrac{1}{4}\tau^2, \qquad \cos\vartheta = 1 - \frac{\tau^2}{2k^2}. \tag{B42}$$

Introducing

$$G_j(k,\tau) = -\frac{\lambda^j}{4\pi}\int e^{i\pi\cdot\mathbf{r}}\, N_j(\mathbf{R} - \tfrac{1}{2}\mathbf{r}, \mathbf{R} + \tfrac{1}{2}\mathbf{r})\, e^{i\tau\cdot\mathbf{R}}\, d\mathbf{r}\, d\mathbf{R}, \quad j = 2, \ldots, 5. \quad \text{(B43)}$$

$$f_B(k,\tau) = -\frac{\lambda}{4\pi}\int e^{i\tau\cdot\mathbf{y}}\, U(y)\, dy, \quad \text{(B44)}$$

[which, from (E19), is seen to be simply the first Born expression for the scattering amplitude], one can write (B39) in the form

$$f(k,\tau) = f_B(k,\tau) + g(k,\tau), \quad \text{(B45)}$$

$$g(k,\tau) = G_2(k,\tau) + G_3(k,\tau) + \frac{G_4(k,\tau)}{\Delta(\gamma^2,k)} + \frac{G_5(k,\tau)}{\Delta(\lambda^2,k)} \quad \text{(B46)}$$

The $G_j(k,\tau)$ have the property

$$G_j(k,\tau) = G_j^*(-k,\tau). \quad \text{(B47)}$$

This follows from the property $N_j(k) = N_j^*(-k)$ for the $N_j$ defined in (B40) by virtue of (B25), (B29), etc.

The analytic properties of $G_j(k,\tau)$, $\Delta(\lambda^2, k)$ as functions of $k$ and the momentum transfer $\tau$ for complex values of $k = k_r + i\kappa$ have been studied by Jost and Pais, and Khuri. The reason for continuing these functions into the (upper half of the) complex $k$-plane is, firstly, because of the fact that the zeroes of $\Delta(\lambda^2, k) = 0$ are connected with the bound states of the system, and for real $\lambda$, the zeroes $k_n$ lie on the positive imaginary axis of the $k$-plane. Secondly, in the study of the "dispersion relations" (Chapter 7, Sect. Z.2), the analytic behavior of $f(k,\tau)$ in the upper half of the complex $k$-plane is of basic importance. These properties are as follows:

**i)** For real $k = k_r$ and for $\frac{1}{2}\tau \leqslant k$ (see (B25)) corresponding to physically possible scattering, if the potential $U(r)$ satisfies the conditions of finiteness

$$\int_0^\infty r|U(r)|\, dr < \infty, \quad \text{(B48a)}$$

$$\int_0^\infty r^2|U(r)|\, dr < \infty, \quad \text{(B48b)}$$

then all the series and integrals in (B46) are uniformly convergent.

**ii)** If $U(r)$ satisfies the conditions

$$|U(r)| \leqslant \frac{M'}{r^2}, \quad \text{(B49a)}$$

$$\int_0^\infty r|U(r)|\, dr \leqslant M < \infty, \quad \text{(B49b)}$$

$$\int_0^\infty e^{\alpha r}\, r^2|U(r)|\, dr \leqslant L < \infty, \qquad \alpha \geqslant 0, \quad \text{(B49c)}$$

then for real $\tau$, and $\frac{1}{2}\tau \leqslant \alpha$, the $G_j(k, \tau)$ are analytic functions of $k$, regular in the upper half of the complex $k$-plane ($\kappa > 0$), and uniformly bounded for $\kappa \geqslant 0$. On the real axis, $G_j(k_r, \tau)$ represents the continuous boundary values of $G_j(k, \tau)$, with branch points $k_r = \pm\frac{1}{2}\tau$. Furthermore, for $\kappa \geqslant 0$,

$$G_j(k, \tau) \to 0 \quad \text{as} \quad |\mathbf{k}| \to \infty, \quad j = 2, \ldots, 5. \tag{B50}$$

If in addition to (B49a, b, c), $U(r)$ also satisfies the following conditions:

$$\int_0^\infty r \left| \frac{dU(r)}{dr} \right| dr < \infty, \tag{B49d}$$

$$\int_0^\infty |U(r)|\, dr < \infty, \tag{B49e}$$

then

$$G_j(k, \tau) \to \frac{1}{|\mathbf{k}|}, \quad \text{as} \quad |\mathbf{k}| \to \infty, \quad j = 3, 4, 5. \tag{B51}$$

The above results also contain the following:

For $\kappa \geqslant 0$, $\Delta(\lambda^2, k; \mathbf{x}, \mathbf{y})$ is continuous in $k$, and

$$\Delta(\lambda^2, k; \mathbf{x}, \mathbf{y}) \to 0 \quad \text{as} \quad |\mathbf{k}| \to \infty. \tag{B52}$$

For $U(r)$ satisfying (B49a), (B49b), it is possible to establish the following:

**iii)** $\Delta(\lambda^2, k)$ is an analytic function of $k$, regular for $\kappa < 0$. For $\kappa \leqslant 0$, $\Delta(\lambda^2, k)$ is uniformly bounded and

$$\Delta(\lambda^2, k) \to 1 \quad \text{as} \quad |\mathbf{k}| \to \infty. \tag{B53}$$

**iv)** Let $\lambda_1, \lambda_2, \ldots$ be the eigenvalues (for a fixed $k$) of the homogeneous integral equation

$$\psi_n(\mathbf{k}, \mathbf{x}) = \lambda_n \int K(\mathbf{x}, \mathbf{y})\, \psi_n(\mathbf{k}, \mathbf{y})\, dy. \tag{B54}$$

Then, for $\lambda = \pm\lambda_n$,

$$\Delta(\lambda^2, k) = 0. \tag{B55}$$

In this case it is possible to factorize an entire function from both $\Delta(\lambda^2, k; \mathbf{x}, \mathbf{y})$ and $\Delta(\lambda^2, k)$ such that

$$\frac{\Delta(\lambda^2, k; \mathbf{x}, \mathbf{y})}{\Delta(\lambda^2, k)} = \frac{\Delta_1(\lambda^2, k; \mathbf{x}, \mathbf{y})}{\Delta_1(\lambda^2, k)}, \tag{B56}$$

where $\Delta_1(\lambda^2, k; \mathbf{x}, \mathbf{y})$, $\Delta_1(\lambda^2, k)$ have no zeroes in common.

**v)** For fixed $\lambda$, $\Delta_1(\lambda, k)$ vanishes for $\mathbf{k}_n$ which is an eigenvalue of

$$\psi_n(\mathbf{k}_n, \mathbf{x}) = \lambda \int K(\mathbf{x}, \mathbf{y})\, \psi_n(\mathbf{k}_n, \mathbf{y})\, dy. \tag{B57}$$

If $k_n$ is real ($k_n = k_{r_n} \neq 0$), $\lambda$ must be complex.

**vi)** *For $\lambda$ real,* $\Delta_1(\lambda, k) = 0$ has no zeroes on the real axis of $k$ (except possibly at $k = 0$). For $\lambda$ real all the zeroes of $\Delta_1(\lambda, k) = \Delta_1(\lambda, \kappa_n) = 0$ are on the positive imaginary axis of the complex $k$-plane, i.e., $k_n = i\kappa_n$.

**vii)** The states $\psi_n(\mathbf{k}_n, \mathbf{x})$ are the bound states of the system, since they are the eigenfunctions of the homogeneous Schrödinger equation (B57).

If $U(r)$ satisfies (B49b), the number of bound states is finite. Then the lowest bound state corresponds to $\kappa_n$ maximum.

**viii)** On the positive imaginary axis of $k$, $K(\mathbf{x}, \mathbf{y})$, $K_2(\mathbf{x}, \mathbf{y})$, $G_j(k, \tau)$, $f_B(k, \tau)$ and hence $f(k, \tau)$ are all real.

## REFERENCES

For an elementary treatment of Green's function, see

1. Schiff, L. I., *Quantum Mechanics.* McGraw-Hill, New York (1955);

and for a more extensive treatment,

2. Courant-Hilbert: *Methods of Mathematical Physics.* Interscience Publ. New York (1953), Vol. **I**, p. 351 ff.

Solution of the integral Schrödinger equation by Fredholm's method:

3. Jost, R. and Pais, A., Phys. Rev. **82**, 840 (1951);
4. Khuri, N. N., Phys. Rev. **107**, 1148 (1957);
5. Cf. Whittaker, E. T., and Watson, C. D. *A course of Modern Analysis,* (Cambridge Univ. Press), 4th ed., Section 11.2 ff.

section

# C Born and Other Approximations

For many problems, a particularly simple approximation, known as the Born approximation, is often used. We shall discuss this and the higher approximations briefly.

## 1. BORN APPROXIMATION FROM THE PHASE SHIFTS

The scattering amplitudes and the phase shifts in the exact treatment of the scattering by a potential field are given by (A2), (A17), (A25), (A26),

$$[\Delta + k^2 - \lambda U(r)]\,\psi(\mathbf{r}) = 0, \tag{C1}$$

$$f(\vartheta) = \frac{1}{2ik}\sum (2l+1)\,[\exp(2i\delta_l) - 1]\,P_l(\cos\vartheta), \tag{C2}$$

$$\sin\delta_l = -k\lambda\int_0^\infty \sqrt{\frac{\pi}{2kr}}\,J_{l+\frac{1}{2}}(kr)\,U(r)\,u_l(r)\,r\,dr, \tag{C3}$$

$$u_l(r) \to r\,\frac{\sin\left(kr - \dfrac{l\pi}{2} + \delta_l\right)}{kr}. \tag{C4}$$

Here, as in (B23), we have introduced a dimensionless parameter $\lambda$ as a measure of the strength of the scattering potential. The function $u_l(r)$, which is a solution of (A13), may be expanded in powers of $\lambda$, the first term of the expansion being the function $v_l(r)$ in (A23). Corresponding to this expansion of $u_l(r)$, we may expand $\delta_l$ in (C2) also in powers of $\lambda$. Thus

$$\delta_l = \lambda\delta_l^{(1)} + \lambda^2\delta_l^{(2)} + \lambda^3\delta_l^{(3)} + \cdots. \tag{C5}$$

Inserting (C5) into (C2) and expanding, we get

$$f(\vartheta) = \frac{1}{k}\sum (2l+1)\, P_l(\cos\vartheta)\{\lambda\delta_l^{(1)} + \lambda^2[\delta_l^{(2)} + i(\delta_l^{(1)})^2]$$
$$+ \lambda^3[\delta_l^{(3)} - \tfrac{4}{3}(\delta_l^{(1)})^3 + 2i\delta_l^{(1)}\delta_l^{(2)}] + \cdots\}. \quad \text{(C6)}$$

The cross section is given, to various powers of $\lambda$, by

$$\sigma = 2\pi \int |f(\vartheta)|^2 \sin\vartheta\, d\vartheta$$
$$= \frac{4\pi}{k^2}\sum_l (2l+1)\{\lambda^2(\delta_l^{(1)})^2 + \lambda^3 2\delta_l^{(1)}\delta_l^{(2)}$$
$$+ \lambda^4[(\delta_l^{(2)})^2 + 2\delta_l^{(1)}\delta_l^{(3)} - \tfrac{5}{3}(\delta_l^{(1)})^4] + \cdots\}. \quad \text{(C7)}$$

If, in this (exact) expression (C6), we drop all quantities of order higher than $\lambda^2$, and approximate $\delta_l^{(1)}$ by $\sin\delta_l^{(1)}$ in (C2) thus,

$$\delta_l^{(1)} \cong \sin\delta_l^{(1)} = -\lambda k \int_0^\infty \sqrt{\frac{\pi}{2kr}}\, J_{l+\frac{1}{2}}(kr)\, U(r)\, v_l(r)\, r\, dr, \quad \text{(C8)}$$

then we obtain what is usually known as the Born approximation. Using $v_l(r)$ in (A23), we obtain

$$\delta_l^{(1)} \cong -\frac{\lambda\pi}{2}\int_0^\infty [J_{l+\frac{1}{2}}(kr)]^2\, U(r)\, r\, dr \quad \text{(C8a)}$$

and

$$f(\vartheta)_B \cong -\frac{\lambda\pi}{2k}\sum_l (2l+1)\, P_l(\cos\vartheta)\int_0^\infty [J_{l+\frac{1}{2}}(kr)]^2\, U(r)\, r\, dr. \quad \text{(C9)}$$

By means of the addition theorem of Bessel functions,

$$\frac{\sin qr}{qr} = \frac{\pi}{2kr}\sum_l (2l+1)[J_{l+\frac{1}{2}}(kr)]^2\, P_l(\cos\vartheta), \quad \text{(C10)}$$

$$q = 2k\sin\frac{\vartheta}{2}, \qquad \mathbf{q} = \mathbf{k} - \mathbf{k}'. \quad \text{(C11)}$$

(C9) becomes

$$f(\vartheta)_B = -\lambda\int_0^\infty U(r)\,\frac{\sin qr}{qr}\, r^2\, dr, \quad \text{(C12)}$$

and

$$d\sigma_B = \lambda^2\left|\int_0^\infty U(r)\,\frac{\sin qr}{qr}\, r^2\, dr\right|^2 \sin\vartheta\, d\vartheta\, d\phi. \quad \text{(C13)}$$

The results (C8a), (C9), (C12), (C13) are called the Born approximation. From (C6) it is seen that it is the first order approximation in a series expansion in powers of the potential.

The Born approximation of the phase shifts can be given in another form. Using the W-K-B method,# we obtain the solution for (A13) in the form

$$v_l(r) \simeq \frac{1}{p^{1/2}} \cos \left[ \int_{r_0}^{r} p\, dr - \frac{\pi}{4} \right], \tag{C14}$$

where

$$p^2 = k^2 - \lambda\, U(r) - \frac{(l + \frac{1}{2})^2}{r^2} > 0, \tag{C15}$$

and $r_0$ is the "distance of closest approach" or the classical "turning point" of the particle, i.e.,

$$p(r_0) = 0. \tag{C16}$$

The phase shift is obviously the difference

$$\delta_l = \lim_{r \to \infty} \left[ \int_{r_0}^{r} p\, dr - \int_{r_0}^{r} \sqrt{k^2 - \frac{(l + \frac{1}{2})^2}{r^2}}\, dr \right]. \tag{C17}$$

If we assume that $k^2$ and $l$ are large so that, except in the neighborhood of $r_0$

$$\lambda U(r) \ll k^2 - \frac{(l + \frac{1}{2})^2}{r^2},$$

we have the Born approximation

$$\delta_l^{(1)} = -\frac{\lambda}{2k} \int_{r_0}^{\infty} \frac{U(r)\, r\, dr}{\sqrt{r^2 - \frac{(l + \frac{1}{2})^2}{k^2}}}, \qquad kr_0 = (l + \tfrac{1}{2}). \tag{C18}$$

This formula could also have been obtained by starting with the Born approximation for the scattered amplitude (C12)

$$f(\vartheta)_B = -\frac{\lambda}{4\pi} \int \exp\,[i(\mathbf{k} - \mathbf{k}')\cdot\mathbf{r}]\; U(r)\, d\mathbf{r}. \tag{C19}$$

Choosing the direction of $\mathbf{k} + \mathbf{k}'$ as the $z$ axis of a cylindrical coordinate system for $\mathbf{r}(z, \rho, \phi)$, and using

$$d\mathbf{r} = dz\, d\rho\, \rho\, d\phi, \qquad q = 2k \sin \frac{\vartheta}{2}, \qquad z^2 = r^2 - \rho^2$$

$$\int_0^{2\pi} \exp \left[ i\left(2k\rho \sin \frac{\vartheta}{2}\right) \cos \phi \right] d\phi = 2\pi\, J_0 \left(2k\rho \sin \frac{\vartheta}{2}\right), \tag{C20}$$

we have

$$f(\vartheta)_B = -\frac{\lambda}{4\pi} \int_{-\infty}^{\infty} dz \int_0^{\infty} \rho\, d\rho\, 2\pi\, J_0\, U(\sqrt{z^2 + \rho^2}). \tag{C21}$$

---

# This is sometimes called Jeffreys' method. The use of half-odd quantum numbers has been justified by R. E. Langer, Phys. Rev. **51**, 669 (1937).

On using the approximate relation

$$P_l(\cos\vartheta) \simeq J_0[(l+\tfrac{1}{2})\vartheta] \quad \text{for small } \vartheta \tag{C22}$$

and setting

$$\begin{aligned} k\rho &\simeq (l+\tfrac{1}{2}) \simeq l && \text{for } k \text{ and } l \text{ large,} \\ 2\sin\frac{\vartheta}{2} &\simeq \vartheta && \text{for } \vartheta \text{ small,} \end{aligned} \tag{C23}$$

we can write

$$f(\vartheta)_B = -\frac{\lambda}{k^2}\int_0^\infty l\,dl \int_{r_0} \frac{U(r)\,r\,dr}{\sqrt{r^2 - \dfrac{(l+\frac{1}{2})^2}{k^2}}} P_l(\cos\vartheta). \tag{C24}$$

On comparison with the expression obtained from (C2), namely,

$$f(\vartheta)_B = -\frac{1}{k}\sum_l (2l+1)\,\delta_l^{(1)} P_l(\cos\vartheta) \tag{C25}$$

it is seen that (C18) holds.

In principle, it is possible to proceed beyond the first approximation (C8a)–(C13) by including the terms of higher order in (C7). Thus, for the second order $\delta_l^{(2)}$ in (C5) we obtain

$$\begin{aligned} \delta_l^{(2)} = (-1)^l \left(\frac{\lambda}{2\pi}\right)^2 \int_0^\infty U(r)\,J^2_{l+\frac{1}{2}}(kr)\,r\,dr \int_0^\infty U(r)\,J_{l+\frac{1}{2}}(kr)\,J_{-l-\frac{1}{2}}(kr)\,r\,dr \\ + (-1)^{l+1}\left(\frac{\lambda}{2\pi}\right)^2 \int_0^\infty U(r)\,J_{l+\frac{1}{2}}(kr)\,r\,dr \int_0^r [J_{l+\frac{1}{2}}(kr)\,J^2_{-l-\frac{1}{2}}(ks) \\ - J_{-l-\frac{1}{2}}(kr)\,J^2_{l+\frac{1}{2}}(ks)]\,U(s)\,s\,ds. \end{aligned} \tag{C26}$$

In actual problems, to attempt a higher order approximation in this way is impractical. For example, an analytical evaluation of $\sum_l (2l+1)\,\delta_l^{(1)}\,\delta_l^{(2)}$ is difficult. One may, in some cases, proceed in the following manner: The first approximation $\delta_l^{(1)}$ is replaced by a "better" approximation $\delta_l'$ and $f(\vartheta)$ in (C2) is evaluated with $\delta_l'$. This $\delta_l'$ may be obtained in one of many ways:

**i)** The variation methods described in Sect. D

**ii)** The W-K-B method: Under certain conditions of the form of the potential $V(r)$ and the energy of the particle, the W-K-B method may prove to be a good approximation. In such cases, $\delta_l'$ can be calculated by means of (C17) which, as distinct from (C18), is *not* the Born approximation.

**iii)** Pais method: Starting from the idea that the phase shift [see (A14) and (A23)] can be related to the order of the Bessel function whose asymptotic form is (A14), Pais obtains the relation

$$\frac{2l+1-(2\delta_l/\pi)}{2l+1-(4\delta_l/\pi)}\,\delta_l = -\frac{\lambda\pi}{2}\int_0^\infty U(r)\,J^2_{l+\frac{1}{2}-(2\delta_l/\pi)}(kr)\,r\,dr. \tag{C27}$$

If $\delta_l$ is small and written in the form (C5), one obtains

$$\delta_l^{(2)} = \frac{2}{\pi}\left(a_l - \frac{\delta_l^{(1)}}{2l+1}\right)\delta_l^{(1)}, \tag{C28}$$

where

$$a_l = -\left(\frac{\partial \delta^{(1)}}{\partial p}\right)_{p=l+\frac{1}{2}}.$$

This method is not valid for $l = 0$ and small values of $l$.

It must be noted that while the procedure of using $\delta_l'$ ("better" than $\delta_l^{(1)}$) in (C2) is justified, it nevertheless does not form a systematic series approximation with respect to the parameter $\lambda$ in (C5).

## 2. BORN APPROXIMATIONS AS SUCCESSIVE APPROXIMATIONS

It is possible to obtain the Born approximation (C12) for the scattering amplitudes $f(\vartheta)$ directly without having to use the phase shifts. We shall solve the Schrödinger equation

$$[\Delta + k^2 - \lambda U(r)]\,\psi(\mathbf{r}) = 0 \tag{C29}$$

by successive approximation. If, in (C29), we regard $\lambda U(r)/k^2$ as "small" in the sense that

$$\frac{\lambda}{k^2}\int \psi^*(\mathbf{r})\,U(r)\,\psi(\mathbf{r})\,d\mathbf{r} = O(\lambda) \ll 1, \tag{C30}$$

we may expand $\psi(r)$ in a series in powers of $\lambda$

$$\psi(\mathbf{r}) = \psi^{(0)}(\mathbf{r}) + \lambda\psi^{(1)}(\mathbf{r}) + \lambda^2\psi^{(2)}(\mathbf{r}) + \cdots. \tag{C31}$$

On putting (C31) onto (C29) and equating quantities of the same order, we obtain

$$(\Delta + k^2)\,\psi^{(0)} = 0 \tag{C32}$$

$$(\Delta + k^2)\,\psi^{(1)} = \lambda\,U(r)\,\psi^{(0)}, \tag{C33}$$

$$(\Delta + k^2)\,\psi^{(2)} = \lambda\,U(r)\,\psi^{(1)}, \tag{C34}$$

$$\cdots\cdots\cdots,$$

a system of (an infinite number of) equations, which is still exact. To the zeroth order approximation, we obtain

$$\psi^{(0)}(\mathbf{r}) = e^{i\mathbf{k}\cdot\mathbf{r}}, \tag{C35}$$

and to the first order approximation, we obtain from the inhomogeneous differential equation (C33),

$$\psi^{(0)}(\mathbf{r}) + \psi^{(1)}(\mathbf{r}) = e^{i\mathbf{k}\cdot\mathbf{r}} - \frac{\lambda}{4\pi}\int \frac{\exp(ik|\mathbf{r}-\mathbf{r}'|)}{|\mathbf{r}-\mathbf{r}'|} U(r')\, e^{i\mathbf{k}\cdot\mathbf{r}'}\, d\mathbf{r}' \qquad \text{(C36)}$$

$$\to e^{i\mathbf{k}\cdot\mathbf{r}} - \frac{\lambda}{4\pi}\frac{e^{ikr}}{r}\int e^{-i\mathbf{k}'\cdot\mathbf{r}'}\, U(r')\, e^{i\mathbf{k}\cdot\mathbf{r}'}\, d\mathbf{r}'$$

$$= e^{i\mathbf{k}\cdot\mathbf{r}} - \frac{\lambda e^{ikr}}{r}\int_0^\infty U(r)\frac{\sin qr}{qr} r^2\, dr$$

$$= e^{i\mathbf{k}\cdot\mathbf{r}} + \frac{e^{ikr}}{r} f(\vartheta), \qquad \text{(C36a)}$$

where $\mathbf{q} = \mathbf{k} - \mathbf{k}'$. It is seen that (C36a) gives the same expression for $f(\vartheta)$ as (C12). $\psi^{(1)}$ is the first Born approximation. It is of historical interest to note that it is from the expression (C36a) that Born was led to the probability interpretation of the $|\psi|^2$ which has become one of the fundamental postulates of quantum mechanics.

In principle, it is possible to proceed to the higher approximations in (C34), etc. Thus, to the second approximation one obtains the expression (C36) from the inhomogeneous equation (C34) in which $\psi^{(1)}$ is given.

$$\psi^{(2)}(\mathbf{r}) = -\frac{\lambda}{4\pi}\int \frac{\exp(ik|\mathbf{r}-\mathbf{r}'|)}{|\mathbf{r}-\mathbf{r}'|} U(r')\, \psi^{(1)}(\mathbf{r}')\, d\mathbf{r}'$$

$$= \left(\frac{\lambda}{4\pi}\right)^2 \int\int \frac{\exp(ik|\mathbf{r}-\mathbf{r}'|)}{|\mathbf{r}-\mathbf{r}'|} U(r') \frac{\exp(ik|\mathbf{r}'-\mathbf{r}''|)}{|\mathbf{r}'-\mathbf{r}''|} U(r'')\, e^{i\mathbf{k}\cdot\mathbf{r}''}\, d\mathbf{r}'\, d\mathbf{r}''$$

$$\to \left(\frac{\lambda}{4\pi}\right)^2 \frac{e^{ikr}}{r}\int\int e^{-i\mathbf{k}'\cdot\mathbf{r}}\, U(r')$$

$$\times \frac{\exp(ik|\mathbf{r}'-\mathbf{r}''|)}{|\mathbf{r}'-\mathbf{r}''|} U(r'')\, e^{i\mathbf{k}\cdot\mathbf{r}''}\, d\mathbf{r}'\, d\mathbf{r}''. \qquad \text{(C37)}$$

Instead of working with the system of differential equations (C32), (C33), . . . , we may also start with the integral equation (B14),

$$\psi(\mathbf{r}) = e^{i\mathbf{k}\cdot\mathbf{r}} - \frac{1}{4\pi}\int \frac{\exp(ik|\mathbf{r}-\mathbf{r}'|)}{|\mathbf{r}-\mathbf{r}'|} U(r')\, \psi(\mathbf{r}')\, d\mathbf{r}' \qquad \text{(C38)}$$

and solve it by iteration. In the first approximation, we replace $\psi(\mathbf{r}')$ in the integrand by the incident wave $e^{i\mathbf{k}\cdot\mathbf{r}'}$, thereby obtaining the same result as in (C36). By inserting the first order solution $\psi^{(1)}$ into $\psi$ in the integrand, we obtain the second Born approximation (C37), etc. Such calculations, because of the rapidly increasing complexities, have been carried out for only a few potentials. In general, there is the question of convergence in the higher approximations that we shall consider in the following pages.

### 3. CONVERGENCE OF THE BORN EXPANSION

Before going into the discussion of the validity of the Born approximation, we shall consider the convergence condition of the Born series and return to the question of the accuracy of the first (and the second) Born approximation in the following subsection.

The $n$th Born approximation wave function $\psi_n(\mathbf{r})$ is, according to (C31), given by

$$\psi_n(\mathbf{r}) = \psi^{(0)}(\mathbf{r}) + \cdots + \lambda^n \psi^{(n)}(\mathbf{r}), \tag{C39}$$

where, as in (C36) and (C37),

$$\psi_n(\mathbf{r}) = e^{i\mathbf{k}\cdot\mathbf{r}} - \lambda \int G^+_{\mathbf{K}}(\mathbf{r}, \mathbf{r}^{(1)})\, U(r^{(1)})\, e^{i\mathbf{k}\cdot\mathbf{r}^{(1)}}\, d\mathbf{r}^{(1)} + \cdots$$
$$+ (-\lambda)^n \int G^+_{\mathbf{K}}(\mathbf{r}, \mathbf{r}^{(1)})\, U(r^{(1)}) \ldots G^+_{\mathbf{K}}(\mathbf{r}^{(n-1)}, \mathbf{r}^{(n)})\, U(r^{(n)})$$
$$\times \exp(i\mathbf{k}\cdot\mathbf{r}^{(n)})\, d\mathbf{r}^{(1)} \ldots d\mathbf{r}^{(n)}, \tag{C40}$$

and $G^+_{\mathbf{K}}(\mathbf{r}, \mathbf{r}')$ is defined in (B13). The scattering amplitude in the $n$th approximation is

$$f_n(\vartheta) = -\frac{\lambda}{4\pi} \int e^{-i\mathbf{k}'\cdot\mathbf{r}}\, U(r)\, \psi_{n-1}(\mathbf{r})\, d\mathbf{r}. \tag{C41}$$

The function $\psi_n(\mathbf{r})$ in (C40) can also be obtained from (B23) by iteration.

The nature of convergence of $\psi_n$ to the true $\psi$ can be analyzed by examining the Fredholm solution of the integral equation (B23). However, as remarked just before (B27), the Fredholm determinants do not exist due to the singularity of the kernel $K(\mathbf{x}, \mathbf{y})$ at $\mathbf{x} = \mathbf{y}$. $K_2(\mathbf{x}, \mathbf{y})$ is, however, non-singular at $\mathbf{x} = \mathbf{y}$. Hence we shall decompose the wave functions into partial waves instead of starting from (B27). From

$$G^+_{\mathbf{K}}(\mathbf{r}, \mathbf{r}') = \sum_{l=0}^{\infty} \frac{(2l+1)}{4\pi r r'}\, G_l(r, r')\, P_l(\cos\Theta), \qquad \Theta = \widehat{\mathbf{r}, \mathbf{r}'},$$
$$G_l(r, r') = -krr'\, j_l(kr_<)[n_l(kr_>) - i j_l(kr_>)],$$

where $j_l$, $n_l$ are the spherical Bessel functions defined in the footnote (A11), and $r_<$, $r_>$ are the smaller and larger one of $r$, $r'$ respectively, we get the integral equation for $\psi_l(\mathbf{r})$ from (B23)

$$\psi(\mathbf{r}) = \sum_{l=0}^{\infty} \frac{1}{kr} (2l+1)\, i^l\, \psi_l(r)\, P_l(\cos\vartheta), \tag{C42}$$

$$\psi_l(r) = kr\, j_l(kr) - \lambda \int_0^{\infty} G_l(r, r')\, U(r')\, \psi_l(r')\, dr' \tag{C43}$$
$$\to \sin\left(kr - \frac{l\pi}{2}\right) + \frac{1}{2i}(S_l - 1) \exp\left[i\left(kr - \frac{l}{2}\pi\right)\right],$$

where $S_l$ is the $S$-matrix element of (A34-5). The scattering amplitude $f(\vartheta)$ is also analyzed into partial amplitudes #

$$f(\vartheta) = \sum_{l=0}^{\infty} \frac{1}{k} (2l + 1) f_l P_l (\cos \vartheta), \tag{C44}$$

$$f_l = -\lambda \int_0^{\infty} r\, j_l(kr)\, U(r)\, \psi_l(r)\, dr = \frac{1}{2i} (S_l - 1). \tag{C45}$$

The decomposed forms of (C40), (C41) are then

$$\begin{aligned} \psi_{n,l}(r) = krj_l(kr) - \lambda k \int_0^{\infty} G_l(r, r^{(1)})\, U(r^{(1)})\, j_l(kr^{(1)})\, r^{(1)}\, dr^{(1)} + \cdots \\ + (-\lambda)^n k \int_0^{\infty} G_l(r, r^{(1)})\, U(r^{(1)}) \ldots G_l(r^{(n-1)}, r^{(n)}) \\ \times U(r^{(n)})\, j_l(kr^{(n)})\, r^{(n)}\, dr^{(1)} \ldots dr^{(n)}, \end{aligned} \tag{C46}$$

$$f_n(\vartheta) = \sum_{l=0}^{\infty} \frac{1}{k} (2l + 1) f_{n,l} P_l (\cos \vartheta), \tag{C47}$$

$$f_{n,l} = -\lambda \int_0^{\infty} rj_l(kr)\, U(r)\, \psi_{n-1,l}(r)\, dr. \tag{C48}$$

From (C42), (C40) and (C46), it is clear that $\psi_n(\mathbf{r})$ converges to $\psi(\mathbf{r})$ if and only if $\psi_{n,l}$ converges to $\psi_l$ for all values of $l$. From (C48), it is seen that $f_{n,l}$ converges to $f_l$ if $\psi_{n,l}$ converges to $\psi_l$. Similarly, the convergence of $\psi_n$ guarantees the convergence of $f_n(\vartheta)$, as seen from (C41).

We shall now consider the case $l = 0$. The integral equation (C43) is then

$$\psi_0(r) = \sin kr - \lambda \int_0^{\infty} G_0(r, r')\, U(r')\, \psi_0(r')\, dr', \tag{C49a}$$

$$G_0(r, r') = \frac{1}{k} \sin kr_< \, e^{ikr_>}. \tag{C49b}$$

Application of the Fredholm theory to (C49a) leads to the solution

$$\psi_0(r; \lambda, k) = \sin kr - \frac{\lambda}{\Delta(\lambda, k)} \int_0^{\infty} \Delta(\lambda, k, r, r') \sin kr'\, dr', \tag{C50}$$

where

$$\Delta(\lambda, k) = 1 + \sum_{n=1}^{\infty} \frac{\lambda^n}{n!} \int_0^{\infty} \begin{vmatrix} K(r_1, r_1) & \ldots & K(r_1, r_n) \\ \cdot & \cdots & \cdot \\ K(r_n, r_1) & \ldots & K(r_n, r_n) \end{vmatrix} dr_1 \ldots dr_n, \tag{C51}$$

# (C42) is (A12). (C43) is (AA-1) and (AA-2) on Pp. 15, 16. $f_l$ in (C44) is $T_l/2i$ of (AA-3).

$$\Delta(\lambda, k, \mathbf{r}, \mathbf{r}') = K(\mathbf{r}, \mathbf{r}') + \sum_{n=1}^{\infty} \frac{\lambda^n}{n!} \int_0^{\infty} \begin{vmatrix} K(r, r') & K(r, r_1) & \dots & K(r, r_n) \\ K(r_1, r') & K(r_1, r_1) & \dots & K(r_1, r_n) \\ \cdot & \cdot & \cdot & \cdot \\ K(r_n, r') & K(r_n, r_1) & \dots & K(r_n, r_n) \end{vmatrix} dr_1 \dots dr_n, \tag{C51a}$$

$$K(r, r') \equiv G_0(r, r')\, U(r').$$

With the condition that $\int_0^{\infty} r|U(r)|\, dr =$ finite, it can be shown that $\Delta(\lambda, k)$ and $\Delta(\lambda, k, r, r')$ are absolutely convergent for any value of $\lambda$ (see refs. 8 and 13). Consider $\Delta(\lambda, k)$ as a function of (complex) $\lambda$, and denote the zeroes of $\Delta(\lambda, k)$ by $\lambda_1 \dots \lambda_n \dots$. Let $\lambda_1$ be the zero with the smallest absolute value. Since the Born expansion of $\psi_0$ is in powers of $\lambda$, $\psi_{n,0}$ in (C46) diverges (converges) if $|\lambda|$ is larger (smaller) than $|\lambda_1|$. $|\lambda_1|$ is the convergence radius, $\lambda_c$, of the expansion (ref. 8). To illustrate the general situation, take the example

$$U(r) = \frac{e^{-r/a}}{a^2(1 - e^{-r/a})}, \tag{C52}$$

for which we have

$$\Delta(\lambda, k) = \prod_{n=1}^{\infty} \left[1 - \frac{\lambda}{n(n - 2ika)}\right]. \tag{C53}$$

The radius of convergence is, from (C53),

$$\lambda_c = |1 - 2ika| = \sqrt{1 + 4(ka)^2}.$$

It may be noted that the value of $\lambda_c$ at $k = 0$ corresponds to a depth of the potential which has just one bound state, with zero binding energy. For the potential $\lambda U(r)$ of (C52), the Born expansion for $l = 0$ converges (diverges) if $k^2a^2 > \frac{1}{4}(|\lambda|^2 - 1)$ $[k^2a^2 < \frac{1}{4}(|\lambda|^2 - 1)]$.

A slightly different Born approximation from (C46) is also used. This starts from the integral equation

$$\varphi_l(r) = kr\, j_l(kr) - \lambda \int_0^{\infty} g_l(r, r')\, U(r')\, \varphi_l(r')\, dr' \tag{C54}$$

instead of from (C43), where

$$g_l(r, r') = -krr'\, j_l(kr_<)\, n_l(kr_>).$$

For $l = 0$, these become

$$\varphi_0(r) = \sin kr - \lambda \int_0^{\infty} g_0(r, r')\, U(r')\, \varphi_0(r')\, dr', \tag{C55a}$$

$$g_0(r, r') = \frac{1}{k} \sin kr_< \cos kr_>. \tag{C55b}$$

Comparing (C49) with (C55), one sees that the choice of the kernel is different in the two cases. The asymptotic forms of $\psi_l$ and $\varphi_l$ are

$$\psi_l(r) \to \sin\left(kr - \frac{l\pi}{2}\right) + \frac{1}{2i}(S_l - 1)\exp\left[i\left(kr - \frac{l}{2}\pi\right)\right], \tag{C56}$$

$$\varphi_l(r) \to \sin\left(kr - \frac{l\pi}{2}\right) + \tan\delta_l \cos\left(kr - \frac{l}{2}\pi\right). \tag{C57}$$

Since both $\psi_l$ and $\varphi_l$ are solutions of the equation

$$\left[\frac{d^2}{dr^2} + k^2 - \lambda U(r) - \frac{l(l+1)}{r^2}\right]\phi = 0, \qquad \phi(0) = 0,$$

the two functions $\psi_l$, $\varphi_l$ should differ only by a constant factor, and in fact

$$\psi_l(r) = \cos\delta_l \exp(i\delta_l)\,\varphi_l(r),$$

where $\varphi_l(r)$ is a real function. The results of the Born approximations based on (C43) and (C54) are, however, different, and accordingly the radii of convergence are also different. The Fredholm theory can be applied to (C54) or (C55) and the convergence condition can be derived as before. We are now dealing with the successive approximation for $\tan\delta_l$ (C57) instead of $(1/2i)(S_l - 1) = \sin\delta_l \exp(i\delta_l)$ (C56),

$$\omega_l \equiv \tan\delta_l = -\lambda \int_0^\infty rj_l(kr)\,U(r)\,\varphi_l(r)\,dr. \tag{C58}$$

In the $n$th approximation

$$\omega_{n,l} = -\lambda \int_0^\infty rj_l(kr)\,U(r)\,\varphi_{n-1,l}(r)\,dr, \tag{C59}$$

where

$$\varphi_{n,l}(r) = kr\,j_l(kr) + \cdots + k(-1)^n \int_0^\infty g_l(r, r^{(1)})\,U(r^{(1)}) \ldots g_l(r^{(n-1)}, r^{(n)}) \times U(r^{(n)})\,j_l(kr^{(n)})\,r^{(n)}\,dr^{(1)} \ldots dr^{(n)}. \tag{C60}$$

Without repeating the details, we shall merely state the result that the poles of $\varphi_l(r)$ lie at $\lambda = \lambda_m$ where $\delta_l = \frac{1}{2}(2m+1)\pi$, at which $\tan\delta_l \to \infty$. The smallest $|\lambda_m|$ is the radius of convergence $\lambda_c'$ of the series (C60).

There are close relations between $f_{n,l}$ and $\omega_{n,l}$. It can easily be verified that

$$f_l = \frac{1}{1+\omega_l^2}(\omega_l + i\omega_l^2), \qquad \omega_l = \frac{1}{1+f_l^2}(f_l - if_l^2),$$

$$f_{n,l} \equiv \frac{1}{k}\sum_{p=1}^{n} A_p\lambda^p, \qquad \omega_{n,l} \equiv \frac{1}{k}\sum_{p=1}^{n} a_p\lambda^p \tag{C61}$$

$$A_1 = a_1, \qquad A_2 = a_2 + \frac{ia_1^2}{k},$$

$$A_p = a_p + \cdots + \left(\frac{i}{k}\right)^m \binom{p-1}{m} a_1^m a_{p-m} + \cdots + \left(\frac{i}{k}\right)^{p-1} a_1^p,$$

$$a_p = A_p + \cdots + \left(-\frac{i}{k}\right)^m \binom{p-1}{m} A_1^m A_{p-m} + \cdots + \left(-\frac{i}{k}\right)^{p-1} A_1^p,$$

where

$$\binom{p}{m} = \frac{p!}{m!(p-m)!}.$$

Thus if $\omega_{n,l}$ has been calculated as a power series of $\lambda$, then $f_{n,l}$ can be found, and vice versa.

Let us estimate the radii of convergence $\lambda_c$, $\lambda_c'$ (Kohn, ref. 13). First consider the expansion (C60). Let the potential strength parameter $\lambda$ be $\lambda_0'(k)$ [or $\lambda_{-1}'(k)$] for which the phase shift is $\pi/2$ (or $-\pi/2$). $\lambda_0'$ $(k = 0) \to \lambda'(k = 0)$ is the depth which houses just one bound state with zero binding energy for a given $l$ [for example, see (E22)], and in general $|\lambda_0'$ $(k = 0)| < |\lambda_{-1}'$ $(k = 0)|$. Therefore the radius of convergence is $\lambda_c' = |\lambda_0|$, i.e., the Born expansion converges if $|\lambda| < |\lambda_0|$, and diverges if $|\lambda| > |\lambda_0|$. The same is true for the expansion (C46) at $k = 0$, because $\lambda_c$ $(k = 0) = \lambda_c'$ $(k = 0)$. At low energies, the value of $\lambda_c$ (and also $\lambda_c'$) increases as $l$ increases. $\lambda_c$ and $\lambda_c'$ are not always monotonically increasing function of energy, and in fact, at $k = 0$,

$$\frac{d\lambda_c}{dk^2} = \frac{d\lambda_c'}{dk^2} < 0 \qquad \text{for } l \geqslant 1.$$

For $k^2 \neq 0$, $\lambda_c$ and $\lambda_c'$ are in general different.

For high energies, the consideration is as follows. Let us assume the potential to be sectionally continuous, to have at most an $r^{-1}$ singularity at the origin, and to satisfy the conditions

$$\lim_{r\to\infty} r^{2+\varepsilon} U(r) = 0 \quad (\varepsilon > 0), \qquad \int_0^\infty U(r)\, dr \neq 0.$$

We shall call $U(r)$ "regular" if $\int_0^\infty |U(r)|\, dr < \infty$, and "singular" if $\lim_{r\to\infty} ar\, U(r) = 1$. For the expansion (C60), it can be shown that, for all $l$, in the limit $k \to \infty$,

$$\text{regular } U: \quad \frac{\lambda_c'(k)}{\pi k} \to \left| \int_0^\infty U(r)\, dr \right|^{-1},$$
$$\text{singular } U: \quad \frac{\lambda_c'(k) \ln ka}{\pi ka} \to 1. \tag{C62}$$

Similarly, for (C46),

$$\text{regular } U: \quad \frac{\lambda_c(k)}{ka} \to \infty,$$
$$\text{singular } U: \quad \frac{\lambda_c(k) \ln ka}{ka} \to \infty. \qquad \text{(C63)}$$

Thus, in the case of finite $\lambda$, the Born approximation always converges for sufficiently high energies.

### 4. VALIDITY OF THE BORN APPROXIMATION

In the preceding subsection, we have studied the condition for the convergence of the Born expansion. We do not know yet how large an error is introduced if the series is broken off after a finite number of terms. [The usual "Born approximation" takes only the first term from the infinite series.] However, it is possible to find an upper and a lower boundary on the error if the radius of convergence $\lambda_c$ or $\lambda_c'$ is known (ref. 14).

Consider the series (C60) with $U(r) \geqslant 0$, and $\omega_{n,l}$ for the $n$th Born approximation for $\omega_l \equiv \tan \delta_l$ of (C58). The usual (first) Born approximation is

$$\omega_{1,l} = -\lambda k \int_0^\infty r^2\, U(r)\, j_l^2(kr)\, dr.$$

The error of $\omega_{n,l}$ is

$$\Delta\omega_{n,l} = \omega_{n,l} - \omega_l.$$

Then, if $|\lambda| < \lambda_c'$, it can be proved that

$$-\frac{|\lambda|^n}{(\lambda_c' - |\lambda|)(\lambda_c')^{n-1}} < \frac{\Delta\omega_{n,l}}{\omega_{1,l}} < 0 \qquad \text{for } n \text{ even,} \qquad \text{(C64)}$$

$$-\frac{|\lambda|^n}{(\lambda_c' - |\lambda|)(\lambda_c')^{n-1}} < \frac{\Delta\omega_{n,l}}{\omega_{1,l}} < \frac{|\lambda|^n}{(\lambda_c' + |\lambda|)(\lambda_c')^{n-1}} \qquad \text{for } n \text{ odd.} \qquad \text{(C65)}$$

In particular, for the first Born approximation,

$$-\frac{|\lambda|}{\lambda_c' - |\lambda|} < \frac{\Delta\omega_{1,l}}{\omega_{1,l}} < \frac{|\lambda|}{\lambda_c' + |\lambda|}. \qquad \text{(C66)}$$

If the radius of convergence $\lambda_c'$ is sufficiently larger than $|\lambda|$, $|\Delta\omega_{n,l}|$ will be $\ll |\omega_{1,l}|$. For the limit $k \to 0$,

$$\omega_{1,l} < \omega_l < \frac{\lambda_c' \omega_{1,l}}{\lambda_c' - |\lambda|}. \qquad \text{(C67)}$$

From (C64), it is seen that $\omega_{2,l}$ gives a lower (upper) bound for $\omega_l$ if the potential $\lambda U$ is attractive (repulsive).

For the total cross section $\sigma_B$ computed by the first Born approximation (C13), we have, for the exact cross section $\sigma$,

$$\frac{\sigma_B}{\left(1 + \frac{|\lambda|}{\lambda_c}\right)^2} \leqslant \sigma \leqslant \frac{\sigma_B}{\left(1 - \frac{|\lambda|}{\lambda_c}\right)^2}. \tag{C68}$$

Consider next the wave function. The improvement in the wave function to a finite order of the Born approximation is good if $|\lambda|/\lambda_c$ (or $|\lambda|/\lambda_c'$) is small. Let the norm of any $\phi$ be defined by $|\phi| \equiv \int \phi^*(\mathbf{r})\, U(r)\, \phi(\mathbf{r})\, d\mathbf{r}$. Then it can be shown that

$$|\psi_{n,l} - \psi_l| \leqq \frac{|\lambda|^{n+1}|\psi_l|}{\lambda_c^{n+1}}, \qquad |\varphi_{n,l} - \varphi_l| \leqq \frac{|\lambda|^{n+1}|\varphi_l|}{(\lambda_c')^{n+1}}. \tag{C69}$$

(If $U(r)$ is not positive definite, the definition of the norm must be modified; but (C69) is valid for the modified norms.)

From (C64)–(C69), it is seen that the validity of the Born approximation is entirely dependent on the ratio $|\lambda|/\lambda_c$ (or $|\lambda|/\lambda_c'$). Several criteria of the validity of the Born approximation have been discussed in the literature on this subject, but some of them are not very accurate from the standpoint mentioned above. Thus one sometimes comes across the statement that the Born approximation is good if the absolute value of the phase shift is small ($|\delta_l| \ll 1$). This is only partly true, since $|\delta|$ can be small at low energies even if $|\lambda|/\lambda_c$ is close to unity, in which case the Born approximation is *not* good. The $|\lambda|/\lambda_0$ (or $|\lambda|/\lambda_c'$) can be estimated, for high energies, from (C63) [or (C62)]. For "regular" potentials, we have

$$\delta_l \to -\frac{\lambda}{2k}\int_0^\infty U(r)\, dr, \qquad \text{for} \quad k \to \infty,$$

the right-hand side being just the Born approximation $\delta_l^{(1)}$ (C8a) in the limit $k \to \infty$.

From (C62)—(C65) and (C69), it is seen that the Born approximation is sufficiently accurate for high energies. An explicit proof of the validity of the approximation is also possible (ref. 13a).

We shall conclude this section with the following remarks:

**i)** It is important to note that the Born approximation for $f(\vartheta)$ given in (C12) is real whereas $f(\vartheta)$ in the exact theory (A17), is essentially complex. Corresponding to the Born approximation (C12) for $f(\vartheta)_B$, one can obtain a first Born approximation, i.e., up to the order $\lambda$, for $\eta(\vartheta)$. See (G10).

**ii)** In general, for small velocities (or $\lambda$ not small) only $\delta_0$ or the first few $\delta_l$ are large ($\gtrsim 1$) and the Born approximation is poor. On the other hand, for high velocities (or small $\lambda$), all the $\delta_l$ are small and the Born approximation, which takes into account all the $\delta_l$ in (C9), is good.

In the case of a Coulomb or screened Coulomb field (also Yukawa field),

$$V(r) = \frac{zZe^2}{r}, \qquad V(r) = \frac{zZe^2}{r} e^{-\lambda r}, \tag{C70}$$

then

$$\delta_l^{(1)} = -\frac{zZe^2\pi}{\hbar v} \int_0^\infty [J_{l+\frac{1}{2}}(kr)]^2 e^{-\lambda r} d(kr). \tag{C71}$$

Thus the Born approximation may not be very good for large values of $z$ and $Z$. Here $f_B$ in (C12) and $\delta_l^{(1)}$ in (C71) can all be calculated simply. Thus

$$\delta_l^{(1)} = -\frac{zZ e^2}{\hbar v} Q_l \left(1 + \frac{\lambda^2}{2k^2}\right), \tag{C72}$$

where $Q_l$ is the Legendre coefficient of the second kind, and (C12) gives

$$f(\vartheta)_B = -\frac{2zZ me^2}{\hbar^2} \cdot \frac{1}{q^2 + \lambda^2}, \qquad q = 2k \sin \frac{\vartheta}{2} \tag{C73}$$

For an unscreened Coulomb field, from (C73), on letting $\lambda \to 0$, one again obtains the classical Rutherford formula

$$|f(\vartheta)_B|^2 = \left(\frac{zZ e^2}{2m v^2}\right)^2 \frac{1}{\sin^4 \frac{\vartheta}{2}} \tag{C74}$$

which is the same as the exact result (A47).

## 5. "HIGH-ENERGY APPROXIMATION" OF MOLIERE

From the expression for the scattering amplitude (C2), one can obtain a "high energy" approximation which is valid under the following conditions:

**i)** For high energies such that a *large* number of partial waves $l$ contribute from 1 up to $10^2$. From the Schrödinger equation (A13), this means

$$k^2a^2 \gtrsim l(l+1), \tag{C75}$$

or

$$ka \gg 1, \tag{C76}$$

where $a$ is of the order of the range of the field $V(r)$.

**ii)** The energy $E(= \hbar^2k^2/2m)$ is large compared with the (absolute) value of $V(r)$, so that the chance of a large angle deflection by $V(r)$ is very small, i.e.,

$$\frac{|V(r)|}{E} \ll 1. \tag{C77}$$

As a consequence of (C77), we have

$$\vartheta \ll 1$$

and we may use (C22). As a consequence of (C76), the summation over $l$ in (C2) may be replaced by integration. Thus, writing $l = k\rho$ as in (C23), for (C2) we have

$$f(\vartheta) = \frac{k}{i}\int_0^\infty \rho\, d\rho\,(e^{2i\delta(\rho)} - 1)\, J_0\,(k\rho\vartheta). \tag{C78}$$

For $\delta(\rho)$, we may use (C18) in the form

$$\delta(\rho) = -\frac{1}{\hbar v}\int_\rho^\infty \frac{V(r)\, r\, dr}{\sqrt{r^2 - \rho^2}}. \tag{C79}$$

The expression (C78) may be "improved" by replacing $J_0(k\rho\vartheta)$ by $J_0[2k\rho \sin(\vartheta/2)]$, which corresponds to representing the "classical trajectory", or the "wave vector $k$", within the region of the field $V(r)$ by a mean direction between initial $\mathbf{k}$ and scattered $\mathbf{k}'$. Thus

$$f(\vartheta) = \frac{k}{i}\int_0^\infty \rho\, d\rho\,\{\exp[2i\delta(\rho)] - 1\}\, J_0\left(2k\rho \sin\frac{\vartheta}{2}\right). \tag{C78a}$$

The expression (C78a) could also have been put in an alternative form which appears in the theory of diffraction in classical optics. By using (C20) and noting that in (C20)

$$2k \sin\frac{\vartheta}{2}\,\rho \cos\phi = (\mathbf{k} - \mathbf{k}')\cdot\boldsymbol{\rho},$$

we have, from (C78a), on writing $\rho\, d\rho\, d\varphi = d^2\rho$,

$$f(\vartheta) = \frac{k}{2\pi i}\int_0^\infty \exp[i(\mathbf{k} - \mathbf{k}')\cdot\boldsymbol{\rho}]\,\{\exp[2i\delta(\rho)] - 1\}\, d^2\rho \tag{C78b}$$

where $\delta(\rho)$, of (C79), may also be given in the form of (C21),

$$\delta(\rho) = -\frac{1}{2\hbar v}\int_{-\infty}^\infty V(\sqrt{z^2 + \rho^2})\, dz. \tag{C79a}$$

On comparing (C78a) with (C21) and (C24), it is seen that (C78b) differs from the Born approximation (i) in having $\exp[2i\delta(\rho)] - 1$ instead of the first nonvanishing term, namely, $2i\delta(\rho)$, in the integrand, and (ii) in replacing the summation over $l$ by an integration.

The first difference (i) is of greater importance in that $f(\vartheta)$ in (C78a, b) is now complex and, unlike the Born approximation in which $f(\vartheta)_B$ is real, it satisfies the general relation (A21)

$$4\pi\, \mathrm{Im}\, f(0) = k\sigma \tag{C80}$$

connecting the *total* cross section $\sigma$ with the coefficient of the imaginary $i$ in the scattered amplitude at $\vartheta = 0$. On putting $\vartheta = 0$ (i.e., $\mathbf{k} = \mathbf{k}'$) in (C78b), we obtain

$$2\pi \operatorname{Im} f(0) = -k \int_0^\infty \{\operatorname{Re} \exp [2i\delta(\rho)] - 1\}\, d^2\rho. \tag{C81}$$

From (C78b) again, we have

$$\begin{aligned} \sigma &= \int |f(\vartheta)|^2\, d\cos\vartheta\, d\phi = \int |f(\vartheta)|^2\, d\Omega_{\mathbf{k}'} \\ &= \left(\frac{k}{2\pi}\right)^2 \int\int_0^\infty \rho'\, d\rho'\, d^2\rho \exp [i(\mathbf{k} - \mathbf{k}')\cdot(\boldsymbol{\rho} - \boldsymbol{\rho}')] \{\exp [2i\delta(\rho)] - 1\} \\ &\qquad\qquad \times \{\exp [-2i\delta(\rho')] - 1\}\, d\Omega_{\mathbf{k}'} \end{aligned}$$

where $|\mathbf{k}| = |\mathbf{k}'|$. The integration over the directions of $\mathbf{k}'$ gives a (two-dimensional) delta function in $\rho - \rho'$ so that

$$\sigma = \int_0^\infty d^2\rho |\exp [2i\delta(\rho)] - 1|^2 = 2 \int_0^\infty \{1 - \operatorname{Re} \exp [2i\delta(\rho)]\}\, d^2\rho. \tag{C82}$$

Thus (C80) is satisfied. Note that, in the Born approximation, $f(\vartheta)$ being real, (C80) is not satisfied.

The expression for $\delta(\rho)$ in (C79) or (C79a) is essentially the first Born approximation $\delta_l^{(1)}$ in (C8a), as can be seen from the way (C24) has been derived from the Born approximation (C19), and also from a direct calculation of $\delta(\rho)$ for the case of screened Coulomb $V(r)$ of (C70). The result is

$$\delta(\rho) = K_0(\lambda\rho),$$

which is in agreement with the Born expression (C72)

$$\lim_{l\to\infty} \delta^{(1)} = \lim_{l\to\infty} Q\left(1 + \frac{\lambda^2}{2k^2}\right) = K_0\left(\frac{\lambda}{k} l\right) = K_0(\lambda\rho)$$

where $K_0$ is a Bessel function (as defined in Watson's Bessel Functions).

Subject to the conditions (C76), (C77), $f(\vartheta)$ as given in (C78a) or (C78b) can be expected to be a good approximation for *small* $\vartheta$. From the point of view, however, of a systematic expansion in powers of "$V(r)$", the discussion given in Sect. C.1, still holds.

### 6. SEMICLASSICAL APPROXIMATION

In some cases of the scattering of a particle, the description in terms of classical dynamics is a good approximation. Since the main difference between the classical and the quantum mechanics lies in the uncertainty principle in the latter, we may examine the conditions under which classical mechanics will give a good approximation, namely, when the effect of the uncertainty principle is not important. To investigate these conditions, let **p** be the initial momentum of the incident particle and $\vartheta$ be the angle of scattering. For the classical description in terms of a trajectory to be meaningful, it is necessary that (i) the uncertainty in the direction of the scattered particle be small compared with $\vartheta$, i.e.,

$$\frac{\Delta p_\perp}{p} \ll \vartheta, \tag{C83}$$

where $\Delta p_\perp$ is the uncertainty in the momentum perpendicular to **p**, and (ii) the uncertainty in the impact parameter $\Delta\rho$ be small compared with the value of the impact parameter itself, i.e.,

$$\Delta\rho \simeq \frac{\hbar}{\Delta p_\perp} \ll \rho. \tag{C84}$$

On combining these two relations, and using $p\rho = l\hbar$ for the angular momentum, it is seen that the condition for the classical description to be valid is

$$l\vartheta \gg 1. \tag{C85}$$

This condition places a restriction on the potentials to which the classical description can be applied. For example, if the potential decreases very fast with distances such that $\vartheta$ decreases with increasing impact parameter faster than $1/\rho$ (or $1/l$), then the classical description is not valid.

Let us assume that the condition (C85) is satisfied, and consider a scattering angle $\vartheta \neq 0$. For $\vartheta \neq 0$, we have the relation#

$$\sum_{l=0} (2l + 1) P_l(\cos\vartheta) = 0. \tag{C86}$$

In the sense of the correspondence principle, according to which the quantum theory passes to the classical theory in the limit of large quantum numbers, we shall take the asymptotic form for large $l$ (ref. 16),

$$P_l(\cos\vartheta) \to \sqrt{\frac{2}{\pi l \sin\vartheta}} \sin\left[\left(l + \frac{1}{2}\right)\vartheta + \frac{\pi}{4}\right]\left\{1 - \frac{1}{8l\vartheta}\cos\left[\left(l + \frac{1}{2}\right)\vartheta + \frac{\pi}{4}\right]\right\}. \tag{C87}$$

Using the condition (C85), and this $P_l$ in (A17), we get for large $l$,

$$f(\vartheta) \to \frac{1}{2ik}\sum\sqrt{\frac{2l}{\pi\sin\vartheta}}\exp(2i\delta_l)\left(\exp\left\{i\left[\left(l + \frac{1}{2}\right)\vartheta - \frac{\pi}{4}\right]\right\} - \exp\left\{\left[-i\left(l + \frac{1}{2}\right)\vartheta - \frac{\pi}{4}\right]\right\}\right). \tag{C88}$$

# This can be found from the way the Legendre coefficients are defined, or, more elegantly, from the expansion of the $\delta$ function

$$\delta(\cos\vartheta - 1) = \tfrac{1}{2}\sum(2l + 1) P_l(\cos\vartheta).$$

On account of the oscillations of the exponential functions with $l\vartheta$, the contribution comes mostly from those $l$ for which the phase in the above exponentials does not change much with $l$, i.e.,

$$2\frac{d\delta_l}{dl} \pm \vartheta = 0, \tag{C89}$$

or

$$\delta_l = \pm \tfrac{1}{2} l\vartheta. \tag{C89a}$$

From this and (C85) it follows that, for the classical approximation to be valid, the phase shifts $|\delta_l|$ must not be small.

It is known that the J-W-K-B method of solving the Schrödinger equation gives a solution in the "classical limit".# From the expression (C17) for the phase shifts,

$$\delta_l = \int_{r_0}^{\infty} \left[\sqrt{k^2 - U(r) - \frac{(l+\frac{1}{2})^2}{r^2}} - k\right] dr - kr_0 + \frac{1}{2} l\pi + \frac{1}{4}\pi, \tag{C90}$$

(C89) gives

$$-\int_{r_0}^{\infty} \frac{(2l+1)}{r^2\sqrt{k^2 - U - \dfrac{(l+\frac{1}{2})^2}{r^2}}}\, dr + \pi \pm \vartheta = 0,$$

which, again on using $l\hbar = p\rho$, and $k^2\hbar^2 = 2\mu E$, $U = (2\mu/\hbar^2)V$, gives

$$\vartheta = \pi - 2\rho \int_{r_0}^{\infty} \frac{p\, dr}{r^2\sqrt{2\mu[E - V(r)] - \left(\dfrac{p\rho}{r}\right)^2}}, \tag{C91}$$

which is the expression in classical dynamics for the deflection angle $\vartheta$ for a particle of initial momentum $p$ and impact parameter $\rho$ in a field $V(r)$. This shows that when the conditions (C85) and (C89) are satisfied, the classical description of the scattering becomes a good approximation.

We have mentioned [in the paragraph following (C85)] that the condition (C85) imposes a condition on the field $U(r)$ for the trajectory picture of classical dynamics to be valid. The Coulomb field is a peculiar case for which both quantum and classical mechanics give the correct scattering cross section, namely, the Rutherford formula. For a physical (in contrast with the mathe-

---

# This follows from the close relation between the Schrödinger equation, with the expansion in the J-W-K-B method, and the Hamilton–Jacobi theory in classical dynamics. For a detailed discussion, see, for example, ref. 17, §§ 19–20.

matical) discussion of the relation between the classical and the quantum theory of scattering, the reader should refer to an article by N. Bohr (ref. 18).

In Chapter III, Sect. M.4.5, we shall see that under proper conditions, the collisions between heavy particles (atoms and molecules) can be treated by the semi-classical (impact parameter) method.

## REFERENCES

The first treatment of the scattering problem in quantum mechanics is given in the classic paper of Born,

**1.** Born, M., Zeits. f. Physik **38**, 803 (1926).

For more extensive treatments, see:

**2.** Mott and Massey, *Theory of Atomic Collisions*. Clarendon Press, Oxford (1949).

**3.** Schiff, L. I., *Quantum Mechanics*. McGraw-Hill, New York (1955).

**4.** Williams, E. J., Rev. Mod. Phys. **17**, 217 (1945).

**5.** Henneberg, W., Zeits. f. Physik **83**, 555 (1933), gives the W-K-B expression (C17); also uses Thomas-Fermi potential for calculating the phase shifts.

**6.** Hellund, E. J., Phys. Rev. **59**, 395 (1941), gives the expression (C26) (a slight error has been corrected here), for the second Born approximation.

The second Born approximation has been treated by a number of authors:

**7.** Wu, T. Y., Phys. Rev. **73**, 934 (1948), treats the case of Gaussian potential.

**7a.** Källen, G., Ark. Fysik **2**, 33 (1950). Yukawa potential.

**8.** Jost, R. and Pais, A., Phys. Rev. **82**, 840 (1951), treats the case of Exponential and Yukawa potential. The condition of convergence of the Born expansion is given.

**9.** Dalitz, R. H., Proc. Roy. Soc. (London) **A206**, 509 (1951), treats the (relativistic) Yukawa potential up to the second Born approximation.

**10.** Pais, A., Proc. Camb. Phil. Soc. **42**, 45 (1946), gives the method of (C27-8).

**11.** Whittaker, E. T. and Watson, G. N., *Modern Analysis* (Cambridge University Press, London, 1940), 4th ed. Chap. XI.

**12.** Bargmann, V., Rev. Modern Physics **21**, 488 (1949).

**13.** Kohn, W., Rev. Mod. Phys. **26**, 293 (1954). On the convergence of Born expansions.

**13a.** Zemach, C. and Klein, A., Nuovo Cimento **10**, 1078 (1958).

**14.** Ohmura (formerly Kikuta), T., Prog. Theore. Phys. **12**, 225, 234 (1954) and a forthcoming paper. Upper and lower bounds of errors in the Born approximations.

**15.** Molière, G., Zeits. Naturforsch. **2a**, 133 (1947), gives (C78)–(C78b).

**16.** Cf. Jahnke, E. and Emde, F., *Tables of Functions*, Dover Publ., New York.

**17.** Breit, G., article in Handbuch der Physik, Vol. **41/1**, Springer (1959).

**18.** Bohr, N., Kgl. Danske Vidensk. Selsk. Mat.-fys Medd. **18**, 1 (1948).

section

# D Variational Methods

In Sect. A, the scattering of a particle by a central field is completely described in terms of the phase shifts. In most actual problems, however, an exact, analytical solution for the phase shifts is not feasible. Unless one resorts to the numerical integration of the Schrödinger equation, some approximate but reasonably accurate methods will be useful.

For the (lowest) discrete eigenvalue $E$ of the Hamiltonian $H$, the variational principle which is equivalent to the Schrödinger equation is

$$\delta \int \psi_E(H - E)\, \psi_E \, d\tau = 0. \tag{D1}$$

By using trial wave functions for $\psi_E$, one approaches $E$ from above, i.e., obtains an upper bound for $E$.

For the continuum states, a similar variational principle exists that gives the stationary property of the phase shifts. The variational principle has been formulated in a few slightly different forms. We shall first describe the form given by Hulthén. The stationary property will be demonstrated in this section by introducing infinitesimal variations. A theory of finite variations based on the Kato identity will be taken up in Sect. E.3, in connection with the theory of the upper bound of the scattering length.

## 1. HULTHÉN'S METHOD

Consider first the scattering of a particle by a field as in Sect. A. The Schrödinger equation is

$$[-\Delta - k^2 + U(r)]\, \psi(\mathbf{r}) = 0. \tag{D2}$$

We seek a solution [see (A12a)]

$$\psi(\mathbf{r}) = \sum \frac{1}{r} u_l(r) P_l(\cos \vartheta) \tag{D3}$$

which asymptotically has the form

$$\psi(\mathbf{r}) \to e^{i\mathbf{k}\cdot\mathbf{r}} + \frac{e^{ikr}}{r} f(\vartheta), \tag{D4}$$

and

$$u_l(0) = 0, \tag{D5}$$

$$u_l(r) \to \sin\left(kr - \frac{l\pi}{2} + \varepsilon_l\right). \tag{D6}$$

For the scattering problem, we wish to know the *exact* phase shift $\varepsilon_l$ (notation changed from $\delta_l$ in Sect. A to avoid confusion with the variation sign here).

The Schrödinger equation for $u_l(r)$ is

$$(L - k^2)\, u_l(r) \equiv \left[-\frac{d^2}{dr^2} + U(r) + \frac{l(l+1)}{r^2} - k^2\right] u_l(r) = 0. \tag{D7}$$

By analogy with (D1), consider the integral

$$I_l \equiv \int_0^\infty \phi_l\, (L - k^2)\, \phi_l(r)\, dr, \tag{D8}$$

where $\phi_l(r)$ is a trial wave function satisfying conditions of the form (D5), (D6).

$$\phi_l(0) = 0, \tag{D5a}$$

$$\phi_l(r) \to \sin\left(kr - \frac{l\pi}{2} + \eta_l\right), \tag{D6a}$$

On varying $\phi_l(r)$ by arbitrary $\delta\phi_l$ subject to the condition ($\eta_l$ being the trial phase shifts)

$$\begin{aligned} \delta\phi_l &= 0, \qquad \text{at } r = 0, \\ \delta\phi_l(r) &\to \cos\left(kr - \frac{l\pi}{2} + \eta_l\right) \delta\eta_l, \end{aligned} \tag{D9}$$

we have

$$\delta I_l = \int \delta\phi_l\, (L - k^2)\, \phi_l\, dr + \int \phi_l\, (L - k^2)\, \delta\phi_l\, dr.$$

As

$$\int_0^\infty \phi_l \frac{d^2}{dr^2} \delta\phi_l\, dr = \int_0^\infty \delta\phi_l \frac{d^2\phi_l}{dr^2} dr + \left[\phi_l \frac{d}{dr}\delta\phi_l - \delta\phi_l \frac{d\phi_l}{dr}\right]_0^\infty,$$

using (D9) we obtain

$$\delta(I_l + k\eta_l) = 2 \int \delta\phi_l \cdot (L - k^2)\, \phi_l(r)\, dr. \tag{D10}$$

From this, it follows that the requirement that $\phi_l$ satisfy (D7) leads to

$$\delta\lambda \equiv \delta(I_l + k\eta_l) = 0, \tag{D11}$$

or, conversely, the requirement that $I_l + k\eta_l \equiv \lambda$ be stationary leads to $\phi_l = u_l$. From

$$\lambda = k\eta_l + \int_0^\infty \phi_l(L - k^2)\,\phi_l\,dr, \tag{D12}$$

it is seen that, when $\phi_l = u_l$ and hence $\eta_l$ is equal to the exact phase shift $\varepsilon_l$ in (D6), we have

$$\lambda = k\varepsilon_l, \tag{D13}$$

i.e., $\lambda$ defined by (D11) is $k$ times the exact $\varepsilon_l$, and the variational equation (D11) is

$$\delta(k\varepsilon_l) = 0. \tag{D14}$$

In the original work, Hulthén imposes a further condition on the trial wave function $\phi_l$, namely,

$$\int_0^\infty \phi_l\,(L - k^2)\,\phi_l\,dr = 0. \tag{D15}$$

In this case, one has, from (D12) and (D14),

$$\delta(k\eta_l) = 0. \tag{D14a}$$

Thus the "trial $\eta_l$" itself is stationary.

One could have expressed the same asymptotic condition (D9) with different normalizations for $\phi_l$. Thus one may write

$$\phi_l(r) \to \cos\left(kr - \frac{l\pi}{2}\right) + \cot\eta_l \sin\left(kr - \frac{l\pi}{2}\right), \tag{D16}$$

or

$$\phi_l(r) \to \sin\left(kr - \frac{l\pi}{2}\right) + \tan\eta_l \cos\left(kr - \frac{l\pi}{2}\right). \tag{D17}$$

The equations corresponding to (D11) and (D12) + (D13) are then

$$\left.\begin{aligned} \delta\lambda &= \delta(I_l + k\cot\eta_l) = 0, \\ k\cot\varepsilon_l &= k\cot\eta_l + \int_0^\infty \phi_l\,(L - k^2)\,\phi_l\,dr \end{aligned}\right\}, \tag{D18}$$

$$\left.\begin{aligned} \delta\lambda &= \delta(I_l - k\tan\eta_l) = 0, \\ k\tan\varepsilon_l &= k\tan\eta_l - \int_0^\infty \phi_l\,(L - k^2)\,\phi_l\,dr \end{aligned}\right\}. \tag{D19}$$

In (D18), the $\phi_l$ is normalized according to (D16); in (D19), according to (D17). Corresponding to (D14), we have in these two cases

$$\delta(k \cot \varepsilon_l) = 0, \tag{D20}$$

$$\delta(k \tan \varepsilon_l) = 0. \tag{D21}$$

(D14), (D20), (D21) are of course entirely equivalent. The form corresponding to (D16), (D18), (D20) is referred to as the second Hulthén method, and that corresponding to (D17), (D19), (D21) as Kohn's method.

In actual calculations, the following procedure may be suggested. We now define $F$ by

$$F \equiv k\lambda + I_l = k\lambda - \int_0^\infty \phi_l (L - k^2) \phi_l \, dr.$$

According to the normalization of the trial function (D6a), (D16) and (D17) we have $\lambda = \eta_l$, $-\cot \eta_l$ and $\tan \eta_l$ respectively. The condition:

$$\delta F = 0 \tag{D22}$$

is equivalent to (D11), (D20) and (D21), respectively. $F$ is the stationary quantity for $k\eta_l$, $-k \cot \eta_l$ and $k \tan \eta_l$ respectively. Starting with a trial wave function $\phi_l$ with an assumed phase $\lambda$ and containing, say, $n$ variational parameters $c_1, \ldots, c_n$, the integral $I_l$ is calculated, leading to $I = I(c_1, \ldots, c_n, \lambda)$. From the variational equation (D22), one obtains $(n + 1)$ equations

$$\frac{\partial F}{\partial \lambda} = 0, \qquad \frac{\partial F}{\partial c_i} = \frac{\partial I_l}{\partial c_i} = 0, \quad (i = 1, \ldots n) \quad \text{(Kohn–Hulthén)} \tag{D23}$$

for the determination of the $(n + 1)$ parameters $\lambda, c_1, \ldots, c_n$. Using the parameter values thus determined, an accurate value for $k\eta_l$, $-k \cot \eta_l$, and $k \tan \eta_l$ respectively is given, by virtue of (D22), by

$$F = k\lambda + I_l(c_1, \ldots, c_n, \lambda).$$

This procedure, as suggested by Hulthén (ref. 4) and Kohn (ref. 3), and often referred to as the Kohn–Hulthén procedure, is preferable to the original method of Hulthén in which $\lambda, c_1, \ldots, c_n$ are to be determined from the following equations

$$I_l(c_1, \ldots, c_n, \lambda) = 0,$$

$$\frac{\partial I_l}{\partial c_i} = 0, \quad i = 1, \ldots n. \quad \text{(Hulthén)} \tag{D24}$$

The reason is that $I_l = 0$ is quadratic in $\lambda$, and $\lambda$ is not uniquely determined. The false $\lambda$ must be discarded on some other arguments.

As an example, we shall take up a simple scattering problem with linear variational parameters to compare the results using different normalization of the wave function (D6a), (D16) and (D17) and different methods (D23), (D24) for the determination of the variational parameters. The differences of the final results on $\eta_l$ are, of course, of the second order with respect to the errors contained in the trial function by virtue of the extremum principle. We may have some idea of the magnitude of these second order quantities from the example (taken from ref. 5c).

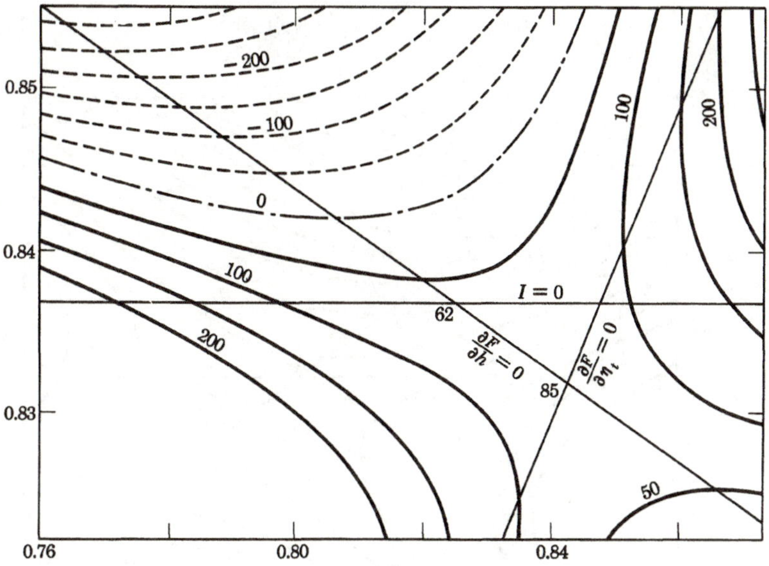


**Figure 1.** Contours of the difference: (exact phase shift − ε) × $10^6$ for various trial phase shift $\eta$ and parameter $h$ in (D25A) (ref. 5c).

Intersection of $\frac{\partial F}{\partial \lambda} = 0$ and $\frac{\partial F}{\partial h} = 0$: (D23)

Intersection of $\frac{\partial F}{\partial h} = 0$ and $I = 0$: (D24)

Intersection of $\frac{\partial F}{\partial \lambda} = 0$ and $I = 0$: (D25)

[From ref. 5c, Turner, J. S., and Makinson, R. E. B., *Proc. Phys. Soc.* (London) **A66**, 866 (1953).] $\lambda = \eta$, normalization (D6a) (Hulthén). The procedure (D24) gives a better result.

The assumed Schrödinger equation and the trial functions are

$$-(L-k^2)u(r) \equiv \left(\frac{d^2}{dr^2} + b\frac{e^{-r}}{r} + k^2\right)u(r) = 0,$$
$$\phi = C\{\sin kr + \tan \eta(1 - e^{-r})(1 + he^{-r}) \cos kr\}, \qquad \text{(D25A)}$$

with the values: $k = 0.8$ and $b = 1.5$. The exact phase shift $\varepsilon = 0.8371$. $C$ is a function of $\eta$ and should be chosen to give the normalization (D6a), (D16) and (D17) for each case. $h$ is the only adjustable parameter besides $\tan \eta$. Corresponding to (D23) and (D24) we have

$$\frac{\partial F}{\partial \lambda} = 0, \qquad \frac{\partial F}{\partial h} = 0. \text{ (Kohn–Hulthén)} \tag{D23}$$

$$I = 0, \qquad \frac{\partial F}{\partial h} = 0. \text{ (Hulthén)} \tag{D24}$$

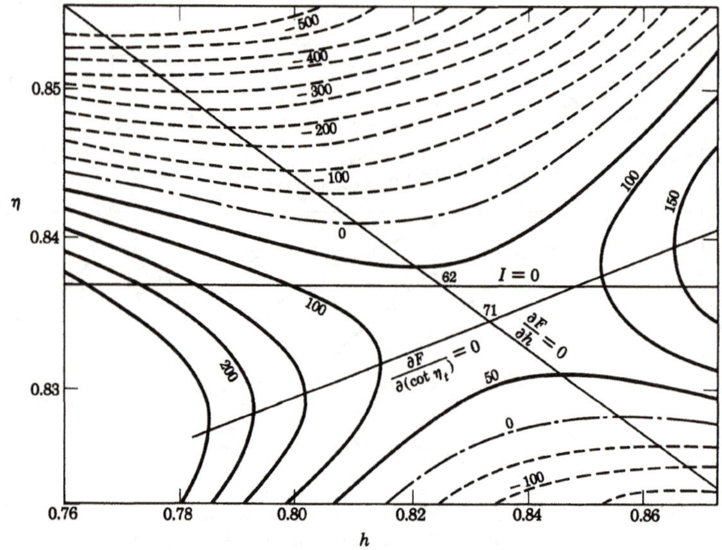


**Figure 2.** Contours of the difference: (exact phase shift $- \varepsilon) \times 10^6$ for various trial phase shift $\eta$ and parameter $h$ in (D25A) (ref. 5c).

Intersection of $\dfrac{\partial F}{\partial \lambda} = 0$ and $\dfrac{\partial F}{\partial h} = 0$: (D23)

Intersection of $\dfrac{\partial F}{\partial h} = 0$ and $I = 0$: (D24)

Intersection of $\dfrac{\partial F}{\partial \lambda} = 0$ and $I = 0$: (D25)

[From ref. 5c, Turner, J. S., and Makinson, R. E. B., *Proc. Phys. Soc.* (London) **A66**, 866 (1953).] $\lambda = -\cot \eta$, normalization (D16) (second Hulthén). The procedure (D24) gives a better result.

We can also impose the following conditions to determine $\lambda$ and $h$.

$$I = 0, \qquad \frac{\partial F}{\partial \lambda} = 0. \tag{D25}$$

We shall calculate the values of the stationary quantity $F$ for various $\eta$ and $h$, and find a more accurate phase shift $\varepsilon$ by the relation: $F = k\varepsilon$, $F = -k \cot \varepsilon$ and $F = k \tan \varepsilon$, for the different normalization of $\phi$ respectively. Figs. 1, 2 and 3 show a contour plot of the difference (exact phase shift—$\varepsilon$) $\times 10^6$ for various $\eta$ and $h$, for the normalization (D6a), (D16) and (D17), respectively. The stationary value of $\varepsilon$ is determined by (D23), and represented by the intersection of the two lines of $\partial F/\partial \lambda = 0$ and $\partial F/\partial h = 0$. The results obtained by (D24) and (D25) are represented by corresponding crosses. The stationary

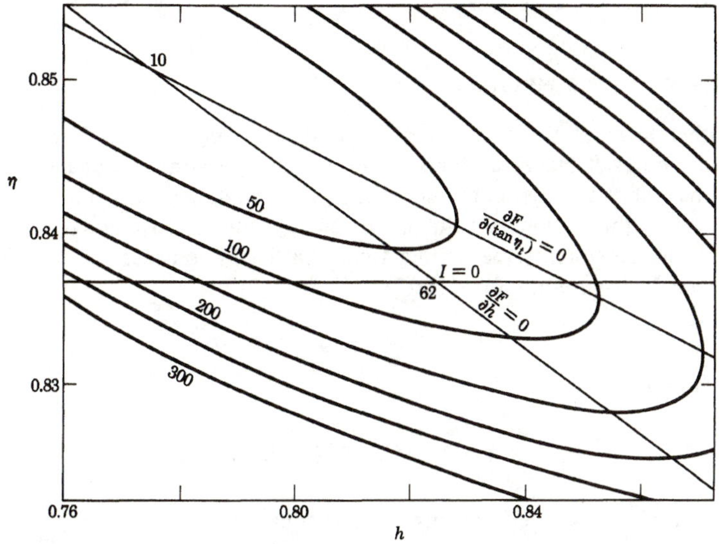


**Figure 3.** Contours of the difference: (exact phase shift $-\ \varepsilon$) $\times 10^6$ for various trial phase shift $\eta$ and parameter $h$ in (D25A) (ref. 5c).

Intersection of $\dfrac{\partial F}{\partial \lambda} = 0$ and $\dfrac{\partial F}{\partial h} = 0$: (D23)

Intersection of $\dfrac{\partial F}{\partial h} = 0$ and $I = 0$: (D24)

Intersection of $\dfrac{\partial F}{\partial \lambda} = 0$ and $I = 0$: (D25)

[From ref. 5c, Turner, J. S., and Makinson, R. E. B., *Proc. Phys. Soc.* (London) **A66**, 866 (1953).] $\lambda = \tan \eta$, normalization (D17) (Kohn). The procedure (D23) gives the best result.

values are saddle points in Figs. 1 and 2, but a maximum in Fig. 3. Thus the variational principles in the scattering problem give neither an upper nor a lower bound on the phase shift in general. The nature of extremum will be further discussed in Sect. E. When $F$ has a maximum, the Kohn–Hulthén procedure (D23) gives the best value attainable by the same trial function. For example, the phase shift determined by (D23) in Fig. 3 is smaller than the true value only by $10^{-5}$ radian. In Fig. 1 and 2 the Hulthén procedure (D24) seems to give the best results. Note that there is a possibility to get the exact phase shift for sets of parameters if the contours have a saddle point. The errors in the phase shift $\varepsilon$ are seen to be generally of the order $\sim 10^{-4}$ radian, a very good result in view of the simplicity of the trial function. The accuracy of the variational calculation entirely depends on the good choice of the trial function.

### 2. SCHWINGER'S METHOD

Another form of the variational method has been given by J. Schwinger (lecture notes at Harvard Univ. in 1947) which gives the stationary value of the phase shift directly, in the same way as the binding energy (discrete state) is given by the stationary value of the energy integral so familiar in atomic problems. To simplify the writing somewhat let us consider $l = 0$ in (D17), and use the asymptotic form (D17)

$$u(r) \to \sin kr + \tan \varepsilon \cos kr. \tag{D26}$$

Equation (D7)

$$\left(\frac{d^2}{dr^2} + k^2\right) u(r) = U(r)\, u(r), \quad U(r) = \frac{2m}{\hbar^2} V(r), \tag{D27}$$

can be replaced by the following integral equation:

$$u(r) = \sin kr + \int_0^\infty G(r, r') \cdot U(r')\, u(r')\, dr', \tag{D28}$$

where $G(r, r')$ is the Green's function chosen to satisfy the asymptotic condition (D26), namely

$$\begin{aligned} G(r, r') &= \frac{1}{k} \sin kr' \cos kr, \qquad \text{for} \quad r' < r, \\ &= \frac{1}{k} \sin kr \cos kr', \qquad \text{for} \quad r < r'. \end{aligned} \tag{D29}$$

Comparing (D28) with (D26), we find

$$\tan \varepsilon = \frac{1}{k} \int_0^\infty \sin kr'\, U(r')\, u(r')\, dr'. \tag{D30}$$

From (D28), one obtains

$$\int_0^\infty U(r)\, u(r) \sin kr\, dr = \int_0^\infty U(r)\, u^2(r)\, dr - \int_0^\infty \int_0^\infty U(r)\, u(r)\, G(r, r')\, U(r')\, u(r')\, dr\, dr'.$$

On dividing both sides of this equation by the square of tan $\varepsilon$ in (D30), one gets

$$k \cot \varepsilon = \frac{\int_0^\infty U(r)\, u^2(r)\, dr - \int_0^\infty \int_0^\infty U(r)\, u(r)\, G(r, r')\, U(r')\, u(r')\, dr\, dr'}{\left[\frac{1}{k}\int_0^\infty U(r)\, u(r) \sin kr\, dr\right]^2} \quad \text{(D31)}$$

From this, the condition for stationary character of $k \cot \varepsilon$ upon arbitrary variations in $u(r)$ leads to

$$0 = \delta(k \cot \varepsilon) = \frac{2\int_0^\infty U(r)\, \delta u\, dr\left[u(r) - \sin kr - \int_0^\infty G(r, r')\, U(r')\, u(r')\, dr'\right]}{\left[\frac{1}{k}\int_0^\infty U(r)\, u(r) \sin kr\, dr\right]^2}$$

which gives exactly the integral equation (D28). Thus (D31) gives a convenient method for calculating $\varepsilon$ by starting with a trial wave function $u(r)$. [As the right-hand side of (D31) is homogeneous (of degree 0) in $u(r)$, we do not even have to normalize the trial wave function $u(r)$.] As to the nature of the extremum in Schwinger's variational principle, see Subsect. 3 (iii) below.

It is seen from (D31) that taking only the first term in the numerator and putting the free particle wave function for $u(r)$ gives

$$k \tan \varepsilon = \int_0^\infty U(r)\, u^2(r)\, dr$$

which is the first Born approximation [see (C8a)]. The second term in the numerator (D31) gives the second Born approximation. Thus (D31) is a very elegant method for the scattering problem. In actual calculations, however, the complexity of the integral involving $G(\mathbf{r}, \mathbf{r}')$ in (D31) is such as to make higher order variational calculations impractical. It is for this reason that the Hulthén method has been more frequently used in the literature.

### 3. GENERAL FORMULATION OF THE VARIATIONAL PRINCIPLE

It is known that the fundamental equations in many branches of physics can be converted into a variational principle of which Euler's equation is the physical law itself. If the stationary quantity is just the quantity we are interested in, the variational principle is useful also in the approximate evaluation of the quantity. For the scattering problem, we have given two types of useful variational principles in the preceding subsections. It will be shown here that these can be extended to more general problems. (Also see Sect. P for the time dependent formulation.)

### *i) Scattering amplitude*

Consider the Schrödinger equation (D2),

$$(L - k^2)\,\psi \equiv [-\Delta - k^2 + U(r)]\,\psi(\mathbf{r}) = 0. \tag{D32}$$

By analogy with (D8), consider the integral

$$I(k_1, k_2) = \int_0^\infty \psi_2^*(L - k^2)\,\psi_1\,d\mathbf{r}. \tag{D33}$$

where $\psi_i(\mathbf{r})$ is a solution of (D32) with the incident wave along the $\mathbf{k}_i$ direction.

$$\psi_i(\mathbf{r}) \to \exp(i\mathbf{k}_i\cdot\mathbf{r}) + f(\vartheta_i)\,\frac{e^{ikr}}{r}, \quad r \to \infty, \tag{D34}$$

$\vartheta_i$ = angle between $\mathbf{k}_i$ and the outgoing direction. By taking an infinitesimal variation of $\psi_i$,

$$\delta\psi_i(r) \to \delta f(\vartheta_i)\,\frac{e^{ikr}}{r}, \quad r \to \infty,$$

we get the following formula in a similar manner as in the derivation of (D10):

$$\delta[I(\mathbf{k}_1, \mathbf{k}_2) - 4\pi f(\widehat{\mathbf{k}_1, \mathbf{k}_2})] = 0. \tag{D35}$$

Therefore, if $\psi_i$ is replaced by an approximate function $\Phi_i$ with the scattering amplitude $f_t$, the quantity

$$f_t(\widehat{\mathbf{k}_1, \mathbf{k}_2}) - \frac{1}{4\pi}\,I(\widehat{\mathbf{k}_1, \mathbf{k}_2}) \tag{D36}$$

is a stationary expression of $f(\widehat{\mathbf{k}_1, \mathbf{k}_2})$, (ref. 3). If the plane wave is used as $\Phi_i$, $[f_t(\widehat{\mathbf{k}_1, \mathbf{k}_2}) = 0]$ (D36) is just the (first) Born approximation.

### *ii) More general cases*

Thus a quadratic integral $I$ leads to a variational principle (D35) for the scattering amplitude. This situation exists in the scattering problems in general. Let the Hamiltonian be $H$, and let the energy of the system be $E$. Consideration of $\delta I(f, i) = \delta[\phi_f^*, (H - E)\phi_i]$ always gives a variational principle for some quantity $T_{fi}$ which essentially determines the transition cross section from the initial state $\phi_i$ to the final state $\phi_f$. The stationary expression $T_{fi}$ has the abstract form of $T_{fi} + I(f, i)$. As an example, the proper choice of the initial and final wave function leads to a variational method on the scattering matrix elements (ref. 3).

### *iii) Relation of the Schwinger method to the other methods*

Schwinger's variation principle (D31) seems at first to be unrelated to the other variational methods. However (D31) can be written in a form similar to the variational expression for the (energy) eigenvalue. By virtue of (D30) we can rewrite (D28) as

$$k \cot \varepsilon \int \frac{\sin kr}{k} \cdot \frac{\sin kr'}{k} U(r') u(r') \, dr' = u(r) - \int G(r, r') U(r') u(r') \, dr'.$$

On multiplying by $u(r) U(r)$ from the left, and integrating over $r$, we get

$$k \cot \varepsilon = \frac{(u, Au)}{(u, Pu)}, \tag{D37}$$

where $A$ and $P$ are hermitian integral operators with the kernel

$$A(r, r') = U(r) \delta(r - r') - U(r) G(r, r') U(r'),$$

$$P(r, r') = U(r) \frac{\sin kr}{k} \cdot \frac{\sin kr'}{k} U(r'),$$

respectively. The form (D37) is nothing but (D31) and a slight extension of the usual Rayleigh–Ritz variational expression. (D37) is equivalent to

$$k \cot \varepsilon = \frac{(u, Au)}{(u, f)^2}, \tag{D38}$$

where $f(r) = k^{-1} \sin kr \, U(r)$. From the abstract form (D37) or (D38), it is seen that Schwinger's method is applicable not only to the phase shift but to other properties. The requirement that the quantity

$$\lambda = \frac{(u, Au)}{(u, Pu)} \quad \text{or} \quad \lambda = \frac{(u, Au)}{(u, f)^2}$$

be stationary with respect to small variations of $u$ leads to the equation for $u$

$$Au = \lambda Pu, \quad \text{or} \quad Au = cf, \quad [c = \lambda \times (u, f)].$$

On the basis of (D37) or (D38) it can be shown that (D31) gives an upper (lower) bound for $k \cot \varepsilon$ if the potential is attractive (repulsive), that is, $V(r) \leqslant 0$ [$V(r) \geqslant 0$], and not too strong so that $|\varepsilon| < \pi$. (ref. 5b.)

## REFERENCES


1. Hulthén, L., Kgl. Fysiograf Sällshapet. Lund Förk,, **14**, No. 21 (1944). The original paper for the Hulthén method.

2. Schwinger, J., Phys. Rev. **78**, 135 (1950); Blatt, J. M. and Jackson, J. D., Phys. Rev. **76**, 18 (1949). Presentation of the Schwinger method.

3. Kohn, W., Phys. Rev. **74**, 1763 (1948). Improvement of the original Hulthén method, also includes variational methods for scattering amplitudes, *S*-matrix etc.

4. Hulthén, L., Arkiv. f. Mat. Astron, Fysik **35A**, No. 25 (1948). Improvement of the original Hulthén method.

5a. Kato, T., Phys. Rev. **80**, 475 (1950). Relation among the Schwinger and Hulthén–Kohn methods.

5b. Kato, T., Progr. Theor. Phys. (Kyoto) **6**, 295 (1951). On Schwinger's method.

5c. Turner, J. S. and Makinson, R. E. B., Proc. Phys. Soc. (London) **A66**, 866 (1953).

6. Marshak, R. E., Phys. Rev. **71**, 688 (1947); Davison, B., Phys. Rev. **71**, 694 (1947); Levine, H. and Schwinger, J., Phys. Rev. **74**, 958 (1948), **75**, 1423 (1949); Levine, H., J. Acous. Soc. Amer. 22, 48 (1950); Papas, C. H., J. Appl. Phys. **21**, 318 (1950); Levitas, A. and Lax, M., J. Acous. Soc. Amer. **23**, 316 (1951). Applications of Schwinger's method to diffraction problems, etc.

7. Lippmann, B. A. and Schwinger, J., Phys. Rev. **79**, 569 (1950); Kato, T., Progr. Theor. Phys. (Kyoto), **6**, 394 (1951); Jackson, J. L., Phys. Rev. **83**, 301 (1951); Kohn, W., Phys. Rev. **84**, 495 (1951); Moses, H. E., Phys. Rev. **92**, 817 (1953); Nuovo Cimento, Supplement Series X, **5**, 120, 144 (1957). Extensions of the method.

section

# E

## Slow Collisions: Theory of Scattering Length and Effective Range

In Sect. A, it is shown that the scattering of a particle by a potential field $V(r)$ is completely described in terms of the phase shift $\delta_l(k)$. For the case of low energies at which only $s$ waves are scattered the amplitude and cross sections are given by

$$f(\vartheta) = \frac{1}{2ik}[\exp(2i\delta_0) - 1].$$

$$\sigma = \frac{4\pi}{k^2}\sin^2\delta_0. \tag{E1}$$

It is found that these can be described, instead of by $\delta_0$, by two quantities that characterize $V(r)$ completely as far as low energy scattering is concerned. These are the "effective range" $r_0$, a concept already evident in the work of Breit *et al.* and introduced explicitly by Schwinger, and Blatt and Jackson, and the "scattering length" $a$ introduced by Fermi, both in connection with nucleon–nucleon scattering.

If the system has a bound state with a small binding energy, the two low energy parameters $r_0$ and $a$ will be completely determined by the bound state wave function. That such a relationship exists between the properties of the bound state and low energy scattering is interesting but not unexpected, since both are determined by the potential $V(r)$. Furthermore, if the bound particle has an electric charge, the low energy parameters also determine the photo-transition rate (from the bound state to the continuum) and the *Bremsstrahlung*, and *vice versa*, as we shall show in the appendices to Sect. E.

We shall first derive, following Bethe, a general formula for a positive energy state.

## I. SCATTERING LENGTH AND EFFECTIVE RANGE FOR SHORT-RANGED POTENTIAL $V(r)$

For $s$ waves, the Schrödinger equation is

$$\frac{d^2u}{dr^2} + [k^2 - U(r)]\,u(r) = 0, \quad U(r) = \frac{2m}{\hbar^2}V(r). \tag{E2}$$

Let $u_1(r)$, $u_2(r)$ be the solutions for two energies $k_1^2$, $k_2^2$. They satisfy

$$u_1(0) = 0, \qquad u_2(0) = 0, \tag{E3}$$

and are normalized such that asymptotically they are

$$\begin{aligned} u_1(r) &\to \frac{1}{\sin\delta_1}\sin(k_1r + \delta_1), \\ u_2(r) &\to \frac{1}{\sin\delta_2}\sin(k_2r + \delta_2). \end{aligned} \tag{E4}$$

From the two equations (E2) for $u_1(r)$, $u_2(r)$, we readily obtain

$$\left[u_2\frac{du_1}{dr} - u_1\frac{du_2}{dr}\right]_0^R = (k_2^2 - k_1^2)\int_0^R u_1u_2\,dr, \tag{E5}$$

where $R$ is an arbitrary radial distance.

Let us take two free-particle solutions

$$\begin{aligned} v_1(r) &= \frac{1}{\sin\delta_1}\sin(k_1r + \delta_1), \\ v_2(r) &= \frac{1}{\sin\delta_2}\sin(k_2r + \delta_2), \end{aligned} \tag{E6}$$

of the equation obtained by putting $U(r) = 0$ in (E2). From the equations for $v_1(r)$, $v_2(r)$, we obtain as in (E5),

$$\left[v_2\frac{dv_1}{dr} - v_1\frac{dv_2}{dr}\right]_0^R = (k_2^2 - k_1^2)\int_0^R v_1v_2\,dr. \tag{E7}$$

On substracting (E6) from (E7), using (E3), (E4), (E6), and letting $R \to \infty$ we get

$$k_2\cot\delta_2 - k_1\cot\delta_1 = (k_2^2 - k_1^2)\int_0^\infty (v_1v_2 - u_1u_2)\,dr. \tag{E8}$$

On defining the "scattering length" $a$ by

$$-\frac{1}{a} = \lim_{k\to 0}[k\cot\delta(k)], \tag{E9}$$

we can write (E8), on letting $k_1 \to 0$ and denoting $k_2$ by $k$

$$k\cot\delta = -\frac{1}{a} + \frac{b}{2}k^2, \tag{E10}$$

where

$$b = 2\int_0^\infty (v_0 v - u_0 u)\, dr. \tag{E11}$$

The factor 2 in (E11) has been introduced so that $b$ [or $r_0$ in (E11a)] has the meaning of the range of the potential. See (E23a). From (E4), (E6), it is seen that the integrand above differs from zero only in the region where $U(r)$ is appreciable. In this region, the wave function $u(r)$ will not depend very much on the energy $k^2$ if $|U(r)|$ is $\gg k^2$. We shall therefore make the approximation of replacing $u$, $v$ by $u_0$, $v_0$ (i.e., for zero energy) in (E11), i.e., we shall take the first two terms in a power series expansion in $k^2$ of $k \cot \delta$:

$$k \cot \delta = -\frac{1}{a} + \frac{r_0}{2} k^2 + O(k^4), \tag{E10a}$$

where

$$r_0 = 2\int_0^\infty (v_0^2 - u_0^2)\, dr \tag{E11a}$$

is defined as the "effective range" of the potential $V(r)$. According to (E4) and (E9), the zero energy $u_0(r)$ has the asymptotic form

$$\begin{aligned} u_0(r) \to v_0(r) &= \lim_{k\to 0} (\cos kr + \cot \delta \sin kr) \\ &= 1 - \frac{r}{a}. \end{aligned} \tag{E12}$$

From (E4), (E6), it is seen that $v_0^2 - u_0^2$ vanishes outside the "range" of $U(r)$.

It is clear that both $r_0$ and $a$ are determined by $U(r)$, that they are insensitive to the exact form of $U(r)$ but depend only on some "integrated" or "average" property of $U(r)$. The meaning of $a$ can be made clearer on the following considerations:

1) For $k^2 \to 0$, we have, from (E1) and (E9)

$$\sigma = \frac{4\pi}{k^2(1 + \cot^2 \delta)} \to 4\pi a^2. \tag{E13}$$

2) From the expression (A25) for $\delta_0$ we have, for $k \to 0$,

$$a = \int_0^\infty r\, U(r)\, u_0(r)\, dr, \tag{E14}$$

where $u_0(r)$ is normalized by the asymptotic behavior (A26) and is hence different from (E12),

$$u_0(r) \to \lim_{k\to 0} \frac{\sin (kr + \delta)}{k} = (r - a) \cos \delta, \tag{E15}$$

where $\cos \delta$ has the value 1 or $-1$.

If the Born approximation is valid, i.e., $u_0$ is replaced by the field free solution $r\sqrt{\pi/2kr}\, J_{1/2}(kr) = (1/k)\sin kr \to r$ as $k \to 0$, (E14) becomes

$$a \cong \int_0^\infty U(r)\, r^2\, dr. \tag{E16}$$

[Note that Born's approximation is not good in both the cases of proton–neutron and electron–H atom scatterings.] (E9), (E14), (E12) or (E15) will characterize the scattering length $a$.

3) The sign of $a$ depends on $U(r)$. The variation of $a$ with $U(r)$ can be brought out explicitly by the following example.

Consider a rectangular well potential

$$U(r) = \begin{cases} -\beta^2, & r < R, \\ 0, & R < r. \end{cases} \tag{E17}$$

If $E = -\gamma^2$, $\gamma > 0$, is a discrete state, then

$$\left(\frac{d^2}{dr^2} - \gamma^2 + \beta^2\right) u_\gamma(r) = 0, \quad r < R,$$

$$\left(\frac{d^2}{dr^2} - \gamma^2\right) u_\gamma(r) = 0, \quad R < r,$$

so that

$$u_\gamma(r) = \begin{cases} A \sin\,(r\sqrt{\beta^2 - \gamma^2}), & r < R, \\ B\, e^{-\gamma r}, & R < r. \end{cases}$$

The continuity condition at $r = R$ leads to

$$\tan\,(\sqrt{\beta^2 - \gamma^2}\, R) = -\frac{\sqrt{\beta^2 - \gamma^2}}{\gamma}. \tag{E18}$$

The condition on $\beta$, $R$ for the existence of 1, 2 or more discrete states is seen to be the following:

$$\begin{aligned} &\text{for only 1 discrete state, } \tfrac{1}{2}\pi < \beta R < \tfrac{3}{2}\pi, \\ &\text{for only 2 discrete states, } \tfrac{3}{2}\pi < \beta R < \tfrac{5}{2}\pi, \text{ etc.} \end{aligned} \tag{E18a}$$

Consider now the scattering by $U(r)$ in (E17). The wave function $u_k$ is now given by

$$\begin{aligned} \left(\frac{d^2}{dr^2} + k^2 + \beta^2\right) u_k = 0, \quad r < R, \\ \left(\frac{d^2}{dr^2} + k^2\right) u_k = 0, \quad R < r, \end{aligned} \tag{E19}$$

$$u_\gamma(r) = \begin{cases} C \sin\,(\sqrt{\beta^2 + k^2}\, r), & r < R, \\ D \sin\,(kr + \delta), & R < r. \end{cases} \tag{E19a}$$

The continuity condition at $r = R$ is

$$k \cot (kR + \delta) = \sqrt{\beta^2 + k^2} \cot (\sqrt{\beta^2 + k^2}\, R), \qquad \text{(E20)}$$

or

$$k \cot \delta = \frac{k \tan (\sqrt{\beta^2 + k^2}\, R) \tan kR + \sqrt{\beta^2 + k^2}}{\tan (\sqrt{\beta^2 + k^2}\, R) - \dfrac{1}{k}\sqrt{\beta^2 + k^2} \tan kR}. \qquad \text{(E20a)}$$

With the definition (E9) for $a$, this gives

$$a = R - \frac{1}{\beta} \tan \beta R. \qquad \text{(E21)}$$

Thus the scattering length $a$ vanishes as $\beta \to 0$. As $\beta R$ increases from 0 ($<\pi/2$, i.e., $U(r)$ has no discrete state), $a$ decreases ($a < 0$). As $\beta R \to \pi/2$, $a$ becomes negatively infinite. The cross section $\sigma$ in (E13) becomes infinite as $\beta R \to \pi/2$, and one has "resonance" at zero energy. The variations of $a$ and $k \cot \delta$ with $\beta R$ for small $k$ are as follows:

| number of discrete states in $U(r)$ | $\beta R$ | $x = k \cot \delta$ | $a$ | $\delta$ | |
|---|---|---|---|---|---|
| 0 | $0 \le \beta R < \pi/2$ | $\infty > x > 0$ | $0 \ge a > -\infty$ | $\simeq 0$ | |
| | $\pi/2$ | $0$ | $-\infty \to +\infty$ | $\pi/2$ | |
| 1 | $\pi/2 < \beta R < 4.49$ | $0 > x > -\infty$ | $\infty > a > 0$ | $\simeq \pi$ | |
| | $4.49 < \beta R < 3\pi/2$ | $\infty > x > 0$ | $0 > a > -\infty$ | $\simeq \pi$ | (E22) |
| | $3\pi/2$ | $0$ | $-\infty \to +\infty$ | $3\pi/2$ | |
| 2 | $3\pi/2 < \beta R < 7.76$ | $0 > x > -\infty$ | $\infty > a > 0$ | $\simeq 2\pi$ | |
| | $7.76 < \beta R < 5\pi/2$ | $\infty > x > 0$ | $0 > a > -\infty$ | $\simeq 2\pi$ | |

The case of $^1S$ proton–neutron scattering corresponds to the first row; while the $^3S$ proton–neutron scattering corresponds to the third row.

The phase shift at zero energy is $n\pi$ where $n$ is the number of bound states. Levinson proved this for central field scatterings (ref. 8). For the more general case, see Sect. L.5,ii.

For the rectangular well potential (E17), $r_0$ can be obtained explicitly. From (E19a) and the $u_k(r)$ for $k \to 0$, $r \to \infty$ in (E12), we get

$$u_0 = \left(1 - \frac{R}{a}\right) \frac{\sin \beta r}{\sin \beta R}, \qquad 0 < r < R,$$

and with (E12) for $v_0$, we have from (E11a)

$$r_0 = 2R - 2\frac{R^2}{a} + \frac{2R^3}{3a^2} + \left(1 - \frac{R}{a}\right)^2 \left(\frac{1}{\beta \tan \beta R} - \frac{R}{\sin^2 \beta R}\right). \qquad \text{(E23)}$$

If $\beta R = (\frac{1}{2} + n)\pi$, $n = 0, 1, 2, \ldots$, $a$ tends to $\pm\infty$ and (E23) simplifies to

$$r_0 = R. \tag{E23a}$$

The total cross section $\sigma$ in terms of $a$ and $r_0$ is

$$\sigma = \frac{4\pi a^2}{1 + a(a - r_0)k^2 + (\frac{1}{2}ar_0)^2 k^4}.$$

The scattering length and effective range for several potentials (Yukawa, exponential, Gaussian, square-well) are given in ref. 2.

## 2. EFFECTIVE RANGE CHARACTERIZED BY THE BOUND STATE

Let the uppermost bound state of (E2) have the energy $E = -(\hbar^2/2m)\gamma^2$, $\gamma^2 > 0$, so that for this bound state,

$$\frac{d^2u_\gamma}{dr^2} - [\gamma^2 + U(r)]\, u_\gamma = 0, \tag{E24}$$

where

$$u_\gamma(0) = 0,$$

and $u_\gamma(r)$ is so normalized that asymptotically,

$$u_\gamma(r) \to e^{-\gamma r}. \tag{E24a}$$

Let us choose, by definition, the following functions

$$v_\gamma(r) = e^{-\gamma r}, \qquad v_k(r) = \frac{\sin(kr + \delta)}{\sin\delta} \tag{E25}$$

which are seen to satisfy the equations

$$\left(\frac{d^2}{dr^2} - \gamma^2\right) v_\gamma(r) = 0, \qquad \left(\frac{d^2}{dr^2} + k^2\right) v_k(r) = 0. \tag{E26}$$

From these two equations, one gets

$$(\gamma^2 + k^2)\int_0^\infty v_\gamma v_k\, dr = \left| v_k \frac{dv_\gamma}{dr} - v_\gamma \frac{dv_k}{dr}\right|_0^\infty$$
$$= k\cot\delta + \gamma. \tag{E27}$$

Using the orthogonality relation between $u_\gamma(r)$ and $u_k(r)$, we have

$$k\cot\delta = -\gamma + \tfrac{1}{2}(\gamma^2 + k^2)\sigma, \tag{E28}$$

where

$$\sigma(E, k) = 2\int_0^\infty (v_\gamma v_k - u_\gamma u_k)\, dr. \tag{E29}$$

If $\gamma^2$ and $k^2$ are small compared with the depth of $U(r)$, we may expand $k \cot \delta$ in powers of $(\gamma^2 + k^2)$ and retain the first two terms only. $v_k(r)$ will be slightly different from $v_\gamma(r)$ and we may expand

$$v_k(r) = cv_\gamma(r) + (\gamma^2 + k^2)f(r) + \cdots.$$

The constant $c$ can be determined as follows. If $k$ is replaced by $-i\gamma$ and $\cot \delta$ by $-i$ (this is the condition for a bound state, see below), $v_k(r)$ in (E25) becomes identical with $v_\gamma(r)$ in (E25). Thus $c = 1$. (E28) then reduces to

$$k \cot \delta = -\gamma + \frac{\rho}{2}(\gamma^2 + k^2) + O[(\gamma^2 + k^2)^2], \tag{E30}$$

where the effective range $\rho$ is characterized by the bound state wave function $u_\gamma$ through#

$$\rho = 2\int_0^\infty (v_\gamma^2 - u_\gamma^2)\, dr. \tag{E31}$$

Suppose that the higher terms in (E10a) and (E30) are exactly zero. Equating (E10a) and (E30), we get

$$\gamma = \frac{1}{a} + \frac{\rho\gamma^2}{2}, \quad r_0 = \rho. \tag{E36}$$

The scattering length $a$ can be calculated from the bound state parameters $\gamma$(E24) and $\rho$ (E31) alone. The effective range $\rho$ at the bound state is equal to the effective range $r_0$ at zero energy in this "shape-independent" approximation.

---

# (E30), (E31) can also be derived from (E8) directly. The asymptotic behavior of $u_k(r)$ is $v_k(r)$ in (E25) that can be written as

$$v_k(r) \to \frac{i\,e^{-i\delta}}{2 \sin \delta}(e^{-ikr} - S\,e^{ikr}), \tag{E32}$$

where $S = e^{2i\delta}$ is the scattering matrix element introduced in (A34)–(A35). If we put $k = -i\gamma$ $(\gamma > 0)$, (E32) becomes

$$v_{-i\gamma}(r) = \frac{i\,e^{-i\delta}}{2 \sin \delta}(e^{-\gamma r} - S\,e^{\gamma r}). \tag{E33}$$

The condition of a bound state is (see Sect. F)

$$S = e^{2i\delta} = 0 \quad \text{at} \quad k = -i\gamma, \quad \gamma > 0. \tag{E34}$$

This is equivalent to one of the following conditions

$$\cot \delta = -i, \qquad \frac{i\,e^{-i\delta}}{2 \sin \delta} = 1, \quad \text{or} \quad \operatorname{Im} \delta = +\infty. \tag{E35}$$

Thus $v_{-\gamma i}(r)$ in (E33) is just the $v_\gamma(r)$ defined in (E25). If we put $k_2 = -i\gamma$ in (E8), then (E8) is identical with (E28)–(E29).

The total cross section $\sigma$ in terms of $\gamma$ and $\rho$ is

$$\sigma = \frac{4\pi}{(\gamma^2 + k^2)[1 - \gamma\rho + \frac{1}{4}\rho^2(\gamma^2 + k^2)]}. \tag{E37}$$

In the appendices to this section, we shall give the relation between the bound state parameters $\gamma$, $\rho$ and the cross section of (1) photo-disintegration and radiative capture, (2) Bremsstrahlung or the absorption of radiation due to free–free transitions. Also we shall indicate the effective range theory for the combined field due to a short-ranged potential plus a long-ranged potential such as the Coulomb or the polarization potential.

## 3. VARIATIONAL PRINCIPLE FOR THE SCATTERING LENGTH

In (E9) or (E36), we have defined the scattering length $a$ by

$$\lim_{k \to 0} (k \cot \varepsilon_l) = -\frac{1}{a}.$$

In (D20), we have an extremal property for $k \cot \varepsilon_l$ [or $k \tan \varepsilon_l$ in (D21)]. From these it follows that the scattering length $a$ has the stationary property.

It will be very convenient to introduce the Kato identity here in order to discuss the errors in variational calculations or the nature of extremum (that is, whether it is a minimum, maximum, or a saddle point). Let the normalization of the exact wave function $u(r)$ and the trial function $\phi(r)$ be specified by

$$\begin{aligned} &u(r) \to \sin(kr + \varepsilon)/\sin(\varepsilon - \vartheta), \quad \text{for } r \to \infty \\ &\phi(r) \to \sin(kr + \eta)/\sin(\eta - \vartheta), \\ &u(0) = \phi(0) = 0, \end{aligned}$$

where $\vartheta$ is an arbitrary constant. The choice of $\vartheta = 0$ (or $\vartheta = -\pi/2$) leads to the normalization (D16) [or (D17)]. On integrating by parts, we have

$$\begin{aligned} \int_0^\infty u(L - k^2)\,\phi\,dr &= k\,[\cot(\varepsilon - \vartheta) - \cot(\eta - \vartheta)] + \int_0^\infty \phi(L - k^2)\,u\,dr \\ &= k\cot(\varepsilon - \vartheta) - k\cot(\eta - \vartheta). \end{aligned} \tag{E38}$$

The left side of (E38) can be rewritten, on putting $\Delta\phi \equiv \phi - u$,

$$\int_0^\infty u(L - k^2)\phi\,dr = \int_0^\infty u(L - k^2)\,\Delta\phi\,dr. \tag{E39}$$

From (E38) and (E39) we have the Kato identity

$$k\cot(\varepsilon - \vartheta) = k\cot(\eta - \vartheta) + \int_0^\infty \phi(L - k^2)\phi\,dr - \int_0^\infty \Delta\phi(L - k^2)\,\Delta\phi\,dr. \tag{E40}$$

Since the last term is quadratic in the small variation $\Delta\phi$, the remaining two terms on the right-hand side constitute a stationary expression of $k \cot (\varepsilon - \vartheta)$. In fact the variation principle in (D18) or (D19) is obtained from (E40) by fixing $\vartheta = 0$ or $\vartheta = -\pi/2$.

The merit of Equation (E40) is to give us a basis for the evaluation of the errors in variational calculations. The integrand of the last term will approach zero as $r$ increases if the trial function $\phi$ is chosen properly. $\int \Delta\phi(L - k^2)\, \Delta\phi\, dr$ is proportional to the energy expectation value (minus the incident energy) of $\Delta\phi$, and is therefore probably positive if the energy of incident particles is small and if the potential $V(r)$ is such that no discrete (bound) state exists. However, unlike the usual variational method for the discrete energy eigenvalue, the variational methods in scattering state give neither an upper nor a lower bound for the phase shift in general. Some useful conclusion will be drawn for the limiting case of $k \to 0$ which is discussed later.

A method for the rigorous estimation of the upper and lower bounds on the phase shift was developed by Kato (ref. 6) on the basis of (E40). The evaluation of the quantity

$$\int_0^\infty \{[(L - k^2)\phi]^2/\rho(r)\}\, dr, \qquad \rho(r) \geqslant 0;\ \rho \to O(r^{-1}) \quad \text{as} \quad r \to \infty,$$

and a rough estimate of some constants are needed in this method.

The variational principle (D12)–(D14) will come from the identity:

$$k \sin(\eta - \varepsilon) = \int_0^\infty \phi(L - k^2)\phi\, dr - \int_0^\infty \Delta\phi(L - k^2)\Delta\phi\, dr,$$

which can be derived similarly with the following asymptotic conditions on $u$ and $\phi$

$$u(r) \to \sin(kr + \varepsilon), \quad u(0) = 0,$$
$$\phi(r) \to \sin(kr + \eta), \quad \phi(0) = 0.$$

Now let us consider the limiting case of $k \to 0$.

If the normalization of the trial wave function $\phi(r)$ at zero energy is specified according to (D16) and (D17), with $l = 0$ and $k \to 0$, we have, respectively,

$$\text{Hulthén:} \quad \phi(r) \to 1 - (r/a_t), \quad \phi(0) = 0. \tag{E41a}$$

$$\text{Kohn:} \quad \phi(r) \to r - a_t, \quad \phi(0) = 0. \tag{E41b}$$

The stationary value $\mu$ for $a$ is given, respectively, by

$$\mu^{-1} = a_t^{-1} - \int_0^\infty \phi(r)\, L\, \phi(r)\, dr, \tag{E42a}$$

$$\mu = a_t + \int_0^\infty \phi(r)\, L\, \phi(r)\, dr. \tag{E42b}$$

Defining $\Delta\phi = \phi - u$, where $u(r)$ is the exact wave function, we get, respectively (ref. 6), on using the Green's theorem

$$\int_0^\infty \phi L\phi\, dr = \int_0^\infty \phi L\Delta\phi\, dr = \int_0^\infty uL\,\Delta\phi\, dr + \int_0^\infty \Delta\phi L\,\Delta\phi\, dr$$

$$= -a^{-1} + a_t^{-1} + \int_0^\infty \Delta\phi L\,\Delta\phi\, dr, \tag{E43a}$$

$$\int_0^\infty \phi L\phi\, dr = a - a_t + \int_0^\infty \Delta\phi L\,\Delta\phi\, dr. \tag{E43b}$$

Using the definition of $\mu$ (E42a, b), we have, respectively,

$$-\mu^{-1} = -a^{-1} + \int_0^\infty \Delta\phi L\,\Delta\phi\, dr, \tag{E44a}$$

$$\mu = a + \int_0^\infty \Delta\phi L\,\Delta\phi\, dr. \tag{E44b}$$

Thus one might assume that $\mu$ gives an upper bound for $a$, because $\int \Delta\phi L\,\Delta\phi\, dr \geqq 0$ if the potential is such that there is no bound state. But a special consideration is needed because $\Delta\phi$ is not a quadratically integrable function and the theory of Hilbert space is not applicable here. We can only say that $\int_0^\infty \Delta\phi\, e^{-\nu r}\, L\,\Delta\phi\, e^{-\nu r}\, dr \geqq 0$ for positive (not zero) $\nu$. Fortunately we can (Spruch and Rosenberg, ref. 5a) verify that

$$\lim_{\nu\to 0} \int_0^\infty \Delta\phi\, e^{-\nu r}\, L\,\Delta\phi\, e^{-\nu r}\, dr = \int_0^\infty \Delta\phi\, L\,\Delta\phi\, dr, \tag{E45}$$

for the normalization (E41b) only. $\mu$ is really greater than the scattering length $a$ in this case. Since (E44) is valid for arbitrary (not too singular) $\Delta\phi$, the upper bound nature of $\mu$ holds for any trial function. If we restrict the trial function by (D15), we get the original Hulthén method. The original method thus may be considered as a special case, and the upper bound theorem is also applicable. The minimization of $\mu$ in the original method [see (D25)] is not complete as in the variational method (D26). Therefore the variational method with the Kohn method (E41b) and (E42b) gives a better result than the original one does.

The proof of (E45) fails for the normalization (E41a); therefore $\mu$ determined by (E42a) does not necessarily give an upper bound for $a$. It can be shown that $\mu^{-1}$ of (E42a) gives a lower bound on $a^{-1}$ only when $a$ is positive.

When the system has bound states, the integral $\int_0^\infty \Delta\phi\, L\,\Delta\phi\, dr$ can be negative, since $L$ is essentially the Hamiltonian. An upper bound for $a$ is, however, still obtainable if the trial wave function satisfies some conditions. (Rosenberg, Spruch and O'Malley, ref. 5b.)

We shall only consider the normalization (E41b) hereafter. Let $v_1(r)$ and $v_2(r)$ be (real) normalized functions. If the $2 \times 2$ matrix $H$ with the elements $H_{ij} = \dfrac{\hbar^2}{2m}\displaystyle\int_0^\infty v_i\, Lv_j\, dr$, is diagonalized, the eigenvalues of $H$, $E_1$ and $E_2$ satisfy the inequality relation

$$E_1 \leqslant H_{11} \leqslant E_2. \tag{E46}$$

Furthermore, if one (and only one) bound state exists in the system, we have

$$E_2 > 0. \tag{E47}$$

We shall assume that $v_1$ is an accurate wave function of the bound state so that

$$H_{11} < 0. \tag{E48}$$

According to (E46), (E47) and (E48) we get

$$E_1 E_2 = H_{11}H_{22} - H_{12}^2 < 0. \tag{E49}$$

Since the inequality sign in (E49) does not depend on the normalization (and also the sign) of $v_1$ and $v_2$, we may replace $v_2$ by $\Delta\phi \cdot e^{-\nu r}$ $(\nu > 0)$ and get

$$\int v_1 L v_1\, dr \cdot \int \Delta\phi\, e^{-\nu r} L\, \Delta\phi\, e^{-\nu r}\, dr < \left(\int v_1 L\, \Delta\phi\, e^{-\nu r}\, dr\right)^2.$$

On taking into account the facts that $\int v_1 L v_1\, dr < 0$, $\int v_1 L\, \Delta\phi\, dr = \int v_1 L\phi\, dr$ and all integrals are continuous at $\nu = 0$, we finally have

$$\int_0^\infty \Delta\phi\, L\, \Delta\phi\, dr > \frac{\left(\int_0^\infty v_1 L\phi\, dr\right)^2}{\int_0^\infty v_1 L v_1\, dr}. \tag{E50}$$

The right side of (E50) can always be evaluated. From (E44b)

$$\mu' \equiv \mu - \left(\int_0^\infty v_1 L\phi\, dr\right)^2 \bigg/ \int_0^\infty v_1 L v_1\, dr > a. \tag{E51}$$

Thus we have found an upper bound for $a$. Furthermore it can now be shown that $\mu'$ is obtainable by the usual variational procedure. Let the new trial function be

$$\phi'(r) = \phi(r) + bv_1.$$

The variational parameter $b$ is determined by using the stationary property of $a_t + \int \phi' L\phi'\, dr$ as

$$b = -\int v_1 L\phi\, dr \bigg/ \int v_1 L v_1\, dr.$$

The resulting stationary value of $a$ is

$$a_t + \int \phi L\phi\, dr - \left(\int v_1 L\phi\, dr\right)^2 \bigg/ \int v_1 L v_1\, dr,$$

which is identical with $\mu'$ in (E51). (It will be noted that the added term is positive.) Therefore the usual variational method gives an upper bound if the trial function is flexible enough so that an approximate wave function $v_1$ for the bound state with a negative energy expectation value can be constructed. This upper bound theorem can be extended to the case where $m$ bound states exist. For details, see ref. 5b. An extension of the method to positive energy scattering is possible if the potential vanishes beyond a certain distance (ref. 5c).

In actual applications, a trial function with $n$ linear variational parameters is often convenient. Let us assume a $\phi$ of this type,

$$\phi_n = u_0 + \sum_{i=1}^{n} b_i u_i,$$

with $u_0(0) = u_i(0) = 0$, $u_0(r) \to r$, $u_1(r) \to -1$, $u_j(r) \to O(r^{-\varepsilon})$, $j \geqq 2$, $\varepsilon > 0$, $r \to \infty$. $b_1$ is an adjustable parameter which is approximately the scattering length. It can be shown (ref. 5d) that the variationally calculated scattering length, $a_n \equiv b_1 + \int_0^\infty \phi_n L \phi_n \, dr$, decreases monotonically as the number $n$ increases from $n = 0$ except for "$m$-jumps" when there exist $m$ bound states. Therefore, after one verifies $m$-jumps, the variational method is assured to give an upper bound on the scattering length. As a special case, if only one bound state exists, and if $\int u_1 L u_1 \, dr < 0$, it can be shown that

$$a_0 \leqslant a_1 \geqslant a_2 \geqslant a_3 \geqslant \cdots \geqslant a \text{ (true)}, \quad a_0 \equiv \int_0^\infty \phi_0 L \phi_0 \, dr.$$

As an illustrative example, let us take the case of an exponential potential well, $L = -(d^2/dr^2) + \lambda e^{-r}$, $u_0 = r$, $u_n = \exp[-(n-1)r](1 - e^{-r})$, $n \geqq 1$. The adjustable parameters $b_i$ are determined by

$$\frac{\partial}{\partial b_i}\left(b_1 + \int_0^\infty \phi_n L \phi_n \, dr\right)$$

$$= 2(L_{i0} + L_{i1}b_1 + \cdots + L_{in}b_n) = 0, \quad (i = 1, \ldots, n),$$

where $L_{ij} = \int_0^\infty u_i L u_j \, dr$. We have used the equality:

$$\int_0^\infty u_0(d^2u_1/dr^2) \, dr = 1 + \int_0^\infty u_1(d^2u_0/dr^2) \, dr.$$

The stationary value $a_n$ is obtained from $a_n = L_{00} + \sum_{i=1}^{n} L_{i0} b_i$.

The calculated values of the scattering length are shown in Table 1. The number of bound states $m$ is zero for $-1.446 < \lambda < +\infty$; one for $-7.618 < \lambda < -1.446$; two for $-18.72 < \lambda < -7.618$, etc. One "jump" of $a_n$ occurs between $a_0$ and $a_1$ for $\lambda = -1.69$; and two jumps occur at $a_1$ and $a_3$

TABLE 1

| $\lambda$ | $-1$ | $-1.69$ | $-7.84$ |
|---|---|---|---|
| $a_0$ | $-2.00$* | $-3.38$ | $-15.68$ |
| $a_1$ | $-5.37$* | 21.99* | 0.68 |
| $a_2$ | $-5.983$* | 17.311* | $-3.36$ |
| $a_3$ | $-6.003$* | 17.310* | 54.28* |
| exact | $-6.007$ | 17.300 | 42.79 |
| $m$ | 0 | 1 | 2 |

for $\lambda = -7.84$. (Since $L_{11} < 0$ for both cases, $a_1 > a_0$.) The values with an asterisk are certain to give an upper bound for the scattering length.

The theorem of upper bound on the scattering length is applicable to more complicated cases such as the scattering of electrons by hydrogen atoms.

It has been shown for the positive energy scattering that the stationary value for the phase shift [see (D18) or (D19)] does not converge smoothly but jumps an infinite number of times as the number of adjustable parameters increases. This is due to the fact that with positive (non-zero) energy E, the number of states of the system with energy smaller than E is infinite (ref. 5d). The variational method proposed in ref. 5c is an exception, namely, after jumps of a finite number of times, the stationary value of the phase shift converges smoothly to the true value.

### 4. VARIATIONAL PRINCIPLE FOR THE EFFECTIVE RANGE

We have seen in (E10a), (E30) that for very low energy particles ($k^2 \to 0$), the scattering is describable in terms of two parameters characterizing the scattering, namely the scattering length $a$ and the effective range $r_0$, according to $k \cot \eta = -(1/a) + \frac{1}{2} r_0 k^2$. The effective range $r_0$ is expressible by the zero energy solution as (E11a) or (E16). In the last section, we have deduced some minimal properties for $a$ under certain conditions. We shall show that $r_0$ can also be given in terms of a variational principle (ref. 7).

We shall again confine ourselves to $s$ waves, for "zero" energy scattering. The Schrödinger equation is

$$(-L + k^2)\, u_k \equiv \left[\frac{d^2}{dr^2} - U(r) + k^2\right] u_k = 0.$$

The solution $u(r)$ is $u(0) = 0$ and is normalized by the asymptotic form (D16)

$$u_k(r) \to \cos kr + \cot \eta \sin kr.$$

For $k^2 = 0$, $u_0(0) = 0,$

$$u_0(r) \to 1 + k(\cot \eta)r \simeq 1 - \frac{1}{a}r, \tag{E52}$$

If we develop $u_k(r)$ in powers of $k^2$ at $k^2 = 0$,

$$u_k(r) = u_0(r) + \tfrac{1}{2}k^2x(r) + \cdots, \tag{E53}$$

$x(r)$ is seen to be given by the equation $Lx(r) = u_0(r)$,

and $$x(0) = Lx(0) = 0,$$

$$x(r) \to r_0 r - r_2 + \frac{1}{3a}r^3. \tag{E54}$$

Consider the variation of the integral $J$

$$\delta J \equiv \delta \int wL^2w\, dr, \tag{E55}$$

where the function $w$ is subject to the conditions

$$w(0) = 0, \qquad Lw(0) = 0,$$

$$w(r) \to r_{0_t}r - r^2 + \frac{1}{3a_t}r^3, \tag{E56}$$

so that $$\delta w(0) = 0, \qquad \delta w \to r\delta r_0 + \frac{r^3}{3}\,\delta\left(\frac{1}{a}\right),$$

$$Lw(r) \to 2 - \frac{2}{a}r, \text{ etc.} \tag{E57}$$

Then it can be seen that

$$\delta \int_0^\infty wL^2w\, dr = -2\delta r_0 + \int_0^\infty \delta w\, L^2w\, dr, \tag{E58}$$

so that the variation principle $\delta(J + 2r_{0_t}) = 0$ leads to

$$L^2w = 0, \tag{E59}$$

which states that $w$ must be actually $x$. Thus a stationary expression for $r_0$ is given by $r_{0_t} + \frac{1}{2}J$, because $J = 0$ for $w = x$.

The above variational principle (E55)–(E59) can be replaced by an equivalent one in which, instead of a fourth order operator $L^2$, we employ only $L$ but two variable functions.

It can be shown that under certain conditions, these methods of calculating $r_0$ lead to an upper bound. If a trial function with $n$ linear variational parameters is assumed (as discussed in the preceding subsection), it can be shown that the variationally calculated effective range $r_{0_t} + \frac{1}{2}J$ decreases monotonically as the number $n$ increases from $n = 0$, except for only one positive jump (see ref. 7a). This minimum principle can be extended to give a maximum principle for the energy derivative of the phase shift (ref. 7b).

#### REFERENCES

**1.** The pioneering work on the analysis of proton–proton (and also proton–neutron) scattering in terms of a potential (nucleus plus Coulomb) is that of Breit, G., Condon, E. U. and Present, D., Phys. Rev. **50**, 825 (1936). Breit and his coworkers investigate in detail the effect of various potential shapes in the analysis of the proton–proton scattering data and bring out what is essentially the effective range theory. Breit, G., Thaxton, H. M. and Eisenbud, L., Phys. Rev. **55**, 1018, 1057 (1939); Hoisengton, L. E., Share, S. S. and Breit, G., Phys. Rev. **56**, 884 (1939), this paper compares various potentials; Breit, G., Broyles, A. A. and Hull, M. H., Phys. Rev. **73**, 869 (1948). Early work on the shape-independent part of the $k \cot \delta$ was also done by Landau, L. and Smorodinsky, J., J. Phys. U.S.S.R. **8**, 154 (1944); Smorodinsky, J., J. Phys. U.S.S.R. **8**, 219 (1944); **11**, 195 (1947).

**2.** Blatt, J. M. and Jackson, J. D., Phys. Rev. **76**, 18 (1949), present Schwinger's proof of the formulas for the effective range (E11a) and (E31) given in his lectures at Harvard in 1947 with the use of the variational method.

**3.** Bethe, H. A., Phys. Rev. **76**, 38 (1949), gives the alternative derivation of the effective range formula. The treatment given in the present section is based on Bethe's work.

**4.** Fermi, E., Ricerca Scient. VII-II, 13 (1936), Introduction of the concept of "scattering length".

**5a.** Spruch, L. and Rosenberg, L., Phys. Rev. **116**, 1034 (1959).

**5b.** Rosenberg, L., Spruch, L. and O'Malley, T. F., Phys. Rev. **118**, 184 (1960). Upper bound of scattering length.

**5c.** Rosenberg, L. and Spruch, L., Phys. Rev. **120**, 474 (1960).

**5d.** Ohmura, T., Phys. Rev., **124**, 130 (1961).

**6.** Kato, T., Progr. Theor. Phys. (Kyoto), **6**, 394 (1951). Upper and lower bounds of phase shift.

**7.** Ohmura, T., J. Math. Phys. **1**, 35 (1960). Stationary expression for effective range.

**7a.** Konno, R. and Kato, T., to be published.

**7b.** Ohmura, T., to be published.

**8.** Levinson, N., Kgl. Danske Vidensk., Mat.-fys. Medd. **25**, No. 9 (1947). Phase shift value at zero energy.

## Appendices to Section E

### E1. PHOTO-TRANSITIONS FROM THE BOUND STATE

We shall show that the cross sections of photo-disintegration processes (electric dipole or magnetic dipole) from a bound $S$ state and of their inverse (radiative capture) processes can be expressed in terms of the bound state parameters $\gamma$ and $\rho$ [(E24), (E31)] and hence related to the low energy scattering data.

Consider a particle of mass $m$ and charge $e$ in a bound $S$ state of binding energy $(\hbar^2/2m)\gamma^2$ in a static, short-range potential $V(r)$. The cross section for photoelectric absorption in an electric dipole transition is given by#

$$\sigma_e = \frac{4\pi e^2}{3\hbar c}\frac{(\gamma^2 + k^2)}{k}|\mathbf{M}_e|^2, \tag{E1-1}$$

where $E = (1/2m)\hbar^2k^2$ is the energy of the ejected particles, $\mathbf{M}_e$ is the matrix element

$$\mathbf{M}_e = N\int_0^\infty u_\gamma(r)\mathbf{r}\psi_p(\mathbf{r})\, d\mathbf{r}. \tag{E1-2}$$

$Nu_\gamma(r)$ is the bound $S$ state wave function such that

$$u_\gamma(r) \to e^{-\gamma r}, \qquad N^2\int_0^\infty u_\gamma^2(r)\, dr = 1. \tag{E1-3}$$

---

# Cf. Heitler: *Quantum Theory of Radiation* (Oxford University Press), Chap. III; or Blatt and Weisskopf: *Theoretical Nuclear Physics* (Wiley and Sons, New York), Chap. XII.

$\psi_p(\mathbf{r})$ is the wave function of the $p$ ($l = 1$) state in the continuum. For *low* energies $E$, $\psi_p(\mathbf{r})$ can be replaced by the $l = 1$ component of a plane wave

$$\phi(r) = \frac{\sin kr}{kr} - \cos kr \tag{E1-4}$$

[see (A8-9)], since $\psi_p(\mathbf{r})$ does not penetrate much into the region where $V(r)$ is appreciable on account of the $l(l + 1)/r^2$ barrier. For small $r$, $u_\gamma(r) \approx r$ and $\phi(r) \approx k^2r^2/3$. Thus the contribution to the integral (E1-2) from the region of small $r$ is small. If the "radius" $(1/\gamma)$ defined by the bound state $u_\gamma(r) \to e^{-\gamma r}$ is very much larger than the range of $V(r)$, one may to a good approximation replace $u_\gamma(r)$ in (E1-2) by the asymptotic form $v_\gamma(r) = e^{-\gamma r}$ (E25).#

To calculate $N$ in (E1-2), we have, from the definition of $\rho$ in (E31),

$$\int_0^\infty u_\gamma^2\,dr = \int_0^\infty v_\gamma^2\,dr - \int_0^\infty (v_\gamma^2 - u_\gamma^2)\,dr = \frac{1}{2\gamma} - \frac{\rho}{2}, \tag{E1-5}$$

so that, from (E1-3),

$$N^2 = \frac{2\gamma}{1 - \gamma\rho}. \tag{E1-6}$$

$M_e$ in (E1-2) is then

$$M_e = \left(\frac{2\gamma}{1 - \gamma\rho}\right)^{1/2} \frac{2k^2}{(\gamma^2 + k^2)^2}, \tag{E1-7}$$

and

$$\sigma_e = \frac{16\pi e^2}{3\hbar c}\left(\frac{2\gamma}{1 - \gamma\rho}\right)\left(\frac{k}{\gamma^2 + k^2}\right)^3, \tag{E1-8}$$

which can be expressed in terms of the threshold frequency $\nu_0$ and the frequency $\nu$ given by

$$h\nu_0 = \frac{\hbar^2\gamma^2}{2m}, \qquad h\nu = \frac{\hbar^2(\gamma^2 + k^2)}{2m}.$$

The formula (E1-8) is not valid for high energies (for $E > 10$ MeV in the case of the deuteron, for which $e$ is half the electronic charge because the neutron has no charge).

---

# In the case of the deuteron, the effective range $\rho \simeq 1.6 \times 10^{-13}$ cm, and $1/\gamma \simeq 4.3 \times 10^{-13}$ cm. The fraction $F$ of the contribution to (E1-2) from the region $r < \rho$ is

$$F < \frac{\int_0^\rho r\left(\frac{k^2r^2}{3}\right)dr}{\int_0^\infty rv_\gamma(r)\phi(r)\,dr} = \frac{(\gamma^2 + k^2)^2}{24}\rho^4,$$

$$< 0.003 \quad \text{for} \quad k^2 = \gamma^2.$$

The same is also true of the photo-ionization of the negative ion of hydrogen where $F < 0.019$. (See ref. 2.)

If the bound particle has a magnetic moment $\mu$ ($\mu$ in units of $e\hbar/2mc$), the cross section of a photo-magnetic disintegration (magnetic dipole transition) is given by [see ref. to (E1-1)]

$$\sigma_m = \frac{\pi}{3\hbar c}\left(\frac{e\hbar}{2mc}\right)^2 \mu^2\left(\frac{\gamma^2 + k^2}{k}\right)|\mathbf{M}_m|^2, \tag{E1-9}$$

where $\mathbf{M}_m$ is the matrix element of the spin operator $\boldsymbol{\sigma}$ between the bound and the continuum state.

For the case of the deuteron,

$$\mathbf{M}_m = N \sin \delta \int_0^\infty u_\gamma(r)\, u_k(r)\, dr, \tag{E1-10}$$

where $N$ is given by (E1-6) and $u_k(r)$ are solutions of (E2) normalized according to (E4). It is easy to verify from (E25) that

$$\int_0^\infty v_\gamma(r)\, v_k(r)\, dr = \frac{\gamma + k \cot \delta}{\gamma^2 + k^2}. \tag{E1-11}$$

To evaluate (E1-10), we write

$$2\int_0^\infty (v_\gamma v_k - u_\gamma u_k)\, dr = \int_0^\infty (v_\gamma^2 - u_\gamma^2)\, dr + \int_0^\infty (v_k^2 - u_k^2)\, dr$$
$$- \int_0^\infty [(v_\gamma - v_k)^2 - (u_\gamma - u_k)^2]\, dr. \tag{E1-12}$$

The first integral on the right is $\frac{1}{2}\rho$, and the second is $\approx\frac{1}{2}r_0$.# The third integral is small since the integrand vanishes outside the range of $V(r)$ and is small within the range of $V(r)$.## Therefore, to a good approximation,

$$\int_0^\infty u_\gamma u_k\, dr = \frac{\gamma + k \cot \delta}{\gamma^2 + k^2} - \frac{1}{4}(\rho + r_0). \tag{E1-13}$$

The magnetic dipole transition cross section from the bound state is then obtained from (E1-9–10) and (E1-13). There are other corrections (exchange magnetic effect, $D$-state effect, etc.) for the deuteron case, but they are not large.

The formation of a bound state by "radiative capture" is the inverse process of photo-disintegration. Its cross section is readily derived from $\sigma_e$ (E1-8) or $\sigma_m$ (E1-9)

$$\sigma_{\text{cap}} = \left(\frac{2\pi v}{kc}\right)^2 \sigma_{\text{disinteg.}}, \tag{E1-14}$$

where $v$ is the velocity of the particle captured.

The photomagnetic process is more important for low energies, $E \lesssim 1$

---

# $\rho$ and $r_0$ can be quite different. For the triplet states of the deuteron, $\rho \simeq 1.6 \times 10^{-13}$ cm, and for the final singlet state, $r_0 \simeq 2.7 \times 10^{-13}$ cm.

## For the case of the deuteron, at $k = 0$, the third integral in (E1-12) is about 2% of the whole integral on the left, which itself is about 20% of the integral (E1-10).

MeV, for the deuteron, whereas the photoelectric process is more important for $E > 1$ MeV.

The photo-disintegration cross section for the deuteron is shown in Fig. 1. The solid line is drawn on the effective range approximation and the experimental points are indicated by the bar. It will be noted here that the

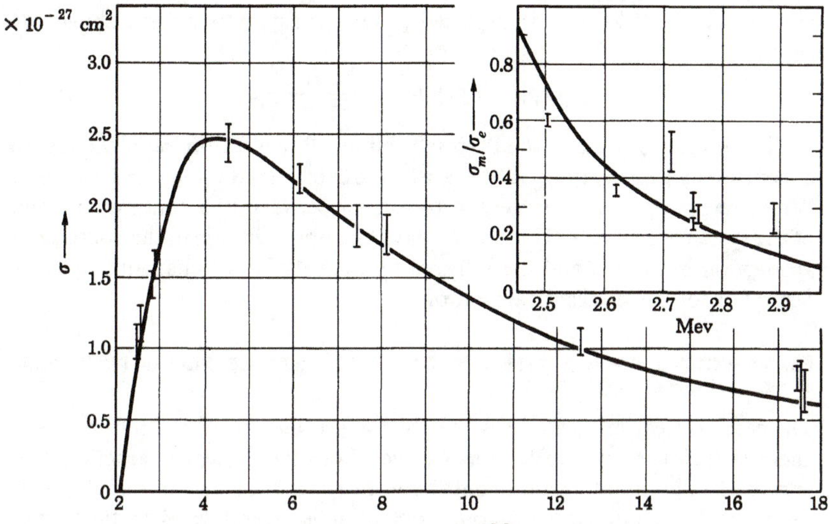


**Figure 1.** The total photodisintegration cross section and the ratio $\sigma_m/\sigma_e$ versus the incident photon energy. The experimental data are indicated by vertical lines representing their errors. Taken from the article by Hulthén, L. and Sugawara, M., in *Handbuch der Physik* **39**, 112 (1957), Springer, Berlin.

ejected particle distribution is proportional to $\sin^2 \vartheta$ ($\vartheta$ is the angle between the directions of the incident photon and the ejected particle) for the electric-dipole process, and is isotropic for the magnetic-dipole transition. Since the final spin states are different for the electric and magnetic transitions, these two transition amplitudes are non-coherent. The differential cross section is just the sum of the individual cross sections.

## E2. BREMSSTRAHLUNG

We have just seen that the photo-transition from the bound state has a close relation to the scattering parameters for low energies. We shall briefly show that the Bremsstrahlung cross section for a low energy charged particle in the field $V(r)$ is also expressible in terms of the $S$ phase shifts alone.

The transition matrix element in the low energy region is proportional to

$$\int r\phi_1(r)\,\phi_2(r)\,dr, \tag{E2-1}$$

where $\phi_1(r)$ is a solution of (E2) with $\phi_1(0) = 0$ and the normalization

$$\phi_1(r) \to \sin(k_1 r + \delta).$$

$\phi_2(r)$ is defined in (E1-4), but with $k = k_2$. Using similar arguments as in the evaluation of (E1-2), we can replace $\phi_1(r)$ by its asymptotic form above and get

$$\int_0^\infty r\phi_1(r)\,\phi_2(r)\,dr = \frac{2k_2^2 \sin\delta(k_1)}{(k_1^2 - k_2^2)^2}. \tag{E2-2}$$

The inverse process of the Bremsstrahlung of an electron in the vicinity of a neutral hydrogen atom is the absorption of radiation by the so-called "free–free transitions", namely, a transition between two continuum states of the system electron + H atom. It plays an important role in the continuous absorption in the red and the infrared by the solar and stellar atmospheres. See ref. 3 for some recent calculations.

### E3. EFFECTIVE RANGE THEORY OF SHORT-RANGE POTENTIAL PLUS COULOMB POTENTIAL

The case of great importance is that of the proton–proton scattering at low energies (below a few MeV). For $S$ waves, both the nuclear interaction $V(r)$ and the Coulomb interaction contribute. For $l \geqslant 1$, on account of the short range of $V(r)$, only the Coulomb potential is operative. The theoretical analysis of the low energy scattering was first made by Breit and his coworkers. A detailed account of the theoretical analysis of the experimental data has been given by Jackson and Blatt (ref. 4). In the following we shall only sketch the treatment of Bethe (ref. 4) on the effective range.

#### 1) *Pure Coulomb scattering*

This case has been treated in Sect. A.6. For proton–proton scattering, we have in (A40a) $Z = Z' = 1$, $m = \frac{1}{2}M$, $M$ being the nucleon mass. From (A34-5) and (A49), we can write the $s$ wave component of $\psi(\mathbf{r})$ of (A2) in the form

$$\psi_0(\mathbf{r}) = \frac{i}{kr}\left[\exp(-ikr + i\alpha \ln 2kr) - \exp(ikr - i\alpha \ln 2kr + 2i\eta_0)\right], \tag{E3-1}$$

where $\eta_0$ is the phase shift due to the Coulomb interaction alone.

#### 2) *V(r) + Coulomb field*

The Schrödinger equation is then

$$\Delta\Psi + \left(k^2 - U(r) - \frac{2k\alpha}{r}\right)\Psi = 0, \quad \alpha = \frac{e^2}{\hbar v}, \tag{E3-2}$$

where $v$ is the relative velocity of the two protons ($mv = \hbar k$). For low energies, only the $s$ wave is affected by the nuclear potential $U(r)$, and its effect is to produce an additional phase shift $\delta_0$. Thus the $s$ wave component of $\Psi(\mathbf{r})$ of (E3-2) is, instead of (E3-1), given by

$$\begin{aligned}\Psi_0(\mathbf{r}) &= \frac{i}{kr}\{\exp(-ikr + i\alpha \ln 2kr) - \exp[ikr - i\alpha \ln 2kr + 2i(\eta_0 + \delta_0)]\} \\ &= \psi_0(\mathbf{r}) - \frac{i}{kr}\exp(ikr - i\alpha \ln 2kr + 2i\eta_0)[\exp(2i\delta_0) - 1]. \qquad \text{(E3-3)}\end{aligned}$$

Since the higher angular momentum waves are assumed to be not affected by $U(r)$, we have for $\Psi(\mathbf{r})$ of (E3-2),

$$\Psi(\mathbf{r}) = \psi(\mathbf{r}) - \psi_0(\mathbf{r}) + \Psi_0(\mathbf{r}). \qquad \text{(E3-4)}$$

For the singlet state scattering amplitude, we have to form the symmetrized expression from (A46), (E3-1) and (E3-3, 4)

$$f_s(\vartheta) = -\left(\frac{e^2}{Mv^2}\right)\exp(ikr - i\alpha \ln 2kr + 2i\eta_0)\left\{\frac{\exp\left[-2i\alpha \ln\left(\sin\frac{\vartheta}{2}\right)\right]}{\sin^2\frac{\vartheta}{2}} + \frac{\exp\left[-2i\alpha \ln\left(\cos\frac{\vartheta}{2}\right)\right]}{\cos^2\frac{\vartheta}{2}} + \frac{i}{\alpha}[\exp(2i\delta_0) - 1]\right\}. \qquad \text{(E3-5)}$$

For the triplet state, there is no $s$ wave on account of the Pauli principle. Since there is no interference between the triplet and the singlet state scattering, the weighted mean of the scattering intensity is, for $\alpha \ll 1$ (i.e., energies higher than about 1 MeV),

$$\sigma(\vartheta) = \left(\frac{e^2}{Mv^2}\right)^2\left[\frac{1}{\sin^4\frac{\vartheta}{2}} - \frac{1}{\sin^2\frac{\vartheta}{2}\cos^2\frac{\vartheta}{2}} + \frac{1}{\cos^4\frac{\vartheta}{2}} - \frac{\sin 2\delta_0}{\sin^2\frac{\vartheta}{2}\cos^2\frac{\vartheta}{2}} + \frac{4\sin^2\delta_0}{\alpha^2}\right]. \qquad \text{(E3-6)}$$

The first three terms are due to the Coulomb scattering (the second or "Mott" term coming from the Pauli principle), the last term is due to the nuclear potential $U(r)$, and the fourth term represents the interference between the nuclear and the Coulomb scattering. It is seen that the relative importance of the scattering due to $U(r)$ increases as $1/\alpha^2$, or as the energy $E = \frac{1}{2}mv^2$. See also the last part of Sect. H.

### 3) *Effective range theory*

For convenience, let us introduce $R$ defined by

$$\frac{1}{R} = \frac{Me^2}{\hbar^2} = 2k\frac{e^2}{\hbar v}. \tag{E3-7}$$

$R$ is the proton Bohr radius. The equation (A40) for $s$ waves in the Coulomb field between the two protons becomes

$$\frac{d^2v}{dr^2} + \left(k^2 - \frac{1}{Rr}\right)v = 0. \tag{E3-8}$$

For the combined field $U(r)$ + Coulomb, (E3-2) for the $s$ waves is

$$\frac{d^2u}{dr^2} + \left[k^2 - \frac{1}{Rr} - U(r)\right]u = 0. \tag{E3-9}$$

We shall refer to Sect. A.6, and construct the following function:

$$\phi(r) = C[G_0(r) + F_0(r)\cot\delta_0], \tag{E3-10}$$

where $C^2 = 2\pi\alpha(e^{2\pi\alpha} - 1)^{-1}$, $\alpha = e^2/\hbar v$, and $\delta_0$ is the phase shift due to the nuclear potential $U(r)$, (see (E3-3)). $\phi(r)$ is a solution of (E3-8) that has the following properties [see (A52-54)]:

For large $r$: $$\phi(r) \to \frac{C}{\sin\delta_0}\sin(kr - \alpha\ln 2kr + \eta_0 + \delta_0), \tag{E3-11}$$

for small $r$: $$\phi(r) \to 1 + \frac{r}{R}\left(\ln\frac{r}{R} + 2\mathcal{C} - 1 + K\right), \tag{E3-12}$$

where

$$K(k^2) = \frac{\pi\cot\delta_0}{e^{2\pi\alpha} - 1} - \ln\alpha - \mathcal{C} + \alpha^2\sum_{n=1}^{\infty}\frac{1}{n(n^2+\alpha^2)}. \tag{E3-13}$$

From (E3-8), for two energies $k_2^2$, $k_1^2$, we obtain

$$\left(\phi_2\frac{d\phi_1}{dr} - \phi_1\frac{d\phi_2}{dr}\right)_{r_1}^{\infty} = (k_2^2 - k_1^2)\int_{r_1}^{\infty}\phi_1\phi_2\,dr. \tag{E3-14}$$

Similarly, from (E3-9), we have

$$\left(u_2\frac{du_1}{dr} - u_1\frac{du_2}{dr}\right)_{r_1}^{\infty} = (k_2^2 - k_1^2)\int_{r_1}^{\infty}u_1u_2\,dr, \tag{E3-15}$$

where $u(r)$ has been normalized according to the asymptotic form

$$u_k(r) \to \frac{C}{\sin\delta}\sin(kr - \alpha\ln 2kr + \eta_0 + \delta_0). \tag{E3-16}$$

On subtracting (E3-15) from (E3-14), using (E3-11), (E3-16) and $u_k(0) = 0$, we get

$$\left[\frac{d\phi_2}{dr} - \frac{d\phi_1}{dr}\right]_{r_1}^{\infty} = (k_2^2 - k_1^2) \int_{r_1}^{\infty} (\phi_1\phi_2 - u_1u_2)\, dr. \tag{E3-17}$$

From (E3-12), we have

$$\frac{d\phi(r)}{dr} = \frac{1}{R} K + \frac{1}{R}\left(\ln \frac{r}{R} + 2\mathcal{C}\right),$$

so that the difference on the left in (E3-17) is independent of the lower limit $r_1$ and we may pass to the limit $r_1 \to 0$. Let $k_1^2 \to 0$ and define the scattering length $a$ by

$$-\frac{1}{a} = \lim_{k\to 0} 2\alpha k K(k^2), \tag{E3-18}$$

where $\alpha = Me^2/2k\hbar$ and $\delta_0 = \delta_0(k)$. Equation (E3-17) finally becomes, on dropping the subscript 2 from $k_2$, $\phi_2$, $u_2$,

$$\frac{1}{R} K(k^2) = -\frac{1}{a} + k^2 \int_0^{\infty} (\phi\phi_0 - uu_0)\, dr. \tag{E3-19}$$

If we expand $\phi(r)$, $u(r)$ in powers of $k^2$ and define the effective range by

$$b = 2 \int_0^{\infty} (\phi_0^2 - u_0^2)\, dr, \tag{E3-20}$$

where the subscript 0 indicates zero energy solutions, then

$$\frac{1}{R} K(k^2) = -\frac{1}{a} + \tfrac{1}{2} b k^2 + O(k^4). \tag{E3-21}$$

For the empirical values of $a$ and $b$, see (K24) in Sect. K below.

## E4. QUANTUM DEFECT METHOD

We have been concerned with the *repulsive* Coulomb potential [plus short-ranged $V(r)$] in the preceding subsection. If the potential is *attractive* (such as that for an electron incident on a positive ion), there is an infinite number of discrete (bound) states. The phase shift $\delta_l$ due to the deviation from a pure Coulomb potential can be estimated from the positions of bound states by the following *quantum defect* method, which has been applied to various electron-ion scattering problems (ref. 7).

Let $I_{nl}$ be the threshold energy for the ejection of the valence electron $n$, $l$ in the neutral atom (or ion) $A$. The effective quantum number $n_l^*$ is defined by

$$I_{nl} = \frac{mZ^2e^4}{2\hbar^2 n_l^{*2}} \quad (Z = \text{residual charge. For neutral atom } Z = 1),$$

and the quantum defect $\mu_l$ is defined by: $\mu_l = n_l^* - n_l$, where $n_l$ is an integer. $\mu_l$ is zero for the pure Coulomb field. It can be shown that $\delta_l(E) = \pi\mu_l(E)$, where $\mu_l(E)$ is the quantum defect for the $n, l$ series extrapolated as a function of incident energy $E$ of electron to the target $A^+$. The radial wave function has the asymptotic form $\sim \sin\left(kr - \frac{l\pi}{2} + \eta_l - \alpha \ln 2kr + \delta_l\right)$. The phase shift $\delta_l$ is zero for the pure Coulomb field. See (A49a). [Note that the $\eta_l$ here is the $\delta_l$ in (A49a).]

The phase shifts thus obtained empirically are often in good agreement with the phase shift values in the Hartree–Fock approximation, for example, the $p$-wave Hartree–Fock phase shift of electron-$Na^+$ ion scattering at zero energy is 2.69 while $\pi\mu_1(0) = 2.63$.

#### REFERENCES TO APPENDICES, SECT. E

1. Bethe, H. A. and Longmire, C., Phys. Rev. **77**, 67 (1950). Effective range theoretical treatment of photo-disintegration of the deuteron.

2. Ohmura, T. and Ohmura, H., Phys. Rev. **118**, 154 (1960). Photo-ionization of the negative hydrogen ion.

3. Ohmura, T. and Ohmura, H., Astrophysical J. **131**, 8 (1960); Phys. Rev. **121**, 513 (1961). Free–free transition of the negative hydrogen ion.

4. Jackson, J. D. and Blatt, J. M., Rev. Mod. Phys. **22**, 77 (1950); Schwinger, J., Phys. Rev. **78**, 135 (1950); Bethe, H. A., Phys. Rev. **76**, 38 (1949). Effective range theory for proton–proton scattering. For earlier work on the Coulomb + nuclear scattering, see Breit, Condon, Present, ref. 1 in Sect. E.

5. Hulthén, L. and Sugawara, M., Handbuch der Physik Vol. **39/1**, Springer-Verlag (1957).

6. Froberg, C. E., Rev. Modern Phys. **27**, 399 (1955), and papers cited there. For full treatment of Coulomb wave functions, see Hull and Breit, ref. 12 in Sect. A.

7. Ham, S. F., Solid State Physics (ed. by Seitz and Turnbull, Vol. 1, p. 127, Academic Press (1955); Seaton, M. J., Comptes Rendu **240**, 1317 (1955); Proc. Phys. Soc. (London) **A70**, 620 (1957); Monthly Notices Roy. Astron. Soc. **118**, 504 (1958); Burgess, A. and Seaton, M. J., Rev. Modern Phys. **30**, 992 (1958).

section

F

# S Matrix and Bound States, Virtual and Decaying States

## 1. BOUND STATES

The $S$ matrix not only gives the scattering amplitudes and cross sections, but is also connected with the bound states of the system. In Sect. B, we have touched upon this point in connection with the solution of the integral equation of scattering by Fredholm's method. The role of the bound states is of basic importance in the problem of the determination of the scattering potential from the phase shifts, as we shall see in the following section. Hence we shall give a more detailed discussion of the bound states below.

For simplicity let us consider the Schrödinger equation for the $s$-wave of a particle scattered by a central potential

$$\left[\frac{d^2}{dr^2} + k^2 - U(r)\right] \psi(r) = 0. \tag{F1}$$

In (A34), we have the solution $\psi(r)$ which satisfies

$$\psi(0) = 0, \tag{F2}$$

$$\psi(r) \to \text{const}\,(e^{-ikr} - S(k)e^{ikr}), \tag{F3}$$

where $S(k) = \exp\,[2i\delta(k)]$ is the $S$ matrix element. If we continue $S(k)$ into the complex $k$-plane, and if

$$S(k) = 0 \quad \text{for} \quad k = -i\kappa, \quad \kappa > 0, \tag{F4}$$

then $\psi(r)$ in (F3) will vanish as $e^{-\kappa r}$ as $r \to \infty$ and will correspond to a bound state, with energy $-\kappa^2\hbar^2/2m$. It would seem as if the zeroes of the $S$ matrix correspond to the bound states of the state. This, however, is not always the case, and it is of importance to study in greater detail the relation between the bound states and the $S$ matrix.

When in 1946 some theoretical physicists were examining the possibility of replacing the present system of quantum mechanics (Hamiltonian, Schrödinger equation, etc.) by a scheme dealing only with the $S$ matrix that describes the directly observable scattering data, it was found by Ma (ref. 1) in a specific example that certain zeroes of the $S$ matrix do not correspond to bound states. He considers the following attractive $U(r)$

$$U(r) = -U_0 e^{-r/a}, \quad U_0 > 0, \quad a > 0, \tag{F5}$$

for which the solution of (F1) is known:

$$\psi(r) = \text{const}\,[J_{-i\rho}(\alpha)J_{i\rho}(\alpha e^{-r/2a}) - J_{i\rho}(\alpha)J_{-i\rho}(\alpha e^{-r/2a})] \tag{F6}$$

where $J$ represents the Bessel functions, and

$$\rho = 2ak, \quad \alpha = 2aU_0^{1/2}.$$

Asymptotically,

$$\psi(r) \rightarrow \text{const}\left[\frac{1}{\Gamma(1+i\rho)}J_{-i\rho}(\alpha)\left(\frac{\alpha}{2}\right)^{i\rho}e^{-ikr} - \frac{1}{\Gamma(1-i\rho)}J_{i\rho}(\alpha)\left(\frac{\alpha}{2}\right)^{-i\rho}e^{ikr}\right]. \tag{F6a}$$

On comparing this with (F3), one gets

$$S(k) = \left(\frac{\alpha}{2}\right)^{-2i\rho} J_{i\rho}(\alpha)\,\Gamma(1+i\rho)/J_{-i\rho}(\alpha)\,\Gamma(1-i\rho). \tag{F7}$$

Let $k = -i\kappa$ ($\kappa$ real and $>0$). Then the zeroes of

$$S(-i\kappa) = 0 \tag{F8}$$

are given by any one of the following equations

$$J_{2a\kappa}(\alpha) = 0, \tag{F9a}$$

$$\Gamma(1 + 2a\kappa) = 0, \tag{F9b}$$

$$J_{-2a\kappa}(\alpha) = \infty, \tag{F9c}$$

$$\Gamma(1 - 2a\kappa) = \infty. \tag{F9d}$$

The first of these has roots $\kappa_n$ and the corresponding $\psi$ are

$$\psi_n(r) = \text{const}\,J_{2a\kappa_n}(\alpha e^{-r/2a}).$$

(F9b), (F9c) have no solutions for $\kappa > 0$. (F9d) has solutions for

$$2a\kappa = 1, 2, 3, \ldots, \tag{F10}$$

but the corresponding $\psi(r)$ vanish identically so that no normalizable eigenfunctions exist for these zeroes of $S(k)$. They have been referred to as redundant zeroes.

The above specific example can be put in a more general form. Consider two independent solutions of (F1), $f(k, r)$ and $f(-k, r)$, which asymptotically behave as

$$f(\pm k, r) \to e^{\mp ikr},$$

or

$$\lim_{r\to\infty} e^{\pm ikr} f(\pm k, r) = 1. \tag{F11}$$

$f(k, r)$ satisfies (F1), and also the integral equation#

$$f(k, r) = e^{-ikr} + \frac{1}{k}\int_r^\infty \sin k(r' - r)\, U(r')\, f(k, r')\, dr', \tag{F12}$$

$f(k, r)$, $f(-k, r)$ do not satisfy (F2), but one can construct the combination

$$\phi(r) = \frac{1}{f(-k)}\,[f(-k)f(k, r) - f(k)f(-k, r)], \tag{F13}$$

where

$$f(k) \equiv f(k, 0), \qquad f(-k) \equiv f(-k, 0). \tag{F14}$$

$\phi(r)$ in (F13) now satisfies

$$\phi(0) = 0$$

and is asymptotically of the form (F13)

$$\phi(r) \to e^{-ikr} - \frac{f(k)}{f(-k)}\, e^{ikr}, \tag{F15}$$

so that the scattering matrix is

$$S(k) = \frac{f(k)}{f(-k)}. \tag{F16}$$

From (F12),

$$f(-k) = f^*(k), \tag{F17}$$

so that from $S(k) = \exp\,[2i\delta(k)]$, one gets

$$\delta(k) = \arg f(k). \tag{F18}$$

Since from (F12), $f(k) \to 1$ as $k \to \infty$, one gets the expected result $\delta(\infty) = 0$.

The above results are for real values of $k$, $(0 \leqslant k < \infty)$. $f(k)$, and hence $S(k)$, do not vanish for any real nonvanishing $k$, and for real $k$, $S(k)$ describes the scattering. The bound states, if there are any, may correspond to the zeroes of $S(k)$ in the complex $k$ plane. To investigate these it is necessary to investigate the continuation of $S(k)$ into the complex $k$-plane. The analytic properties of $S(k)$ in the complex plane depend on the nature of the potential $U(r)$. [Obviously, for a repulsive potential, there are no bound

# That $f(k, r)$, $f(-k, r)$ are linearly independent solutions can be seen from the non-vanishing of the Wronskian $f(k, r)f'(-k, r) - f(-k, r)f'(k, r) = 2ik \neq 0$. They are complex conjugate of each other.

states and hence probably no zeroes corresponding to them.] The problem has been studied by Jost, Bargmann and others. If the potential $U(r)$ is such that

$$\int_0^\infty r|U(r)|\,dr < \infty, \tag{F19}$$

and, on writing $k = k_r - i\kappa$ where $k_r$, $\kappa$ are real, then $f(k)$ is regular in the lower half of the $k$-plane, i.e., $\kappa > 0$; $f(k)$ is continuous on the real axis; and $f(k)$ has singularities in the upper half of the $k$-plane, $\kappa < 0$. If $U(r)$ is such that

$$\int_0^\infty e^{2\mu r}|U(r)|\,dr < \infty, \quad \mu > 0, \tag{F20}$$

then $f(k)$ is regular in the region-$\kappa < \mu$ of the $k$-plane.

To come back to the relation between $S(k)$ and the bound states, it is seen from (F12) that if $k = -i\kappa$ is a zero of $f(k)$, i.e.,

$$f(k) = f(-i\kappa) = 0, \quad \kappa > 0, \tag{F21}$$

then $f(k, r)$ in (F12) will behave as $e^{-\kappa r}$ and will describe a bound state. This bound state will correspond to a zero of $S(k) = S(-i\kappa)$, $\kappa > 0$, provided [see (F16)]

$$f(-k) = f(+i\kappa) \neq 0. \tag{F22}$$

On the other hand, the zeroes of $S(-i\kappa) = 0$#

$$S(-i\kappa) = \frac{f(-i\kappa)}{f(i\kappa)} = 0, \quad \kappa > 0, \tag{F23}$$

will correspond to bound states provided

$$|f(i\kappa)| \neq \infty. \tag{F25}$$

The redundant zeroes of $S(-i\kappa)$ found by Ma in (F10) correspond to $|f(i\kappa)| = \infty$, and hence do not correspond to (F21) for bound states, as shown by Jost. It is also clear that $S(-i\kappa)$ does not necessarily vanish at the bound state if the denominator $f(i\kappa)$ vanishes as well as the numerator $f(-i\kappa)$.

To summarize, it is seen that if the potential satisfies (F20), then

$$f(i\kappa) \text{ is regular for } 0 < \kappa < \mu,$$

and hence the zeroes of $S(k)$ at $k = -i\kappa_n$, $0 < \kappa_n < \mu$,

$$S(-i\kappa_n) = \frac{f(-i\kappa_n)}{f(i\kappa_n)} = 0$$

have a one–one correspondence with the bound states with energy $-\kappa_n^2$.

---

# From (F12), it can be verified that $f(-k)$, $f^*(k^*)$ satisfy the same integral equation, i.e.,

$$f(-k) = f^*(k^*). \tag{F24}$$

## 2. VIRTUAL STATES

In the case of proton–neutron scattering, the total cross section for triplet and for singlet states scatterings

$$\sigma = \tfrac{1}{4}\,{}^1\sigma + \tfrac{3}{4}\,{}^3\sigma \tag{F26}$$

for unpolarized particles. Consider the triplet state scattering at very low energies at which only the $s$ wave contribution is important. The deuteron is known to be a bound state of energy $-\gamma^2$ $(\gamma > 0)$. $f(k)$ has a zero at $k = -i\gamma$ and $f(-i\gamma) = 0$. Since $\gamma$ is small, for small energies one can expand $f(k)$ in the neighborhood of $k = -i\gamma$

$$f(k) = c(k + i\gamma) + c'(k + i\gamma)^2 + \cdots. \tag{F27}$$

Using only the first term in (F27) in (F16), (F24)

$$\exp\,[2i\delta(k)] = S(k) = \frac{f(k)}{f(-k)}, \tag{F28}$$

one obtains

$$\cot\delta = -\frac{\gamma}{k}, \qquad {}^3\sigma(k) \simeq \frac{4\pi}{k^2 + \gamma^2}. \tag{F29}$$

From the binding energy of the deuteron $\gamma^2\hbar^2/2m = 2.23$ MeV, or $\gamma = 2.32 \times 10^{12}\ \text{cm}^{-1}$, one gets, for $k \to 0$, a value for ${}^3\sigma \simeq 3.63$ barns, which is much smaller than the experimentally measured value $\sigma \simeq 20.5$ barns.

The solution of this "discrepancy" was suggested by Wigner, namely, that the singlet ${}^1\sigma$ cross section might be very large on account of the "accidental" existence of a "virtual" state of very small positive energies (due to the potential ${}^1U(r)$ for the singlet $S$ state). Putting $f(-i\gamma') = 0$, $\gamma' < 0$, in (F21) will not lead to a bound state since $\psi(r)$ in (F15) will behave as $e^{-\gamma' r}$ for large $r$. On expanding $f(k)$ about $-i\gamma'$, $\gamma' < 0$, as in (F27), and using again the first term, one gets by (F28)–(F29)

$${}^1\sigma(k) \simeq \frac{4\pi}{k^2 + (\gamma')^2}. \tag{F30}$$

The observed value of $\sigma$ in (F26) can be accounted for by

$$\frac{\hbar^2}{2m}\gamma'^2 = 0.0664\ \text{MeV}, \quad \text{or} \quad \gamma' = -0.40 \times 10^{12}\ \text{cm}^{-1}. \tag{F31}$$

The "state" with energy 0.0664 MeV has been called the "virtual state" of the deuteron. The name "virtual" has caused not infrequent confusions. Its definition, from the point of view of the $S$ matrix, is that

$$f(k) = f(-i\gamma') = 0, \quad \text{with} \quad \gamma' < 0. \tag{F32}$$

It is contrasted with the bound state which is defined by (F21)

$$f(k) = f(-i\gamma) = 0, \quad \text{with} \quad \gamma > 0. \tag{F33}$$

If one considers a potential $\lambda U(r)$ (attractive) such that for a certain value of the parameter $\lambda$ (real, $>0$) there exists a bound state, then by continuously decreasing $\lambda$, the energy $-\gamma^2$ increases (i.e., $\gamma$ decreases) and the bound state disappears as $\gamma \to 0$. Further decreasing of $\lambda$ is accompanied by $\gamma$ going into $\gamma' < 0$ and one talks of a "virtual" state of energy $\gamma'^2 > 0$. See also Sect. A.4.

For a fuller discussion of the many other definitions of the "virtual" state of the deuteron, see ref. 4.

### 3. DECAYING STATES

In addition to the zeroes of $f(k)$ on the imaginary axis of the complex $k$-plane there may also be zeroes for complex values of $k$ for which

$$f(k) = f(-k_0 + ik_1) = 0, \quad k_0, k_1 \text{ real.} \tag{F34}$$

The corresponding wave function (F13) has the asymptotic form

$$\phi(r) = f(-k_0 + ik_1, r) \to c \exp(ik_0 r) \exp(k_1 r) \tag{F35}$$

which will be an outgoing wave if $k_0 > 0$. The number $N$ of particles going through the surface of a sphere of radius $a$ per unit time is

$$N = \frac{\hbar}{2im}\left(\frac{\partial\phi}{\partial r}\phi^* - \frac{\partial\phi^*}{\partial r}\phi\right) 4\pi a^2$$

$$= \frac{4\pi a^2 \hbar |c|^2 k_0}{m} \exp(2k_1 a). \tag{F36}$$

At $r = a$,

$$|\psi(r)|_a^2 = |c|^2 e^{2k_1 a}.$$

(F36) shows that the particles come out with the velocity $v = \frac{\hbar k_0}{m}$. The time-dependent wave function has the factor

$$\psi(r)\, e^{-iEt/\hbar},$$

where

$$E = \frac{\hbar^2 k^2}{2m} = \frac{\hbar^2}{2m}[k_0^2 - k_1^2 - 2ik_0 k_1].$$

On writing

$$E = E_0 - i\frac{\Gamma}{2}, \quad \Gamma = \frac{2k_0 k_1 \hbar^2}{m},$$

it is seen that

$$\left|\psi(r) \exp(-iEt/\hbar)\right|^2_{r=a} = |c|^2 \exp(2k_1 a) \exp[-(\Gamma/\hbar)t], \tag{F37}$$

i.e., if $k_0 > 0$, $k_1 > 0$, the system decays in time with the mean lifetime $\hbar/\Gamma$. If $\hbar/\Gamma$ is long compared with the characteristic (periodic) time of the system, the "decay state" above may be regarded as quasi-stable. The decaying state differs from the "virtual state" in that for the latter, $\Gamma = 0$ as seen from (F34) and (F32) for their respective definitions.

The scattering by a potential having such decaying states has been treated in Sect. 6 A.4.

## REFERENCES

For references to the $S$ matrix, see Sect. W.

1. Ma, S. T., Phys. Rev. **69**, 668 (1946); **71**, 195 (1947). Redundant zeroes of the $S$ matrix.

2. ter Haar, D., Physica **12**, 509 (1946). Redundant zeroes.

3. Jost, R., Helv. Phys. Acta **20**, 256 (1947). Analysis on the zeroes of the $S$ matrix.

4. Ma, S. T., Rev. Modern Phys. **25**, 853 (1953). Interpretation of the virtual level of the deuteron in terms of the $S$ matrix.

section

# G Determination of $V(r)$ from the Scattering Data

The theory of scattering of the preceding sections enables us to evaluate the cross section $\sigma(\vartheta, E)$ as a function of the angle and the energy from any given potential $V(r)$. In the field of nuclear collisions (nucleon–nucleon, nucleon–nucleus), the interaction law $V(\mathbf{r}_1 - \mathbf{r}_2)$ between the two colliding particles is not known from first principles. Thus the problem presented to us is that of trying to deduce, with the help of the mathematical analysis, as much information about the interaction law $V(\mathbf{r}_1 - \mathbf{r}_2)$ as possible from appropriate experimental data on the scattering cross sections $\sigma(\vartheta)$ [also polarization data, see Sect. J in the following chapter].

To study the problem of finding $V$ from the observed $\sigma(\vartheta)$, it will be convenient to separate this problem into two parts, namely, (i) the determination of the values of $\delta_l$ from the experimental $\sigma(\vartheta)$, and (ii) the construction of $V(r)$ from $\delta_l$. We shall be concerned with (i) in the present subsection, and (ii) in the following subsection.

## 1. DETERMINATION OF $\delta_l$ FROM $\sigma(\vartheta)$

### i) *Exact calculation of $\delta_l(k)$ from $\sigma(\vartheta,k)$*

From (A20), we obtain

$$\sin^2 \delta_l = \frac{k}{2} \int_0^\pi \sqrt{\sigma(\vartheta)} \sin \eta(\vartheta)\, P_l (\cos \vartheta) \sin \vartheta \, d\vartheta. \tag{G1}$$

This relation affords a simple method of determining values of $\delta_l$ from the experimental values of $\sigma(\vartheta)$, provided $\eta(\vartheta)$ can also be determined from the experimental value of $\sigma(\vartheta)$ by a relation other than (G1). Such a relation for $\eta(\vartheta)$ can be found as follows.

Let $\mathbf{k}$, $\mathbf{k}'$, $-\mathbf{k}'$ be three wave vectors corresponding to the same energy

$$|\mathbf{k}|^2 = |\mathbf{k}'|^2,$$

and let $\psi_{\mathbf{k}}(\mathbf{r})$, $\psi_{\mathbf{k}'}(\mathbf{r})$, $\psi_{-\mathbf{k}'}(\mathbf{r})$ be the wave functions describing the scattering of a particle incident with $\mathbf{k}$, $\mathbf{k}'$, $-\mathbf{k}'$ respectively. From (A2), we obtain

$$\begin{aligned} \psi_{-\mathbf{k}'}\,\Delta\psi_{\mathbf{k}} - \psi_{\mathbf{k}}\,\Delta\psi_{-\mathbf{k}'} &= 0, \\ \psi^*_{\mathbf{k}'}\,\Delta\psi_{\mathbf{k}} - \psi_{\mathbf{k}}\,\Delta\psi^*_{\mathbf{k}'} &= 0. \end{aligned} \tag{G2}$$

On integrating over a volume of a sphere around the scattering center and using Green's theorem, we obtain the surface integrals

$$\iint_S \left(\psi_{-\mathbf{k}'}\frac{\partial}{\partial \mathbf{r}}\psi_{\mathbf{k}} - \psi_{\mathbf{k}}\frac{\partial}{\partial \mathbf{r}}\psi_{-\mathbf{k}'}\right) d\mathbf{S} = 0, \tag{G3a}$$

$$\iint_S \left(\psi^*_{\mathbf{k}'}\frac{\partial}{\partial \mathbf{r}}\psi_{\mathbf{k}} - \psi_{\mathbf{k}}\frac{\partial}{\partial \mathbf{r}}\psi^*_{\mathbf{k}'}\right) d\mathbf{S} = 0. \tag{G3b}$$

We have, from (A3),

$$\psi_{\mathbf{k}}(\mathbf{r}) \to e^{i\mathbf{k}\cdot\mathbf{r}} + \frac{e^{ikr}}{r} f(\widehat{\mathbf{k}_r, \mathbf{k}}), \text{ etc.,} \tag{G4}$$

where $\mathbf{k}_r$ is in the direction of the observed scattered wave, and $\widehat{\mathbf{k}_r, \mathbf{k}}$ the angle between $\mathbf{k}_r$ and $\mathbf{k}$. If we make the sphere large enough, we may use the asymptotic wave function (G4) for $\psi_{\mathbf{k}}$, $\psi_{\mathbf{k}'}$, $\psi_{-\mathbf{k}'}$, in (G3a) and (G3b), and obtain, respectively,

$$f(\widehat{\mathbf{k}', \mathbf{k}}) = f(\widehat{-\mathbf{k}, -\mathbf{k}'}), \tag{G5}$$

$$\frac{1}{2i}[f(\widehat{\mathbf{k}', \mathbf{k}}) - f^*(\widehat{\mathbf{k}, \mathbf{k}'})] = \frac{k}{4\pi}\int f^*(\widehat{\mathbf{k}'', \mathbf{k}'})\, f(\widehat{\mathbf{k}'', \mathbf{k}})\, d\Omega_{\mathbf{k}''}, \tag{G6}$$

where the integration is over the directions of $\mathbf{k}''$. (G5) shows a microscopic reversibility in the scattering process. On putting $\mathbf{k}' = \mathbf{k}$ in (G6), one obtains again the optical theorem (A21).

Since $V(r)$ is central, reversing all the directions of $\mathbf{k}$, $\mathbf{k}'$ should leave the $(\vartheta)$ invariant, i.e.,

$$f(\mathbf{k}', \mathbf{k}) = f(-\mathbf{k}', -\mathbf{k}). \tag{G7}$$

On combining this with (G5), we obtain

$$f(\widehat{\mathbf{k}', \mathbf{k}}) = f(\widehat{\mathbf{k}, \mathbf{k}'}) \tag{G8}$$

which shows that $f(\mathbf{k}', \mathbf{k})$ depends only on the numerical value of the angle between $\mathbf{k}$ and $\mathbf{k}'$ [i.e., $f(\vartheta)$ is invariant under rotation and inversion].

On using (G8) and (A19) in (G6), we obtain the relation#

$$\sqrt{\sigma(\vartheta)}\,\sin\eta(\vartheta) = \frac{k}{4\pi}\int\sqrt{\sigma(\vartheta')\,\sigma(\vartheta'')}\,\cos\,[\eta(\vartheta') - \eta(\vartheta'')]\,d\Omega_{\mathbf{k}''}, \tag{G9}$$

where $\vartheta$ is the angle between $\mathbf{k}$ and $\mathbf{k}'$, $\vartheta'$ the angle between $\mathbf{k}$ and $\mathbf{k}''$, and $\vartheta''$ the angle between $\mathbf{k}'$ and $\mathbf{k}''$. If the cross section $\sigma(\vartheta)$ as a function of $\vartheta$ is known experimentally, this relation (G9) may be regarded as an integral equation for the function $\eta(\vartheta)$. With the $\eta(\vartheta)$ determined from this equation, one can determine the phase shifts $\delta_l$ from (G1).

In actual practice, the procedure usually adopted is somewhat as follows. Starting with the low energy scattering data, one determines the $\delta_l$ for small $l$. Additional $\delta_l$ for higher $l$ are introduced to fit the $\sigma(\vartheta)$ for higher energies. In the case of nucleon–nucleon scattering, information from the polarization experiments is needed for the determination of the $\delta_l$.

### *ii) Qualitative relations between the phase shifts $\delta_l$ and $\sigma(\vartheta)$*

From the exact expression (A17) for the scattered amplitude which can be written

$$f(\vartheta) = \frac{1}{2k}\sum_{l=0}^{\infty}(2l+1)\,(\sin 2\delta_l + i\,2\sin^2\delta_l)\,P_l\,(\cos\vartheta)$$
$$\equiv \sum (d_l + ie_l)\,P_l\,(\cos\vartheta)$$

and from

$$\sigma(\vartheta) = |f(\vartheta)|^2,$$

it is simple to show that a relation such as

$$\sigma(0) \geqslant \sigma(\vartheta), \quad 0 \leqslant \vartheta \leqslant \pi \tag{i}$$

is ensured by the following conditions on the coefficients $d_l$, $e_l$, namely,

A) All $d_l$ have the same sign; and all $e_l$ have the same sign.
Similarly, the following inequality relations

$$\sigma(0) \geqslant \sigma(\pi) \tag{ii}$$

$$\sigma(0) \geqslant \tfrac{1}{2}\sigma(\pi) \tag{iii}$$

$$\sigma(0) \geqslant \sigma\left(\frac{\pi}{2}\right) \tag{iv}$$

---

# One may obtain a Born approximation for $\eta(\vartheta)$ by replacing the $\sqrt{\sigma(\vartheta)}$ by the Born approximation for the amplitude $f(\vartheta)_B$ so that (G9) becomes

$$4\pi f(\vartheta)_B\,\eta^{(1)}(\vartheta) = k\int f(\vartheta'')_B f(\vartheta')_B\,d\Omega_{\mathbf{k}''}. \tag{G10}$$

$$\sigma(0) + \sigma(\pi) > 2\sigma\left(\frac{\pi}{2}\right) \tag{v}$$

$$\sigma(0) \geqslant \frac{2}{3m-1}\left[\sum_{n=1}^{m=1} \sigma\left(\frac{n\pi}{m}\right) + \frac{1}{2}\sigma(\pi)\right] \tag{vi}$$

$$\frac{1}{\pi}\int_0^\pi \sigma(\vartheta)\, d\vartheta \leq \frac{3}{\alpha}\int_0^\alpha \sigma(\vartheta)\, d\vartheta\# \tag{vii}$$

imply certain conditions on the $d_l$ and $e_l$ such as A) and the following:

**B)** All $\left(\sum_{l=m}^{\infty} d_{2l}\, C_{2l,\,l-m}\right)$, $\left(\sum_{l=m}^{\infty} d_{2l+1}\, C_{2l+1,\,l-m}\right)$, $m > 0$, have the same sign; similarly for the corresponding sums in $e_l$, where $C_{2l,\,j}$ are the coefficients in

$$P_l(\cos\vartheta) = \sum_{j=0}^{l} C_{2l,\,j} \cos(2l - 2j)\vartheta, \text{ etc.}$$

**C)**
$$\left(\sum d_{2l}\right)\left(\sum d_{2l+1}\right) + \left(\sum e_{2l}\right)\left(\sum e_{2l+1}\right) \geqslant 0,$$

where all the summations are from $l = 0$ to $\infty$.

**D)** All $d_{2l}$ have the same sign, and all $e_{2l}$ have the same sign.

The dependence of the validity of (i)–(vii) on the conditions A)–D) is given in the following table:

| Condition | Inequality relations | | | | | | |
|---|---|---|---|---|---|---|---|
| A) | (i) | (ii) | (iii) | (iv) | (v) | (vi) | (vii) |
| B) | (i) | (ii) | (iii) | (iv) | | (vi) | (vii) |
| C) | | (ii) | (iii) | (iv) | | | |
| D) | | | | (iv) | (v) | | |

From a comparison of the empirical data on $\sigma(\vartheta)$ with the relations (i)–(vii), one can see which of the relations **A)** . . . **D)** are satisfied, and from this one can draw some qualitative conclusions about the $\delta_l$. From the general properties of the $\delta_l$ for various types of potential (attractive, repulsive, exchange, existence of bound states, etc.) discussed in Sects. A.3, (i)–(vi), and E.1.(3), one can make some inferences about $V(r)$. It is clear, however, that on account of the nature of the conditions **B)–C)** in which only the *sums* of a series appear, the usefulness of the above table as a criterion for ascertaining $\delta_l$ and $V(r)$ is rather limited.

# The proof of the inequality (vii) rests on a theorem of Erdös and Fuchs. For details, see ref. 3.

## 2. DETERMINATION OF $V(r)$ FROM THE PHASE SHIFTS

The preceding subsection shows how the theory enables us to determine the $\delta_l$ from the observed $\sigma(\vartheta)$ for a given energy $E$. Our next problem is to investigate the possibility of determining the potential $V(r)$ from the given $\delta_l$. This leads to the mathematical theory of a problem which is the inverse of the Sturm–Liouville problem. In the latter, given a differential equation (for example, a Schrödinger equation with a given known $V(r)$ in our scattering problem) and the prescribed boundary (or asymptotic) conditions, one determines the whole spectrum of discrete eigenvalues and eigenfunctions and the phase shifts of the continuum eigenfunctions. The inverse problem is to see whether, from some appropriately given spectra of the phase shifts $\delta_l(E)$ (the $\delta_l$ for all energies $0 \leqslant E < \infty$ for one or more $l$), it is possible to determine a unique $V(r)$ that will reproduce the given $\delta_l(E)$. A considerable literature has grown up in the last ten years on both the mathematical and the physical aspects of this problem. We shall not be able to deal extensively with these studies. Instead, we shall only sketch the basic ideas of the theory and state the main conclusions.

### *i) Brief statements of results*

The mathematical theory of the inverse Sturm–Liouville problem has been studied by G. Borg, and, with special reference to the scattering problem, by Gel'fand and Levitan (ref. 5). The answer as to whether the determination of a unique $V(r)$ can be made from the $\delta_l$ alone is in the negative in the general case (Bargmann, ref. 4). Only when $V(r)$ has certain special properties is this possible.

**1)** Consider first the corresponding problem in classical mechanics of determining $V(r)$ from the observed cross section $\sigma(\vartheta)$. There the relation between the angle $\vartheta$ of scattering and the impact parameter $\rho$ (which is related to $l$ by $\rho mv = l\hbar$), namely, $\vartheta = \vartheta(\rho, V(r))$, is given by the classical equation of motion of the particle in the field $V(r)$,

$$\vartheta(\rho, V(r)) = \pi - 2\int_{r_0}^{\infty} \frac{\rho\, dr}{r^2\sqrt{1 - \left(\frac{\rho}{r}\right)^2 - \frac{V(r)}{E}}}, \tag{G11}$$

where $r_0$ is the largest root of the radicand. The relation between the differential cross section $\sigma(\vartheta)$ and $\rho$ is, for a central field,

$$2\pi\sigma(\vartheta) \sin \vartheta\, d\vartheta = -\rho\, d\rho.$$

It has been shown by Keller, Kay and Schmoys (ref. 20) that from $\sigma(\vartheta)$ for all $\vartheta$, (i.e., for all $\rho$), classical mechanics enables $V(r)$ to be determined for all $r > r_{\text{minimum}}$, where $r_{\text{min}}$ is the distance of closest approach for that energy $E$, i.e., $V(r_{\text{min}}) = E$.

**2)** Consider next the case for which the Born approximation for the phase shifts in the form (C18) is valid, namely,

$$\delta_l = -\frac{\lambda}{2k}\int_{r_0}^{\infty}\frac{U(r)\,r\,dr}{\sqrt{r^2-\left(\frac{l+\frac{1}{2}}{k}\right)^2}}, \quad kr_0 = l+\tfrac{1}{2}. \tag{G12}$$

If we regard $l$ as a continuous variable, we can solve this equation for $V(r)$. This integral equation can be transformed by simple substitutions

$$\xi = \frac{1}{r^2}, \qquad x = \left(\frac{k}{l+\frac{1}{2}}\right)^2, \qquad D(x) = -\frac{2k}{\lambda}\left(\frac{l+\frac{1}{2}}{k}\right)\delta_l,$$

$$W(\xi) = \tfrac{1}{2}r^3U(r),$$

into

$$D(x) = \int_0^x \frac{W(\xi)\,d\xi}{\sqrt{x-\xi}}$$

which is the Abel equation. The solution can be shown to be

$$\lambda U(r) = \frac{4k}{\pi}\frac{d}{dr}\left[r\int_{kr-\frac{1}{2}}^{\infty}\frac{\delta_l\,dl}{(l+\frac{1}{2})\sqrt{\left(\frac{l+\frac{1}{2}}{k}\right)^2-r^2}}\right], \tag{G13}$$

which enables $V(r)$ to be calculated if $\delta_l$ is a known function of $l$. It is to be noted from (G12) that $(k\delta_l)$ is a function of $(l+\frac{1}{2})/k$ so that if we write

$$\eta = \frac{l+\frac{1}{2}}{k},$$

(G13) becomes

$$\lambda U(r) = \frac{4}{\pi}\frac{d}{dr}\left[r\int_r^{\infty}\frac{(k\delta_l)\,d\eta}{\eta\sqrt{\eta^2-r^2}}\right] \tag{G13a}$$

which is independent of $k$.

In an exact theory, the situation is very much more complicated.

**3)** Consider the simplest case when $V(r)$ is a repulsive potential [$V(r)$ is a monotonically decreasing function with $V(\infty)\to 0$, say]. In this case it is possible to determine $V(r)$ *uniquely* from a spectrum of $\delta_l(E)$ (i.e., the phase shifts for *all* energies and one given angular momentum $l$). Two methods of constructing the potential $V(r)$ from the $\delta_l$ have been given by Jost and Kohn, one based on the Gel'fand–Levitan equation to be described below and one on an earlier method (ref. 9).

Alternatively, it is also possible to determine $V(r)$ from all the phase shifts $\delta_l$ (i.e., for all $l$) for one single energy $E$ (ref. 13).

**4)** Consider next the case of a potential of the "short range" type [i.e., $V(r)$, while attractive, decreases in numerical value with increasing $r$ so fast that there are only a finite number of discrete eigenvalues or bound states in the field]. From an exact knowledge of the whole (energy) spectrum of the phase shifts $\delta_l(E)$ of a given $l$, provided that there exists no discrete eigenvalues of that same $l$, it is possible to determine $V(r)$ uniquely.

5) If for that $l$ [for which the spectrum $\delta_l(E)$ is given] there exist $m$ discrete eigenvalues $\lambda_l$, then it is not possible to determine $V(r)$ uniquely, but an $m$-fold variety of potentials $V(r)$ can be found that are all consistent with the $\delta_l(E)$ and $\lambda_l$ [or, a $2m$-fold variety $V(r)$ that are all consistent with the $\delta_l$(E)].

This may be understood on the following considerations. Consider the states of a given angular momentum. Let $\psi_j(r)$ be the normalized bound state functions and $\psi(k, r)$ the continuum state functions normalized to sin $[kr + \delta(k)]$ at large distances. If the potential $U(r)$ is varied by an infinitesimal amount $\Delta U(r)$, the eigenvalues $\lambda_j$ of the bound states and the phase shifts $\delta(k)$ will undergo the following changes

$$\Delta\lambda_j = \int_0^\infty \Delta U(r)|\psi_j(r)|^2\, dr, \quad j = 1, \ldots, m,$$

$$\Delta\delta(k) = -\frac{1}{k}\int_0^\infty \Delta U(r)[\psi(k,r)]^2\, dr.$$

Therefore all the eigenvalues $\lambda_j$ and the phase shifts $\delta(k)$ will remain unchanged if $\Delta U(r)$ is chosen to be orthogonal to the squares of all the bound state and continuum eigenfunctions. It can be readily verified that the functions $\psi_j(r)(d\psi_j/dr)$ have this property. Thus an $m$-fold variety of potentials $U(r)$ can be constructed by forming $\Delta U(r)$ out of linear combinations of $m\psi_j(d\psi_j/dr)$, which will all have the same bound state eigenvalues $\lambda_j$ and the same phase shifts $\delta(k)$.

It may be of interest to present an explicit example which forms a family of phase-equivalent potentials (Bargmann, ref. 4). Let $\phi(k, r)$ be the wave function with angular momentum zero,

$$\left(\frac{d^2}{dr^2} + k^2\right)\phi = U(r)\phi, \tag{G14}$$

where it will be assumed that

$$\int_0^\infty r^n|U(r)|\, dr < \infty, \quad n = 1 \text{ and } 2. \tag{G15}$$

The two independent solutions $f(\pm k, r)$ of the above equation can be chosen as [see Section F, from (F11), for more details]

$$f(\pm k, r) \rightarrow e^{\mp ikr}, \quad (r \rightarrow \infty). \tag{G16}$$

The solution of (G14) satisfying the initial conditions

$$\phi(k, r = 0) = 0, \qquad \left.\frac{\partial\phi(k, r)}{\partial r}\right|_{r=0} = 1, \tag{G17}$$

is expressed by the linear combination of $f(\pm k, r)$ as

$$\phi = \frac{1}{2ik}\,[f(k)f(-k, r) - f(-k)f(k, r)], \tag{G18}$$

which asymptotically

$$\phi \to -\frac{1}{2ik}f(-k)\left[e^{-ikr} - \frac{f(k)}{f(-k)}f(k, r)\right],$$

where $f(k) \equiv f(k, 0)$. From (G16) we get

$$f(-k, r) = f^*(k, r) \quad \text{and} \quad f(-k) = f^*(k).$$

Then the phase shifts $\delta$ are related to $f(k)$ by (F16) and (A35)

$$e^{2i\delta} = f(k)/f^*(k). \tag{G19}$$

For the potential $U_1(r)$

$$U_1(r) = \frac{\rho\sigma\{4\rho\sigma + (\rho - \sigma)^2\cosh[(\rho + \sigma)r - 2\vartheta] - (\rho + \sigma)^2\cosh(\rho - \sigma)r\}}{[\sigma\sinh(\rho r - \vartheta) - \rho\sinh(\sigma r - \vartheta)]^2}, \tag{G20}$$

the equation (G14) can be solved analytically. The solution is given by

$$f_1(k, r) = \frac{e^{-ikr}}{[2k - i(\rho - \sigma)][2k - i(\rho + \sigma)]}\left(4k^2 + \rho^2 + \sigma^2 - \frac{4ikw_1'}{w_1} - \frac{2w_1''}{w_1}\right),$$

where

$$w_1(r) = (\sigma e^{\rho r} - \rho e^{\sigma r}) + e^{2\vartheta}(\rho e^{-\sigma r} - \sigma e^{-\rho r}), \quad \vartheta > 0, \quad \rho > \sigma > 0,$$

and the phase shift is readily calculated as

$$\delta(k) = \tan^{-1}\frac{(\rho - \sigma)}{2k} + \tan^{-1}\frac{(\rho + \sigma)}{2k} \tag{G21}$$

$\delta(k)$ does not depend on $\vartheta$, therefore $U_1(r)$ forms a family of phase-equivalent potentials for various values of $\vartheta$. This ambiguity of potentials comes from the existence of the bound state at $k = -i\frac{1}{2}(\rho + \sigma)$. The binding energy is $\frac{1}{4}(\rho + \sigma)^2$. For potential $U_2(r)$, which is obtained by merely substituting $-\sigma$ for $\sigma$ in $U_1(r)$ but with $\vartheta < 0$, the solution $f_2(k, r)$ is again obtained by the same substitution. The phase shift $\delta(k)$ is, however, exactly the same as for $U_1(r)$. Thus $U_2(r)$ is another phase equivalent potential to $U_1(r)$. The position of the bound state is, however, different and the binding energy is $\frac{1}{4}(\rho - \sigma)^2$.

### *ii) Elementary theory of Gel'fand and Levitan*

In the following we shall give a simple version of the Gel'fand–Levitan technique of obtaining the potential $U(r)$ from $|f(k)|$, the discrete eigenvalues $\lambda_j = -k_j^2$, $k_j^2 > 0$, and the normalization constants $C_j$ of the eigenfunctions corresponding to the discrete spectrum. At the same time the procedure will enable us to find the eigenfunctions $\phi(k, r)$ of equation (G18) and the eigenfunctions $\phi_j(r)$ corresponding to the discrete eigenvalues $\lambda_j$.

Again, we take the equations (G14)–(G19). From (G18) the asymptotic form of $\varphi$ is

$$\varphi(k, r) \to \frac{|f(k)|}{k} \sin(kr + \delta). \tag{G22}$$

From (G17),

$$\varphi(k, 0) = 0.$$

Let the number $m$ of discrete states be finite, say

$$\lambda = -k_1^2, \ldots, -k_m^2, \tag{G23}$$

[$m$ is finite under the condition (G15)]. The corresponding eigenfunctions are given by $\varphi(-ik_1, r), \ldots, \varphi(-ik_m, r)$. Let $C_j^{-1/2}\varphi(-ik_j, r)$ be the normalized eigenfunctions.

It will first be shown that the phase shift $\delta(k)$ and the binding energies $k_j^2$ determine $f(k)$ completely.# From (G19) we have

$$\delta(k) = \text{Im}\,[\ln f(k)] = \arg f(k). \tag{G24}$$

Consider first the case of no bound states. $f(k)$ is regular for $\text{Im}\,(k) < 0$ and continuous for $\text{Im}\,(k) \leqslant 0$. With the condition of (G15) for $n = 1$, we get

$$\lim_{|k|\to\infty} f(k) = 1. \quad \text{Im}\,(k) \leqslant 0. \tag{G25}$$

Therefore for $\text{Im}\,(p) < 0$, by the Cauchy integral theorem and (G25),

$$\ln f(p) = \frac{i}{2\pi}\int_{-\infty}^{\infty} \frac{\ln f(k)}{k - p}\,dk,$$

$$0 = \frac{i}{2\pi}\int_{-\infty}^{\infty} \frac{[\ln f(k)]^*}{k - p}\,dk.$$

Using (G24) we get, for real $p$,

$$\ln f(p) = \frac{1}{\pi}\int_{-\infty}^{\infty} \frac{\delta\,dk}{p - k - i\epsilon}. \tag{G26}$$

---

# This statement is correct provided the potential is short-ranged, i.e., satisfies the condition (G15).

If the system has bound states, the binding energies are required for the construction of $f(k)$, which has zeroes at $-ik_j$, and hence $\log f(k)$ has logarithmic branch points there. But the function $\bar{f}(k)$

$$\bar{f}(k) \equiv f(k) \prod_j^m \frac{(ik + k_j)}{(ik - k_j)} \tag{G27}$$

has no zeroes for $\text{Im}\,(k) < 0$ and has the same asymptotic form as (G25). Therefore $\bar{f}(k)$ is expressed by a modified phase shift $\bar{\delta}$,

$$\log \bar{f}(p) = \frac{1}{\pi} \int_{-\infty}^{\infty} \frac{\bar{\delta}\, dk}{p - k - i\epsilon}, \tag{G28}$$

where $\bar{\delta}(k) = \delta(k) - 2 \sum_{j=1}^{m} \tan^{-1}\left(\frac{k_j}{k}\right)$. Thus $f(k)$ has been determined by $\delta(k)$ and $k_j$ (ref. 12).

We shall now give the Gel'fand–Levitan theorem:#

Let us define $D(r, r')$ by

$$D(r, r') = \frac{2}{\pi} \int_0^{\infty} \sin kr \sin kr' [|f(k)|^{-2} - 1]\, dk + \sum_{j=1}^{m} \frac{\sinh k_j r \sinh k_j r'}{C_j} \tag{G29}$$

and assume the (Gel'fand–Levitan) equation

$$K(r, r') = -D(r, r') - \int_0^r K(r, r'')\, D(r'', r')\, dr'' \tag{G30}$$

has a unique solution for $K(r, r')$. Then the potential $U(r)$ which reconstructs the phase shift $\delta(k)$ and point eigenvalues $\lambda_j$ is given by

$$U(r) = 2 \frac{d}{dr} K(r, r) \tag{G31}$$

provided $U(r)$ satisfies the short-range requirement (G15). Furthermore, the eigenfunctions of $-(d^2/dr^2) + U(r)$ corresponding to the eigenvalues $k^2$ and $\lambda_j = -k_j^2$ are

$$\begin{aligned} \phi(k, r) &= \frac{\sin kr}{k} + \int_0^r K(r, r') \frac{\sin kr'}{k}\, dr', \\ \phi_j(r) &= \sinh k_j r + \int_0^r K(r, r') \sinh k_j r'\, dr'. \end{aligned} \tag{G32}$$

These functions satisfy the completeness relation

$$\frac{2}{\pi} \int_0^{\infty} \phi(k, r) \frac{k^2}{|f(k)|^2} \phi(k, r')\, dk + \sum_{j=1}^{m} \frac{\phi_j(r)\, \phi_j(r')}{C_j} = \delta(r - r'), \tag{G33}$$

# The following presentation is an adaptation by H. E. Moses of the original method of Gel'fand and Levitan (ref. 5, and Jost and Kohn, ref. 12). The authors are indebted to Dr. Moses for the presentation below in private communications.

and the orthonormality relations

$$\int_0^\infty \phi(k, r)\,\phi(k', r)\,dr = \frac{\pi}{2}\frac{|f(k)|^2}{k^2}\,\delta(k - k'),$$

$$\int_0^\infty \phi(k, r)\,\phi_i(r)\,dr = 0, \tag{G33a}$$

$$\int_0^\infty \phi_i(r)\,\phi_j(r)\,dr = C_i\delta_{ij}.$$

Before proving part of the theorem we shall discuss equation (G30) briefly. For a fixed value of $r$, it is a Fredholm integral equation for $K(r, r')$. Hence from the theory of Fredholm equations, conditions on $D(r, r')$ can be given to obtain unique solutions for $K(r, r')$.

We shall now prove part of the theorem: We shall show that if $\phi(k, r)$ as given by (G32) is an eigenfunction of $-(d^2/dr^2) + U(r)$, it is the *same* as $\phi(k, r)$ given by (G18). This fact is shown simply by noting that $\phi(k, 0) = 0$, $\phi'(k, 0) = 1$ using either form, and that these boundary conditions determine $\phi(k, r)$ uniquely as a solution of the eigenvalue equation. Hence $f(k)$ can be reconstructed from the asymptotic form of $\phi(k, r)$ [equation (G22)].

Let us now show that $\phi(k, r)$ as given by (G32) satisfies

$$\left[-\frac{d^2}{dr^2} + U(r)\right]\phi(k, r) = k^2\,\phi(k, r), \tag{G34}$$

where $U(r)$ is given by (G31).

We note that

$$\left(\frac{d^2}{dr^2} + k^2\right)\phi(k, r) = \frac{d^2}{dr^2}\int_0^r K(r, r')\frac{\sin kr'}{k}\,dr' + \int_0^r K(r, r')\,k^2\frac{\sin kr'}{k}\,dr'$$

$$= \frac{d^2}{dr^2}\int_0^r K(r, r')\frac{\sin kr'}{k}\,dr' - \int_0^r K(r, r')\left(\frac{d^2}{dr'^2}\frac{\sin kr'}{k}\right)dr'.$$

But,

$$\frac{d^2}{dr^2}\int_0^r K(r, r')\frac{\sin kr'}{k}\,dr' = \left[\frac{d}{dr}K(r, r)\right]\frac{\sin kr}{r} + K(r, r)\cos kr$$

$$+ \left[\frac{\partial}{\partial r}K(r, r')|_{r'=r}\right]\frac{\sin kr}{r} + \int_0^r\left[\frac{\partial^2}{\partial r'^2}K(r, r')\right]\frac{\sin kr'}{k}\,dr'.$$

Integrating twice by parts,

$$\int_0^r K(r, r')\left(\frac{d^2}{dr'^2}\frac{\sin kr'}{k}\right)dr' = K(r, r)\cos kr$$

$$- \left[\frac{\partial}{\partial r'}K(r, r')|_{r'=r}\right]\frac{\sin kr}{k} + \int_0^r\left[\frac{\partial^2}{\partial r'^2}K(r, r')\right]\frac{\sin kr'}{k}\,dr'.$$

Hence

$$\left(\frac{d^2}{dr^2}+k^2\right)\phi(k,r) = 2\left[\frac{d}{dr}K(r,r)\right]\frac{\sin kr}{k} + \int_0^r\left[\frac{\partial^2}{\partial r^2}K(r,r') - \frac{\partial^2}{\partial r'^2}K(r,r')\right]\frac{\sin kr'}{k}\,dr'. \quad \text{(G35)}$$

In the above equation we have used the relation

$$\frac{\partial}{\partial r}K(r,r')|_{r'=r} + \frac{\partial}{\partial r'}K(r,r')|_{r'=r} = \frac{d}{dr}K(r,r).$$

We now wish to evaluate $(\partial^2/\partial r^2)\,K(r,r') - (\partial^2/\partial r'^2)\,K(r,r')$ which appears in equation (G35).

From the Gel'fand–Levitan equation (G30)

$$\frac{\partial^2}{\partial r^2}K(r,r') = -\frac{\partial^2}{\partial r^2}D(r,r') - \left[\frac{d}{dr}K(r,r)\right]D(r,r') - K(r,r)\frac{\partial}{\partial r}D(r,r') - \int_0^r\left[\frac{\partial^2}{\partial r^2}K(r,r'')\right]D(r'',r')\,dr'' - \left[\frac{\partial}{\partial r}K(r,r')|_{r'=r}\right]D(r,r').$$

Also using the fact that

$$\frac{\partial^2}{\partial r^2}D(r,r') = \frac{\partial^2}{\partial r'^2}D(r,r'),$$

and integrating twice with respect to $r''$

$$\begin{aligned}\frac{\partial^2}{\partial r'^2}K(r,r') &= -\frac{\partial^2}{\partial r^2}D(r,r') - \int_0^r K(r,r'')\frac{\partial^2}{\partial r''^2}D(r'',r')\,dr'' \\ &= -\frac{\partial^2}{\partial r^2}D(r,r') - K(r,r)\frac{\partial}{\partial r}D(r,r') \\ &\quad + \left[\frac{\partial}{\partial r'}K(r,r')|_{r'=r}\right]D(r,r') - \int_0^r\left[\frac{\partial^2}{\partial r''^2}K(r,r'')\right]D(r'',r')\,dr''.\end{aligned}$$

Hence

$$\frac{\partial^2}{\partial r^2}K(r,r') - \frac{\partial^2}{\partial r'^2}K(r,r') = -2\left[\frac{d}{dr}K(r,r)\right]D(r,r') - \int_0^r\left[\frac{\partial^2}{\partial r^2}K(r,r'') - \frac{\partial^2}{\partial r''^2}K(r,r'')\right]D(r'',r')\,dr''. \quad \text{(G36)}$$

We shall regard (G36) as an integral equation for

$$\frac{\partial^2}{\partial r^2}K(r,r') - \frac{\partial^2}{\partial r'^2}K(r,r').$$

A solution of (G36) is

$$\frac{\partial^2}{\partial r^2} K(r, r') - \frac{\partial^2}{\partial r'^2} K(r, r') = 2\left[\frac{d}{dr} K(r, r)\right] K(r, r'), \tag{G37}$$

as can be verified by substitution into (G36) and using the Gel'fand–Levitan equation (G30). The solution of (G36) is unique because we have assumed the Gel'fand–Levitan equation has a unique solution. On substituting (G37) into (G35) and using (G32) we find

$$\left(\frac{d^2}{dr^2} + k^2\right) \phi(k, r) = U(r)\, \phi(k, r)$$

as required. A similar proof shows that

$$\left[\frac{d^2}{dr^2} + \lambda_j\right] \phi_j(r) = U(r)\, \phi_j(r).$$

The proofs of the completeness and orthonormality relations (G33) and (G33a) will not be given because they are rather long.

To summarize, the determination of the scattering potential $U(r)$ from the phase shifts proceeds as follows. From the phase shifts $\delta_l(k)$ (for a given $l$, but for all $k^2 > 0$) and the energies of the (finite number of) bound states $-k_j^2$, the $|f(k)|$ can be uniquely determined. Then, with the normalization constants $C_j$ in (G29) the Gel'fand–Levitan integral equation (G30) determines $K(r, r_1)$ uniquely. From $K(r, r_1)$, a unique potential $U(r)$ is obtained by means of (G31), which reproduces the phase shifts $\delta_l(k)$ and the discrete eigenvalues $-k_j^2, j = 1, \ldots, m$.

It is to be noted that the eigenvalues $-k_j^2$ of bound states and the constant $C_j$ are necessary independent data in addition to the phase shifts $\delta(k)$. If there are $m$ bound states (for $l = 0$, say), then a $2m$-parameter family of $U(r)$ can be found generally that are "phase-shift equivalent", i.e., give the same phase shifts, or $S(k)$.

We have discussed the so-called effective range theory in Sect. E on the basis of the belief that the $S$ matrix must vanish at the discrete spectrum. (The binding energy has been determined by the scattering length and the effective range.) This statement has now turned out to be not true in general. Only in the case of sufficiently short-range potentials will the zeroes of $S$ and the bound states have a one–one correspondence. It has been shown in ref. 12 (and Sect. F, ref. 3) that the one–one correspondence holds within the region $-\mu < \text{Im}\,(k) < 0$ if the potential $U(r)$ satisfies the condition,

$$\lim_{r \to \infty} e^{2\mu r}\, U(r) = 0. \tag{G38}$$

For the neutron–proton system, $\mu$ can be determined from meson theory, and $k_1$ (which corresponds to the binding energy $-k_1^2$ of the deuteron) is smaller than $\mu$. Therefore the zero of $\exp[2i\delta(k)]$ corresponds to the deuteron state. For this fixed point of the zero, the constant $C$ is uniquely determined, because the constructed potential has the property of (G38) only for the special value of $C$. All other values of $C$ are eliminated by the condition (G38). If we know that $U(r)$ satisfies (G38) with $\mu = \infty$, all the binding energies and the potential are uniquely determined by $\delta(E)$.

Since the early work of Breit and his coworkers in 1939, many attempts have been made to deduce the law of interaction between two nucleons from the empirical data on the two-nucleon system, such as the properties of the ground state of the deuteron, low and high energy nucleon–nucleon scattering (up to *ca.* 400 MeV). In this case the problem is complicated by the interaction being dependent not only on spin and parity, but also being non-central, since tensor and two-body $\mathbf{L}\cdot\mathbf{S}$ interactions are definitely indicated. For the determination of $V(|\mathbf{r}_i - \mathbf{r}_j|)$, the differential $\sigma(\vartheta)$ and polarization data are needed. [For the theory of scattering by non-central fields, see Chapter 2 below.] The general procedure has been to represent the observed cross sections and polarization data (at each energy) by as many phase shifts as are warranted by the accuracies of these data, and this analysis is carried out for all the energies at which the scattering experiments have been performed. Then potentials of various assumed forms are tried to reproduce these empirical phase shifts. It is clear that, with only a finite number, in fact of the order of 20, of phase shifts for a finite number of scattering energies of limited precision, one does not expect to be able to obtain a unique law of interaction. This is in fact found to be the case. Also there is the possibility that the two-nucleon interaction may not be expressible by a potential in a simple closed form at all. For a summary of the nucleon–nucleon potentials, see Chapter 2, Sect. K, below.

# Appendix I

## Generalized Theory of Gel'fand-Levitan: Method of Spectral Weight Function

The theory of Gel'fand and Levitan described in Sect. G. 2 above in its essence has been formulated in a generalized form by Kay and Moses. In their formulation, the spectral weight operator is used. In the elementary case, the weight operator is $|f(k)|^2$. We shall give below their theory for the case in which the system has only a continuous spectrum. Our object here is rather to present the point of view and the method of the theory than to give the theory in all its generalizations and details. For these the reader is referred to their published papers.

The general idea of the theory is as follows. When the particle is at large (infinite) distances from the scattering potential, it is in an eigenstate $\Phi_a$ of $H_0$. At finite distances, when the scattering potential $V(r)$ is operative, the state becomes an eigenstate $\psi_a$ of $H = H_0 + \varepsilon V$. The relation between these two states is expressed by an operator $U$ (in a unique way), and from this $U$, a spectral weight function $W$ can be constructed. We are, however, interested in the inverse problem, namely, given the boundary condition (i.e., the state $\Phi_a$ to which $\psi_a$ asymptotically approaches as $r \to \infty$) and $U$ and therefore $W$, to find the conditions under which the Hamiltonian $H$ [i.e., the potential $V(r)$] can be uniquely determined.

### *i) Spectral weight function W*

We shall confine ourselves to the case when the Hamiltonians $H_0$ and $H_0 + \varepsilon V$ have the same spectrum which is continuous (i.e., no bound states).

Let $A_0$ denote all other observables that together with $H_0$ form a complete commuting set. Let the eigenstate of $H_0$, $A_0$ corresponding to the eigenvalues $E$, $a$

$$H_0|H_0, A_0; E, a\rangle = E|H_0, A_0; E, a\rangle \tag{G39}$$

be denoted for brevity by#

$$|Ea\rangle_0 \equiv |E, a\rangle_0 \equiv |H_0A_0; Ea\rangle \equiv |H_0, A_0; E, a\rangle. \tag{G40}$$

Since $H$, $H_0$ have the same spectrum (i.e., the same range $E_a \leqslant E \leqslant E_b$, and the same degeneracy), the eigenstates of $H$ can be labelled by the same eigenvalues $E$, $a$, and

$$H|HA; Ea\rangle = E|HA; Ea\rangle. \tag{G39A}$$

Similarly, let

$$|Ea\rangle \equiv |HA; Ea\rangle. \tag{G40A}$$

Let $U$ be an operator such that

$$|Ea\rangle = U|Ea\rangle_0. \tag{G41}$$

Using (G39), (G39A), we readily get (since $|Ea\rangle_0 \neq 0$),

$$HU = UH_0, \quad \text{or} \quad UH_0 - H_0U = \varepsilon VU. \tag{G42}$$

We shall show later [footnote to (G78)] that an inverse $U^{-1}$ exists, so that $H$ is expressible by a canonical transformation in terms of $H_0$

$$H = UH_0U^{-1}. \tag{G42A}$$

Now, given $H_0$ and $H = H_0 + \varepsilon V$, (G42A) does not determine $U$ uniquely. In fact any $U$ satisfying the operator equation

$$U = L + \varepsilon P \int \frac{1}{E - H_0} VU\delta(E - H_0)\, dE, \tag{G43}$$

provided

$$LH_0 - H_0L = 0, \tag{G44}$$

satisfies (G42A).## To make $U$ (and $L$) unique, we must specify the state $|Ea\rangle$

---

# For simplicity, we shall omit the comma between $H_0$ and $A_0$ and that between $E$ and $a$, etc., in the rest of the present section.

## $\delta(E - H_0)$ is defined by

$$\delta(E - H_0) = \int |Ea\rangle_0\, da\, {}_0\langle Ea|, \tag{G45}$$

and has the properties

$$H_0\delta(E - H_0) = \delta(E - H_0)H_0 = E\delta(E - H_0), \tag{G46}$$

$$\int \delta(E - H_0)\, dE = \mathbf{I} \quad \text{(unity operator)}. \tag{G47}$$

Operating on $H_0$ by (G43) on the left and on the right, using (G44) and (G46), we obtain

$$UH_0 - H_0U = \varepsilon \int VU\delta(E - H_0)\, dE = \varepsilon VU$$

which is (G42).

of $H$ in (G41). It can be shown that if the $|Ea\rangle_-$ in (G41) is to represent outgoing wave and $|Ea\rangle_0$ the initial state,

$$|Ea\rangle_- = U_-|Ea\rangle_0, \tag{G48}$$

then#

$$U_- = \mathbf{I} + \varepsilon \int \gamma_-(E - H_0)\, VU_-\delta(E - H_0)\, dE, \tag{G49}$$

where

$$\gamma_\pm(E - H_0) = \pm i\pi\delta(E - H_0) + \frac{P}{E - H_0} = \lim_{\zeta\to 0^+} \frac{1}{E - H_0 \mp i\zeta}. \tag{G50}$$

Similarly, if $|Ea\rangle_+$ is to represent ingoing wave

$$|Ea\rangle_+ = U_+|Ea\rangle_0, \tag{G48A}$$

then

$$U_+ = \mathbf{I} + \varepsilon \int \gamma_+(E - H_0)\, VU_+\delta(E - H_0)\, dE. \tag{G49A}$$

Here and in the following, **I** is the unity operator.

The $U_+$, $U_-$ are unitary, as can be seen as follows. The $|Ea\rangle_-$ form a complete set. A state $|\phi\rangle$ can be expanded

$$|\phi\rangle = \int\int |Ea\rangle_-\, dE\, da\, _-\langle Ea|\phi\rangle,$$

$$\overset{\text{(G48)}}{=\!=} \int\int U_-|Ea\rangle_0\, dE\, da\, _0\langle Ea|U_-^\dagger|\phi\rangle.$$

By means of the completeness of $|Ea\rangle_0$, one gets

$$U_-U_-^\dagger = \mathbf{I}. \tag{G51A}$$

Similarly,

$$U_+U_+^\dagger = \mathbf{I}, \tag{G51B}$$

i.e.,

$$U_\pm^\dagger = U_\pm^{-1}. \tag{G51}$$

The operators $U$ are, however, generally not unitary.

Now we define an operator $W$ by

$$WU^\dagger U = \mathbf{I}, \quad \text{or} \quad W = U^{-1}U^{\dagger -1}. \tag{G52}$$

This $W$ has the following properties:

i) $W$ has an inverse $W^{-1}$. This follows from (G52), which gives $W = (U^\dagger U)^{-1}$ and hence $W^{-1} = U^\dagger U$. (G53)

ii) $W$ is hermitian. From (G52), one has for the transposed $\tilde{W} = \tilde{U}^{\dagger -1}\tilde{U}^{-1} = U^{*-1}\tilde{U}^{-1}$. Also $W^* = U^{*-1}U^{\dagger * -1} = U^{*-1}\tilde{U}^{-1}$ $\therefore$ $\tilde{W} = W^*$. (G54)

# It is seen that, with $U_\mp$ given by (G49), (G49A), and $\gamma_\mp$ in (G50), equations (G48), (G48A) are just the Schrödinger equations (B20).

**iii)** $W$ is positive definite. From (G52), for any state $|\phi\rangle$, we have

$$\langle\phi|W|\phi\rangle = \langle\phi|U^{-1}\,U^{\dagger-1}|\phi\rangle \equiv \langle\vartheta|\vartheta\rangle,$$

where

$$|\vartheta\rangle = U^{\dagger-1}|\phi\rangle, \quad \text{or} \quad U^{\dagger}|\vartheta\rangle = |\phi\rangle.$$

Now $|\vartheta\rangle \neq 0$ for any non-zero $|\phi\rangle$. Hence $\langle\vartheta|\vartheta\rangle\cdot > 0$, i.e.,

$$\langle\phi|W|\phi\rangle\cdot > 0. \tag{G55}$$

**iv)** $WH_0 = H_0W$. From (G42A), and (G52)

$$H = UH_0U^{-1} = UH_0WU^{\dagger},$$
$$H^{\dagger} = U^{-1\dagger}\,H_0^{\dagger}U^{\dagger} = UWH_0U^{\dagger}.$$

Since $H$, $H_0$ are self-adjoint, we get

$$WH_0 - H_0W = 0. \tag{G56}$$

For any state $|\phi\rangle$, we have by (G52), $|\phi\rangle = UWU^{\dagger}|\phi\rangle$, or

$$|\phi\rangle = \int\int\int\int U|Ea\rangle_0\, dE\, da\, {}_0\langle Ea|\, W|Fb\rangle_0\, dF\, db\, {}_0\langle Fb|U^{\dagger}|\phi\rangle.$$

Since $W$ commutes with $H_0$, $W$ is diagonal with $H_0$, and by (G41),

$$|\phi\rangle = \int dE \int\int |Ea\rangle\, da\, {}_0\langle a|W(E)|b\rangle_0\, db\, \langle bE|\phi\rangle.$$

This expression takes a simpler form if there is no degeneracy,

$$|\phi\rangle = \int dE\, |E\rangle W(E) \langle E|\phi\rangle. \tag{G57}$$

Thus $W(E)$ is a weight factor. This suggests the name spectral weight function (operator) for $W$.

### *ii) Inverse problem: determination of potential V from W*

*α) Boundary conditions:*

From (G42), it is seen that if $U$ is known, the potential $V$ can be obtained from

$$\varepsilon VU = UH_0 - H_0U. \tag{G42}$$

The problem then is to obtain a *unique* $U$ from a given spectral weight function $W$ according to

$$WU^{\dagger}U = \mathbf{I} \quad \text{or} \quad UWU^{\dagger} = \mathbf{I}, \tag{G52}$$

such that $U|Ea\rangle_0$ is an eigenfunction of $H = UH_0U^{-1}$ corresponding to some prescribed boundary conditions on

$$|Ea\rangle = U|Ea\rangle_0. \tag{G58}$$

For convenience in the following analysis, we shall replace the single equation $WU^+U = \mathbf{I}$ in (G52) by the equivalent pair

$$WU^{\dagger} = U_0, \tag{G59}$$

$$U_0 = U^{-1}. \tag{G59A}$$

From (G52)–(G56), it is readily seen that the necessary conditions (G58) to have a solution $U$ that satisfies (G59) are just the properties (i)–(iv) in (G53)–(G56). These conditions, however, are not sufficient for $U$ to be unique. To obtain a unique $U$, we must specify the *boundary conditions* for $|Ea\rangle$ (as stated just after (G44)). To do this, let us introduce a complete set of hermitian operators $Q$ whose eigenvalues $q$ lie in the range $q_0 \leqslant q \leqslant q_1$ ($q$ may be taken to be the spatial coordinates). The boundary conditions on $|Ea\rangle$ may be expressed in terms of the representatives of $|Ea\rangle$, $|Ea\rangle_0$ in the $Q$-representation as

$$\lim_{q\to q_0} [\langle q|Ea\rangle - \langle q|Ea\rangle_0] = 0, \tag{G60}$$

i.e., $\langle q|Ea\rangle$ asymptotically approaches $\langle q|Ea\rangle_0$.

This boundary condition (G60) can best be expressed, instead of $U$, $U_0$, in terms of $K$, $K_0$ defined by

$$U = \mathbf{I} + \varepsilon K, \qquad U_0 = \mathbf{I} + \varepsilon K_0. \tag{G60a}$$

Then (G58) can be expressed in terms of the representatives in the $Q$-representation

$$\langle q|Ea\rangle = \langle q|Ea\rangle_0 + \varepsilon \int_{q_0}^{q_1} \langle q|K|q'\rangle\, dq'\, \langle q'|Ea\rangle_0, \tag{G61}$$

$$\langle q|Ea\rangle_0 = \langle q|Ea\rangle + \varepsilon \int_{q_0}^{q_1} \langle q|K_0|q'\rangle\, dq'\, \langle q'|Ea\rangle. \tag{G61A}$$

If $K$, $K_0$ have the following properties

$$\langle q|K|q'\rangle = \begin{cases} 0 & \text{for } q < q', \\ \text{bounded} & \text{for } q' \leqslant q, \end{cases} \tag{G62}$$

then (G61) becomes

$$\langle q|Ea\rangle = \langle q|Ea\rangle_0 + \varepsilon \int_{q_0}^{q} \langle q|K|q'\rangle\, dq'\, \langle q'|Ea\rangle_0$$

which thus satisfies (G60). Similarly for (G61A). Any operator satisfying (G62) is said to be triangular.

To sum up, we introduce a $Q$-representation in which $K$, $K_0$ are triangular to ensure the boundary condition (G60).# The next problem is the investiga-

---

# If $q_0$ is $-\infty$, one would then have to add further conditions on $\langle q|K|q'\rangle$, $\langle q|K_0|q'\rangle$ to ensure (G62). A sufficient, but stringent, one is to require these matrix elements to vanish for $q < q_{\min}$, when $q_{\min}$ is finite.

tion of the conditions under which $U$, $U_0$ can be uniquely determined [and from them, $H$ by (G42)]. These conditions can be expressed in terms of the following theorems:

I. A necessary condition is that $\langle q|W - \mathbf{I}|q'\rangle$ be a bounded function of $q'$ at $q' = q$.
II. A sufficient condition for a solution of $WU^\dagger = U_0$ is that $W - \mathbf{I}$ be bounded $< 1$.
III. A sufficient condition for the uniqueness of $U$, $U_0$ of $WU^\dagger = U_0$ is that $K$, $K_0$ be triangular.
IV. A sufficient condition for $U^{-1} = U_0$ is that either $\langle q|K|q'\rangle$ or $\langle q|K_0|q'\rangle$ be a bounded continuous function of its arguments.
V. An alternative sufficient condition for the uniqueness of $U$, $U_0$ and for $U^{-1} = U_0$ is that $W - \mathbf{I}$ be bounded less than 1 and $W - \mathbf{I}$ have an inverse.

*β) Determination of U (or K): Gel'fand–Levitan equation:*

To obtain $U$ from (given) $W$, we now introduce the operator $\Omega$ by

$$W = \mathbf{I} + \varepsilon\Omega. \tag{G63}$$

From (G59) and (G53), we have

$$WU^\dagger = U_0 \to UW^\dagger = U_0^\dagger \to UW = U_0^\dagger.$$

Substituting (G61) and (G63) into the last equation, we get

$$K = K_0^\dagger - \Omega - \varepsilon K\Omega \tag{G64}$$

which gives $K$ in terms of $\Omega$ (i.e., $U$ in terms of $W$). This equation is the generalized Gel'fand–Levitan equation. In the $Q$-representation, in which $K$, $K_0$ are triangular, this equation is

$$\begin{aligned}\langle q|K|q'\rangle &= \langle q|K_0^\dagger|q'\rangle - \langle q|\Omega|q'\rangle \\ &\qquad - \varepsilon\int_{q_0}^{q} \langle q|K|q''\rangle\, dq''\langle q''|\Omega|q'\rangle, \quad \text{for } q' < q, \\ &= 0 \quad \text{for } q < q'.\end{aligned} \tag{G65}$$

By (G62),

$$\langle q|K_0^\dagger|q'\rangle = \langle q'|K_0|q\rangle^* = 0 \quad \text{for } q' < q.$$

Hence (G65) becomes

$$\langle q|K|q'\rangle = -\langle q|\Omega|q'\rangle - \varepsilon\int_{q_0}^{q} \langle q|K|q''\rangle\, dq''\, \langle q''|\Omega|q'\rangle. \tag{G66}$$

This generalized Gel'fand–Levitan equation is of the Fredholm type. One may solve it by iteration, or by the following procedure.

Define $\delta(q - Q)$ by

$$\delta(q - Q)|q'\rangle = \delta(q - q')|q'\rangle, \tag{G67}$$

which, on using the completeness relation $\int_{q_0}^{q_1} |q\rangle\, dq\langle q| = \mathbf{I}$, gives

$$\delta(q - Q) = |q\rangle\langle q|. \tag{G67a}$$

Define $\eta(q - Q)$ by

$$\eta(q - Q)|q'\rangle = \eta(q - q')|q'\rangle, \tag{G68}$$

where

$$\eta(x) = \begin{cases} 1 & \text{for } 0 < x, \\ 0 & \text{for } x < 0. \end{cases}$$

Then

$$\langle q|K|q'\rangle = \langle q|\bar{K}|q'\rangle\, \eta(q - q'), \tag{G69}$$

where on the right, the value of $\langle q|\bar{K}|q'\rangle$ for $q' > q$ is arbitrary and $\langle q|\bar{K}|q'\rangle = \langle q|K|q'\rangle$ for $q' < q$. Using this function $\eta$, we can write (G66) in the form

$$\langle q|K|q'\rangle = -\langle q|\Omega|q'\rangle\eta(q - q') - \varepsilon \int_{q_0}^{q_1} \langle q|K|q''\rangle\eta(q - q'')\, dq''\langle q''|\Omega|q'\rangle\eta(q - q'). \tag{G70}$$

On using $\delta$ in (G67), this equation can be put in the operator form

$$K = -\int_{q_0}^{q_1} \delta(q - \Omega)\, \Omega\eta(q - \Omega)\, dq - \varepsilon \int_{q_0}^{q_1} \delta(q - \Omega)\, K\eta(q - \Omega)\Omega\eta(q - \Omega)\, dq. \tag{G71}$$

We now assert that if $[\mathbf{I} + \varepsilon\Omega\eta(q - \Omega)]^{-1}$ exists, the solution of the Gel'fand–Levitan equation is#

$$K = -\int_{q_0}^{q_1} \delta(q - Q)[\mathbf{I} + \varepsilon\Omega\eta(q - \Omega)]^{-1}\Omega\eta(q - Q)\, dq. \tag{G72}$$

Now if $\Omega$ is bounded and $< 1/|\varepsilon|$, then $\Omega\eta(q - Q)$ is certainly bounded and $< 1/|\varepsilon|$, and

$$[\mathbf{I} + \varepsilon\Omega\eta(q - Q)]^{-1} = \sum_n \varepsilon^n(-1)^n\, [\Omega\eta(q - Q)]^n \tag{G74}$$

exists. Substituting this into (G72), we may obtain $K$ by iteration.

Having found $K$, we can find $U$, $U_0$ from (G61), (G67A), (G72), (G73),

$$U = \int_{q_0}^{q_1} \delta(q - Q)[\mathbf{I} + \varepsilon\Omega\eta(q - Q)]^{-1}\, dq, \tag{G75}$$

---

# Proof: If $[\mathbf{I} + \varepsilon\Omega\eta(q - \Omega)]^{-1}$ exists, then

$$\mathbf{I} = \frac{1}{\mathbf{I} + \varepsilon\Omega\eta}[\mathbf{I} + \varepsilon\Omega\eta] \rightarrow \frac{\mathbf{I}}{\mathbf{I} + \varepsilon\Omega\eta} = \mathbf{I} - \varepsilon[\mathbf{I} + \varepsilon\Omega\eta]^{-1}\Omega\eta. \tag{G73}$$

Using this in (G72), substituting this $K$ from (G72) into (G71), and using (G73) again, one gets back (G72), thus proving (G72) satisfies (G71).

$$U_0 = WU^\dagger \overset{(G63)}{=\!=} (\mathbf{I} + \varepsilon\Omega) \int_{q_0}^{q_1} [\mathbf{I} + \varepsilon\Omega\eta(q - Q)]^{-1}\, \delta(q - Q)\, dq$$
$$= \int_{q_0}^{q_1} \{[\mathbf{I} + \varepsilon\eta(q - Q)\Omega] + \varepsilon\eta(Q - q)\Omega\}[\mathbf{I} + \varepsilon\eta(q - Q)\Omega]^{-1}\, \delta(q - Q)\, dq^{\#}$$
$$= \mathbf{I} + \varepsilon \int_{q_0}^{q_1} \eta(Q - \eta)\Omega\, [\mathbf{I} + \varepsilon\eta(q - Q)\Omega]^{-1}\, \delta(q - Q)\, dq. \tag{G76}$$

Hence from (G61), as the companion to (G72),

$$K_0 = \frac{1}{\varepsilon}(U_0 - \mathbf{I}) = \int_{q_0}^{q_1} \eta(Q - q)\Omega\, [\mathbf{I} + \varepsilon\eta(q - Q)\Omega]^{-1}\, \delta(q - Q)\, dq. \tag{G77}$$

Having obtained $K$, $K_0$ from $\Omega$ (i.e., from $W$), and $U$, $U_0$ from (G61), or in $Q$-representation,

$$\begin{aligned} \langle q|U|q'\rangle &= \delta(q - q') + \varepsilon\langle q|K|q'\rangle, \\ \langle q|U_0|q'\rangle &= \delta(q - q') + \varepsilon\langle q|K_0|q'\rangle, \end{aligned} \tag{G78}$$

we have to complete the program by proving that the solution $U$, $U_0$ of $WU^\dagger = U_0$ in (G78) is unique,## and $U^{-1} = U_0$.###

---

\# Note that $\eta(q - Q) + \eta(Q - q) = \mathbf{I}$ from the definition (G68).

\## The proof that $U$, $U_0$ is unique is as follows:

Assume that $U' = \mathbf{I} + \varepsilon K'$, $U_0' = \mathbf{I} + \varepsilon K_0'$ is another solution. From $WU^\dagger = U_0$, $WU'^\dagger = U_0'$, we get

$$W(K^\dagger - K'^\dagger) = K_0 - K_0',$$

or

$$WR^\dagger = R_0, \qquad R^\dagger = K^\dagger - K'^\dagger, \qquad R_0 = K_0 - K_0'.$$

By hypothesis, $K$, $K_0$, $K'$, $K_0'$ are triangular; so are $R^\dagger$ and $R_0$, and $WR^\dagger = R_0$ can be written

$$\int_q^{q_1} \langle q|W|q''\rangle\, dq'' \langle q''|R^\dagger|q'\rangle = \langle q|R_0|q'\rangle,$$

or

$$\int_{q_0}^{r} dq \int_{q_0}^{q'} dq'' \langle r|R|q\rangle\langle q|W|q''\rangle\langle q''|R^\dagger|q'\rangle = \int_{q'}^{r} \langle r|R|q\rangle\, dq\, \langle q|R_0|q'\rangle,$$

where $q' < r$. On letting $r \to q'$, since $R$, $R_0$ are bounded, the integral $\to 0$. On writing

$$\langle\Phi| = \int dq\, \langle q'|R|q\rangle\langle q|, \quad \text{and hence} \quad \int |q''\rangle\langle q''|R^\dagger|q'\rangle\, dq'' = |\Phi\rangle,$$

we obtain

$$\langle\Phi|W|\Phi\rangle = 0.$$

Since by (G55), $W$ is positive definite, it follows that $|\Phi\rangle = 0$, i.e., $R = R_0 = 0$. Hence $U' = U$, $U_0' = U_0$, and the solution $U$, $U_0$ is unique.

\### The proof that $U^{-1} = U_0$ can be given in two steps.

(a) Proof that $U^{-1}$ exists.

A necessary and sufficient condition for $U$ to have an inverse $U^{-1}$ is that the relation $U|\Phi\rangle = 0$ implies $|\Phi\rangle = 0$.

$WU^+$

$\gamma$) *Asymptotic form for the eigenfunction $|Ea\rangle$ of $H$:*

In (G60), we have imposed the asymptotic condition for $q \to q_0$ (in the $Q$-representation)

$$\langle q|Ea\rangle \xrightarrow{q-q_0} \langle q|Ea\rangle_0. \tag{G81}$$

We have shown that a unique $U'$ can be found such that

$$|Ea\rangle = U|Ea\rangle_0 \tag{G82}$$

satisfies the above asymptotic condition. The question now is: what is the asymptotic form of this $|Ea\rangle$ as $q$ approaches the other limit $q_1$ of the range $q_0 \leq q \leq q_1$.

---

Now, from $U = \mathbf{I} + \varepsilon K$, the equation $U|\Phi\rangle = 0$ in the $Q$-representation is

$$\langle q|\Phi\rangle + \varepsilon \int_{q_0}^{q} \langle q|K|q'\rangle\, dq' \,\langle q'|\Phi\rangle = 0.$$

According to Volterras' theory, if $\langle q|K|q'\rangle$ is continuous and bounded (which it is, by hypothesis), the only continuous $\langle q|\Phi\rangle$ that can satisfy the above equation is $\langle q|\Phi\rangle = 0$. Hence $U^{-1}$ exists.

(b) Proof that $U_0 = U^{-1}$.

From (G59), we have $UW^\dagger = U_0^\dagger$, and by (G54) $W$ is hermitian so that $UW = U_0^\dagger$. Hence

$$UWU^\dagger = U_0^\dagger U^\dagger.$$

But by (G59), $UWU^\dagger = UU_0$. Hence

$$UU_0 = U_0^\dagger U^\dagger. \tag{G79}$$

By (G61), we obtain

$$K + K_0 + \varepsilon KK_0 = K^\dagger + K_0^\dagger + \varepsilon K_0^\dagger K^\dagger,$$

or in the $Q$-representation,

$$\langle q|K|q'\rangle + \langle q|K_0|q'\rangle + \varepsilon\langle q|KK_0|q'\rangle = \langle q|K^\dagger|q'\rangle + \langle q|K_0^\dagger|q'\rangle + \varepsilon\langle q|K^\dagger K_0^\dagger|q'\rangle.$$

Now, from the triangularity of $K$, $K_0$, it follows (from the left-hand side) that

$$\langle q|K^\dagger|q'\rangle + \langle q|K_0^\dagger|q'\rangle + \varepsilon\langle q|K^\dagger K_0^\dagger|q'\rangle = 0, \quad \text{for} \quad q \leq q'.$$

But $\langle q|K^\dagger|q'\rangle = \langle q'|K|q\rangle^*$ etc. Hence

$$\langle q|K^\dagger|q'\rangle = \langle q|K_0^\dagger|q'\rangle = \langle q|K^\dagger K_0^\dagger|q'\rangle = 0, \quad \text{for} \quad q' \leq q.$$

Hence

$$K + K_0 + \varepsilon KK_0 = K^\dagger + K_0^\dagger + \varepsilon K^\dagger K_0^\dagger = 0,$$

and

$$UU_0 = \mathbf{I} + \varepsilon(K + K_0 + \varepsilon KK_0) = \mathbf{I}. \tag{G80}$$

Since we have proved $U^{-1}$ exists, hence $U_0 = U^{-1}$.

In the proof above, we have tacitly assumed that $Q$ has a discrete spectrum. If $Q$ has a continuous spectrum, it is only necessary to leave the diagonal $\langle q|K|q\rangle$, $\langle q|K_0|q\rangle$ etc. finite (instead of zero), and $\langle q|K|q'\rangle$, etc. $= 0$ for $q \lessgtr q'$. The diagonal elements have a measure zero and the same result

$$UU_0 = \mathbf{I}$$

holds.

Let us define a state $|H_0B_0; Ea\rangle$ by

$$|H_0B_0; Ea\rangle \equiv W^{-1}|H_0A_0; Ea\rangle = W^{-1}|Ea\rangle_0. \tag{G83}$$

[see (G40) for notation]. From (G56), $WH_0 - H_0W = 0$, it follows that $W^{-1}H_0 - H_0W^{-1} = 0$. Hence

$$\begin{aligned} H_0|H_0B_0; Ea\rangle &= H_0W^{-1}|Ea\rangle_0 = W^{-1}E|Ea\rangle_0 \\ &= EW^{-1}|H_0A_0; Ea\rangle = E|H_0B_0; Ea\rangle, \end{aligned} \tag{G84}$$

i.e., $|H_0B_0; Ea\rangle$ is an eigenstate of $H_0$ of energy $E$. Hence $|H_0B_0; Ea\rangle$ differs from $|H_0A_0; Ea\rangle = |Ea\rangle_0$ only on account of the degeneracy. $|H_0B_0; Ea\rangle$ is in general *not* an eigenfunction of $A_0$.

It can be proved that, if $q_1$ is finite, as $q \to q_1$,

$$\langle q|HA; Ea\rangle \xrightarrow{q\to q_1} \langle q|H_0B_0; Ea\rangle, \tag{G85}$$

i.e., the state $|HA; Ea\rangle = U|H_0A_0; Ea\rangle$ in (G82) that satisfies (G81) goes into the state $W^{-1}|H_0A_0; Ea\rangle = |H_0B_0; Ea\rangle$ as $q \to q_1$.#

### *iii) Remarks*

We shall now summarize the theory as follows: We are given the boundary (asymptotic) conditions for the eigenstate $|HA; Ea\rangle$ of $H$, such that, in terms of the representatives in the $Q$-representation,

$$\langle q|W^{-1}|H_0A_0; Ea\rangle \xleftarrow{q\to q_1} \langle q|HA; Ea\rangle \xrightarrow{q\to q_0} \langle q|H_0A_0; Ea\rangle. \tag{G87}$$

The theory enables $K$ in $U = \mathbf{I} + \varepsilon K$ to be uniquely determined from the Gel'fand–Levitan equation, in terms of the $\Omega$ in $W = \mathbf{I} + \varepsilon\Omega$. Then the potential $V$ in $H = H_0 + \varepsilon V$ is uniquely determined by

$$\varepsilon VU = UH_0 - H_0U.$$

---

# The proof is as follows: from (G52),

$$W = U^{-1}U^{\dagger-1} = U_0U_0^\dagger.$$

$$\begin{aligned} |HA; Ea\rangle &= U|H_0A_0; Ea\rangle \overset{(G83)}{=\!=} UW|H_0B_0; Ea\rangle \\ &= U^{\dagger-1}|H_0B_0; Ea\rangle = U_0^\dagger|H_0B_0; Ea\rangle. \end{aligned} \tag{G86}$$

On account of the triangularity of $K$, $K_0$,

$$\langle q|K_0|q'\rangle = 0, \quad q \le q'; \qquad \langle q|K_0^\dagger|q'\rangle = 0, \quad q' \le q.$$

Using $U_0 = \mathbf{I} + \varepsilon K_0$, we have from (G86),

$$\langle q|HA; Ea\rangle = \langle q|H_0B_0; Ea\rangle + \varepsilon \int_q^{q_1} \langle q|K_0^\dagger|q'\rangle \, dq' \, \langle q'|H_0B_0; Ea\rangle.$$

As $q \to q_1$, we arrive at (G85).

For $q_1 \to +\infty$, we must impose further restrictions on $\langle q|K_0^\dagger|q'\rangle$ to make the integral above vanish as $q \to q_1$.

It is of interest to note that in this time-independent theory, the operator $W^{-1}$ relates the *asymptotic states* in *space*. It is analogous to the $S$ operator (see Chapter 4, Sects. R, S, T; Chapter 6, Sect. W) which relates the *asymptotic states* in *time* in the time-dependent theory:

$$\Psi(+\infty) = S\Psi(-\infty).$$

They are, however, not really equivalent, $W$ being hermitian and $S$ being unitary.

It is mentioned at the beginning of (i) of this Appendix that we have confined ourselves to the case in which $H_0$ and $H$ have the same spectrum which is entirely continuous. [This restriction is reflected in the transformation $H = UH_0U^{-1}$ in (G42A)]. The extension of the theory, however, can be made to the more general case in which $H$ may have bound states in addition to a continuous spectrum (see Sect. G.2). For further details and extensions of the theory sketched above, the reader is referred to the papers by Kay and Moses.

# Appendix 2

## Amdur's Determination of Interatomic Potential from Measured Scattering Cross Sections

In this connection, we might discuss the extensive work of Amdur and his school on the determination of the interatomic potential $V(r)$ from the measured scattering cross sections (ref. 19).

The method employed can be briefly summarized as follows: A beam of atoms (say helium) of energy $E$ (500–2000 eV) after passing through a layer of gas (scatterer, say, also helium), falls on a detector of a width such that atoms scattered through angles $\vartheta$ greater than a value $\vartheta_{\min}$ are not registered. By comparing the intensities of the collected beams with and without the scattering layer, the cross section for scattering angles $\vartheta > \vartheta_{\min}$, i.e.,

$$S(\vartheta_{\min}; E) = 2\pi \int_{\vartheta_{\min}}^{\pi} \sigma[\vartheta, E; V(r)] \sin \vartheta \, d\vartheta$$

can be deduced. $S(\vartheta_{\min}; E)$ is then measured for various energies $E$ of the incident beam.

The analysis of the results is based on classical mechanics, according to which the minimum angle $\vartheta_{\min}$ is related to an impact parameter $\rho_m$ through Equation (G11). In terms of this $\rho_m$, the $S(\vartheta_{\min}; E)$ is given by

$$S(\vartheta_{\min}; E) = \pi \rho_m^2[\vartheta_{\min}, E; V(r)],$$

where $V(r)$ indicates the functional dependence of the $\rho_m$ calculated from (G11) on the potential. The procedure of analysis is as follows: (i) A certain analytic form, such as

$$V(r) = \frac{K}{r^s}, \quad K, s = \text{constants}$$

is assumed, and (ii) the relation between $\vartheta_{\min}$ and $\rho_m$ is calculated from (G11) with this assumed $V(r)$. (iii) The constants $K$, $s$ are then adjusted until the function relation between the *calculated* $\rho_m$ and the energy $E$ fits the relation between the observed $S(\vartheta_{\min}; E)$ and $E$. The results of this analysis are then given in the form of a function $V(r)$ (i.e., $K$ and $s$) together with a range of $r$ for its validity

$$\rho_m(E_2) \leqslant r \leqslant \rho_m(E_1),$$

where

$$\rho_m(E_1) = \sqrt{S(\vartheta_{\min}; E)/\pi}$$

and $E_1$, $E_2$ are the lower and upper values of the range of energy $E$ employed in the experiments.

It is to be noted that, without using any information about the *differential* cross section, to determine $V(r)$ in the manner described above suffers from the following two incompatible conditions. On the one hand, it is clear that the wider the range of $E$, the better the determination of $V(r)$ [since in the limit of a very narrow range in $E$, the observed $S[\vartheta_{\min}, E; V(r)]$ can be fitted by a variety of functions for $V(r)$]. On the other hand, unless the actual potential is expressible by a form, such as $V(r) = K/r^s$, over a wide range of distance $r$, the use of the relation (G11) (which depends on the value of $V(r)$ from $r_0$ to $r \to \infty$) can only lead to an approximate relation between $\vartheta_{\min}$ and $\rho_m$.

It must also be noted that for the lighter atoms (helium) at these energies (2000 eV), the effect of the uncertainty principle begins to be appreciable so that the use of classical mechanics may not be satisfactory. (See Sect. C.6.)

## REFERENCES

The expressions (G5–8) are given by:

1. Glauber, R. and Schomaker, V., Phys. Rev. **89**, 667 (1953). [The method (G2–7) is an extension of that of Feenberg, Phys. Rev. **40**, 40 (1932).]

2. Przikov, L., Ryndin, R. and Smorodinsky, J., Nuclear Physics **3**, 436 (1957). This paper gives the relation (G9).

3. Reeves, H., Annals of Phys. **3**, 386 (1958). (The inequality relations in Subsection 1.ii.)

The classic papers on the "inverse problem" are:

4. Bargmann, V., Phys. Rev. **75**, 301 (1949); Rev. Mod. Phys. **21**, 488 (1949). (Examples of "equivalent potentials", i.e., potentials giving the same phase shifts.)

5. Gel'fand, I. M. and Levitan, B. M., Doklady Akad. Nauk. S.S.S.R., **77**, 557 (1951); Isvest. Akad. Nauk S.S.S.R., **15**, 309 (1951), formulate the "inverse problem" and the integral equation (G30) or (G66).

5a. Marchenko, V. A., Doklady Akad. Nauk S.S.S.R. **104**, 695 (1955).

A review article has recently appeared:

6. Feddeyev, L. D., Uspekhi Matem. Nauk, Tom **XIV-4** (88), P. 57 (1959), which gives a fairly complete review of the subject and a comprehensive bibliography (translated into English by B. Sacklar, Inst. Math. Sci., New York Univ.).

7. Levinson, N., Phys. Rev. **89**, 755 (1953). This paper presents the methods of Gel'fand-Levitan in a slightly different form from that given in the text (G29)–(G37).

8. Kay, I. and Moses, H. E., Nuovo Cimento **X,2**, 917 (1955). (Generalized form of Gel'fand-Levitan theory in terms of spectral measure functions.)

Further papers on the subject are:

9. Levinson, N., Phys. Rev. **75**, 1445 (1949). [$U(r)$ is determined by the phase shifts $\delta_l(k)$ for all $l$]; Kgl. Danske Vidensk. Selsk. Mat.-fys. Medd. **25**, No. 9 (1949) (uniqueness).

10. Borg, G., Acta Math. **78**, 1 (1946). Mathematical theory of uniqueness problem from given discrete eigenvalues.

11. Holmberg, B., Nuovo Cimento **9**, 597 (1952). Equivalent potential.

12. Jost, R. and Kohn, W., Phys. Rev. **87**, 977 (1952) (construction of potential); Phys. Rev. **88**, 382 (1952) (equivalent potentials); Kgl. Danske Vidensk. Mat.-fys. Medd. **27**, No. 9 (1953) (application of Gel'fand-Levitan equation).

13. Wheeler, J. A., Phys. Rev. **99**, 630 (1955).

14. Corinaldesi, E., Nuovo Cimento **11**, 468 (1954). Construction of potentials from phase shifts and binding energies of relativistic equations.

15. Newton, R. G. and Jost, R., Nuovo Cimento, Ser. **X,1**, 590 (1955). Construction of potentials from $S$-matrix for systems of differential equations.

16. Ohmura, T., Prog. Theor. Phys. (Kyoto) **16**, 231 (1956). Stationary expression for potential curve itself.

17. Kay, I. and Moses, H. E., Nuovo Cimento, Ser. **X,3**, 66; 276 (1956); **5**, 230 (1957). Determination of scattering potential from spectral measure function; one-dimensional Schrödinger equation.

18. Ekstein, H., Phys. Rev. **117**, 1590 (1960). Equivalent Hamiltonian.

19. Amdur, I., J. Chem. Phys. **9**, 503 (1941); **17**, 844 (1949); Amdur, I. and Harkness, A. L., J. Chem. Phys. **22**, 664 (1954); Amdur, Glick and Perlman, Proc. Am. Acad. Arts Sci. **76**, 101 (1948).

20. Keller, J. B., Kay, I. and Shmoys, J., Phys. Rev. **102**, 557 (1956). The determination of the potential from the scattering data in classical mechanics.

chapter

2

# Scattering of a Particle by a Non-Central Field

The study of the scattering of a particle by a non-central field has been stimulated mostly by the problem of nucleon–nucleon scattering. Historically, the first empirical evidence for the existence of a deviation of the interaction between the proton and the neutron in the deuteron from central symmetry came with the discovery of the presence of a quadrupole moment in the ground state of the deuteron from the molecular beam experiments of Kellogg, Rabi, Ramsey and Zacharias. This quadrupole moment has since been explained by the presence of a "tensor" interaction of the form

$$TV_T(r) = \left[\frac{3(\boldsymbol{\sigma}_1 \cdot \mathbf{r})(\boldsymbol{\sigma}_2 \cdot \mathbf{r})}{r^2} - (\boldsymbol{\sigma}_1 \cdot \boldsymbol{\sigma}_2)\right] V_T(r), \qquad \text{(H1)}$$

where

$$\mathbf{r} = \mathbf{r}_2 - \mathbf{r}_1, \qquad r = |\mathbf{r}_2 - \mathbf{r}_1|, \tag{H1a}$$

and $\boldsymbol{\sigma}_1$, $\boldsymbol{\sigma}_2$ are the (vector) Pauli matrix operators for the two particles. Further definite evidence for the presence of such a tensor interaction between two nucleons came from the magnetic moment of the deuteron, and the differential cross sections of nucleon–nucleon scatterings at high energies ($>100$ MeV).

From the measurements of the differential cross sections and the polarization effects of the nucleon–nucleon scatterings at still higher energies ($E \simeq 300$ MeV), and their theoretical analysis, there has also been very strong evidence that in addition to the central and the tensor interactions, there is an interaction of the type

$$V_{so} = V_{so}(r)\mathbf{L}\cdot\mathbf{S}, \tag{H2}$$

where

$$\begin{aligned} \mathbf{r} &= \mathbf{r}_2 - \mathbf{r}_1, \qquad & \mathbf{p} &= \tfrac{1}{2}(\mathbf{p}_2 - \mathbf{p}_1), \\ \mathbf{L} &= [\mathbf{r} \times \mathbf{p}], \qquad & \mathbf{S} &= \tfrac{1}{2}(\boldsymbol{\sigma}_1 + \boldsymbol{\sigma}_2). \end{aligned} \tag{H2a}$$

Many attempts have been made in recent years to construct, from the empirical data on the deuteron system and on the proton–proton and proton–neutron scattering, a phenomenological law of interaction between two nucleons. In the present chapter we shall describe the theory of scattering for these two types ($T$ and $\mathbf{L}\cdot\mathbf{S}$) of non-central forces on which the analysis of the scattering cross section and the polarization data is based, and shall summarize the results on the phenomenological interaction law.

section

# H Scattering by Tensor and L·S Potential: Partial Wave Analysis

## 1. PRELIMINARIES

The Pauli spin operators $\boldsymbol{\sigma}(\sigma_x, \sigma_y, \sigma_z)$ for a spin $\frac{1}{2}$ particle are represented by the matrices

$$\sigma_x = \begin{pmatrix} 0 & 1 \\ 1 & 0 \end{pmatrix}, \quad \sigma_y = \begin{pmatrix} 0 & -i \\ i & 0 \end{pmatrix}, \quad \sigma_z = \begin{pmatrix} 1 & 0 \\ 0 & -1 \end{pmatrix},$$

$$\begin{aligned} \sigma_x^2 &= \sigma_y^2 = \sigma_z^2 = 1, \\ \sigma_x\sigma_y &= -\sigma_y\sigma_x = i\sigma_z, \text{ etc.} \end{aligned} \tag{H3}$$

It is well known that $\mathbf{L}\cdot\mathbf{S}$ in (H2) commutes with each of the following operators

$$\mathbf{J}^2 = (\mathbf{L} + \mathbf{S})^2, \qquad \mathbf{L}^2, \qquad \mathbf{S}^2, \quad M = L_z + \tfrac{1}{2}(\sigma_1 + \sigma_2)_z.$$

Hence $J$, $L$, $S$, $M$ are all exact quantum numbers of $V_{so}$ in (H2). The eigenvalues of $\mathbf{L}\cdot\mathbf{S}$ in a state $(J, L, S, M)$ are given by

$$\tfrac{1}{2}[J(J+1) - L(L+1) - S(S+1)].$$

For singlet states, $S = 0$ and $J = L$, so that $V_{so}$ in (H2) does not contribute to any scattering in the singlet states.

From the relation

$$(\boldsymbol{\sigma}\cdot\mathbf{A})(\boldsymbol{\sigma}\cdot\mathbf{B}) = (\mathbf{A}\cdot\mathbf{B}) + i(\boldsymbol{\sigma}\cdot[\mathbf{A}\times\mathbf{B}]),$$

where $\mathbf{A}$, $\mathbf{B}$ commute with each other and also with $\boldsymbol{\sigma}$, and the relations

$$\begin{aligned} (\boldsymbol{\sigma}_1 + \boldsymbol{\sigma}_2)^2 &= 6 + 2(\boldsymbol{\sigma}_1\cdot\boldsymbol{\sigma}_2), \\ (\boldsymbol{\sigma}_1\cdot\boldsymbol{\sigma}_2)^2 &= 3 - 2(\boldsymbol{\sigma}_1\cdot\boldsymbol{\sigma}_2), \end{aligned}$$

one readily obtains the following relations

$$
\begin{aligned}
(\boldsymbol{\sigma}_1 \cdot \boldsymbol{\sigma}_2)\, T - T(\boldsymbol{\sigma}_1 \cdot \boldsymbol{\sigma}_2) &= 0, \\
(\boldsymbol{\sigma}_1 + \boldsymbol{\sigma}_2)\, T - T(\boldsymbol{\sigma}_1 + \boldsymbol{\sigma}_2) &= 0, \\
PT = TP &= T, \\
T^2 = 6 + 2(\boldsymbol{\sigma}_1 \cdot \boldsymbol{\sigma}_2) &- 2T,
\end{aligned}
\tag{H4}
$$

where $P$ is the Bartlett operator

$$P = \tfrac{1}{2}[1 + (\boldsymbol{\sigma}_1 \cdot \boldsymbol{\sigma}_2)]. \tag{H5}$$

If we denote by ${}^3\chi^{M_s}$ and ${}^1\chi^{M_s}$ the symmetric (triplet) and antisymmetric (singlet) spin wave functions

$$
\begin{aligned}
{}^3\chi^1 &= \begin{pmatrix}1\\0\end{pmatrix}_1 \begin{pmatrix}1\\0\end{pmatrix}_2, \qquad {}^3\chi^0 = \frac{1}{\sqrt{2}}\left[\begin{pmatrix}1\\0\end{pmatrix}_1 \begin{pmatrix}0\\1\end{pmatrix}_2 + \begin{pmatrix}1\\0\end{pmatrix}_2 \begin{pmatrix}0\\1\end{pmatrix}_1\right], \\
{}^3\chi^{-1} &= \begin{pmatrix}0\\1\end{pmatrix}_1 \begin{pmatrix}0\\1\end{pmatrix}_2, \qquad {}^1\chi^0 = \frac{1}{\sqrt{2}}\left[\begin{pmatrix}1\\0\end{pmatrix}_1 \begin{pmatrix}0\\1\end{pmatrix}_2 - \begin{pmatrix}1\\0\end{pmatrix}_2 \begin{pmatrix}0\\1\end{pmatrix}_1\right],
\end{aligned}
\tag{H6}
$$

we obtain the following relations

$$
\begin{aligned}
(\boldsymbol{\sigma}_1 + \boldsymbol{\sigma}_2)^2\, {}^3\chi &= 8\, {}^3\chi, \qquad & (\boldsymbol{\sigma}_1 + \boldsymbol{\sigma}_2)^2\, {}^1\chi &= 0, \\
P\, {}^3\chi &= {}^3\chi, & P\, {}^1\chi &= -{}^1\chi, \\
& & T\, {}^1\chi &= 0,
\end{aligned}
\tag{H7}
$$

$$
\begin{aligned}
T\,{}^3\chi^1 &= 4\sqrt{\tfrac{3}{5}}\, Y_{2,2}\, {}^3\chi^{-1} - 2\sqrt{\tfrac{6}{5}}\, Y_{2,1}\, {}^3\chi^0 + 2\sqrt{\tfrac{2}{5}}\, Y_{2,0}\, {}^3\chi^1, \\
T\,{}^3\chi^0 &= 2\sqrt{\tfrac{6}{5}}\, Y_{2,1}\, {}^3\chi^{-1} - 4\sqrt{\tfrac{2}{5}}\, Y_{2,0}\, {}^3\chi^0 + 2\sqrt{\tfrac{6}{5}}\, Y_{2,-1}\, {}^3\chi^1, \\
T\,{}^3\chi^{-1} &= 2\sqrt{\tfrac{2}{5}}\, Y_{2,0}\, {}^3\chi^{-1} - 2\sqrt{\tfrac{6}{5}}\, Y_{2,-1}\, {}^3\chi^0 + 4\sqrt{\tfrac{3}{5}}\, Y_{2,-2}\, {}^3\chi^1,
\end{aligned}
\tag{H8}
$$

where the $Y_{L,M_L}$ are the normalized spherical harmonics with the phase as defined in Condon and Shortley's book (ref. 3), namely,

$$
\begin{aligned}
Y_{2,0} &= \sqrt{\tfrac{5}{2}}\tfrac{1}{2}(3\cos^2\vartheta - 1), \\
Y_{2,\pm 1} &= \mp\tfrac{3}{2}\sqrt{\tfrac{5}{3}}\sin\vartheta\cos\vartheta\, e^{\pm i\varphi}, \\
Y_{2,\pm 2} &= \tfrac{3}{4}\sqrt{\tfrac{5}{3}}\sin^2\vartheta\, e^{\pm 2i\varphi},
\end{aligned}
$$

and

$$
\begin{aligned}
(\mathbf{L}\cdot\mathbf{S})\,\chi^1 &= \frac{1}{\sqrt{2}}(L_x - iL_y)\,\chi^0 + L_z\chi^1, \\
(\mathbf{L}\cdot\mathbf{S})\,\chi^0 &= \frac{1}{\sqrt{2}}(L_x - iL_y)\,\chi^{-1} + \frac{1}{\sqrt{2}}(L_x + iL_y)\,\chi^1, \\
(\mathbf{L}\cdot\mathbf{S})\,\chi^{-1} &= -L_z\chi^{-1} + \frac{1}{\sqrt{2}}(L_x + iL_y)\,\chi^0.
\end{aligned}
\tag{H9}
$$

Since we shall be concerned mainly with the triplet states in this and the next section, we shall omit, as in (H9), the superscript 3 for triplet states from the ${}^3\chi^{M_s}$ in the following.

Consider the tensor interaction (H1). That the total spin $\mathbf{S}^2 = \frac{1}{4}(\boldsymbol{\sigma}_1 + \boldsymbol{\sigma}_2)^2$ commutes with $T$ [see (H4)] means that $S$ is an exact quantum number of the Hamiltonian containing $V_T(r)T$. Other exact quantum numbers are#

$$J, \quad M, \quad \text{and} \quad \text{parity}. \tag{H10}$$

The orbital angular momentum, however, does not commute with $T$ so that $L$ is not an exact quantum number. Thus we cannot talk of $S, P, D, F, \ldots$ states. It is, however, convenient to retain these spectroscopic notations and

---

# The concept of "parity" of a quantum mechanical state, first introduced by Wigner, refers to a property of the wave function with respect to the space inversion operation in which all coordinate vectors are reversed

$$\mathscr{I}: \quad \mathbf{x} \to -\mathbf{x}, \qquad \mathbf{x} = (\mathbf{x}_1, \mathbf{x}_2, \ldots, \mathbf{x}_n). \tag{H10A}$$

Consider the system whose Hamiltonian $H = K + V$, where $K$, $V$ are the kinetic and potential energy, is invariant under the inversion $\mathscr{I}$. Then solutions $\psi(\mathbf{x})$ and $\psi(-\mathbf{x})$ of the Schrödinger equation

$$(H - E)\psi(\mathbf{x}) = 0, \qquad (H - E)\psi(-\mathbf{x}) = 0,$$

can be found such that

$$\psi(-\mathbf{x}) = \mathscr{I}\psi(\mathbf{x}) \quad = c\psi(\mathbf{x}),$$

and

$$\psi(\mathbf{x}) = \mathscr{I}\psi(-\mathbf{x}) = c\psi(-\mathbf{x}),$$

where $c$ is a constant. From these, it follows that $c = 1$ or $-1$. The parity of $\psi$ is then defined as even or odd according as $c = 1$ or $-1$. Let the parity be denoted by the superscripts $+$, $-$. Then $\psi^+$, $\psi^-$ are solutions with the same energy $E$, and any linear combinations of $\psi^+$, $\psi^-$ are also solutions of $H$. In the scattering problem, the solution is in general such a combination, consisting of even $l$ and odd $l$ partial waves.

If the interaction $V$ is not invariant under $\mathscr{I}$, then the eigenstates $\psi$ of $K + V = H$ will not have a definite parity. Suppose $H$ is made up of two parts; namely $H^+$ which is symmetric, and $H^-$ which is antisymmetric, with respect to the inversion $\mathscr{I}$. Then obviously an even $\psi^+$, or an odd $\phi^-$, is not a solution of $(H^+ + H^-)\psi = E\psi$. The solution must have mixed parity, such as $\psi = \psi^+ + \phi^-$. For then the equation $(H^+ + H^- - E)(\psi^+ + \phi^-) = 0$ is satisfied by

$$(H^+ - E)\psi^+ = H^- \frac{1}{H^+ - E} H^-\psi^+,$$

and

$$(H^+ - E)\phi^- = -H^-\psi^+.$$

In the scattering problems dealt with in the present volume, we shall have no occasion to come across such parity non-conserving interactions.

to describe the possible states as combinations of $S, P, \ldots$ states, as shown in the following table:

| $J \backslash L$ | Even parity 0 | 2 | 4 | 6 | Odd parity 1 | 3 | 5 |
|---|---|---|---|---|---|---|---|
| 0 | | | | | $^3P_0$ | | |
| 1 | $^3S_1$ | $+\ ^3D_1$ | | | $^3P_1$ | | |
| 2 | | $^3D_2$ | | | $^3P_2$ | $+\ ^3F_2$ | |
| 3 | | $^3D_3$ | $+\ ^3G_3$ | | | $^3F_3$ | |
| 4 | | | $^3G_4$ | | | $^3F_4$ | $+\ ^3H_4$ |
| 5 | | | $^3G_5$ | $+\ ^3I_5$ | | | $^3H_5$ |

(H11)

Any two triplet states having the same $J$ and the same parity are "mixed up" by the tensor operator $T$. Thus the ground state of the deuteron is described by the linear combination of a $^3S_1$ and a $^3D_1$ state. Let us denote the radial wave functions of any three states in a given row of (H11) as follows:

$$\begin{array}{ccc} L = J & L = J - 1 & L = J + 1 \\ \hline \dfrac{1}{r}\, v_J(r) & \dfrac{1}{r}\, u_{J-1}(r) & \dfrac{1}{r}\, w_{J+1}(r) \end{array}. \tag{H12}$$

Let the interaction of the system be represented by the following spin and parity dependent potential (multiplied by $M/k$)

$$\begin{aligned} {}^3U^{\text{even}} &= -U_c^+(r) - U_T^+(r)T - U_{so}^+(r)\,\mathbf{L}\cdot\mathbf{S}, \\ {}^3U^{\text{odd}} &= -U_c^-(r) - U_T^-(r)T - U_{so}^-(r)\,\mathbf{L}\cdot\mathbf{S}, \\ {}^1U^{\text{even}} &= -{}^1U_c^+(r), \qquad {}^1U^{\text{odd}} = -{}^1U_c^-(r), \end{aligned} \tag{H13}$$

where the superscripts $+$, $-$ indicate even and odd parity respectively, $c$ denotes a central field, and the various $U(r)$ are functions of $r$ alone. The Schrödinger equation of the system leads, for each group of three states of a given $J$ [namely those in a row in (H11)], to one separate equation and two coupled equations. For example, for a given $J$ = an odd integer, we have

$L = J$:

$$\frac{d^2v_L}{dr^2} + \left[k^2 - \frac{L(L+1)}{r^2} + U_c^-(r) + 2U_T^-(r) - U_{so}^-(r)\right] v_L(r) = 0, \tag{H14}$$

$L = J - 1$:

$$\begin{aligned} \frac{d^2u_L}{dr^2} + \left[k^2 - \frac{L(L+1)}{r^2} + U_c^+(r) - \frac{2L}{2L+3}\, U_T^+(r) + LU_{so}^+(r)\right] u_L(r) \\ = -6\, \frac{\sqrt{(L+1)(L+2)}}{2L+3}\, U_T^+(r)\, w_{L+2}(r), \end{aligned} \tag{H15}$$

$L = J + 1$:

$$\frac{d^2 w_L}{dr^2} + \left[k^2 - \frac{L(L+1)}{r^2} + U_c^+(r) - \frac{2L+2}{2L-1} U_T^+(r) - (L+1)\, U_{so}^+(r)\right] w_L(r) = -6\,\frac{\sqrt{L(L-1)}}{2L-1}\, U_T^+(r)\, u_{L-2}(r).$$

For $J = 1$, (H15) give the equations for the coupled state ${}^3S_1 + {}^3D_1$, and (H14) gives ${}^3P_1$. For $J =$ an even integer, we obtain a similar set of equations with the parity indices $+$ and $-$ in (H14-5) interchanged.

## 2. PARTIAL WAVE AND PHASE SHIFT ANALYSIS

Since the spin $S$ is an exact quantum number, the scatterings from the triplet and the singlet states contribute incoherently to the scattering intensity. Therefore we may consider the singlet state and the triplet state scatterings separately.

The scattering amplitudes ${}^1f(\vartheta)^+$, ${}^1f(\vartheta)^-$ in the singlet-even and the singlet-odd states are determined by the singlet state potentials in (H13). They are given by the theory of Sect. A for central fields, namely,

$$\begin{aligned} {}^1f(\vartheta)^+ &= \frac{1}{2ik}\sum_{L\ \mathrm{even}} (2L+1)[\exp(2i\delta_L^+) - 1]\, P_L(\cos\vartheta), \\ {}^1f(\vartheta)^- &= \frac{1}{2ik}\sum_{L\ \mathrm{odd}} (2L+1)[\exp(2i\delta_L^-) - 1]\, P_L(\cos\vartheta), \end{aligned} \tag{H16}$$

where $\delta_L^+$, $\delta_L^-$ are determined by

$$\frac{d^2 v_L}{dr^2} + \left[k^2 - \frac{L(L+1)}{r^2} + {}^1U_c^{\pm}(r)\right] v_L(r) = 0.$$

For the triplet state scattering, a partial wave analysis can be carried out as follows.

Consider the initial state of a nucleon–nucleon scattering represented by a plane wave (in the relative coordinate $\mathbf{r} = \mathbf{r}_2 - \mathbf{r}_1$) with the wave vector $\mathbf{k}$ (taken to be the $Z$-axis),

$$e^{ikz}\, {}^3\chi^{M_S^0},$$

where $M_S^0$ is the sum of the spin angular momentum components of the two nucleons along the direction of $\mathbf{k}$. Since $M = M_L + M_S$ is an exact quantum number and since the initial $M_L^0 = 0$, the scattered waves will have three possible $M_L$ values

$$M_L = M_S^0 + 1, \qquad M_S^0, \qquad M_S^0 - 1. \tag{H17}$$

The scattered wave corresponding to a given $M$ $(=M_S^0)$ (i.e., with $S = 1$; $J$ unspecified; $L = J - 1, J, J + 1$) is expressible in the form

$$\Psi(\mathbf{r}, \sigma) = \sum_J \sum_L A_{JL} \frac{1}{r} \psi_L(r) F_{JML}(\vartheta, \varphi, \sigma), \tag{H18}$$

where the summation in $L$ is taken over $L = J - 1, J, J + 1$ [i.e., over a row in (H11)] and the summation in $J$ is taken over all rows in (H11) with $J \geqslant |M|$. The angular and spin part $F_{JML}(\vartheta, \varphi, \sigma)$ for a given $J$, $M$, $L$ is given by the familiar transformation from the $(m_L, m_S)$- to the $(J, M)$-representation#

$$F_{JML}(\vartheta, \varphi, \boldsymbol{\sigma}) = \sum_{m_l} C_{m_l}^{JML} Y_{L,m_l}(\vartheta, \varphi) \chi^{m_S}, \tag{H19}$$

where the summation taken over $m_l$ is such that $m_l + m_S = M_S^0$ as in (H17). The $Y_{l,m_l}$ are the normalized harmonics with phases as defined in Condon–Shortley (ref. 3). The triplet index for ${}^3\chi$ has been dropped, since in the following we are always only concerned with the triplet states. The coefficients $C$ are well known. For each $M$, $L$, there are nine coefficients $C$ corresponding to $J = L + 1, L, L - 1$, and $m_l = M - 1, M, M + 1$.

The scattering problem now is to find a solution $\Psi_M$ of the Schrödinger equation that represents an incident plane wave (H16) and an outgoing wave, i.e.,

$$\Psi_M(\mathbf{r}, \boldsymbol{\sigma}) \to e^{ikz}\chi^M + \frac{e^{ikr}}{r} f_M(\vartheta, \varphi, \boldsymbol{\sigma}). \tag{H20}$$

The radial wave functions $\psi_L(r)$ in (H18) are solutions of (H14), (H15) that vanish at $r = 0$ and have the asymptotic forms

$$\begin{aligned} v_L(r) &\to A_L \sin\left(kr - \frac{l\pi}{2} + \delta_l\right), \\ u_L(r) &\to B_L \sin\left(kr - \frac{l\pi}{2} + \varepsilon_l\right), \\ w_L(r) &\to C_L \sin\left(kr - \frac{l\pi}{2} + \zeta_l\right). \end{aligned} \tag{H21}$$

The $A_L$, $B_L$, $C_L$ are determined by the asymptotic condition (H20) for outgoing waves in terms of the phases $\delta_l$, $\varepsilon_l$, $\zeta_l$. These phases are determined by the equations (H14), (H15). Thus for $M = 1$ (for $M = -1$, the $C_{m_l}^{JML}$ are similar to the following, with $m_l$ replaced by $-m_l$, and a $-$ sign for the first row)

---

# The Clebsch-Gordan coefficients $C_{m_l}^{JML}$ are written as $(LSm_lm_s|LSJM)$ in Condon and Shortley's *Theory of Atomic Spectra*, p. 75 (ref. 3). The symbols $C_{LS}(JMm_lm_s)$ and $S_{Jm_lm_s}^{LS}$ are also sometimes used.

$$C_0^{L,1,L} : C_1^{L,1,L} : C_2^{L,1,L} = \frac{1}{\sqrt{2}} : -\frac{1}{\sqrt{L(L+1)}} : -\sqrt{\frac{(L-1)(L+2)}{2L(L+1)}},$$

$$C_0^{L+1,1,L} : C_1^{L+1,1,L} : C_2^{L+1,1,L} = \sqrt{\frac{L+2}{2(2L+1)}} : \sqrt{\frac{L(L+2)}{(L+1)(2L+1)}} : \sqrt{\frac{L(L-1)}{2(L+1)(2L+1)}}, \tag{H22}$$

$$C_0^{L-1,1,L} : C_1^{L-1,1,L} : C_2^{L-1,1,L} = \sqrt{\frac{L-1}{2(2L+1)}} : -\sqrt{\frac{(L-1)(L+1)}{L(2L+1)}} : \sqrt{\frac{(L+1)(L+2)}{2L(2L+1)}}.$$

For $M = 0$,

$$C_{-1}^{L,0,L} : C_0^{L,0,L} : C_1^{L,0,L} = \frac{1}{\sqrt{2}} : 0 : -\frac{1}{\sqrt{2}},$$

$$C_{-1}^{L+1,0,L} : C_0^{L+1,0,L} : C_1^{L+1,0,L} = \sqrt{\frac{L}{2(2L+1)}} : \sqrt{\frac{L+1}{2L+1}} : \sqrt{\frac{L}{2(2L+1)}}, \tag{H23}$$

$$C_{-1}^{L-1,0,L} : C_0^{L-1,0,L} : C_1^{L-1,0,L} = \sqrt{\frac{L+1}{2(2L+1)}} : -\sqrt{\frac{L}{2L+1}} : \sqrt{\frac{L+1}{2(2L+1)}}.$$

Using the asymptotic forms (H21) and that of $e^{ikz}$ in (H20), we obtain, for $M = 1$, for example,

$$\begin{aligned}
2ikr(\Psi_{M=1} - e^{ikz}\chi^M) &= \sum_{L=2}\Big[C_2^{L,1,L}A_L\{\exp[i(\delta_l)] - \exp[-i(\delta_l)]\} \\
&\qquad + C_2^{L+1,1,L}B_L\{\exp[i(\varepsilon_l)] - \exp[-i(\varepsilon_l)]\} \\
&\qquad + C_2^{L-1,1,L}C_L\{\exp[i(\zeta_l)] - \exp[-i(\zeta_l)]\}\Big]\, Y_{L,2}\chi^{-1} \\
&\quad + \sum_{L=1}\Big[C_1^{L,1,L}A_L\{\exp[i(\delta_l)] - \exp[-i(\delta_l)]\} \\
&\qquad + C_1^{L+1,1,L}B_L\{\exp[i(\varepsilon_l)] - \exp[-i(\varepsilon_l)]\} \\
&\qquad + C_1^{L-1,1,L}C_L\{\exp[i(\zeta_l)] - \exp[-i(\zeta_l)]\}\Big]\, Y_{L,1}\chi^0
\end{aligned}$$

$$+ \sum_{L=0} \Bigg[ C_0^{L,1,L} A_L \{\exp[i(\delta_l)] - \exp[-i(\delta_l)]\}$$
$$+ C_0^{L+1,1,L} B_L \{\exp[i(\varepsilon_l)] - \exp[-i(\varepsilon_l)]\}$$
$$+ C_0^{L-1,1,L} C_L \{\exp[i(\zeta_l)] - \exp[-i(\zeta_l)]\}$$
$$- \sqrt{\frac{2}{2L+1}}(2L+1)\, i^L \Bigg\{\exp\left[i\left(kr - \frac{L\pi}{2}\right)\right]$$
$$- \exp\left[-i\left(kr - \frac{L\pi}{2}\right)\right]\Bigg\}\Bigg] Y_{L,0}\, \chi^1, \quad \text{(H24)}$$

where we use the notation

$$(\delta_l) \equiv kr - \frac{L\pi}{2} + \delta_L, \text{ etc.} \quad \text{(H25)}$$

In the summations over $L$ above, it is necessary to remember that $J \geqslant |M|$ and hence to exclude from the summation such non-existent states as ${}^3S_0$, ${}^3S_{-1}$, ${}^3P_0$ for $M = 1$. For $L = 0$, from the condition that the coefficient of $e^{-ikr}$ be zero, and $C_0^{110} = 1$ from (H22), we find

$$B_0 = \sqrt{2}\exp(i\varepsilon_0). \quad \text{(H26a)}$$

For $L = 1$, the condition for the coefficient of $e^{-ikr}$ to vanish together with the values of $C_1^{111}$, $C_1^{211}$, $C_0^{111}$, $C_0^{211}$ from (H22), gives

$$A_1 = \sqrt{3}\,i\exp(i\delta_1), \qquad B_1 = \sqrt{3}\,i\exp(i\varepsilon_1), \qquad C_1 = 0. \quad \text{(H26b)}$$

For $L \geqslant 2$, we obtain similarly

$$A_L = \sqrt{2L+1}\,i^L\exp(i\delta_L), \qquad B_L = \sqrt{L+2}\,i^L\exp(i\varepsilon_L),$$
$$C_L = \sqrt{L-1}\,i^L\exp(i\zeta_L). \quad \text{(H26c)}$$

Finally, for the scattered wave for $M = 1$, we have

$$\Psi'_{M=1}(\mathbf{r}, \boldsymbol{\sigma}) - e^{ikz}\chi^1(\sigma)$$
$$= \frac{e^{ikr}}{2ikr}\Bigg[\sum_{L=2}\left\{\begin{matrix}-(2L+1)\exp(2i\delta_L) + L\exp(2i\varepsilon_L) \\ + (L+1)\exp(2i\zeta_L)\end{matrix}\right\}$$
$$\times \sqrt{\frac{(L-1)(L+2)}{2L(L+1)(2L+1)}}\, Y_{L,2}\,\chi^{-1}$$
$$+ \sum_{L=1}\left\{\begin{matrix}-(2L+1)\exp(2i\delta_L) + L(L+2) \\ \times \exp(2i\varepsilon_L) - (L^2-1)\exp(2i\zeta_L)\end{matrix}\right\}$$
$$\times \sqrt{\frac{1}{L(L+1)(2L+1)}}\, Y_{L,1}\,\chi^0$$
$$+ \sum_{L=1}\left\{\begin{matrix}(2L+1)\exp(2i\delta_L) + (L+2)\exp(2i\varepsilon_L) \\ + (L-1)\exp(2i\zeta_L) - 2(2L+1)\end{matrix}\right\}$$
$$\times \sqrt{\frac{1}{2(2L+1)}}\, Y_{L,0}\,\chi^1$$
$$+ [\exp(2i\varepsilon_0) - 1]\,\chi^1\Bigg]. \quad \text{(H27)}$$

For $M = -1$, an entirely similar calculation shows that $\Psi_{M=-1}$ is obtainable from $\Psi_{M=1}$ above by simply changing $m_l$ into $-m_l$ and $M_S$ into $-M_S$. For $M = 0$, we obtain

$$L = 0, \qquad B_0 = \sqrt{2}\exp(i\varepsilon_0),$$
$$L \geqslant 1, \qquad A_L = 0, \qquad B_L = \sqrt{2L+2}\,i^L \exp(i\varepsilon_L),$$
$$C_L = -\sqrt{2L}\,i^L \exp(i\zeta_L), \tag{H28}$$

$$\Psi_{M=0}(\mathbf{r}, \boldsymbol{\sigma}) - e^{ikz}\chi^0(\boldsymbol{\sigma})$$
$$= \frac{e^{ikr}}{2ikr}\left[\sum_{L=1}[\exp(2i\varepsilon_L) - \exp(2i\zeta_L)]\sqrt{\frac{L(L+1)}{2L+1}}\,Y_{L,1}\chi^{-1}\right.$$
$$+ \sum_{L=0}\{(L+1)\exp(2i\varepsilon_L) + L\exp(2i\zeta_L) - (2L+1)\}\sqrt{\frac{2}{2L+1}}\,Y_{L,0}\chi^0$$
$$\left. + \sum_{L=1}[\exp(2i\varepsilon_L) - \exp(2i\zeta_L)]\sqrt{\frac{L(L+1)}{2L+1}}\,Y_{L,-1}\chi^1\right]. \tag{H29}$$

The expressions (H27), (H29) are the exact formulas for the scattering amplitudes by a non-central field of the type in (H13). They are the generalizations of the Faxen–Holtzmark formula (A17) for a central field.

We may note the following interrelation between the ratio of the amplitudes $B_{L-2}/C_L$ and the phase shifts. Because of the coupling of the equations for $u_L$ and $w_{L+2}$, their solutions depend on the ratio of the amplitudes $B_L/C_{L+2}$. For a given ratio $B_{L-2}/C_L$, the phases $\varepsilon_{L-2}$, $\zeta_L$ are uniquely determined. The ratio $B_{L-2}/C_L$ is, however, arbitrary and is determined only by the requirement that $\Psi_M(\mathbf{r}, \boldsymbol{\sigma}) - e^{ikz}\chi^M$ represent an outgoing wave alone. See Sect. I, (I6) in the following.

Note that, since the ratio $B_{L-2}/C_L$ is the same for $M = 1$ and $-1$, the phase shifts are the same for $M = \pm 1$. We may introduce the notation

$$M = \pm 1: \quad \varepsilon_L \equiv \varepsilon_L^{JM} = \varepsilon_L^{L+1,1}, \qquad \delta_L \equiv \delta_L^{JM} = \delta_L^{L,1}, \qquad \zeta_L \equiv \zeta_L^{JM} = \zeta_L^{L-1,1}$$
$$M = 0: \quad \varepsilon_L \equiv \varepsilon_L^{JM} = \varepsilon_L^{L+1,0}, \qquad \delta_L \equiv \delta_L^{JM} = \delta_L^{L,0}, \qquad \zeta_L \equiv \zeta_L^{JM} = \zeta_L^{L-1,0}. \tag{H30}$$

In the absence of the tensor and **L·S** interactions, the coefficients of $\chi^{-1}$ and $\chi^0$ in (H27) vanish and there are then no spin flips in the scattering. Similarly, the spin flip terms with $\chi^{-1}$ and $\chi^1$ in (H29) vanish.

### 3. SCATTERING MATRIX REPRESENTATION: REAL PHASE SHIFTS

In the foregoing treatment, the phase shifts $\varepsilon_l$, $\zeta_l$ in (H21) of the coupled wave functions $u_{J-1}(r)$, $w_{J+1}(r)$ are complex. A treatment using a different representation is possible in which the phase shifts are real. The method, suggested by Schwinger and worked out by Rohrlich and Eisenstein and by Blatt and Biedenharn, has often been used in the analysis of nucleon–nucleon scattering data. It will be briefly described below.

We recall that the coupled equations (H15) for $u_{J-1}$, $w_{J+1}$ are each of the second order, so that there are four independent solutions in general. The solution satisfying the desired asymptotic condition (H20) for outgoing waves is completely determined by the initial conditions $u_{J-1}(0) = 0$, $w_{J+1}(0) = 0$ and the ratio of the amplitudes $B_{L-2}/C_L$ in (H21). [See paragraph following (H29).] One may proceed by first discarding the asymptotic condition and taking two arbitrary independent solutions which vanish at $r = 0$. The asymptotic condition is then satisfied by taking a linear combination of the two solutions.

Thus let the radial wave functions be $u_{J-1}$, $w_{J+1}$ which asymptotically are

$l = J - 1$:

$$u_{J-1}(r) \to A_1 \exp\{-i[kr - \tfrac{1}{2}(J-1)\pi]\} - B_1 \exp\{i[kr - \tfrac{1}{2}(J-1)\pi]\} \tag{H31}$$

$l = J + 1$:

$$w_{J+1}(r) \to A_2 \exp\{-i[kr - \tfrac{1}{2}(J+1)\pi]\}\ B_2 \exp\{i[kr - \tfrac{1}{2}(J+1)\pi]\}$$

and therefore have ingoing and outgoing waves. Let the wave function with total angular momentum $J$ and parity $(-1)^{J+1}$ be written, in the notation of (H18),

$$\Psi = \frac{1}{r} u_{J-1}(r)\, F_{J,M,J-1}(\vartheta, \varphi, \sigma) + \frac{1}{r} w_{J+1}(r)\, F_{J,M,J+1}(\vartheta, \varphi, \sigma). \tag{H31a}$$

In a generalization of the scattering matrix $S$ of Chapter 1 (A34-5), one has the following relations

$$\begin{aligned} B_1 &= S_{11}A_1 + S_{12}A_2, \\ B_2 &= S_{21}A_1 + S_{22}A_2, \end{aligned} \tag{H32}$$

or in matrix notation, with $B$, $A$ as column matrices,

$$B = SA. \tag{H32a}$$

The $S$ matrix can be diagonalized by a unitary transformation $U$,

$$USU^{-1} = \begin{pmatrix} S_1 & 0 \\ 0 & S_2 \end{pmatrix}. \tag{H33}$$

With this transformation,

$$UB = USA = \begin{pmatrix} S_1 & 0 \\ 0 & S_2 \end{pmatrix} UA. \tag{H34}$$

The eigen-phase shifts $\delta_\alpha$, $\delta_\beta$ are defined by the eigenvalues of the $S$ matrix (see A35)

$$\exp(2i\delta_\alpha) = S_1, \qquad \exp(2i\delta_\beta) = S_2. \tag{H35}$$

On account of the unitary property of the $S$ matrix (see Chapter 6, Sect. W), $\delta_\alpha$, $\delta_\beta$ are real numbers. The two eigenvectors $UA$ are given by

$$UA_\alpha = \begin{pmatrix}1\\0\end{pmatrix}, \qquad UA_\beta = \begin{pmatrix}0\\1\end{pmatrix}. \tag{H36}$$

Since the $S$ matrix is symmetric,# the number of independent parameters contained in $S_{ij}$ is only $3[=2(2+1)/2]$. Two of these are $\delta_\alpha$ and $\delta_\beta$. Hence the unitary matrix $U$ must be expressible by one real number, say $\varepsilon$,

$$U = \begin{pmatrix}\cos\varepsilon & \sin\varepsilon\\ -\sin\varepsilon & \cos\varepsilon\end{pmatrix}. \tag{H37}$$

From (H36), we have

$$\begin{pmatrix}A_{1\alpha}\\A_{2\alpha}\end{pmatrix} \equiv A_\alpha = U^{-1}\begin{pmatrix}1\\0\end{pmatrix} = \begin{pmatrix}\cos\varepsilon\\ \sin\varepsilon\end{pmatrix}, \qquad \begin{pmatrix}A_{1\beta}\\A_{2\beta}\end{pmatrix} \equiv A_\beta = \begin{pmatrix}-\sin\varepsilon\\ \cos\varepsilon\end{pmatrix} \tag{H38}$$

and from (H32a), (H35)

$$\begin{aligned} B_\alpha &= U^{-1}\begin{pmatrix}\exp(2i\delta_\alpha)\\0\end{pmatrix} = \begin{pmatrix}\exp(2i\delta_\alpha)\cos\varepsilon\\ \exp(2i\delta_\alpha)\sin\varepsilon\end{pmatrix}\\ B_\beta &= U^{-1}\begin{pmatrix}0\\ \exp(2i\delta_\beta)\end{pmatrix} = \begin{pmatrix}-\exp(2i\delta_\beta)\sin\varepsilon\\ \exp(2i\delta_\beta)\cos\varepsilon\end{pmatrix}. \end{aligned} \tag{H39}$$

We now have two independent solutions of the form (H31), which are the eigenvectors of the $S$ matrix. The asymptotic forms of these are given by

$$\begin{aligned} u_\alpha(r) &\to \cos\varepsilon\sin\left(kr - \frac{(J-1)\pi}{2} + \delta_\alpha\right),\\ w_\alpha(r) &\to \sin\varepsilon\sin\left(kr - \frac{(J+1)\pi}{2} + \delta_\alpha\right), \end{aligned} \tag{H40}$$

$$\begin{aligned} u_\beta(r) &\to -\sin\varepsilon\sin\left(kr - \frac{(J-1)\pi}{2} + \delta_\beta\right),\\ w_\beta(r) &\to \cos\varepsilon\sin\left(kr - \frac{(J+1)\pi}{2} + \delta_\beta\right). \end{aligned} \tag{H40a}$$

To sum up, in order to find the $\alpha$- and $\beta$-solution, one first starts with an arbitrary set of two independent solutions $u$, $w$ of (H15) satisfying the initial condition that they vanish at $r = 0$. The $\alpha$ and $\beta$ solutions are then so constructed by forming two independent linear combinations of the initial two solutions that the resulting combinations have the same phase shifts in $u$ and

# See Chapter 6 for general discussion of $S$ matrix. In the present case, it will be seen below that the boundary condition is independent of $M$, and Equations (H15) do not involve $M$ because of the rotational invariance of the Hamiltonian. Therefore the scattering is really independent of $M$. This makes the scattering matrix symmetric.

$w$. In the low-energy limit, these two $\alpha$, $\beta$ solutions behave as if they were the partial waves corresponding to $l = J - 1$ and $l = J + 1$ [we called the solution the $\alpha$ solution whose $w$ component has the smaller value as $k^2 \to 0$]. In this limit, $\varepsilon$ is proportional to the energy.

Since the differential cross section is expressible in terms of the $S$ matrix alone, and since the eigen-phase shifts $\delta_\alpha$, $\delta_\beta$ and the "mixing" parameter $\varepsilon$ specify the $S$ matrix completely, the cross section for the triplet system (for which tensor force is operative) can be expressed in terms of $\delta_{J\alpha}$, $\delta_{J\beta}$, $\varepsilon_J$. The expression, however, is not simple.

On using (H32a), (H38) and (H39) we get the expression for $S$ matrix elements

$$S = \begin{pmatrix} \cos^2 \varepsilon \exp(2i\delta_\alpha) + \sin^2 \varepsilon \exp(2i\delta_\beta), & \tfrac{1}{2} \sin(2\varepsilon) [\exp(2i\delta_\alpha) - \exp(2i\delta_\beta)] \\ \tfrac{1}{2} \sin(2\varepsilon) [\exp(2i\delta_\alpha) - \exp(2i\delta_\beta)], & \sin^2 \varepsilon \exp(2i\delta_\alpha) + \cos^2 \varepsilon \exp(2i\delta_\beta) \end{pmatrix}. \quad \text{(H41)}$$

In the recent literature another way of parametrization is sometimes used besides (H41). This consists in using three real parameters, $\bar{\delta}_{J-1,J'}$, $\bar{\delta}_{J+1,J}$, $\bar{\varepsilon}_J$ instead of $\delta_{\alpha J}$, $\delta_{\beta J}$ and $\varepsilon_J$.

$$S = \begin{pmatrix} \exp(i\bar{\delta}_{J-1}), & 0 \\ 0, & \exp(i\bar{\delta}_{J+1}) \end{pmatrix} \begin{pmatrix} \cos 2\bar{\varepsilon}, & i \sin 2\bar{\varepsilon} \\ i \sin 2\bar{\varepsilon}, & \cos 2\bar{\varepsilon} \end{pmatrix} \begin{pmatrix} \exp(i\bar{\delta}_{J-1}), & 0 \\ 0, & \exp(i\bar{\delta}_{J+1}) \end{pmatrix}$$

$$= \begin{pmatrix} \cos 2\bar{\varepsilon} \exp(2i\bar{\delta}_{J-1}), & i \sin 2\bar{\varepsilon} \exp[i(\bar{\delta}_{J+1} + \bar{\delta}_{J-1})] \\ i \sin 2\bar{\varepsilon} \exp[i(\bar{\delta}_{J+1} + \bar{\delta}_{J-1})], & \cos 2\bar{\varepsilon} \exp(2i\bar{\delta}_{J+1}) \end{pmatrix}.$$

The "bar" quantities are introduced by Stapp (ref. 5) and are convenient when Coulomb effects are discussed.

We shall merely write down the differential cross section formula for proton–proton scattering by neglecting the nuclear phase shifts higher than $D$-waves. The coupling between $^3P_2$ and $^3F_2$ is also omitted.

$$d\sigma_{pp}(\vartheta) = d\sigma_{\text{Mott}} + d^1\sigma_N + d^1\sigma_{\text{int}} + d^3\sigma_N + d^3\sigma_{\text{int}},$$

where 1 and 3 denote the singlet and triplet states, $N$ and int denote, respectively, the nuclear scattering and the interference between the nuclear and the Coulomb scattering amplitudes.

$$d\sigma_{\text{Mott}} = k^{-2}(\eta/2)^2 \{\operatorname{cosec}^4(\vartheta/2) + \sec^4(\vartheta/2) - \operatorname{cosec}^2(\vartheta/2)\sec^2(\vartheta/2) \times \cos[2\eta \ln \tan(\vartheta/2)]\}\, d\Omega, \qquad \eta = e^2/\hbar v,$$

$$d^1\sigma_N = k^{-2}\{(\sin {}^1\delta_0)^2 + 25(\sin {}^1\delta_2)^2[P_2(\cos\vartheta)]^2 + 10 \sin {}^1\delta_0 \sin {}^1\delta_2 \cos[{}^1\delta_0 - {}^1\delta_2 + 2(\sigma_0 - \sigma_2)]\, P_2(\cos\vartheta)\}\, d\Omega,$$

$$d^1\sigma_{\text{int}} = k^{-2}(\eta/2)\{[\operatorname{cosec}^2(\vartheta/2)\cos\phi_0^s + \sec^2(\vartheta/2)\cos\phi_0^c]\sin {}^1\delta_0 + 5[\operatorname{cosec}^2(\vartheta/2)\cos\phi_2^s + \sec^2(\vartheta/2)\cos\phi_2^c]\sin {}^1\delta_2\, P_2(\cos\vartheta)\}\, d\Omega,$$

where

$$\phi_l^s = 2\eta \ln \sin(\vartheta/2) + 2(\sigma_l - \sigma_0) + {}^1\delta_l,$$
$$\phi_l^c = 2\eta \ln \cos(\vartheta/2) + 2(\sigma_l - \sigma_0) + {}^1\delta_l,$$
$$\sigma_l = \arg \Gamma\,(l+1+i\eta) \text{ as in (A49b)}$$

and also, with the notation ${}^{2s+1}\delta_J^l$,

$$\begin{aligned} d^3\sigma_N = k^{-2}\{&[(\sin {}^3\delta_0^1)^2 + (9/4)(\sin {}^3\delta_1^1)^2 + (13/4)(\sin {}^3\delta_2^1)^2 \\ &- 2 \sin {}^3\delta_0^1 \sin {}^3\delta_2^1 \cos({}^3\delta_0^1 - {}^3\delta_2^1) \\ &- (9/2) \sin {}^3\delta_1^1 \sin {}^3\delta_2^1 \cos({}^3\delta_1^1 - {}^3\delta_2^1)] \\ &+ [(9/4)(\sin {}^3\delta_1^1)^2 + (21/4)(\sin {}^3\delta_2^1)^2 \\ &+ 6 \sin {}^3\delta_0^1 \sin {}^3\delta_2^1 \cos({}^3\delta_0^1 - {}^3\delta_2^1) \\ &+ (27/2) \sin {}^3\delta_1^1 \sin {}^3\delta_2^1 \cos({}^3\delta_1^1 - {}^3\delta_2^1)] \cos^2\vartheta\}\, d\Omega, \end{aligned}$$

$$\begin{aligned} d^3\sigma_{\text{int}} = -k^{-2}(\eta/2)\cos\vartheta \sum_{J=0,1,2} \{&[\operatorname{cosec}^2(\vartheta/2)\cos\phi_{1J}^s \\ &- \sec^2(\vartheta/2)\cos\phi_{1J}^c](2J+1)\sin {}^3\delta_J^1\}\, d\Omega, \end{aligned} \quad \text{(H42)}$$

where

$$\phi_{1J}^s = 2\eta \ln \sin(\vartheta/2) + 2(\sigma_1 - \sigma_0) + {}^3\delta_J^1,$$
$$\phi_{1J}^c = 2\eta \ln \cos(\vartheta/2) + 2(\sigma_1 - \sigma_0) + {}^3\delta_J^1.$$

For neutron–proton scattering the formula is more complicated, and the reader is referred to refs. 4 and 5.

## REFERENCES


1. Rarita, W. and Schwinger, J., Phys. Rev. **59**, 436 (1941); probably treat the scattering by tensor potential for the first time.
2. Ashkin, J. and Wu, T. Y., Phys. Rev. **73**, 973 (1948); treat the tensor force scattering by the method of partial waves and also in Born approximation. Subsection 2 is based on this work.
3. Condon, E. U. and Shortley, G. H., *Theory of Atomic Spectra.* Cambridge University Press (1935).

**4a.** Blatt, J. M. and Biedenharn, L. C., Phys. Rev. **86**, 399 (1952); Real phase shift representation for tensor forces. Also **4b.** Rohrlich, F. and Eisenstein, J., Phys. Rev. **75**, 705 (1949).

5. Stapp, H. P., Phys. Rev., **103**, 425 (1956); Stapp, H. P. *et al.*, Phys. Rev. **105**, 302 (1957).
6. Matsumoto, M., Progr. Theor. Phys. **15**, 329 (1955). The misprints in this paper as well as ref. 4b are corrected in Taketani *et al.*, Prog. Theor. Phys. Supplement **3**, p. 92 (1956). For general formulas (including *p–p*), see Breit, G. *et al.*, Phys. Rev. **97**, 1047, 1051 (1955); Wright, S. C., Phys. Rev. **99**, 996 (1955); Stapp, H. P., Ypsilantis, T. and Metropolis, N., Phys. Rev. **105**, 302 (1957).

section I

# Scattering by Tensor and L·S Fields: Born Approximation

The exact treatment of the preceding section can be replaced by the Born approximation for high energies ($>200$ MeV in the case of nucleon–nucleon scattering). One way to obtain the Born approximation is to start from the exact expressions (H27), (H29) for the scattering amplitudes and to replace the exact phase shifts therein by their Born approximations. This procedure is parallel to that given in Sect. C.1 for the case of central fields. Another way is to obtain the scattering amplitudes with plane wave approximations in a manner similar to that given in Sect. C.2, without going through the phase shifts. It seems instructive to give both procedures.

## 1. BORN APPROXIMATION FROM THE EXACT THEORY

Since the tensor and the **L·S** forces do not operate on singlet states, the scattering amplitudes for singlet states are given by the central field alone [see (H16), (C8-12), (H13)]

$$
\begin{aligned}
{}^1f^+(\vartheta) &= \int_0^\infty U_c^+(r)\,\frac{\sin qr}{q}\,r\,dr, \\
{}^1f^-(\vartheta) &= \int_0^\infty U_c^-(r)\,\frac{\sin qr}{q}\,r\,dr,
\end{aligned}
\qquad \text{(I1)}
$$

where

$$q = 2k \sin\frac{\vartheta}{2}.$$

For the triplet state scattering, the phase shifts must be obtained by solving the radial wave equations (H14-15) for $v_L(r)$, $u_L(r)$ and $w_L(r)$. For the Born approximation, we shall assume

$$\begin{aligned} v_L(r) &= f_L(r)[1 + \varphi_L(r)], \\ u_L(r) &= g_L(r)[1 + \psi_L(r)], \\ w_L(r) &= h_L(r)[1 + \chi_L(r)], \end{aligned} \tag{I2}$$

where

$$\varphi_L \ll 1, \qquad \psi_L \ll 1, \qquad \chi_L \ll 1. \tag{I3}$$

Corresponding to the asymptotic forms (H21) for $v_L(r)$, $u_L(r)$, $w_L(r)$, we must take the asymptotic forms of the initial approximation to be

$$\begin{aligned} f_L(r) &\to A_L kr\, j_L(kr), \\ g_L(r) &\to B_L kr\, j_L(kr), \\ h_L(r) &\to C_L kr\, j_L(kr), \end{aligned} \tag{I4}$$

where $j_L(kr)$ is defined in (A11). The usual perturbation method of finding $\varphi_L(r)$, $\psi_L(r)$, $\chi_L(r)$ and meeting the required asymptotic forms gives, corresponding to (H14), (H15) for odd $J$#

$$\begin{aligned} \delta_L &= k \int_0^\infty (U_c^- + 2U_T^-)[j_L(kr)]^2 r^2\, dr \\ \varepsilon_L &= k \int_0^\infty \Big[\Big(U_c^+ - \frac{2L}{2L+3} U_T^+\Big) j_L(kr) \\ &\qquad + 6\frac{\sqrt{(L+1)(L+2)}}{2L+3} \frac{C_{L+2}}{B_L} U_T^+ j_{L+2}(kr)\Big] j_L(kr)\, r^2\, dr, \\ \zeta_L &= k \int_0^\infty \Big[\Big(U_c^+ - \frac{2(L+1)}{2L-1} U_T^+\Big) j_L(kr) \\ &\qquad + 6\frac{\sqrt{L(L-1)}}{2L-1} \frac{B_{L-2}}{C_L} U_T^+ j_{L-2}(kr)\Big] j_L(kr)\, r^2\, dr, \end{aligned} \tag{I5}$$

where the ratios $C_{L+2}/B_L$, $B_{L-2}/C_L$ have the values given in (H26) and (H28), namely,

$$\begin{aligned} M = \pm 1, \quad C_{L+2}/B_L &= -\left(\frac{L+1}{L+2}\right)^{1/2} \exp i\,(\zeta_{L+2} - \varepsilon_L), \\ M = 0, \qquad\qquad &= \left(\frac{L+2}{L+1}\right)^{1/2} \exp i(\zeta_{L+2} - \varepsilon_L). \end{aligned} \tag{I6}$$

Substitution of these into the expressions for $\varepsilon_L$, $\zeta_L$ gives two simultaneous equations for the determination of $\varepsilon_L$ and $\zeta_L$. It is seen that in general, the phase shifts $\varepsilon_L$, $\zeta_L$ are complex.

---

# For simplicity in writing, we shall leave out the spin-orbit $U_{so}^+(r)$, $U_{so}^-(r)$ in the following.

In the approximation (I3), the phase shifts are assumed to be small so that, for consistency, we shall replace the factors $\exp i(\zeta_{L+2} - \varepsilon_L)$ in (I6) by 1. Thus we obtain, for $M = 0$,

$$\varepsilon_L = k\int_0^\infty \left[\left(U_c^+ - \frac{2L}{2L+3}U_T^+\right)j_L(kr) + 6\frac{L+2}{2L+3}U_T^+\,j_{L+2}(kr)\right]j_L(kr)\,r^2\,dr,$$

$$\zeta_L = k\int_0^\infty \left[\left(U_c^+ - \frac{2L+2}{2L-1}U_T^+\right)j_L(kr) + 6\frac{L-1}{2L-1}U_T^+\,j_{L-2}(kr)\right]j_L(kr)\,r^2\,dr, \tag{I7}$$

and for $M = \pm 1$,

$$\varepsilon_L = k\int_0^\infty \left[\left(U_c^+ - \frac{2L}{2L+3}U_T^+\right)j_L(kr - 6\frac{L+1}{2L+3}U_T^+\,j_{L+2}(kr)\right]j_L(kr)\,r^2\,dr,$$

$$\zeta_L = k\int_0^\infty \left[\left(U_c^+ - \frac{2L+2}{2L-1}U_T^+\right)j_L(kr) - 6\frac{L}{2L-1}U_T^+\,j_{L-2}(kr)\right]j_L(kr)\,r^2\,dr. \tag{I8}$$

These are the expressions corresponding to the Born approximation for the phase shift $\delta_l$ for central fields in (C8a).

It can now be shown that, with these approximate formulas for the phase shifts, the series expressions (H27), (H29) for the scattering amplitudes can be summed in closed forms. To illustrate this, take the amplitude for $M = 0$, $M_s = -1$, i.e., the initial spin state $M_s^0 = 0$ is changed by the tensor interaction into the spin state $M_s = -1$ after the scattering.

$$2ikf^{M=0}_{M_s=-1}(\vartheta, \varphi, \sigma) = 2i\frac{e^{i\varphi}}{\sqrt{2\pi}}\frac{1}{\sqrt{2}}\sum_{L=1}(\varepsilon_L - \zeta_L)\,P_L^1(\cos\vartheta)\,\chi^{-1}$$

$$= 12i\frac{e^{i\varphi}}{\sqrt{2\pi}}\frac{1}{\sqrt{2}}\chi^{-1}\sum_{L=1}\int_0^\infty\left[-\frac{(L-1)(2L+1)}{(kr)}j_L(kr)\right.$$

$$\left. + (2L+1)\,j_{L+1}(kr)\right] \times j_L(kr)\,U_T^+(r)\,r\,dr\cdot P_L^1(\cos\vartheta).$$

On differentiating with respect to $x$ the following expression

$$\frac{\sin Q}{Q} = \sum_{L=0}(2L+1)\,P_L(x)[j_L(kr)]^2,$$

where

$$Q = 2kr\sin\frac{\vartheta}{2}, \qquad x = \cos\vartheta,$$

one obtains

$$-\left(\cos Q - \frac{\sin Q}{Q}\right)\left(\frac{1+x}{1-x}\right)^{1/2} = 2\sum_{L=1}(2L+1)\,P_L^1(x)[j_L(kr)]^2.$$

Differentiating this expression with respect to $(kr)$, one gets

$$-\cos\frac{\vartheta}{2}\left[-\sin Q - \frac{\cos Q}{Q} + \frac{\sin Q}{Q^2}\right]$$

$$= 2\sum_{L=1}(2L+1)\,P_L^1(x)\,j_L(kr)\left[\frac{L}{kr}j_L(kr) - j_{L+1}(kr)\right].$$

With these expressions, we obtain

$$f_{M_s=-1}^{M=0}(\vartheta,\varphi,\sigma) = \frac{e^{i\varphi}}{\sqrt{2\pi}}\,3\sqrt{\pi}\,\chi^{-1}k^2\sin\vartheta(1-\cos\vartheta)\int_0^\infty \frac{J_{5/2}(Q)}{Q^{5/2}}\,U_T^+(r)\,r^4\,dr. \tag{I9}$$

Entirely similar calculations give the other scattering amplitudes for $M = 0$, $M_s = 0, +1$.

## 2. BORN APPROXIMATION FROM PLANE WAVES

For the scattering by the potential (H13), the singlet state scattering amplitudes are given in (I1).

For the triplet state scattering, in order to describe the dependence of the scattering amplitude on the parity through the different $U^{\text{even}}$ and $U^{\text{odd}}$ and to make the result immediately applicable to the cases of proton–proton and of proton–neutron scattering, it is convenient to introduce the isobaric spin $\mathbf{T}(T_x, T_y, T_z)$ and the isobaric spin wave functions $\tau_1, \tau_2$. In this formalism, the operators $T_x$, $T_y$, $T_z$ for each nucleon are represented by matrices similar to the spin operators $\sigma_i$ in (H3), and the proton, neutron states, represented by $\tau^+$, $\tau^-$, are the two eigenstates of $T_z$. For the two-nucleon system, the symmetric and antisymmetric states are as follows:

| | $p, p$ | $p, n$ | $n, n$ |
|---|---|---|---|
| $\tau_s$ | $\tau_1^+\tau_2^+$ | $\frac{1}{\sqrt{2}}(\tau_1^+\tau_2^- + \tau_2^+\tau_1^-)$ | $\tau_1^-\tau_2^-$ |
| $\tau_a$ | | $\frac{1}{\sqrt{2}}(\tau_1^+\tau_2^- - \tau_2^+\tau_1^-)$ | |

The operator $P_H$ defined by

$$-P_H = \tfrac{1}{2}(1 + \mathbf{T}_1\cdot\mathbf{T}_2)$$

and known as the Heisenberg operator, has the following eigenvalues

$$(-P_H)\,\tau_s = \tau_s, \qquad (-P_H)\,\tau_a = -\tau_a.$$

To represent the fact that $U^{\text{even}}$, $U^{\text{odd}}$ in (H13) are the even, odd state interactions respectively, we may associate an operator $\frac{1}{4}(1 - \mathbf{T}_1\cdot\mathbf{T}_2)$ with

${}^3U^{\text{even}}$, and operator $\frac{1}{4}(3 + \mathbf{T}_1\cdot\mathbf{T}_2)$ with ${}^3U^{\text{odd}}$, since these operators have the eigenvalues

$$\begin{aligned} \tfrac{1}{4}(1 - \mathbf{T}_1\cdot\mathbf{T}_2)\begin{Bmatrix}\tau_s\\ \tau_a\end{Bmatrix} &= \begin{Bmatrix}0\tau_s\\ 1\tau_a\end{Bmatrix}, \\ \tfrac{1}{4}(3 + \mathbf{T}_1\cdot\mathbf{T}_2)\begin{Bmatrix}\tau_s\\ \tau_a\end{Bmatrix} &= \begin{Bmatrix}1\tau_s\\ 0\tau_a\end{Bmatrix}, \end{aligned} \tag{I13}$$

and the isobaric spin wave function for triplet, even and triplet, odd state must be $\tau_a, \tau_s$ respectively according to the Pauli principle. For the $p, p$ and the $n, n$ system, only the triplet, odd and the singlet, even states are allowed so that, of (H13), only ${}^3U^{\text{odd}}$ contributes to the scattering.

For a two-nucleon system, with a given $z$ component of total angular momentum $M$ $(=M_L + M_S)$, we may write the wave functions of even and odd parities in the form

$$\Psi_+ = \sum \phi^+_{M_L}(\mathbf{r})\,\chi^{M_s}\tau_a, \qquad \Psi_- = \sum \phi^-_{M_L}(\mathbf{r})\,\chi^{M_s}\tau_s, \tag{I14}$$

where the summations are taken over all $M_L$, $M_S$ for the same $M$, and $\phi^+$, $\phi^-$ are the spatial wave functions of even and odd parity respectively. These may be expressed as

$$\begin{aligned} \phi^+(\mathbf{r}) &= \psi(\mathbf{r}_1, \mathbf{r}_2) + \psi(\mathbf{r}_2, \mathbf{r}_1), \\ \phi^-(\mathbf{r}) &= \psi(\mathbf{r}_1, \mathbf{r}_2) - \psi(\mathbf{r}_2, \mathbf{r}_1), \end{aligned} \tag{I15}$$

$\psi(\mathbf{r}_1, \mathbf{r}_2)$ being neither even nor odd (for example, a plane wave in the relative coordinate $\mathbf{r}$).

The Schrödinger equations are, from (I14) and (H13),

$$\begin{aligned} (\Delta + k^2)\left(\sum \phi^+\,\chi^{M_s}\tau_a\right) &= {}^3U^{\text{even}}\,\Psi_+, \\ (\Delta + k^2)\left(\sum \phi^-\,\chi^{M_s}\tau_s\right) &= {}^3U^{\text{odd}}\,\Psi_-. \end{aligned} \tag{I16}$$

Multiplying these two equations by $\overset{*}{\chi}{}^{M_s'}\tau_a^*$, $\overset{*}{\chi}{}^{M_s'}\tau_s^*$ respectively and integrating over the spin and isobaric spin coordinates, we get, on using (I15),

$$(\Delta + k^2)\,\psi(\mathbf{r}_1, \mathbf{r}_2) = A\tfrac{1}{2}[\psi(\mathbf{r}_1, \mathbf{r}_2) + \psi(\mathbf{r}_2, \mathbf{r}_1)] + B\tfrac{1}{2}[\psi(\mathbf{r}_1, \mathbf{r}_2) - \psi(\mathbf{r}_2, \mathbf{r}_1)], \tag{I17}$$

where

$$A = -U_c(r)\,\delta_{M_s', M_s} - U_T^+(r)\langle M_s'|T|M_s\rangle - U_{so}^+(r)\langle M_s'|\mathbf{L}\cdot\mathbf{S}|M_s\rangle, \tag{I18}$$

and $B$ given by a similar expression with $+$ replaced by $-$. The matrix elements of $T$ and $\mathbf{L}\cdot\mathbf{S}$ have been given in (H8) and (H9).

Let us choose the direction of the incident $\mathbf{k}$ as the direction of spatial quantization, i.e., initially,

$$M_S^0 = M, \qquad M_L^0 = 0. \tag{I19}$$

For a given $M\,(=1, 0, -1)$, the scattering amplitudes for a given $M_S\,(=1, 0, -1)$ are, from (I17),

$$f^M_{M_S} = \tfrac{1}{2}[f^M_{M_S}(k, \vartheta)^+ + f^M_{M_S}(-k, \vartheta)^+ + f^M_{M_S}(k, \vartheta)^- - f^M_{M_S}(-k, \vartheta)^-] \times \frac{1}{\sqrt{2\pi}} \exp\,[i(M - M_S)\varphi]\,\chi^{M_S}, \quad \text{(I20)}$$

where the first two terms are from the even, and the last two terms from the odd states. All the above expressions are still exact.

Let us now make the plane wave approximation by putting, on the right-hand side of (I17),

$$f(\mathbf{r}_1, \mathbf{r}_2) = e^{i\mathbf{k}\cdot\mathbf{r}}, \qquad f(\mathbf{r}_2, \mathbf{r}_1) = e^{-i\mathbf{k}\cdot\mathbf{r}}. \quad \text{(I21)}$$

Then, with (H8), (H9), we obtain

$$\begin{aligned}
f^1_1(k, \vartheta)^+ &= -\frac{1}{4\pi}\int e^{-i\mathbf{k}'\cdot\mathbf{r}}\Big[{}^3U^+_c(r) + 2\sqrt{\tfrac{2}{5}}\,Y_{2,0}(\xi, \eta)\,{}^3U^+_T(r) + {}^3U^+_{so}(r)L_z\Big]\,e^{i\mathbf{k}\cdot\mathbf{r}}\,d\mathbf{r},\\
f^1_0(k, \vartheta)^+ &= -\frac{1}{4\pi}\int e^{-i\mathbf{k}'\cdot\mathbf{r}}\Big[-2\sqrt{\tfrac{6}{5}}\,Y_{2,1}(\xi, \eta)\,{}^3U^+_T(r) + \frac{1}{\sqrt{2}}\,{}^3U^+_{so}(r)(L_x - iL_y)\Big]\,e^{i\mathbf{k}\cdot\mathbf{r}}\,d\mathbf{r}, \qquad \text{(I22)}\\
f^1_{-1}(k, \vartheta)^+ &= -\frac{1}{4\pi}\int e^{-i\mathbf{k}'\cdot\mathbf{r}}\,4\sqrt{\tfrac{3}{5}}\,Y_{2,2}(\xi, \eta)\,{}^3U^+_T(r)\,e^{i\mathbf{k}\cdot\mathbf{r}}\,d\mathbf{r},
\end{aligned}$$

with the corresponding $f^1_{M_S}(-k, \vartheta)^+$ obtained by replacing $e^{i\mathbf{k}\cdot\mathbf{r}}$ in (I22) by $e^{-i\mathbf{k}\cdot\mathbf{r}}$, the $f^1_{M_S}(k, \vartheta)^-$ by replacing the $U^+$ in (I22) by $U^-$, etc. $\vartheta$ is the scattering angle between $\mathbf{k}'$ and $\mathbf{k}$. $\xi$, $\eta$ are the polar angles of $\mathbf{r}$.

For $M = 0$, similarly,

$$\begin{aligned}
f^0_1(k, \vartheta)^+ &= -\frac{1}{4\pi}\int e^{-i\mathbf{k}'\cdot\mathbf{r}}\Big[2\sqrt{\tfrac{6}{5}}\,{}^3U^+_T(r)\,Y_{2,-1}(\xi, \eta) + \frac{1}{\sqrt{2}}\,{}^3U^+_{so}(r)(L_x + iL_y)\Big]\,e^{i\mathbf{k}\cdot\mathbf{r}}\,d\mathbf{r},\\
f^0_0(k, \vartheta)^+ &= -\frac{1}{4\pi}\int e^{-i\mathbf{k}'\cdot\mathbf{r}}\Big[{}^3U^+_c(r) - 4\sqrt{\tfrac{2}{5}}\,{}^3U^+_T(r)\,Y_{2,0}(\xi, \eta)\Big]\,e^{i\mathbf{k}\cdot\mathbf{r}}\,d\mathbf{r}, \qquad \text{(I23)}\\
f^0_{-1}(k, \vartheta)^+ &= -\frac{1}{4\pi}\int e^{-i\mathbf{k}'\cdot\mathbf{r}}\Big[2\sqrt{\tfrac{6}{5}}\,Y_{2,1}(\xi, \eta)\,{}^3U^+_T(r) + \frac{1}{\sqrt{2}}\,{}^3U^+_{so}(r)(L_x - iL_y)\Big]\,e^{i\mathbf{k}\cdot\mathbf{r}}\,d\mathbf{r},
\end{aligned}$$

with $f^0_{M_s}(k, \vartheta)^-$ given by (I23) with $U^+$ replaced by $U^-$, etc.

For $M = -1$, we have similarly

$$f^{-1}_{1}(k, \vartheta)^{+} = -\frac{1}{4\pi}\int e^{-i\mathbf{k}'\cdot\mathbf{r}}\, 4\sqrt{\frac{3}{5}}\, Y_{2,-2}(\xi, \eta)\, {}^{3}U_{T}^{+}(r)\, e^{i\mathbf{k}\cdot\mathbf{r}}\, d\mathbf{r},$$

$$f^{-1}_{0}(k, \vartheta)^{+} = -\frac{1}{4\pi}\int e^{-i\mathbf{k}'\cdot\mathbf{r}}\Big[-2\sqrt{\frac{6}{5}}\, Y_{2,-1}(\xi, \eta)\, {}^{3}U_{T}^{+}(r) + \frac{1}{\sqrt{2}}\, {}^{3}U_{so}^{+}(r)(L_x + iL_y)\Big]\, e^{i\mathbf{k}\cdot\mathbf{r}}\, d\mathbf{r}, \tag{I24}$$

$$f^{-1}_{-1}(k, \vartheta)^{+} = -\frac{1}{4\pi}\int e^{-i\mathbf{k}'\cdot\mathbf{r}}\Big[{}^{3}U_{c}^{+}(r) + 2\sqrt{\frac{2}{5}}\, Y_{2,0}(\xi, \eta)\, {}^{3}U_{T}^{+}(r) - {}^{3}U_{so}^{+}(r)\, L_z\Big]\, e^{i\mathbf{k}\cdot\mathbf{r}}\, d\mathbf{r},$$

with $f^{-1}_{M_S}(-k, \vartheta)^{+}$ given by replacing $e^{i\mathbf{k}\cdot\mathbf{r}}$ by $e^{-i\mathbf{k}\cdot\mathbf{r}}$ in (I24), etc.

The 36 integrals in (I22-24) are not all different. We may choose $\mathbf{k}$ as the polar axis of $\mathbf{r}(\xi, \eta)$ so that

$$\mathbf{k}\cdot\mathbf{r} = kr \cos \xi,$$
$$\mathbf{k}'\cdot\mathbf{r} = kr\,(\cos \xi \cos \vartheta + \sin \xi \sin \vartheta \cos \eta).$$

The integration over $\eta$ leads to Bessel functions

$$\int_0^{2\pi} \exp(-iz \cos \eta)\, d\eta = 2\pi J_0(z),$$

$$\int_0^{2\pi} \exp[i(\eta - z \cos \eta)]\, d\eta = -2\pi i J_1(z), \tag{I25}$$

$$\int_0^{2\pi} \exp[i(2\eta - z \cos \eta)]\, d\eta = -2\pi J_2(z).$$

The integrals over $\xi$ are the Sonine and Gegenbauer finite integrals (ref. 2). Thus for example,

$$\int_0^{\pi} \exp(\pm ix \cos \xi)\, J_0(y \sin \xi) \sin \xi\, d\xi = \sqrt{\frac{2\pi}{z}}\, J_{1/2}(z), \quad z^2 = x^2 + y^2.$$

To illustrate the possible simplifications of the results in (I22-24), let us leave out the $\mathbf{L}\cdot\mathbf{S}$ interaction and calculate the contribution $S^{M}_{M_s}(k, \vartheta)^{\pm}$ of the tensor interactions to the $f$ in (I22-24). On carrying out the integrations over the angles as indicated above, we obtain

$$S^{1}_{1}(k, \vartheta)^{\pm} = \sqrt{\frac{\pi}{2}}\int_0^{\infty}\Big[-\frac{J_{1/2}(Q)}{Q^{1/2}} + 3\frac{J_{3/2}(Q)}{Q^{3/2}} - \frac{3Q^4}{(2kr)^2}\frac{J_{5/2}(Q)}{Q^{5/2}}\Big]\, {}^{3}U_{T}^{\pm}(r)\, r^2\, dr,$$

$$S^{1}_{0}(k, \vartheta)^{\pm} = \frac{3}{2}\sqrt{\pi}\int_0^{\infty} \sin \vartheta\, \frac{J_{5/2}(Q)}{Q^{1/2}}\, {}^{3}U_{T}^{\pm}(r)\, r^2\, dr, \tag{I26}$$

$$S^{1}_{-1}(k, \vartheta)^{\pm} = -3\sqrt{\frac{\pi}{2}}\, k^2 \sin^2 \vartheta \int_0^{\infty} \frac{J_{5/2}(Q)}{Q^{5/2}}\, {}^{3}U_{T}^{\pm}(r)\, r^4\, dr,$$

where

$$Q = 2kr \sin \frac{\vartheta}{2}.$$

By means of (H8) and (I25), and the fact that the Bessel Function $J_n(z)$ is even or odd in $z$ according as $n$ is even or odd, we obtain the following relations

$$\begin{aligned} S_1^0(k, \vartheta)^\pm &= -S_0^1(k, \vartheta)^\pm, & S_1^{-1}(k, \vartheta)^\pm &= S_{-1}^1(k, \vartheta)^\pm, \\ S_0^0(k, \vartheta)^\pm &= -2S_1^1(k, \vartheta)^\pm, & S_0^{-1}(k, \vartheta)^\pm &= S_0^1(k, \vartheta)^\pm, \\ S_{-1}^0(k, \vartheta)^\pm &= -S_0^1(k, \vartheta)^\pm, & S_{-1}^{-1}(k, \vartheta)^\pm &= S_1^1(k, \vartheta)^\pm, \end{aligned} \tag{I27}$$

$$\left.\begin{aligned} S_1^1(-k, \vartheta)^\pm &= S_1^1(k, \vartheta)^\pm \\ S_0^1(-k, \vartheta)^\pm &= S_0^1(k, \vartheta)^\pm \\ S_{-1}^1(-k, \vartheta)^\pm &= S_{-1}^1(k, \vartheta)^\pm \end{aligned}\right\}, \quad \text{with } Q \text{ replaced by } 2kr \cos \frac{\vartheta}{2}, \tag{I28}$$

with (I27) again for $-k$ replacing $k$. Thus there are only three integrals, namely, those in (I26), as far as the tensor force part of the scattering amplitude is concerned. For the **L·S** part, one would obtain a similar simplification.

To compare this direct calculation of the amplitude with the use of plane waves with that from the exact theory (see preceding subsection), let us again take the tensor force part $S_{-1}^0(k, \vartheta)^\pm$ of $f_{-1}^0(k, \vartheta)^\pm$. By means of (I26) and (I27), we obtain exactly the same result as in (I9).

The differential cross section for the triplet state scattering is finally obtained from

$$\sigma(\xi) = \frac{1}{3} \sum_{M=-1}^{1} \left\{ \sum_{M_S} |f_{M_S}^M(\xi)|^2 \right\}. \tag{I29}$$

## REFERENCES


1. Ashkin, J. and Wu, T. Y., Phys. Rev. **73**, 973 (1948).
2. Watson, G. N., *A Treatise on the Theory of Bessel Functions*, Cambridge Univ. Press, 2nd ed., Chap. XII.
3. Rubinow, S. I., Phys. Rev. **98**, 183 (1955), variational principle for the scattering by tensor forces.
4. Biedenharn, L. C. and Blatt, J. M., Phys. Rev. **93**, 1387 (1954), variational principle based on the integral equation.

section J

# Polarization Effects

In Sects. H and I, the effects of spin-dependent interactions, such as the tensor and the $\mathbf{L}\cdot\mathbf{S}$, on the scattering cross sections have been treated. We shall show that, from measurements of the polarization (spin) of the scattered particles, additional useful information can be derived concerning the interaction of the particles.

In the case of electron scattering, the possibility of detecting the polarization, due to the small spin-orbit interaction (which is a relativistic effect), in a double-scattering experiment was first treated by Mott. Because of the smallness of the effect, successful experiments were done considerably much later. The possibility of polarizing nucleons by scattering from $He^4$ nucleus was suggested by Schwinger. Since then extensive experiments have been carried out on the scattering of nucleons from various atomic nuclei, especially nucleons. These have pointed to the presence of spin-dependent interactions, and in the case of nucleon–nucleon, the polarization data for high energy (mainly above 100 MeV) scatterings indicate the presence of an $\mathbf{L}\cdot\mathbf{S}$ interaction in addition to the spin- and parity-dependent central and tensor interactions. We shall in the following describe briefly the theory on which an analysis of the polarization effects can be based.

## 1. FORMAL DESCRIPTION OF POLARIZATION EFFECT

### *i) Single scattering of spin $\frac{1}{2}$ particles*

For simplicity, let us consider the scattering of a spin $\frac{1}{2}$ particle by a target of spinless particles. The asymptotic form of the scattered wave can be written

$$\Psi \to e^{ikz}\,\chi^{m_{si}} + \frac{e^{ikr}}{r}\sum_{m_{sf}} f_{if}(\vartheta, \varphi)\,\chi^{m_{sf}}. \tag{J1}$$

$\chi^{m_{si}}, \chi^{m_{sj}}$ ($m_{si}, m_{sj} = \pm\frac{1}{2}$) denote the spin state of the incident and scattered particle respectively. The scattering amplitudes $f_{ij}$ (i.e., $f_{m_{si},m_{sj}}$) are the matrix elements of a $2 \times 2$ matrix $f$ which can be expressed in terms of the Pauli matrices $\sigma_x, \sigma_y, \sigma_z$ of (H3) and the unit matrix **E**.

$$f = g\mathbf{E} + (\mathbf{h}\cdot\boldsymbol{\sigma}). \tag{J2}$$

It is clear that $g$ corresponds to the spin-independent and $\mathbf{h}$ to the spin-flip causing part of the interaction. (J2) can be put in a general, parity conserving form on the following consideration. For the scattering process, a relevant unit axial vector is

$$\mathbf{n} = \frac{1}{k^2 \sin\vartheta}[\mathbf{k}\times\mathbf{k}'], \tag{J3}$$

where $\mathbf{k}$, $\mathbf{k}'$ are the wave vector of the incident and elastically scattered wave (in the center of mass system) respectively, and $\vartheta$ is the angle between $\mathbf{k}$ and $\mathbf{k}'$. ($|\mathbf{n}| = 1$, $|\mathbf{k}| = |\mathbf{k}'|$). $f$ in (J2) can be expressed in the general form

$$f = g(\vartheta)\,\mathbf{E} + h(\vartheta)(\mathbf{n}\cdot\boldsymbol{\sigma}). \tag{J4}$$

Consider an incident particle whose spin state is $\chi^c$ representing a polarization in the direction of the unit vector $\mathbf{c}$. This $\chi^c$ can be expressed as

$$\chi^c = a\chi^{1/2} + b\chi^{-1/2} \tag{J5}$$

and

$$(\mathbf{c}\cdot\boldsymbol{\sigma})\,\chi^c = \chi^c,$$

where

$$a = \sqrt{\frac{1+c_z}{2}}, \qquad b = \frac{c_x + ic_y}{\sqrt{2(1+c_z)}}, \qquad |\mathbf{c}^2| = 1.$$

If for a moment we use the notation

$$\chi^{1/2} = \begin{pmatrix}1\\0\end{pmatrix}, \qquad \chi^{-1/2} = \begin{pmatrix}0\\1\end{pmatrix}, \qquad \chi^c = \begin{pmatrix}a\\b\end{pmatrix},$$

the scattered wave of the beam with $\chi^c$, if we choose $\mathbf{n}$ to be the direction of the $y$-axis (namely $\varphi = 0$), can be obtained from (J1) and (J4),

$$\Psi' \rightarrow e^{ikz}\begin{pmatrix}a\\b\end{pmatrix} + \frac{e^{ikr}}{r}\begin{pmatrix}f_c^+(\vartheta)\\f_c^-(\vartheta)\end{pmatrix}, \tag{J6}$$

where

$$f_c^+ = ag - ibh, \qquad f_c^- = bg + iah. \tag{J7}$$

The cross section is then

$$\begin{aligned}\sigma(\vartheta,\phi) &= |f_c^+|^2 + |f_c^-|^2\\ &= |g(\vartheta)|^2 + |h(\vartheta)|^2 + 4\,\mathrm{Re}\,(g^*h)\,\mathrm{Im}\,(a^*b)\\ &= |g(\vartheta)|^2 + |h(\vartheta)|^2 + 2c_y\,\mathrm{Re}\,(g^*h).\end{aligned} \tag{J8}$$

From our choice of **n** as the $y$-axis, this can be expressed in the form

$$\sigma(\vartheta, \phi) = (1 + P\,\mathbf{n}\cdot\mathbf{c})\, I(\vartheta) = \{1 + P(\vartheta)\cos\phi\}\, I(\vartheta), \tag{J9}$$

where

$$I(\vartheta) = |g(\vartheta)|^2 + |h(\vartheta)|^2,$$
$$P(\vartheta) = 2\,\frac{\mathrm{Re}\,(g^*h)}{I(\vartheta)}. \tag{J10}$$

Thus $\sigma(\vartheta, \phi)$ depends on $\vartheta$ (angle between **k** and **k**′) and the angle, denoted by $\phi$, between the direction **c** of the polarization of the incident particle, and the direction **n** in (J3). It is seen that the spin component normal to **n** contributes nothing to the scattering cross section.

For an unpolarized incident beam (such as usually produced by a high-energy accelerator), we can find the expectation value $\langle\boldsymbol{\sigma}\rangle$ of the spin of the scattered particle (in the direction **k**′) by averaging over the two cases $\chi_1^c = \chi^{1/2}$ and $\chi_2^c = \chi^{-1/2}$. Analogously to (J6), we have

$$f_1 = \begin{pmatrix} f_1^+(\vartheta) \\ f_1^-(\vartheta) \end{pmatrix} = \begin{pmatrix} g \\ ih \end{pmatrix}, \qquad f_2 = \begin{pmatrix} f_2^+ \\ f_2^- \end{pmatrix} = \begin{pmatrix} -ih \\ g \end{pmatrix}.$$

From

$$\langle\boldsymbol{\sigma}\rangle = (f_1^*\boldsymbol{\sigma}f_1 + f_2^*\boldsymbol{\sigma}f_2)/(f_1^*f_1 + f_2^*f_2),$$

one gets

$$\langle\sigma_x\rangle = 0, \qquad \langle\sigma_z\rangle = 0,$$

and

$$\langle\sigma_y\rangle = \langle\sigma_n\rangle = 2\,\frac{\mathrm{Re}\,(g^*h)}{|g|^2 + |h|^2} = P(\vartheta). \tag{J11}$$

Thus the scattered particle in the direction **k**′ has a non-vanishing expectation value of the spin only for the component along **n**, i.e., normal to **k** and **k**′, and the value is $P(\vartheta)$ of (J9). We may call $P$ the "polarizing factor". (J11) can be put in the form

$$\langle\boldsymbol{\sigma}\rangle = P\mathbf{n}, \tag{J11a}$$

where **n** is a unit vector $\perp$ to **k** and **k**′.

### *ii) Double-scattering*

Suppose an unpolarized beam in the direction **k** is scattered by a target A (of spinless particles) into the direction **k**′, the angle between **k** and **k**′ being $\vartheta_1$. This scattered beam is polarized in the direction $\mathbf{n}_1$ ($\mathbf{n}_1 \perp$ to the plane **k**, **k**′). Let this scattered beam fall on a second (similar) target B and let us measure the scattered intensities from B in directions **k**″ ($\vartheta_2$, $\phi_{12}$), where $\vartheta_2$ is the angle between **k**″ and **k**′ and $\phi_{12}$ the angle between $\mathbf{n}_1$ and $\mathbf{n}_2$, $\mathbf{n}_2$ being $\perp$ to plane **k**′, **k**″.

Since, after the scattering by A, the expectation value of the spin is $\langle\boldsymbol{\sigma}\rangle_1 = P_1\mathbf{n}_1$, the "initial" spin for the second scattering, by B, is $\langle\boldsymbol{\sigma}\rangle_1$. Hence by (J9), the differential cross section of the double scattering is

$$\begin{aligned}\sigma(\vartheta_1, \vartheta_2, \phi_{12}) &= (1 + P_2\mathbf{n}_2\cdot\langle\boldsymbol{\sigma}\rangle_1)\, I_2(\vartheta_2)\, I_1(\vartheta_1) \\ &= \{1 + P_2(\vartheta_2)\, P_1(\vartheta_1)\cos\phi_{12}\}\, I_1(\vartheta_1)\, I_2(\vartheta_2). \qquad \text{(J12)}\end{aligned}$$

Thus a convenient way to measure $P(\vartheta)$ is to measure the scattered intensities from $B$ in the directions $\phi_{12} = 0$ and $\pi$ and form the ratio.

$$\frac{\sigma(\phi = 0) - \sigma(\phi = \pi)}{\sigma(\phi = 0) + \sigma(\phi = \pi)} = P_1(\vartheta_1)\, P_2(\vartheta_2). \qquad \text{(J13)}$$

The absolute sign of $P(\vartheta)$ is, however, not determined. All the formulas are also valid even if $\mathbf{k}$, $\mathbf{k}'$, $\mathbf{k}''$ are considered as wave vectors in the laboratory system.

It will be noted from (J10) that $P$ becomes zero when either $g$ or $h$ vanishes. Since the central force does not give rise to a spin–flip scattering (that is $h = 0$), $P$ is necessarily zero if the scattering is caused by a central force only.

## 2. SCATTERING AMPLITUDES IN TERMS OF PHASE SHIFTS

Let us rewrite $g(\vartheta)$ and $h(\vartheta)$ in terms of phase shifts. The wave function describing the scattering of a spin $\frac{1}{2}$ particle by a spinless target may be decomposed into partial waves in a manner closely analogous to that of Sect. H. We write (H18) for the present case as follows

$$\Psi_M(\mathbf{r}, \boldsymbol{\sigma}) = \sum_{J \geqslant 1/2} A_{JL} \frac{1}{r} \psi_{JL}(r)\, F_{JML}(\vartheta, \varphi, \boldsymbol{\sigma}). \qquad \text{(J14)}$$

$F_{JML}$ is constructed of $Y_{L,M_l}(\vartheta, \varphi)$ and $\chi^{\pm 1/2}$. The associated transformation coefficients $C_{m_l}^{JML}$ are

$$C_0^{L+1/2,\,1/2,\,L} = \sqrt{\frac{L+1}{2L+1}}, \qquad C_1^{L+1/2,\,1/2,\,L} = \sqrt{\frac{L}{2L+1}},$$

$$C_0^{L-1/2,\,1/2,\,L} = -\sqrt{\frac{L}{2L+1}}, \qquad C_1^{L-1/2,\,1/2,\,L} = \sqrt{\frac{L+1}{2L+1}},$$

$$C_{-1}^{L+1/2,\,-1/2,\,L} = \sqrt{\frac{L}{2L+1}}, \qquad C_0^{L+1/2,\,-1/2,\,L} = \sqrt{\frac{L+1}{2L+1}},$$

$$C_{-1}^{L-1/2,\,-1/2,\,L} = -\sqrt{\frac{L+1}{2L+1}}, \qquad C_0^{L-1/2,\,-1/2,\,L} = \sqrt{\frac{L}{2L+1}}.$$

Using the asymptotic form for $\psi_{LJ}$,

$$\psi_{LJ} \to \frac{1}{k}\sin\left(kr - \frac{L\pi}{2} + \delta_{JL}\right), \qquad \text{(J15)}$$

and the notation (H25)

$$(\delta_{JL}) \equiv \left(kr - \frac{L\pi}{2} + \delta_{JL}\right), \tag{J16}$$

we have

$$\begin{aligned}
&2ikr(\Psi_M - e^{ikz}\chi^+)_{M=\frac{1}{2}} \\
&\quad\to \sum_{J\geq|M|}\sum_{L=J-\frac{1}{2}}^{J+\frac{1}{2}} A_{JL}\{\exp[i(\delta_{JL})] - \exp[-i(\delta_{JL})]\} \\
&\qquad\qquad \times \sum_{m_l} C^{JML}_{m_l}\, Y_{Lm_l}(\vartheta,\varphi)\,\chi^{m_s} \\
&\qquad - \sqrt{\frac{2}{2L+1}}(2L+1)\, i^L \left\{\exp\left[i\left(kr - \frac{L\pi}{2}\right)\right]\right. \\
&\qquad\qquad \left. - \exp\left[-i\left(kr - \frac{L\pi}{2}\right)\right]\right\} Y_{L0}\,\chi^+ \\
&\quad = \sum_L \left[A_{L+\frac{1}{2},L}\sqrt{\frac{L+1}{2L+1}}\{\exp[i(\delta_{L+\frac{1}{2},L})] - \exp[-i(\delta_{L+\frac{1}{2},L})]\}\right. \\
&\qquad - A_{L-\frac{1}{2},L}\sqrt{\frac{L}{2L+1}}\{\exp[i(\delta_{L-\frac{1}{2},L})] - \exp[-i(\delta_{L-\frac{1}{2},L})]\} \\
&\qquad - \sqrt{2(2L+1)}\, i^L \left\{\exp\left[i\left(kr - \frac{L\pi}{2}\right)\right]\right. \\
&\qquad\qquad \left.\left. - \exp\left[-i\left(kr - \frac{L\pi}{2}\right)\right]\right\}\right] Y_{L0}(\vartheta)\,\chi^+ \\
&\quad + \sum_L \left[A_{L+\frac{1}{2},L}\sqrt{\frac{L}{2L+1}}\{\exp[i(\delta_{L+\frac{1}{2},L})] - \exp[-(\delta_{L+\frac{1}{2},L})]\}\right. \\
&\qquad \left. + A_{L-\frac{1}{2},L}\sqrt{\frac{L+1}{2L+1}}\{\exp[i(\delta_{L-\frac{1}{2},L})] - \exp[-i(\delta_{L-\frac{1}{2},L})]\}\right] \\
&\qquad\qquad \times Y_{L1}(\vartheta,\varphi)\chi^-.
\end{aligned} \tag{J17}$$

The coefficient of $e^{-ikr}$ must vanish because of the outgoing wave boundary condition. This gives

$$\begin{aligned}
&-A_{L+\frac{1}{2},L}\sqrt{\frac{L+1}{2L+1}}\exp(-i\delta_{L+\frac{1}{2},L}) \\
&\qquad + A_{L-\frac{1}{2},L}\sqrt{\frac{L}{2L+1}}\exp(-i\delta_{L-\frac{1}{2},L}) + \sqrt{2(2L+1)}\, i^L = 0,
\end{aligned}$$

$$A_{L+\frac{1}{2},L}\sqrt{\frac{L}{2L+1}}\exp(-i\delta_{L+\frac{1}{2},L}) + A_{L-\frac{1}{2},L}\sqrt{\frac{L+1}{2L+1}}\exp(-i\delta_{L-\frac{1}{2},L}) = 0,$$

leading to

$$A_{L+\frac{1}{2},L} = \sqrt{2(L+1)}\, i^L \exp(i\delta_{L+\frac{1}{2},L}),$$

$$A_{L+\frac{1}{2},L} = -\sqrt{2L}\, i^L \exp(i\delta_{L-\frac{1}{2},L}).$$

Comparing (J1) with (J17), we obtain

$$f_{11} = \frac{1}{2ik} \sum_L \left[\sqrt{\frac{2(L+1)^2}{2L+1}} \exp(2i\delta_{L+\frac{1}{2},L}) + \sqrt{\frac{2L^2}{2L+1}} \exp(2i\delta_{L-\frac{1}{2},L}) - \sqrt{2(2L+1)}\right] Y_{L0}(\vartheta),$$

$$f_{12} = \frac{1}{2ik} \sum_{L\geqslant 1} \left\{\sqrt{\frac{2L(L+1)}{2L+1}} [\exp(2i\delta_{L+\frac{1}{2},L}) - \exp(2i\delta_{L-\frac{1}{2},L})]\right\} Y_{L1}(\vartheta, \varphi), \text{ etc.} \tag{J18}$$

The determination of the phase shifts in terms of the potential is analogous to that described in Sects. H, I.

We get finally, by noticing $f = g\mathbf{E} + h\sigma_y$ in the coordinate system adopted here,

$$g(\vartheta) = f_{11} = f_{22}, \qquad h(\vartheta) = if_{12} = -if_{21}, \qquad (\varphi = 0).$$

To examine the qualitative dependence of $P(\vartheta)$ on $\vartheta$ and energy, only the $s$- and $p$-wave contributions will be retained in (J18). The polarization parameter $P(\vartheta)$ is given by

$$I(\vartheta)P(\vartheta) = -\frac{2}{k^2} \sin\vartheta \sin(\delta_+ - \delta_-) \times [\sin\delta_0 \sin(\delta_+ + \delta_- - \delta_0) + 3\cos\vartheta \sin\delta_+ \sin\delta_-], \tag{J19}$$

where $\delta_0 = \delta_{1/2,0}$, $\delta_+ = \delta_{3/2,1}$ and $\delta_- = \delta_{1/2,1}$. If the $p$-wave phase shifts have small magnitudes, we have a simple expression for $P$.

$$P = 2(\delta_+ - \delta_-)\sin\vartheta. \tag{J20}$$

If the $p$-wave scattering is due to two split resonances with energies $E_{3/2}$ and $E_{1/2}$ and a width $\Gamma$ for both levels, the phase shifts are expressed by

$$\tan\delta_+ = \frac{1}{\varepsilon + x}, \qquad \varepsilon = \left(\frac{E_{3/2} + E_{1/2}}{2} - E\right)\Big/\left(\frac{\Gamma}{2}\right),$$

$$\tan\delta_- = \frac{1}{\varepsilon - x}, \qquad x = (E_{3/2} - E_{1/2})/\Gamma. \tag{J21}$$

$P$ is shown in Fig. 1 and Fig. 2 as a function of $\varepsilon$, $x$ and the $s$-phase shift $\delta_0$ for $\vartheta = 90°$. These figures are taken from a work of Wolfenstein (ref. 5). If the sign of $x$ is changed the sign of $P$ changes, and if the sign of $\delta_0$ is changed

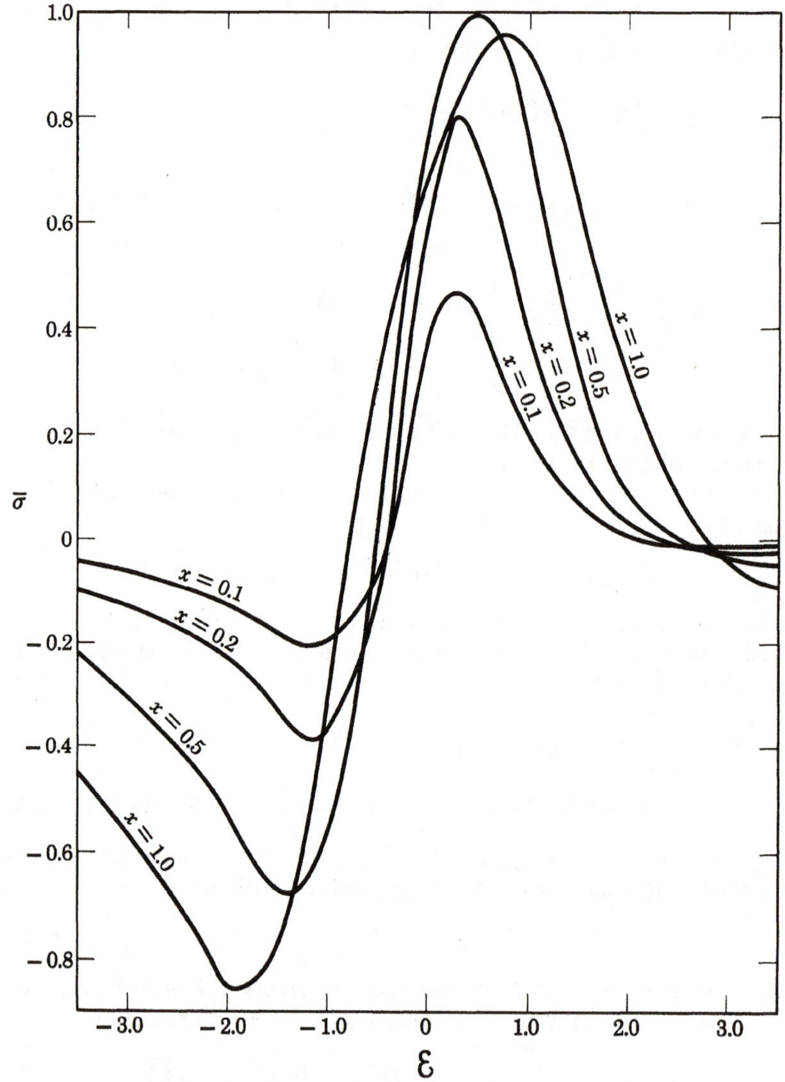


**Figure 1.** Polarization $\bar{\sigma}$ (denoted as $P$ in the text) of protons scattered from a doublet $P$ resonance as a function of energy $\epsilon$ and splitting $x$. $\delta_0 = 45°$, $\eta = 0$, $\vartheta = 90°$ [Ref. 5, Wolfenstein, L., *Phys. Rev.* **75**, 1664 (1949)].

only the sign of $\varepsilon$ changes. We can observe that a large amount of polarization can be obtained for $x$ greater than 0.2 by an interference effect of two resonance levels. For smaller values of $x$, the polarization is approximately

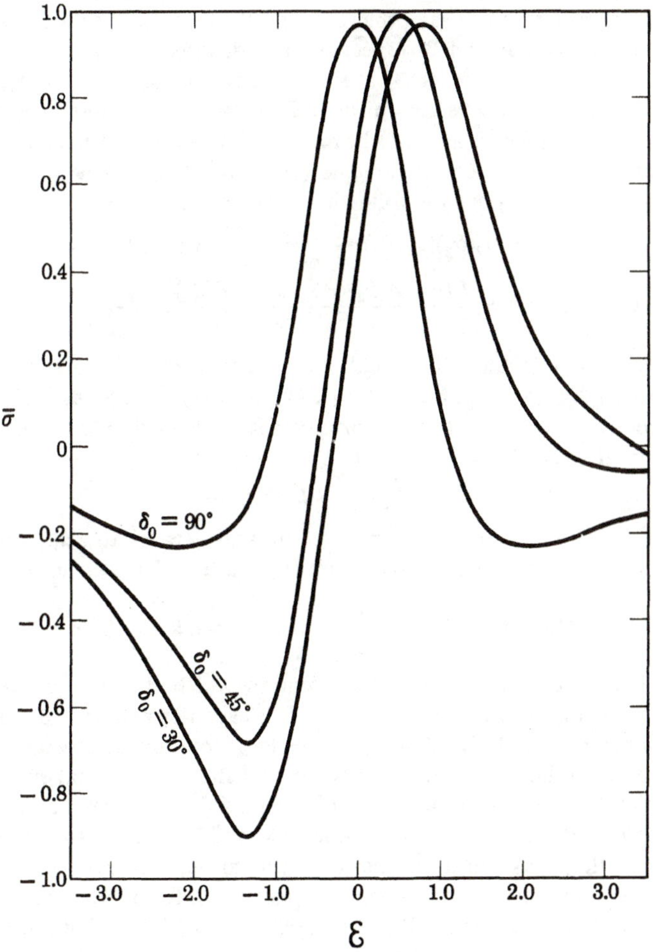


**Figure 2.** Polarization $\bar{\sigma}$ (denoted as $P$ in the text) of protons scattered from a doublet $P$ resonance as a function of energy $\epsilon$ and $s$ phase shift $\delta_0$. $x = 0.5$, $\eta = 0$, $\vartheta = 90°$ [Ref. 5, Wolfenstein, L., *Phys. Rev.* **75**, 1664 (1949)].

proportional to $x$. Such a predicted nature of $P$ was first confirmed by the Minnesota group (ref. 6) in the double scattering experiment of protons for the helium target.

### 3. TRIPLE SCATTERING EXPERIMENT: SPIN $\frac{1}{2}$ PARTICLES

Further information can be obtained from a triple-scattering experiment. All quantities will be referred to the laboratory system below.

Let the expectation values of the spin after the first and the second scattering (by the scatterer A and B respectively) be denoted by $\langle\boldsymbol{\sigma}\rangle_1$, $\langle\boldsymbol{\sigma}\rangle_2$. It can be shown that $\langle\sigma_x\rangle_2$, $\langle\sigma_y\rangle_2$, $\langle\sigma_z\rangle_2$ are linear functions of $\langle\sigma_x\rangle_1$, $\langle\sigma_y\rangle_1$, $\langle\sigma_z\rangle_1$. Let $\mathbf{k}$, $\mathbf{k}'$, $\mathbf{k}''$ be the wave vectors of the initial, the once scattered (by A) and the twice scattered (by B after A) beam. $\mathbf{k}$, $\mathbf{n}\times\mathbf{k}$ are polar vectors, $\boldsymbol{\sigma}$, $\mathbf{k}\times\mathbf{k}'$ axial vectors, $(\langle\boldsymbol{\sigma}\rangle_2\cdot\mathbf{n}_2)$ is a scalar, and $(\langle\boldsymbol{\sigma}\rangle_2\cdot[\mathbf{n}_2\times\mathbf{k}''])$ is pseudoscalar. The expectation value $\langle\boldsymbol{\sigma}\rangle_2$ has the following general form

$$\langle\boldsymbol{\sigma}\rangle_2\cdot\mathbf{n}_2 = I_2(\vartheta_2)\{P_2(\vartheta_2) + D(\vartheta_2)\langle\boldsymbol{\sigma}\rangle_1\cdot\mathbf{n}_2\}/\sigma_2(\vartheta_2, \phi_{12}), \tag{J22}$$

$$\langle\boldsymbol{\sigma}\rangle_2\cdot[\mathbf{n}_2\times\mathbf{k}''] = \frac{I_2(\vartheta_2)\{A(\vartheta_2)\langle\boldsymbol{\sigma}\rangle_1\cdot\mathbf{k}' + R(\vartheta_2)\langle\boldsymbol{\sigma}\rangle_1\cdot[\mathbf{n}_2\times\mathbf{k}']\}}{\sigma_2(\vartheta_2, \phi_{12})}, \tag{J23}$$

where $\sigma_2(\vartheta_2, \phi_{12})$ is the differential cross section of the polarized beam characterized by $\langle\boldsymbol{\sigma}\rangle_1$, and $\cos\vartheta_2 = \mathbf{k}'\cdot\mathbf{k}''$, $\cos\phi_{12} = \langle\boldsymbol{\sigma}\rangle_1\cdot\mathbf{n}_2/|\langle\boldsymbol{\sigma}\rangle_1|$. The $P_2(\vartheta_2)$ can be shown to be the polarization factor defined in (J8) and (J9) by noting that if $\langle\boldsymbol{\sigma}\rangle_1 = 0$, (J22) reduces to

$$\langle\boldsymbol{\sigma}\rangle_2\cdot\mathbf{n}_2 = P_2(\vartheta_2).$$

If $|\langle\boldsymbol{\sigma}\rangle_1|$ equals unity and $\mathbf{n}_2$ is parallel or anti-parallel to $\langle\boldsymbol{\sigma}\rangle_1$, the spin expectation values of the scattered beam (by scatterer B) is given by

$$\langle\boldsymbol{\sigma}\rangle_2\cdot\mathbf{n}_2 = \frac{(P_2 \pm D)}{(1 \pm P_2)}, \qquad \langle\boldsymbol{\sigma}\rangle_2\cdot(\mathbf{n}_2\times\mathbf{k}'') = 0. \tag{J24}$$

Thus the parameter $D$ exhibits the extent to which the second scattering depolarizes a completely polarized beam. $D$ is called sometimes "depolarization parameter". Similarly, $R$ and $A$ are called "rotation parameters". Such a terminology is, however, only suggestive, and should not be taken literally.

Now let the scattered beam $\mathbf{k}''$ from B fall on a third scatterer C, and let us consider the scattering by C in the plane of $\mathbf{k}'$ and $\mathbf{k}''$, i.e., $\mathbf{n}_3$ is parallel to $\mathbf{n}_2$ or $-\mathbf{n}_2$. Let the measured intensities of scattered beam be $I_3^n(+)$ and $I_3^n(-)$ respectively. This sort of experiment supplies the information about the $\mathbf{n}_2$ component of $\langle\boldsymbol{\sigma}\rangle_2$ just as in (J13). Using (J12) and (J22) one obtains

$$\frac{I_3^n(+) - I_3^n(-)}{I_3^n(+) + I_3^n(-)} = \frac{P_3(\vartheta_3)\{P_2(\vartheta_2) + D(\vartheta_2)\,P_1(\vartheta^1)\cos\phi_{12}\}}{1 + P_1(\vartheta_1)\,P_2(\vartheta_2)\cos\phi_{12}}, \tag{J25}$$

where $\phi_{12}$ is the angle between $\mathbf{n}_1$ and $\mathbf{n}_2$. From similar measurement of the intensity $I_3'(+)$ and $I_3'(-)$ corresponding to $\mathbf{n}_3\cdot(\mathbf{n}_2\times\mathbf{k}'') = 1$ and $-1$ respectively, we get

$$\frac{I_3'(+) - I_3'(-)}{I_3'(+) + I_3'(-)} = \frac{P_3(\vartheta_3)\,P_1(\vartheta_1)\,R(\vartheta_2)\sin\phi_{12}}{1 + P_1(\vartheta_1)\,P_2(\vartheta_2)\cos\phi_{12}}. \tag{J26}$$

The formulas (J25) and (J26) will enable $D(\vartheta)$ and $R(\vartheta)$ to be determined from the observed $I$'s and the $P_1P_2$, $P_2P_3$ and $P_3P_1$ which are already determined from the double scattering experiment. $A(\vartheta)$ can not be measured by this sort of experiment, because $\langle\boldsymbol{\sigma}\rangle_1$ is perpendicular to $\mathbf{k}'$ after the first scattering. The $\mathbf{k}''$ component of $\langle\boldsymbol{\sigma}\rangle_2$ can not be detected by means of the differential cross section measurement of the third scattering. One should rotate the spin direction by a magnetic field in order to measure these quantities.

$P(\vartheta)$ has been expressed in terms of $g(\vartheta)$ and $h(\vartheta)$. $R(\vartheta)$ and $D(\vartheta)$ are also expressible in terms of $g$ and $h$. Let us take $\mathbf{n}_2 \times \mathbf{k}'$, $\mathbf{n}_2$ and $\mathbf{k}'$ as $\mathbf{x}$-, $\mathbf{y}$-, and $\mathbf{z}$-axes respectively, and suppose that $\langle\sigma_x\rangle_1 = 1$, and $\langle\sigma_y\rangle = \langle\sigma_z\rangle = 0$. The initial and final spin state of the beam is described by

$$\chi_i = \begin{pmatrix} 1/\sqrt{2} \\ 1/\sqrt{2} \end{pmatrix} \quad \text{and} \quad \chi_f = \begin{pmatrix} (g - ih)/\sqrt{2} \\ (g + ih)/\sqrt{2} \end{pmatrix} \Big/ (|g|^2 + |h|^2)^{1/2}.$$

The formula (J23) is reduced to

$$\begin{aligned} R(\vartheta) &= \langle\boldsymbol{\sigma}\rangle_2 \cdot (\mathbf{n}_2 \times \mathbf{k}'') \\ &= \langle\chi_f, (\sigma_x \sin\vartheta + \sigma_z \cos\vartheta)\chi_f\rangle \\ &= [(\cos\vartheta + \sin\vartheta)|g|^2 + (\cos\vartheta - \sin\vartheta)|h|^2]/(|g|^2 + |h|^2). \end{aligned} \tag{J27}$$

Similarly assuming $\langle\sigma_y\rangle = 1$, $\langle\sigma_x\rangle = \langle\sigma_z\rangle = 0$, we get from (J22)

$$P_2(\vartheta_2) + D(\vartheta_2) = \frac{\sigma_2(\vartheta, \phi)}{I_2(\vartheta)} \langle\sigma_y\rangle_2. \tag{J28}$$

The initial and final states of the beam are characterized by

$$\chi_i = \begin{pmatrix} 1/\sqrt{2} \\ i/\sqrt{2} \end{pmatrix}, \qquad \chi_f = \begin{pmatrix} (g + h)/\sqrt{2} \\ i(g + h)/\sqrt{2} \end{pmatrix} \Big/ (g + h) = \chi_i.$$

Thus we get $\langle\sigma_y\rangle_2 = 1$, and from (J28), $D(\vartheta) = 1$. The experimentally measured values of $D$ for proton beam and for various nuclei of which the spins are not zero are not substantially different from unity. A complex nucleus behaves as if it had zero spin (ref. 7b). The important exception is found in nucleon–nucleon scattering. If $g$ and $h$ are expressed by the phase shifts, (J27) gives a formula for $R$ in terms of the phase shift.

## 4. MORE COMPLICATED CASES

The discussion has been limited to elastic scattering of spin $\frac{1}{2}$ particles by a spinless target. Some of the extensions to more complicated cases are straightforward. The general forms for $P$, $D$, $R$ etc. given by (J22) and (J23) are valid even if the target spin is not restricted to be zero, providing that the targets are unpolarized. These parameters are, however, more complicated functions of the scattered amplitude, which can be expressed in terms of the phase shifts and other scattering parameters in principle, but actually not in a simple way. We shall first mention the nucleon–nucleon scattering.

The scattering amplitude $f$ for nucleon–nucleon scattering is represented by a $4 \times 4$ matrix in the spin space. The basic spin functions may be taken to be product spin functions of the incident and target nucleons, or more conveniently, ${}^1\chi^0$, ${}^3\chi^1$, ${}^3\chi^0$, ${}^3\chi^{-1}$ employed in (H6). We have seen that the four matrix elements defined in (J1) are characterized by two independent functions, $g(\vartheta)$ and $h(\vartheta)$. The sixteen ($=4 \times 4$) matrix elements for nucleon–nucleon scattering are again not completely independent. Considerations similar to those leading to (J4) lead to the following general form for the matrix

$$\begin{aligned} f = BS + C(\boldsymbol{\sigma} + \boldsymbol{\sigma}_t)\cdot\mathbf{n} + G(\boldsymbol{\sigma}\cdot\mathbf{q}\boldsymbol{\sigma}_t\cdot\mathbf{q} + \boldsymbol{\sigma}\cdot\mathbf{p}\boldsymbol{\sigma}_t\cdot\mathbf{p})T \\ + H(\boldsymbol{\sigma}\cdot\mathbf{q}\boldsymbol{\sigma}_t\cdot\mathbf{q} - \boldsymbol{\sigma}\cdot\mathbf{p}\boldsymbol{\sigma}_t\cdot\mathbf{p})T + N\boldsymbol{\sigma}\cdot\mathbf{n}\boldsymbol{\sigma}_t\cdot\mathbf{n}T + Q(\boldsymbol{\sigma} - \boldsymbol{\sigma}_t)\cdot\mathbf{n}, \quad \text{(J29)} \end{aligned}$$

where $t$ refers to the target, and $\mathbf{q}$ and $\mathbf{p}$ are unit vectors in the directions $(\mathbf{k}' - \mathbf{k})$ and $(\mathbf{k}' + \mathbf{k})$ respectively. $S$ and $T$ are singlet and triplet projection operators. Thus the nucleon–nucleon scattering is represented by six quantities $B$–$Q$, which are functions of $\vartheta$ and energy. For proton–proton scattering $f$ must be invariant for the interchange of two protons, therefore $Q = 0$. The explicit formulas for $I$, $P$, $D$ and $R$ for proton–proton scattering have been given by Wolfenstein (ref. 8) in terms of the coefficients $B$, $C$, $G$, $H$, $N$. Since $f$ can be expressed in terms of the phase shift parameters, $B, \ldots, N$ can be expressed by the phase shifts and other parameters (ref. 7c). For the explicit formulas of $I$, $P$, $D$, $R$ and $A$, see Stapp *et al.*, ref. 7a.

There are other types of useful experiments for the case of nucleon–nucleon scattering. One is the correlation experiment which measures the polarization of both the scattered and the recoiled nucleons after the first scattering. Since $\mathbf{k}'$ and $\mathbf{k}_t'$ ($t$ referring to the target) are orthogonal, we shall have the orthogonal coordinate axes: $\mathbf{n}$, $\mathbf{k}'/k' \equiv \mathbf{P}$ and $-\mathbf{k}'/k = \mathbf{K}$. ($\mathbf{P}$ and $\mathbf{K}$ follow the conventional notation.) The expectation value of the dyadic $\boldsymbol{\sigma}\boldsymbol{\sigma}_t$ is expressed in general by

$$\langle\boldsymbol{\sigma}\boldsymbol{\sigma}_t\rangle = C_{nn}\mathbf{nn} + C_{KP}\mathbf{KP},$$

by assuming unpolarized incident beam. The terms proportional to $\mathbf{nP}$ and $\mathbf{nK}$ are eliminated on the basis of spatial inversion invariance, while the components of $\mathbf{PK}$, $\mathbf{KK}$ and $\mathbf{PP}$ vanish because $\boldsymbol{\sigma}$ and $\boldsymbol{\sigma}_t$ do not have the $\mathbf{P}$ and $\mathbf{K}$ components, respectively. The spin correlation coefficients, $C_{nn}$ and $C_{KP}$, are determined by the coincidence measurements of both the scattered and the recoiled nucleons in the second scattering with combinations of azimuthal angles with respect to $\mathbf{n}$. $C_{nn}$ and $C_{KP}$ can be expressed by $B \sim N$ and also by the phase shift parameters.

An extensive phase shift analysis using more than ten phase shifts has been carried out for proton–proton scattering at 310 MeV. The measured values of the five parameters $I_0$–$A$ (and $C_{nn}$, $C_{KP}$) determine the phase shifts quite well (ref. 7a). For proton–proton scattering, let $l_{\max}$ be the maximum orbital angular momentum of a partial wave that is altered by nuclear interaction. The differential cross section,

$$I(\vartheta) = \sum_{n=0}^{2l_{\max}} C_n P_n (\cos \vartheta),$$

has $l_{\max} + 1$ coefficients. (We have taken into account the fact $C_n = 0$ for odd $l$ because of the Pauli principle.) For even $l$, only the singlet states are allowed, and for odd $l$, the triplet states. Therefore we have $2l_{\max} + 1$ (or $2l_{\max} + 2$) phase shifts $\delta_J^l$ if $l_{\max}$ is even (or odd). We have $(l_{\max} - 2)/2$ [or $(l_{\max} - 1)/2$] mixing parameters $\varepsilon_J$ besides the phase shifts. Thus the measurement of (ordinary) differential cross section does not determine $\delta_J^l$ and $\varepsilon_J$. The measurement of $P(\vartheta)$ provides additional $l_{\max}$ parameters, but $I(\vartheta)$ and $P(\vartheta)$ are still not enough to determine $\delta_J^l$ and $\varepsilon_J$. Each measurement of $D(\vartheta)$, $R(\vartheta)$ and $A(\vartheta)$ will give us new $2l_{\max} + 1$ parameters. The phase shift analysis has been able to fix only two possible sets of $\delta_J^l$ and $\varepsilon_J$, although the experimental uncertainties prevent a unique determination at present. Similar analyses are being performed for other energies. From these analyses, the phase shifts with $l > 3$ are in excellent agreement with the predicted ones from the one-pion exchange meson potential (see ref. 7a and Sect. K).

Now consider particles of higher spins. The expectation values of the three spin components are not enough to specify the average spin properties of the beam. For spin 1 particle, for example, we must consider another expectation value of five tensor operators, such as $(S_x \pm S_y)^2$, together with the spin components $S_{x,y,z}$ (ref. 1). A general review is found in the work by L. Wolfenstein (ref. 1), which covers more general treatment of polarization of the beam.

If one can use a polarized target, one may dispense with the first scattering E as a polarizer. Experimentally this calls for very low temperatures for the target.

## 5. POLARIZATION IN BORN APPROXIMATION. SCATTERING BY SPIN-ORBIT POTENTIAL

To see the qualitative features of polarization of scattered particles by a spin orbit potential, we shall employ the Born approximation. This kind of treatment for nucleon polarization from complex nuclei is due to Fermi (ref. 10). The potential is assumed to have the following form

$$V(r) + W(r)\frac{\boldsymbol{\sigma}}{2}\cdot\frac{\mathbf{L}}{\hbar},$$

where **L** is the relative orbital angular momentum and $\boldsymbol{\sigma}$ is the Pauli matrix for the incident nucleon. In the case of elastic scattering of nucleons by nucleus, the nucleus may be replaced by an optical potential $V(r)$ in the first approximation, namely, the square well potential (of rounded edge) with an imaginary part (see Sect. U). The radial shape of $W(r)$ is not known well. There seem, however, to be empirical inference and theoretical arguments that the potential is most effective at the surface of the nucleus. Since the angular momentum **L** depends on the position of the nucleon (from the center of the nucleus), $W(r)$ should be small within the nucleus. If $W(r)$ is taken to be proportional to $dV(r)/dr$, the qualitative feature discussed above may be satisfied. On the other hand, if a charged particle is scattered by electromagnetic interactions, $V(r)$ and $W(r)$ are proportional to $1/r$ and $[1/r\, d/dr\,(1/r)]$ respectively. The spin-orbit coupling in the Dirac theory of a particle in the static potential $V(r)$ is given by

$$W(r) = \frac{-1}{2}\left(\frac{\hbar}{mc}\right)^2 \frac{1}{r}\frac{dV(r)}{dr}.$$

We shall introduce $Y(r)$, by which $W(r)$ is expressed, as follows,

$$W(r) = -\left(\frac{\hbar}{mc}\right)^2 \frac{1}{r}\frac{dY(r)}{dr}. \tag{J30}$$

The scattering amplitude by (J30) from wave vector **k** to **k**′ is given in the Born approximation as,

$$\begin{aligned} f &= \frac{m}{2\pi\hbar^2}\left(\frac{\hbar}{mc}\right)^2 \int \exp[-i(\mathbf{k}' - \mathbf{k})\cdot\mathbf{r}] \frac{1}{2ir}\frac{dY}{dr}(\boldsymbol{\sigma}\cdot\mathbf{r}\times\boldsymbol{\nabla})\, d\tau \\ &= +\frac{m}{2\pi\hbar^2}\left(\frac{\hbar}{mc}\right)^2 \int \exp[-i(\mathbf{k}' - \mathbf{k})\cdot\mathbf{r}]\frac{1}{2}(\boldsymbol{\sigma}\cdot\boldsymbol{\nabla} Y\times\mathbf{k})\, d\tau. \end{aligned}$$

Using the vector identity $\mathbf{A}\cdot\mathbf{B}\times\mathbf{C} = \mathbf{B}\cdot\mathbf{C}\times\mathbf{A}$, and integrating by parts,

$$\begin{aligned} f &= \frac{m}{2\pi\hbar^2}\left(\frac{\hbar}{mc}\right)^2 \frac{\boldsymbol{\sigma}}{2}\cdot\mathbf{k}\times \int \boldsymbol{\nabla}\{\exp[-i(\mathbf{k}' - \mathbf{k})\cdot\mathbf{r}]\}\, Y(r)\, d\tau, \\ &= \frac{-im}{2\pi\hbar^2}\left(\frac{\hbar}{mc}\right)^2 \frac{\boldsymbol{\sigma}}{2}\cdot\mathbf{k}\times(\mathbf{k}' - \mathbf{k}) \int \exp[-i(\mathbf{k}' - \mathbf{k})\cdot\mathbf{r}]\, Y(r) d\tau \\ &= \frac{i}{2}\left(\frac{\hbar k}{mc}\right)^2 \sin\vartheta\cdot y(|\mathbf{k}' - \mathbf{k}|)\, \boldsymbol{\sigma}\cdot\mathbf{n}, \end{aligned} \tag{J31}$$

where we have used the definition of $y$,

$$y(|\mathbf{k}' - \mathbf{k}|) = -\frac{m}{2\pi\hbar^2}\int \exp[-i(\mathbf{k}' - \mathbf{k})\cdot\mathbf{r}]\, Y(r)\, d\tau \tag{J31a}$$

and $\mathbf{k} \times \mathbf{k}' = k^2 \sin \vartheta \mathbf{n}$. By comparing (J30) with (J2), we have the Born approximation expression for $g$ and $h$,

$$g(\vartheta) = -\frac{m}{2\pi\hbar^2} \int \exp\left[-i(\mathbf{k}' - \mathbf{k})\cdot\mathbf{r}\right] V(r)\, d\tau,$$

$$h(\vartheta) = i\eta^2 \sin \vartheta\, y(|\mathbf{k}' - \mathbf{k}|)/2, \qquad \eta = \hbar k/(mc).$$

If the imaginary part of $V(r)$ is taken to be proportional to its real part, and $Y(r)$ proportional to $U(r)$,

$$V(r) = (a + ib)\, U(r), \qquad Y(r) = dU(r),$$

the differential cross section $I(\vartheta)$ and the polarization parameter $P(\vartheta)$ are given by

$$I(\vartheta) = u^2(|\mathbf{k}' - \mathbf{k}|)[a^2 + b^2 + (d^2\eta^4 \sin^2 \vartheta)/4], \tag{J32a}$$

$$I(\vartheta)P(\vartheta) = \eta^2 \frac{bd}{a} \sin \vartheta \times u^2(|\mathbf{k}' - \mathbf{k}|), \tag{J32b}$$

where $u(|\mathbf{k}' - \mathbf{k}|)$ is a Fourier transform of $U(r)$ corresponding to (J31a). It is observed from (J32a, b) that the polarization $P$ vanishes if the potential $V(r)$ and $W(r)$ are both real (i.e., $b = 0$). This fact is due to the use of the Born approximation, and is generally not valid for more accurate treatments. A particular feature of the formulas (J32a) and (J32b) is that the polarization $P$ is independent of the shape of the potential and the size of the well, but depends only on the relative magnitude of $a$, $b$, and $d$. The factor $\eta^2$ $(=\hbar^2k^2/m^2c^2)$ in (J32b) shows that $P$ is proportional to the energy of incident particles in the low energy scattering. The maximum value of the polarization is given by

$$P_{\max} = \frac{b}{\sqrt{a^2 + b^2}} \tag{J33a}$$

at an angle $\vartheta_{\max}$,

$$\sin \vartheta_{\max} = \frac{2\sqrt{a^2 + b^2}}{d\eta}. \tag{J33b}$$

The scattering data of protons on various nucleus up to 400 MeV suggest that $a \sim b$, and $d \sim 15a$, by comparing $P(\vartheta)$ with (J33a) and (J33b). The spin-orbit coupling postulated in the nuclear shell model of Mayer and Jensen is also of the same order, that is, $d \sim 15a$. In the case of a square well potential,

$$U(r) = \begin{cases} -1; & r < R, \\ \phantom{-}0; & r > R, \end{cases}$$

we have

$$u(|\mathbf{k}' - \mathbf{k}|) = \frac{2mR^3}{\hbar^2}\left\{\frac{\sin q}{q^3} - \frac{\cos q}{q^2}\right\},$$

where $q = 2kR \sin(\vartheta/2)$. The exact calculation using optical potentials with rounded edge will give a better fit to experimental polarization curves. See also Sect. U.

## REFERENCES


1. Wolfenstein, L., Annual Review of Nuclear Science **6**, 43 (1956). Review article on nucleon polarization.

2. Mott, N. F., Proc. Roy. Soc. **A124**, 425 (1929); **A135**, 429 (1932). Proposal on electron polarization by scattering.

3. Shull, C. G., Chase, C. T. and Myers, F. E., Phys. Rev. **63**, 29 (1943). Experiment on electron polarization.

4. Schwinger, J., Phys. Rev. **69**, 681 (1946); **73**, 407 (1947). Proposal on nucleon polarization.

5. Wolfenstein, L., Phys. Rev. **75**, 1664 (1949). Further theoretical development on nucleon polarization.

6. Critchfield, C. L. and Dodder, D. C., Phys. Rev. **76**, 602 (1949). Heusinkveld, M. and Freier, G., Phys. Rev. **85**, 80 (1952). Earliest experiments on nucleon polarization.

**7a.** Chamberlain, O., Segré, E., Tripp, R., Wiegand, C. and Ypsilantis, T., Phys. Rev. **105**, 288 (1957); Stapp, H. P., Ypsilantis, T. J. and Metropolis, N., Phys. Rev. **105**, 302 (1957); MacGregor, M. H., Moravcsik, M. J. and Stapp, H. P., Phys. Rev. **116**, 1248 (1959). Phase shift analysis of 310 MeV proton–proton scattering. See also ref. 24 of Sect. K for more recent work.

**7b.** Chamberlain, *et al.*, Phys. Rev. **102**, 1659 (1956). Proton–nucleus scattering.

**7c.** Wright, S. C., Phys. Rev. **99**, 996 (1955).

8. Wolfenstein, L., Phys. Rev. **96**, 1654 (1954). Theory of triple scattering.

9. Blin-Stoyle, R. J., Grace, M. A. and Halban, H. H., Progress in Nuclear Physics **3**, 63 (1953). Stationary polarized nuclei.

10. Fermi, E., Nuovo Cimento [9] **11**, 407 (1954); [10] **2**, 84 (1955).

section

# K Nucleon-Nucleon Scattering

The study of the interaction between two nucleons is one of the most basic problems in nuclear physics, and the most direct method for this study is that of the nucleon–nucleon scattering experiments. The physical theory of nuclear interactions is the meson theory of Yukawa which has furnished the basis for the qualitative understanding of many phenomena. While no completely satisfactory meson field theory of nuclear forces has yet been found, it is nevertheless of some interest to explore certain qualitative features of the theory, and to have some meson-theoretic basis for the empirical potential.

Now in the symmetric meson theory, to the second order in the (renormalized) coupling constant $f$ between a pion and a nucleon and on neglecting the recoil of the nucleon, one obtains the static, nucleon–nucleon potential for one-pion exchange (between two physical, or "dressed", not bare, nucleons)

$$V_2(x) = \tfrac{1}{3}V(x)(\boldsymbol{\tau}_1\cdot\boldsymbol{\tau}_2)[(\boldsymbol{\sigma}_1\cdot\boldsymbol{\sigma}_2) + X], \qquad \text{(K1)}$$

$$V(x) = \frac{mc^2f^2}{4\pi}\cdot\frac{e^{-x}}{x}, \qquad X = T\left(1 + \frac{3}{x} + \frac{3}{x^2}\right),$$

$$x = r/\lambda, \qquad \lambda = \frac{\hbar}{mc} = 1.42 \times 10^{-13}\ \text{cm},$$

$m$, $\lambda$ being the pion mass and Compton wavelength respectively, and $T$ the tensor operator (H1). This $V_2(x)$ is singular at $x = 0$ (no bound state exists for the $1/x^3$ potential), and a fourth order (in $f$) calculation (Taketani *et al.* ref. 14) leads to even larger potential for small $x$. Thus such meson theories (no cut-off of energies of virtual mesons, neglect of higher order meson exchange, nucleon recoil, radiative corrections, etc.) are not expected to give a reliable nucleon–nucleon potential for small distances (say $x < 1$). For large distances

($x > 1$), however, $V_2(x)$ above, or the fourth order $V_4(x)$, may be expected to be meaningful. Since, for high angular momentum waves (large $l$) only the $V(x)$ from the region of large distances contributed significantly to the scattering, one may use $V_2(x)$, or $V_4(x)$, from the meson theory for the estimation of the phase shifts of these waves. Also, from (K1), we obtain

$$\begin{aligned} {}^3V^{\text{even}} &= -V(x)(1+X), & {}^3V^{\text{odd}} &= \tfrac{1}{3}V(x)(1+X), \\ {}^1V^{\text{even}} &= -V(x), & {}^1V^{\text{odd}} &= 3V(x). \end{aligned} \tag{K2}$$

We shall see later below how the meson theory correctly describes many qualitative features of the two-nucleon system. There are authors (many of whom are Japanese) who study the nucleon–nucleon scattering in the light of the meson theory as far as possible. For small distances, the meson-theoretic potential is replaced by empirical potentials with adjustable parameters which are chosen to fit the experimental cross sections and polarization data.

On the other hand, there are authors who take a more phenomenological approach in which the observed differential cross sections and polarization data are reduced to phase shifts, and potentials of certain assumed forms and parameters are adjusted to reproduce these phase shifts.

Since the number of phase shifts (for various angular momenta $l$, $J$) that enter significantly into the scattering increases with the energy of the collision, as experimental data for higher and higher energies become available (up to the threshold for pion production by nucleon collisions), more phases will be available for the construction of the nucleon–nucleon potential.

While the general theoretical method of relating the observed scattering (cross section and polarization) data is that described in the preceding Sects. H, I, J, the actual analyses are difficult and lengthy. We shall not be able to discuss these analyses here, but summarize the results of some recent work.

For convenience, let us write the interaction between two nucleons in a general spin- and parity-dependent form, and use the notation of (H13)

$$\begin{aligned} &(p, n) && {}^3V^{\text{even}},\ {}^3V^{\text{odd}},\ {}^1V^{\text{even}},\ {}^1V^{\text{odd}}, \\ &(p, p);\ (n, n) && {}^3V^{\text{odd}},\ {}^1V^{\text{even}}. \end{aligned} \tag{K3}$$

If we assume a central-field interaction between two nucleons#

$$V_{ij}^c = [a_0 + a_\sigma(\boldsymbol{\sigma}_i\cdot\boldsymbol{\sigma}_j) + a_\tau(\boldsymbol{\tau}_i\cdot\boldsymbol{\tau}_j) + a_{\sigma\tau}(\boldsymbol{\sigma}_i\cdot\boldsymbol{\sigma}_j)(\boldsymbol{\tau}_i\cdot\boldsymbol{\tau}_j)]\, J^c(r_{ij}), \tag{K4}$$

where $J^c(r_{ij})$ is a function of $|\mathbf{r}_i - \mathbf{r}_j|$, and may be taken as a Yukawa potential

$$J^c(r) = \frac{r_c}{r}\exp\left(-\frac{r}{r_c}\right), \tag{K5}$$

# A more general form will be one having different radial dependences in each of the four terms in (K4).

then the coefficients of $J^c(r)$ for the various states in (K3) are

$$\begin{aligned} &\text{Triplet, even:} \quad a_0 + a_\sigma - 3a_\tau - 3a_{\sigma\tau}, \\ &\text{Triplet, odd:} \quad a_0 + a_\sigma + a_\tau + a_{\sigma\tau}, \\ &\text{Singlet, even:} \quad a_0 - 3a_\sigma + a_\tau - 3a_{\sigma\tau}, \\ &\text{Singlet, odd:} \quad a_0 - 3a_\sigma - 3a_\tau + 9a_{\sigma\tau}. \end{aligned} \tag{K6}$$

For the tensor interaction $T$ (H1) and the spin-orbit interaction $V_{SO}$ (H2), in view of the first relation in (H4) [from which one gets $(\boldsymbol{\sigma}_1 \cdot \boldsymbol{\sigma}_2) T = T(\boldsymbol{\sigma}_1 \cdot \boldsymbol{\sigma}_2) = T$] and the relation $(\boldsymbol{\sigma}_i \cdot \boldsymbol{\sigma}_j)(\boldsymbol{\sigma}_i + \boldsymbol{\sigma}_j) = (\boldsymbol{\sigma}_i + \boldsymbol{\sigma}_j)$, we have, corresponding to (K4),

$$\begin{aligned} V_1^T &= [b_0 + b_\tau(\boldsymbol{\tau}_i \cdot \boldsymbol{\tau}_j)]\, J^T(r)\, T, \\ V^{LS} &= [c_0 + c_\tau(\boldsymbol{\tau}_i \cdot \boldsymbol{\tau}_j)]\, J^{LS}(r)\, \mathbf{L} \cdot \mathbf{S}., \end{aligned} \tag{K7}$$

where $\mathbf{L} = \mathbf{r} \times \mathbf{p}$ (H2a), and $\mathbf{S} = \frac{1}{2}(\boldsymbol{\sigma}_i + \boldsymbol{\sigma}_j)$, and we may assume

$$J^T(r) = \frac{r_T}{r} \exp\left(-\frac{r}{r_T}\right), \qquad J^{LS}(r) = \frac{r_{LS}}{r} \exp\left(-\frac{r}{r_{LS}}\right). \tag{K8}$$

In addition to $V^c$, $V^T$, $V^{LS}$, there are evidences for a repulsive core which may be represented by a "hard" sphere

$$V_{\text{core}}(r) = \begin{cases} \infty, & r \leqslant r_0, \\ 0, & r > r_0, \end{cases} \tag{K9}$$

or by some other functions. The nucleon–nucleon interaction may then be taken to be the sum of

$$V_{ij} = V_{ij}^c + V_{ij}^T + V_{ij}^{LS} + V_{\text{core}} + \left(\frac{e^2}{r_{ij}}\right), \tag{K10}$$

where the last term is the Coulomb interaction in the case of two protons. The constants $a_0$, $a_\sigma$, $a_\tau$, $a_{\sigma\tau}$, $b_0$, $b_\tau$, $c_0$, $c_\tau$, and the range parameters $r_c$, $r_T$, $r_{LS}$, $r_0$ are constants to be adjusted to fit the experimental data.

## 1. LOW ENERGY DATA

### *i) Deuteron, ground state properties*

The binding energy, the electric quadrupole moment and the magnetic moment of the deuteron provide three relations for the constants appearing in the triplet-even potential. For the ground state (mixture of $^3S$ and $^3D$), the Schrödinger equation is the coupled equations in (H15), which take the form

$$\begin{aligned} \frac{d^2u}{d\rho^2} - \left[1 - A\frac{e^{-a\rho}}{\rho}\right]u &= -\sqrt{8}\,B\frac{e^{-b\rho}}{\rho}\,w \\ \frac{d^2w}{d\rho^2} - \left[1 + \frac{6}{\rho^2} - A\frac{e^{-a\rho}}{\rho} + 2B\frac{e^{-b\rho}}{\rho} + 3C\frac{e^{-c\rho}}{\rho}\right]w &= -\sqrt{8}\,B\frac{e^{-b\rho}}{\rho}\,u, \end{aligned} \tag{K11}$$

where

$$\alpha^2 = \frac{M}{\hbar^2}|E|, \qquad E = -2.226 \text{ MeV},$$

$$a = \frac{1}{\alpha r_c}, \qquad b = \frac{1}{\alpha r_T}, \qquad c = \frac{1}{\alpha r_{LS}}, \tag{K12}$$

$$Aa|E| = |V_c|, \qquad Bb|E| = |V_T|, \qquad Cc|E| = |V_{LS}|,$$

and

$$^3V^{\text{even}}(r) = V_c J^c(r) + V_T J^T(r) T + V_{LS} J^{LS}(r)(\mathbf{L}\cdot\mathbf{S}), \tag{K13}$$

$J^c$, $J^T$, $J^{LS}$ being the functions in (K5), (K8).

The equations (K11) are solved under the boundary conditions

$$\begin{aligned} &u(\rho_0) = w(\rho_0) = 0 \quad \text{at} \quad \rho_0 = \alpha r_0 \\ &u(\rho) \to A_s\, e^{-\rho}, \qquad w(\rho) \to A_d\left(1 + \frac{3}{\rho} + \frac{3}{\rho^2}\right) e^{-\rho}, \quad \rho \to \infty, \end{aligned} \tag{K14}$$

$r_0$ being the radius of the repulsive core in (K9). The solution of (K11) subject to (K14) gives one relation among $A$, $B$, $C$, $a$, $b$ and $c$.

If the $u(r)$ and $w(r)$ are normalized according to

$$\int_0^\infty [u^2(r) + w^2(r)]\, dr = 1,$$

the quadrupole moment of the deuteron is then

$$Q = \frac{\sqrt{2}}{10}\int_0^\infty \left(u(r) - \frac{1}{2\sqrt{2}}\, w(r)\right) w(r)\, r^2\, dr. \tag{K15}$$

On equating this to the observed value $Q = 2.73 \times 10^{-27}$ cm², one obtains a second relation among $A$, $B$, $C$, $a$, $b$ and $c$.

The magnetic moment of the deuteron is, in units of the nuclear magneton $e\hbar/2Mc$,

$$\mathscr{M} = \mu_P \sigma_P + \mu_N \sigma_N + \tfrac{1}{2}\vec{L},$$

where $\mu_P = 2.790$, $\mu_N = -1.913$, and the factor $\frac{1}{2}$ arises from the fact that only the proton has a charge. From

$$L = J - S = J - \tfrac{1}{2}(\sigma_P + \sigma_N),$$

one obtains

$$\mathscr{M} = \tfrac{1}{2}J + (\mu_P + \mu_N - \tfrac{1}{2})\, S + \tfrac{1}{2}(\mu_N - \mu_P)(\sigma_N - \sigma_P).$$

For triplet state, $\sigma_N = \sigma_P$. Only $S \cos \widehat{\mathbf{SJ}}$ contributes to the average magnetic moment,

$$S \cos \widehat{\mathbf{SJ}} = J \cdot \frac{S^2 + J^2 - L^2}{2J^2} = J\,\frac{2 + 2 - \overline{L^2}}{2\cdot 2},$$

so that

$$\overline{\mathscr{M}} = J[\mu_P + \mu_N - \tfrac{1}{4}(\mu_P + \mu_N - \tfrac{1}{2})\, \overline{L^2}].$$

For the ground state, only $^3D$ contributes to $\overline{L^2}$,

$$\overline{L^2} = l(l+1)\int w^2(r)\,dr = 6\int_0^\infty w^2\,dr,$$

and the magnetic moment of the deuteron is

$$\mu_D = \overline{\mathscr{M}} = \mu_P + \mu_N - \tfrac{3}{2}(\mu_P + \mu_N - \tfrac{1}{2})\int w^2(r)\,dr. \tag{K16}$$

From the observed $\mu_D = 0.857$, $\mu_P + \mu_N = 0.877$, we obtain

$$\int w^2(r)\,dr = 0.04, \tag{K16a}$$

which gives another relation among the constants $A, B, \ldots, c$. (K16a), however, is not a very accurate relation on account of the many meson-theoretic corrections.

### *ii) Proton–neutron scattering*

For low energies (for which only $S$ wave scattering is important), the scattering can be completely described in terms of the scattering length $a$ and the effective range $r_0$, as shown in Sect. E. All potentials are equivalent as far as low energy scattering is concerned (below $\sim$10 MeV), provided they correspond to the same $a$ and $r_0$. This was noted by Breit and his coworkers.

Let $a_t$ and $a_s$ be the scattering length in the triplet and singlet (even) system respectively. The total cross section is, on using (E13),

$$\begin{aligned}\sigma &= 4\pi(\tfrac{3}{4}a_t^2 + \tfrac{1}{4}a_s^2)\\ &= 20.4 \pm 0.2 \text{ barns for zero energy.}\end{aligned} \tag{K17}$$

From the measured value of the coherent scattering amplitude $f$ (from the total reflection of thermal neutrons from a liquid mirror containing protons), one has another relation

$$f = 2(\tfrac{3}{4}a_t + \tfrac{1}{4}a_s) = -3.78\,(1 \pm 0.006) \times 10^{-13}\text{ cm}. \tag{K18}$$

From (K17), (K18), we get#

$$\begin{aligned}a_t &= 0.538 \pm 0.004 \times 10^{-12}\text{ cm},\\ a_s &= -2.37 \pm 0.01 \times 10^{-12}\text{ cm}.\end{aligned} \tag{K19}$$

For the deuteron, the "radius" $1/\gamma$, defined by (E24a) and the binding energy $E = -2.226$ MeV, is

$$\frac{1}{\gamma} = 4.314\,(1 \pm 0.001) \times 10^{-13}\text{ cm}. \tag{K20}$$

---

# The values of the low energy parameters in (K17), (K19) are taken from the article of Hulthén and Sugawara (1957), but the errors indicated here are taken to be somewhat larger than their values.

From (K19) and (E36), we get the effective range of the triplet-even potential

$$\rho = r_{0t} = 1.72\,(1 \pm 0.02) \times 10^{-13}\ \text{cm}. \tag{K21}$$

For the singlet-even potential, $r_{0S}$ can not yet be determined as accurately as $r_{0t}$. A consistent value from a comparison of the total proton–neutron cross sections below $\sim 10$ MeV, and the photo-disintegration of the deuteron (see Appendix 1 to Sect. E) is

$$r_{0S} = 2.7\,(1 \pm 0.2) \times 10^{-13}\ \text{cm}. \tag{K22}$$

A remark may be made here concerning the meson-theoretic potential (K1). The quadrupole moment $Q$ in (K15), depending on $w(r)$, is very sensitive to the strength of the tensor force, namely, $f^2$ in (K1). Also the effective range which for the case of the general potential (K13) is

$$r_0 = 2\int_0^\infty \{e^{-2\gamma r} - [u^2(r) + w^2(r)]\}\,dr, \tag{K23}$$

depends also on $f^2$. It is shown by Iwadare *et al.* (ref. 21) that for all plausible potentials for the region $x < 1$ (together with (K1) for $x > 1$) that reproduce the observed $r_0$ and $Q$, the value of $f^2/4\pi$ must be taken between 0.06 and 0.09. This is in satisfactory agreement with the generally accepted value 0.08 derived from the pion-nucleon scattering.

For the singlet-even states, (K2) gives an attractive potential. Using (K2) and $f^2/4\pi = 0.08$, we do not get the correct scattering length $a_s$ nor the effective range $r_{0s}$. But (K1) is not expected to be valid for $x < 1$. Hence, if one assumes

$${}^1V^{\text{even}} = \begin{cases} -0.08\,mc^2\,\dfrac{e^{-x}}{x} & \text{for } x > 1, \\ V_0 & \text{for } x_0 < x < 1, \\ \infty & \text{for } 0 < x < x_0, \end{cases}$$

it is then found that to get the $a_s = -2.37 \times 10^{-12}$ cm in (K19), the corresponding values of $x_0$, $V_0$ and $r_{0s}$ are (Taketani *et al.*, ref. 5):

| $x_0$ | $V_0$ in MeV | $r_{0s}$ in $10^{-13}$ cm |
|---|---|---|
| 0.37 | −114 | 2.50 |
| 0.49 | −175 | 2.65 |
| 0.54 | −225 | 2.72 |

Thus, to get $r_{0s} = 2.7 \times 10^{-13}$ cm of (K22), $x_0$ cannot be smaller than 0.4 say, *i.e.*, the core in (K9) has $r_0 \sim 0.6 \times 10^{-13}$ cm. This conclusion is not much altered if $V_4(x)$ instead of $V_2(x)$ is used for the region $x > 1$. This, in a way, gives an indirect argument for the repulsive core introduced by Jastrow.

### *iii) Proton–proton scattering*

The accurate data in the 0.2–4 MeV range are from the van de Graaf and in the 4 ~ 12 MeV range are from the cyclotron measurements. Here the effect of the Coulomb field must be included, and the theory has been given in Appendix 3 to Sect. E. If we write (E3–21)

$$\frac{K}{R} = -\frac{1}{a} + \frac{r_0}{2}k^2, \tag{K24a}$$

the data for $E \lesssim 5$ MeV give[#, ##]

$$\begin{aligned} a &= (-7.69 \pm 0.02) \times 10^{-13}\,\text{cm}, \\ r_0 &= (2.65 \pm 0.10) \times 10^{-13}\,\text{cm}. \end{aligned} \tag{K24}$$

The close agreement between the value $r_{0s}$ for $p$–$n$ in (K22) and $r_0$ for $p$–$p$ in (K24) indicates the charge symmetry of nucleon–nucleon interaction. If we assume the same range for $p$–$n$ and $p$–$p$, then the small difference expressed in terms of the potential amplitude (or strength, or well depth) is as follows: for the singlet-even potential, the depth of a rectangular well for $p$–$n$ is 3.3 per cent greater than that for $p$–$p$; for the Yukawa potential, 1.6 per cent. This small difference may be explained by taking into account such effects as the electromagnetic interactions between the dipoles of the nucleons and the mass difference between $\pi^\pm$ and $\pi^0$.

For $p$–$p$ scattering at low energies (< 10 MeV so that the nuclear interaction affects only the $s$ wave) the scattered amplitude is given by (E3–5), and the approximate cross section $\sigma(\vartheta)$ (for $\alpha = e^2/\hbar v < 1$, i.e., $E > 1$ MeV) is given by (E3–6). The pure Coulomb part $\sigma_c$ of $\sigma(\vartheta)$ is peaked in the forward and the backward directions and symmetrical about $\vartheta = \pi/2$ (in the center of mass system). The pure nuclear scattering is isotropic ($s$-waves) and increases with the energy $E$. The ratio $\sigma/\sigma_c$ is $\simeq 1$ for $\vartheta \simeq 0$, and increases with $\vartheta$ ($\sigma/\sigma_c \simeq 7$ at $\vartheta = \pi/2$ for 1 MeV). (For $E < 0.5$ MeV, however, $\sigma/\sigma_c$ decreases

---

# The values in (K24) are taken from Hulthén and Sugawara (ref. 4). If the experimental data are fitted to an expression containing higher power terms, namely, $-Pr_0^3k^4 + Qr_0^5k^6$, in (K24a) to take into account the effect of the shape of the potential, it has been found that the values of $r_0$ and $a$ are somewhat different depending on the shape of the potential (square well, exponential and Yukawa). This unsatisfactory state of affairs is probably due to the insufficiently accurate data at low energies. For a fuller discussion, see Hulthén and Sugawara.

## An approximate neutron–proton scattering length $a_n$ is obtainable from the relation $a_n^{-1} \cong a_p^{-1} - [\ln(r_0/R) + 0.33]/R$ where $a_p$ is the corresponding $p$–$p$ scattering length for the same potential, and $R = 2.88 \times 10^{-12}$ cm, as defined in (E 3–7). If we use the values in (K24) for $a_p$ and $r_0$, we shall get $a_n \cong -1.75 \times 10^{-12}$ cm.

slightly with increasing $\vartheta$.) The deviation of this ratio from unity at $\vartheta = \pi/2$ is noticeable at about $E = 0.2$ MeV.

Let us again consider the meson-theoretic potential (K1), which for $p$–$p$, reduces to ${}^3V^{\text{odd}}$ and ${}^1V^{\text{even}}$ of (K2). For energies less than 4 MeV, the "distance of closest approach" for $p$-wave is $\kappa \gtrsim 4$ so that (K1) may hold. For the ${}^3P_{0,1,2}$ waves, the phases $\delta_0$, $\delta_1$, $\delta_2$ are small, and on expanding the cross section in the $\delta$, these shifts appear in the interference term through the average [see (H42)]

$$\delta_{\text{av}} = \tfrac{1}{9}(\delta_0 + 3\delta_1 + 5\delta_2). \tag{K25}$$

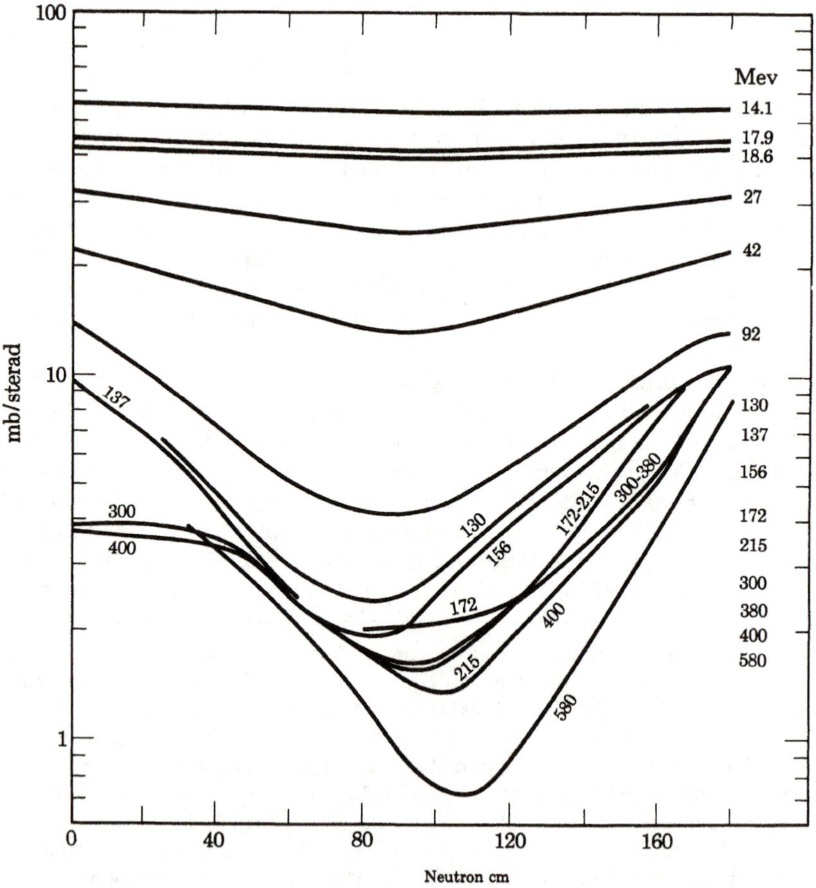


**Figure 1.** Experimental differential cross section of proton–neutron scattering at various energies [Ref. 1, Hess, W. N., *Review Mod. Phys.* **30**, 368 (1958)].

The contribution of the tensor force part of (K2) to these δ are, to the Born approximation (which is good for the *p*-wave), proportional to the matrix elements of $T$ in the $^3P_0$, $^3P_1$, $^3P_2$, states, namely, $-4$, 2, $-0.4$ respectively. On multiplying these by the weights 1, 3, 5, it is seen that the tensor force contributes nothing to the average $\delta_{av}$, and $\delta_{av}$ comes only from the central force part of (K2). Furthermore, the analysis of the *p–p* scattering data ($E < 5$ MeV) indicates a repulsive central potential. This is in agreement with $^3V^{odd}$ in (K2) (ref. 26).

## 2. HIGH ENERGY NUCLEON–NUCLEON SCATTERING

The differential cross sections for a few energies for *p–n* scattering are given in Fig. 1, and for *p–p* in Fig. 2.

An immediately noticeable feature from Fig. 1 is the approximate symmetry of $\sigma(\vartheta)$ about $\vartheta = \pi/2$. This indicates small contributions from the

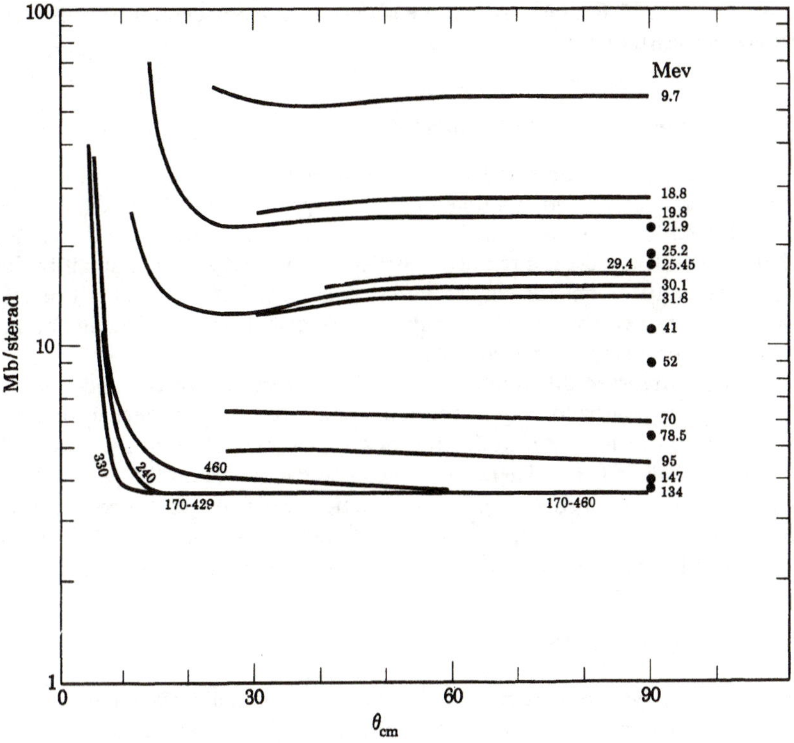


**Figure 2.** Experimental differential cross section of proton–proton scattering at various energies [Ref. 1, Hess, W. N., *Review Mod. Phys.* **30**, 368 (1958)].

odd partial waves (i.e., $l = 1, 3, \ldots$), or, small ${}^1V^{\text{odd}}$, ${}^3V^{\text{odd}}$. On the other hand, the appreciable dip at $\vartheta \simeq \pi/2$ for $E \gtrsim 40$ MeV indicates that $d$-waves, with a positive phase shift, are already important at $E \simeq 40$ MeV. The contributions from the tensor interaction are important at higher energies and tend to flatten out the $\sigma(\vartheta)$ curve.

For the $p$–$p$ scattering, we first note that on account of the identity of the two protons, (i) the $\sigma(\vartheta)$ curve should be symmetric about $\vartheta = \pi/2$, and (ii) only triplet-odd and singlet-even states exist and hence only ${}^3V^{\text{odd}}$, ${}^1V^{\text{even}}$ determine the scattering. At 32 MeV (Fig. 2), $\sigma(\vartheta)$ shows the large Coulomb scattering for $\vartheta < 20°$; the interference between the Coulomb and nuclear scatterings for $20° < \vartheta < 50°$ and the predominantly nuclear scattering for $50° < \vartheta < \pi/2$.

For energies 70–400 MeV, $\sigma(\vartheta)$ is almost isotropic, apart from the Coulomb scatterings at very small angles ($\vartheta < 5°$ for 300 MeV). Since the total cross section ~4 mb/steradian at 340 MeV exceeds the maximum $s$-wave cross section of 2.6 mb/steradian, the observed isotropy cannot be attributed to $s$-wave scattering alone.

## 3. NUCLEON–NUCLEON POTENTIAL

To account for the observed $\sigma(\vartheta)$ for $p$–$n$ and $p$–$p$ on the assumption of "charge symmetry" and of a potential of the type $V^C + V^T$ (K2, K5) has met with difficulties. One very fruitful suggestion was made by Jastrow (ref. 13) that as in (K10), there is a repulsive core of the type (K9) with a small radius ($r_0 \simeq 0.6 \times 10^{-13}$ cm for singlet states), which makes its effect felt mostly in high energy collisions. The repulsive core also helps to account for the saturation property of nuclear interactions.

But the observed differential and total cross sections for $p$–$n$ and $p$–$p$ and especially the polarization (polarization, depolarization, etc., Sect. J) data for $p$–$p$ seem to call for the introduction of an $\mathbf{L}\cdot\mathbf{S}$ interaction of the type (K7) (Case and Pais, ref. 12). There has been some reluctance on the part of some authors to believe such a large $\mathbf{L}\cdot\mathbf{S}$ interaction necessary; but a recent study of the depolarization seems to indicate that tensor force is not sufficient and $\mathbf{L}\cdot\mathbf{S}$ force is necessary to account for the high energy data (Nigam, ref. 22).

### *i) Gammel–Thaler potential*

While the analyses of scattering and polarization data and the determination of empirical nucleon–nucleon potential are still being carried out in a few laboratories, two such sets of potentials are at present available. One is that of Signell and Marshak (ref. 16), and the other is that of Gammel and Thaler. They are different in the functional forms of the various potentials in (K8), but they seem about equal in an overall fitting of the data. The Gammel–Thaler (ref. 17) potential is given below:

TABLE 1. *Constants in the Gammel–Thaler potential. For notation of* $V_c$, $V_T$ $V_{LS}$, *see* (K11), (K3), (K5), (K6), (K7).

| | Triplet even | Triplet odd | Singlet even | Singlet odd |
|---|---|---|---|---|
| Central $V_c$ in MeV | $-100.7$ | 0 | $-425.5$ | 100 |
| Tensor $V_T$ in MeV | $-257$ | 22 | | |
| **L·S** $V_{LS}$ in MeV | $(-5000)$ | $-7317.5$ | | |
| Core $r_0$ in $10^{-13}$ cm | 0.4 | 0.4 | 0.4 | 0.5 |
| Central $1/r_c$ in $10^{13}$ cm$^{-1}$ | 1.23 | | 1.45 | 1.0 |
| Tensor $1/r_T$ in $10^{13}$ cm$^{-1}$ | 1.203 | 0.8 | | |
| **L·S** $1/r_{LS}$ in $10^{13-}$ cm$^{-1}$ | 3.7 | 3.7 | | |

The distinctive feature of the Gammel–Thaler potential is the presence of strong **L·S** interaction with a very short range. However, the $V_{LS} = -5000$ MeV for the triplet-even states has been added to account for the high energy data but has not been adjusted to the low energy (deuteron) data [see Gammel and Thaler (ref. 3), also Tauber and Wu (ref. 23)]. In their most recent work, Gammel and Thaler (1960) state that the introduction of an **L·S** potential for the triplet-even states does not improve the fitting at high energies. This is in agreement with the result of Signell and Marshak. Hence probably the **L·S** term, if any, is very small for triplet-even states.

### *ii) Signell–Marshak potential*

Instead of (K1) and its fourth-order $V_4(x)$, Gartenhaus, following Chew, uses a pseudoscalar meson theory with a cut-off (in the energy of the virtual mesons) and obtains a static two-nucleon potential to the fourth order in the renormalized coupling constant $f$. This potential is not readily expressible in simple analytic forms, and is best shown in Fig. 3 which also shows the Gammel–Thaler potential for comparison. We must note that this Gartenhaus potential is also not valid for very small distances on account of the approximate nature of the theory.

In order to fit all the two-nucleon data up to ~150 MeV, Signell and Marshak have found it necessary to modify the Gartenhaus potential and to add an **L·S** term to the triplet-odd states, namely,

$$(\mathbf{L}\cdot\mathbf{S})\,\frac{V_0}{x_c}\,\frac{d}{dx}\left(\frac{e^{-x}}{x}\right)\bigg|_{r=r_c}, \qquad r \leqslant r_c$$

and

$$(\mathbf{L}\cdot\mathbf{S})\,\frac{V_0}{x}\,\frac{d}{dx}\left(\frac{e^{-x}}{x}\right), \qquad r > r_c,$$

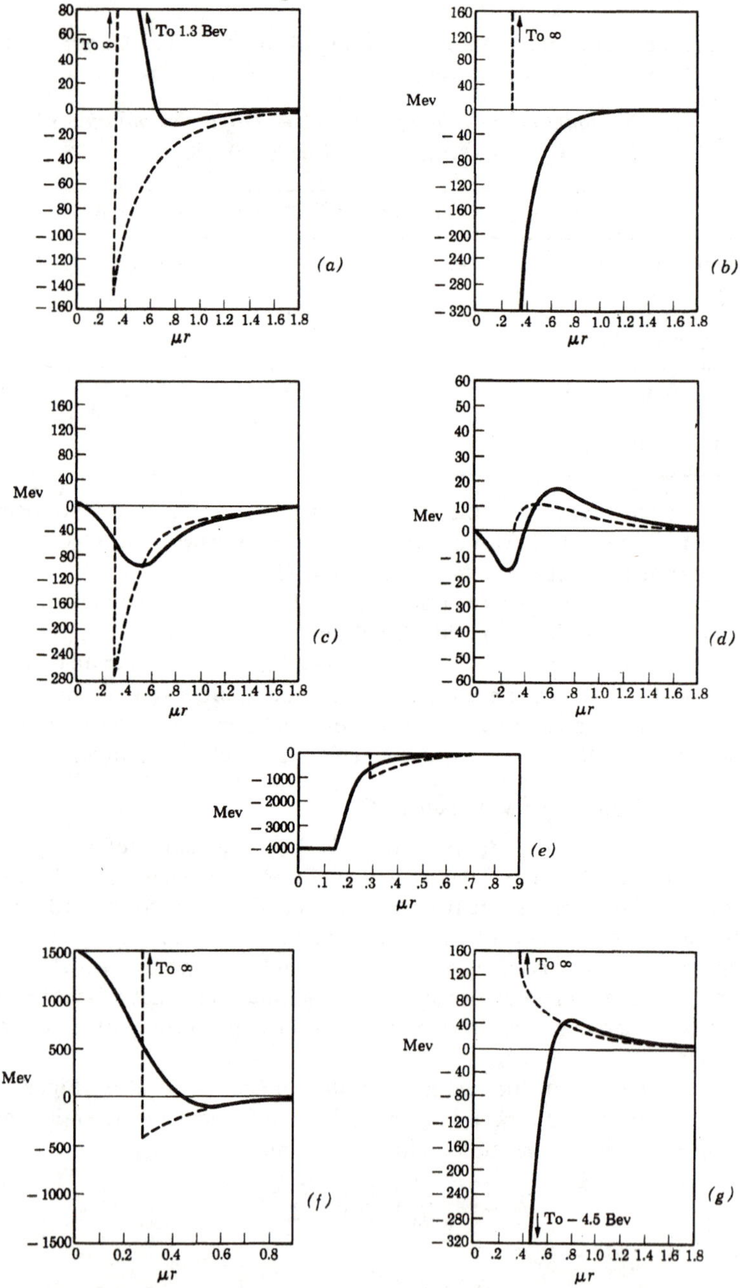


**Figure 3.** Two-nucleon potentials. Dashed curve: Gammel–Thaler (ref. 17); Solid curve: Gartenhaus (ref. 15). (a) Triplet-even, central potential; (b) Triplet-odd, central potential; (c) Triplet-even, tensor potential; (d) Triplet-odd, tensor potential; (e) Triplet-odd, L·S potential; (f) Singlet-even, central potential; (g) Singlet-odd, central potential. [From ref. 16, Signell, P., and Marshak, R., *Phys. Rev.* **109**, 1229 (1958).]

where $x = r/r_0$, $x_c = r/r_c$, $V_0$, $r_0$, $r_c$ being empirical parameters ($V_0$ = 30 MeV, $r_0 = 1.07 \times 10^{-13}$ cm, $r_c = 0.21 \times 10^{-13}$ cm). The modifications of the Gartenhaus potentials in Fig. 3 are (i) for ${}^1V^{\text{even}}$, the core potential has to be raised to make it more repulsive, (ii) for ${}^1V^{\text{odd}}$, the attractive core well must be removed, (iii) for the central potential of ${}^3V^{\text{even}}$, the height of the core potential must be raised, (iv) for the central potential of ${}^3V^{\text{odd}}$, the attractive well has to be removed. Although these modifications look rather drastic, it is perhaps to be noted that they all pertain to the core region in which the meson-theoretic (Gartenhaus) potential is not reliable anyway. More recently, a phenomenological potential in analytic form for the triplet-odd and singlet-even has been given by Bryan (ref. 19) for the proton–proton data. See also Hamada, ref. 19.

For further details of the analysis of the experimental data and the construction of potentials, the reader must be referred to the recent review articles.

## REFERENCES


The literature on the subject is a very extensive one. The following is a list of the most recent review articles:

1. Hess, W. N., Rev. Mod. Phys. **30**, 368 (1958). Experimental data above 10 MeV.
2. Phillips, R. J. N., Reports on Progress in Physics, **12**, 562 (1959). Both phenomenology and meson theory.
3. Gammel, J. L. and Thaler, R. M., Proc. Elementary Particles and Cosmic Ray Physics, Vol. **5**, 99 (1960). Phenomenological potentials and analysis of data.
4. Hulthén, L. and Sugawara, M., Handbuch der Physik. Vol. **39**, 1, Springer-Verlag (1957). Both phenomenology and meson theory.
5. Taketani, M. *et al.*, Supplement **3** to Progr. Theor. Phys. (1956). Summary of the meson theoretical studies in Japan; the necessity of good methodology in attacking the problem is emphasized.
6. Mather, K. B. and Swan, P., *Nuclear Scattering*, Chapters 6, 7, 9, and 10 (Cambridge at the University Press) (1958).

Some of the theoretical and more recent papers are:

7. Yukawa, H., Proc. Phys. Math. Soc. Japan, **17**, 48 (1935). Original paper on the meson theory of nuclear forces.
8. Schwinger, J. and Teller, E., Phys. Rev. **52**, 286 (1937). Suggestion of ortho- and para-hydrogen scattering to verify the spin dependence suggested by Wigner.

9. Eisenbud, L. and Wigner, E., Proc. Nat. Acad. Sc. **27**, 281 (1941). General form of the potential, including spin dependence.

10. Rarita, W. and Schwinger, J., Phys. Rev. **59**, 436, 556 (1941). Pioneer work in the phenomenological analysis, including tensor forces.

11. Breit, G. *et al.*, Phys. Rev. **50**, 825 (1936); **55**, 1018 (1939). Pioneer work on *p–p* scattering analysis. See also references in Sect. E.

12. Case, K. M. and Pais, A., Phys. Rev. **80**, 203 (1950). Consideration of two-body **L·S** potential.

13. Jastrow, R., Phys. Rev. **81**, 165 (1951). Introduction of the hard core.

14. Taketani, M., Machida, S. and Onuma, S., Progr. Theor. Phys. (Kyoto), **6**, 638 (1951); **7**, 45 (1952). Fourth order static, meson-theoretic potential.

15. Gartenhaus, S., Phys. Rev. **100**, 900 (1955). Meson theoretical potential with cut-off.

16. Signell, P. and Marshak, R., Phys. Rev. **109**, 1229 (1958); Signell, Zinn and Marshak, Phys. Rev. Letters, **1**, 416 (1958). Semiphenomenological potential which reasonably fits the experimental data up to 150 MeV.

17. Gammel, J. and Thaler, R., Phys. Rev. **107**, 291, 1337 (1957). Phenomenological potential which reasonably fits the experimental data up to 310 MeV.

18. Sugawara, M. and Okubo, S., Phys. Rev. **117**, 605, 611 (1960). Fourth order potential with recoil correction.

19. Bryan, R. A., Nuovo Cimento, **16**, 895 (1960). Proposed potential for *p–p* scattering; Hamada, T., Prog. Theor. Phys. (Tokyo) **24**, 1033 (1960); **25**, 247 (1961).

20. Konuma, M., Miyazawa, H. and Otsuki, S., Progr. Theor. Phys. **19**, 17 (1958). Static potential obtained by Miyazawa's method from dispersion relation. One- and two-pion exchange, and radiative corrections are included.

21. Iwadare, Otsuki, Tamagaki and Watari, Progr. Theor. Phys. **15**, 86; **16**, 455 (1956). Determination of the coupling constant.

22. Nigam, B. P., Progr. Theor. Phys. (Kyoto), **23**, 61 (1960).

23. Tauber, G. E. and Wu, T. Y., Nuclear Physics, **16**, 545 (1960); **22**, 339 (1961). Hartree–Fock calculation of binding energy and $^2P$ splitting due to **L·S** interaction of $O^{15}$ with Gammel–Thaler potential.

24. Breit, Hull, Lassila and Pyatt, Phys. Rev. **120**, 2227 (1960); Hull, Lassila, Ruppal, McDonald and Breit, Phys. Rev. **122**, 1606 (1961); MacGregor, M. H. *et. al.*, Phys. Rev. **113**, 1559 (1959); Phys. Rev. Letters **4**, 524, 2154 (1960); Phys. Rev. **123**, 1835 (1961). Extensive phase shift analysis of *p–p* and *n–p* data from low energy to 350 MeV.

25. Breit, G. *et. al.* Nuclear Phys. **15**, 216 (1960); Phys. Rev Letters, **4**, 79; **5**, 274 (1960). On the validity of one- and two-pion exchange potentials.

26. Otsuki, Taketani, Tamegaki and Watari, Prog. Theor. Phys. (Kyoto), **25**, 427 (1961).

chapter

# 3

# Collision between Composite Particles

In the previous chapters, we have dealt with the scattering of a particle by a potential field. The theory includes the problem of the collision between two particles having no internal degrees of freedom except the spin. While the theory adequately covers the very important case of nucleon–nucleon scattering, it does not apply to a large variety of problems. We shall just mention a few of these in the order of their complexities. Let $A, B, \ldots$ denote particles having no internal degrees of freedom (such as the electron, proton, neutron, etc.), and let $X, Y, \ldots$ denote composite particles (for example, $X$ an atom, $XY$ a molecule, etc.). Let an excited state be denoted by an asterisk. Then we have

i) Elastic collisions

$$\begin{aligned} A + X &\to A + X \\ A + XY &\to A + XY \end{aligned} \tag{L1}$$

ii) Inelastic collisions

$$A + X \to A + X^* \tag{L2}$$

$$A + XY \to A + XY^* \tag{L3}$$

$$X + YZ \to X + YZ^* \tag{L4}$$

$$WX + YZ \to WX^* + YZ^* \tag{L5}$$

iii) Rearrangement collisions

$$X + YZ \to XY + Z \tag{L6}$$

$$WX + YZ \to WY + XZ \tag{L7}$$

(L1) covers all elastic collisions between an electron and an atom or a molecule, etc. (L2) represents the excitation or deexcitation of an atom by an electron for example. (L3) covers the excitation or deexcitation by an electron of the electronic, vibrational or rotational state, including the dissociation, of a molecule. (L4, 5) represent the transfer of the translational energy of atoms and molecules into the electronic, vibrational or rotational motion of molecules. (L6, 7) represent the large class of processes forming the subject of chemical reactions. The above list of examples is by no means complete, but it is clear that many of the processes mentioned are of basic importance in many fields of physics, astrophysics, aerophysics and chemistry. A knowledge, theoretical or experimental, about many such processes is most desirable, if not absolutely necessary, in many researches.

Now, in principle, ordinary quantum mechanics already contains the first principles for treating any system consisting of atoms and molecules. In practice, however, very few *accurate* theoretical calculations have been carried out for any of the processes mentioned above. Only very recently has the simplest of three-body processes, namely, the scattering of an electron by a hydrogen atom, been calculated with fair accuracy by the variational method. For most other processes, the available theoretical treatments usually consist of various approximate estimates.

The difficulties involved in the study of the probabilities or cross sections of these processes are of two kinds, namely, i) the formulation of the appropriate interactions for the scattering problem, and ii) the calculation. To clarify these, consider the process (L6) for example. It is obvious that to start with a Hamiltonian containing explicitly all the electron–electron, electron–nuclei interactions in the systems $X + YZ$ and $XY + Z$ would lead to prohibitively complicated calculations. It is natural to try to "average out" the electronic motions and to work with some interaction $V(X, Y, Z)$ among the three particles. It is exactly the lack of sufficiently well grounded knowledge about this interaction $V$ (in six-dimensional configuration space) that makes the formulation of the collision problem most uncertain. The second difficulty, ii) above, is of a technical nature. With a given $V(X, Y, Z)$, the problem becomes one of solving the Schrödinger equation with reliable accuracy. The situation may be summarized by saying that even with the aid of high speed computing machines, very little work has been carried out for most of the processes mentioned above.

In the present chapter, it is not our hope to attempt to deal with the actual solution of these problems. We shall rather treat in detail the formulation of the method for the simplest of three-body collisions, and to give the general, but *formal*, method for formulating the problem of rearrangement collisions. Brief references will be made to some approximate methods and calculations on this subject.

section

L

# Scattering of an Electron by Hydrogen Atom

We shall first consider the incident electron 1 and the "atomic electron" 2 as distinguishable and find the probability of "direct scattering" in which 1 goes away, and that of "exchange scattering" in which 1 is captured to form the atom and 2 is ejected. The Pauli principle for the nondistinguishability is then included for the correct treatment of the problem.

The Hamiltonian of the system (electron + H atom) is, after separating off the part due to the motion of the center of mass,

$$H = -\tfrac{1}{2}(\Delta_1 + \Delta_2) + U(r_1) + U(r_2) + U(r_{12}), \tag{L8}$$

and the Schrödinger equation is

$$(H - E)\Psi(\mathbf{r}_1, \mathbf{r}_2) = 0. \tag{L9}$$

$\mathbf{r}_1$, $\mathbf{r}_2$ are the coordinates referred to the "heavy" nucleus of the atom. For convenience, we express length in units of the Bohr radius and energy in units of rydbergs.

## 1. "DIRECT" SCATTERING

For "direct" (elastic or inelastic) scattering in which particle 1 is incident with momentum $\mathbf{k}\hbar$ upon the atom in a state $W_n$, and emerges with momentum $\mathbf{k}_m\hbar$, we shall write $H$ in the form

$$\begin{aligned} H &= H_0' + U', \\ H_0' &= -\Delta_1 + \mathscr{H}(2), \quad \mathscr{H}(2) = -\Delta_2 + U(r_2) \equiv \mathscr{H}_0(2) + U(r_2), \\ U' &= U(r_1) + U(\mathbf{r}_1, \mathbf{r}_2). \end{aligned} \tag{L10}$$

The meaning for grouping $H$ in this form and using the prime is to have $H_0'$

correspond to the initial state of particle 1 being free and 2 in the atomic field $U(r_2)$ The eigenvalues $W_n$ and eigenfunctions of $\mathscr{H}$ are the well-known hydrogenic ones, but we shall choose to express the wave equation

$$(\mathscr{H} - W_m)\,\xi_m(\mathbf{r}) = [\mathscr{H}_0 - W_m + U(r)]\,\xi_m(\mathbf{r}) = 0 \tag{L11}$$

in the form of an integral equation (B14)

$$\xi_n(\mathbf{r}) = -\frac{1}{4\pi}\int \frac{\exp(-\sqrt{-W_n}|\mathbf{r}-\mathbf{r}'|)}{|\mathbf{r}-\mathbf{r}'|}\,U(r')\,\xi_n(\mathbf{r}')\,d\mathbf{r}', \quad W_n < 0, \tag{L12}$$

and#

$$\xi_{\mathbf{p}}(\mathbf{r}) = e^{i\mathbf{p}\cdot\mathbf{r}} - \frac{1}{4\pi}\int \frac{\exp(ip|\mathbf{r}-\mathbf{r}'|)}{|\mathbf{r}-\mathbf{r}'|}\,U(r')\,\xi_{\mathbf{p}}(\mathbf{r}')\,d\mathbf{r}', \quad W_p = p^2 > 0. \tag{L13}$$

Next consider the equation

$$(H_0' - E_a')\,\phi'_{\mathbf{K},m}(\mathbf{r}_1, \mathbf{r}_2) = 0 \tag{L15}$$

which by (L10) is

$$[-\Delta_1 + \mathscr{H}(\mathbf{r}_2) - E_a']\,\phi'_{\mathbf{K},m}(\mathbf{r}_1, \mathbf{r}_2) = 0. \tag{L16}$$

From (L11) and (L16), we have

$$\phi'_{\mathbf{K},n}(\mathbf{r}_1, \mathbf{r}_2) = \exp(i\mathbf{k}\cdot\mathbf{r}_1)\,\xi_n(\mathbf{r}_2), \quad \text{2 in bound state,} \tag{L17}$$

or

$$\phi'_{\mathbf{K},p}(\mathbf{r}_1, \mathbf{r}_2) = \exp(i\mathbf{k}\cdot\mathbf{r}_1)\,\xi_{\mathbf{p}}(\mathbf{r}_2), \quad \text{2 in continuous state,} \tag{L18}$$

with, respectively,

$$E_a' = W_n + k^2, \quad W_n < 0, \tag{L17a}$$

or

$$= W_p + k^2, \quad W_p > 0. \tag{L18a}$$

Finally, consider the equation

$$(H - E_a')\,\Psi'_{\mathbf{K},n}(\mathbf{r}_1, \mathbf{r}_2) = [H_0' + U'(\mathbf{r}_1, \mathbf{r}_2) - E_a']\,\Psi'_{\mathbf{K},n} = 0. \tag{L19}$$

This equation can again be put in the form of an integral equation which automatically ensures an outgoing wave in the particle 1 (or, we say, with

---

# In the form (B20), this equation is

$$\xi_{\mathbf{p}}(\mathbf{r}) = e^{i\mathbf{p}\cdot\mathbf{r}} + \frac{1}{W_a - \mathscr{H}_0 + i\varepsilon}\,U(r)\xi_{\mathbf{p}}(\mathbf{r}), \tag{L13a}$$

where the operator $1/(W - \mathscr{H}_0 + i\varepsilon)$ has the meaning (B21), namely,

$$\frac{1}{W_a - \mathscr{H}_0 + i\varepsilon}\,U(r)\xi_{\mathbf{p}}(\mathbf{r}) = (2\pi)^{-3}\lim_{\varepsilon\to 0}\frac{1}{W_a - W_p + i\varepsilon} \times \int \exp[i\mathbf{p}\cdot(\mathbf{r}-\mathbf{r}')]\,U(r')\,\xi_{\mathbf{p}}(\mathbf{r}')\,d\mathbf{r}'. \tag{L14}$$

respect to $H_0'$). If the system is initially in the state $\phi'_{\mathbf{k},n}(\mathbf{r}_1, \mathbf{r}_2)$ in (L17) where $\xi_n(\mathbf{r}_2)$ may be either a bound or a continuous state, we have#

$$\Psi'_{\mathbf{k},n}(\mathbf{r}_1, \mathbf{r}_2) = \phi'_{\mathbf{k},n}(\mathbf{r}_1, \mathbf{r}_2) - \frac{1}{4\pi} \sum_m \int\!\!\int\!\!\int \frac{\exp(ik_m|\mathbf{r}_1 - \mathbf{r}_1'|)}{|\mathbf{r}_1 - \mathbf{r}_1'|}$$
$$\times\, \xi_m(\mathbf{r}_2)\, \xi_m^*(\mathbf{r}_2')\, U'(\mathbf{r}_1', \mathbf{r}_2')\, \Psi'_{\mathbf{k},n}(\mathbf{r}_1', \mathbf{r}_2') \cdot d\mathbf{r}_1'\, d\mathbf{r}_2', \quad \text{(L20)}$$

where

$$k_m^2 + W_m = k^2 + W_n = E_a'. \quad \text{(L21)}$$

Here $\sum_m \int$ is to be taken over all states $m$, bound or unbound. Consider those states

$$k_m^2 > 0.$$

We have, asymptotically for large $r_1$,

$$\Psi'_{\mathbf{k},n}(\mathbf{r}_1, \mathbf{r}_2) \to \exp(i\mathbf{k}\cdot\mathbf{r}_1)\, \xi_n(\mathbf{r}_2) + \sum_m \int \frac{\exp(ik_m r_1)}{r_1}\, f_m(\vartheta_1, \varphi_1)\, \xi_m(\mathbf{r}_2) \quad \text{(L22)}$$

$$f_m(\vartheta_1, \varphi_1) = -\frac{1}{4\pi} \int\!\!\int \exp(-i\mathbf{k}_m\cdot\mathbf{r}_1')\, \xi_m^*(\mathbf{r}_2')$$
$$[U(r_1') + U(\mathbf{r}_1', \mathbf{r}_2')]\, \Psi'_{\mathbf{k},n}(\mathbf{r}_1', \mathbf{r}_2')\, d\mathbf{r}_1'\, d\mathbf{r}_2'. \quad \text{(L23)}$$

Thus $f_m(\vartheta_1, \varphi_1)$ gives the scattering amplitude of 1 being scattered with a momentum $\mathbf{k}_m(\vartheta_1, \varphi_1)$ and 2 being excited (or deexcited) from the initial state $\xi_n(\mathbf{r}_2)$ to the state $\xi_m(\mathbf{r}_2)$. For $m = n$, then, by (L21), $|\mathbf{k}_m|$ has the same value as the initial $\mathbf{k}$, $|\mathbf{k}_m| = |\mathbf{k}|$, i.e., $f_n(\vartheta_1)$ gives the elastically scattering amplitude. On putting in (L23) the explicit Coulomb potentials, one has

$$f_m(\vartheta_1, \varphi_1) = -\frac{1}{4\pi} \int\!\!\int \exp(-i\mathbf{k}_m\cdot\mathbf{r}_1')\, \xi_m^*(\mathbf{r}_2')$$
$$\times \left[-\frac{2}{r_1'} + \frac{2}{|\mathbf{r}_1' - \mathbf{r}_2'|}\right] \Psi'_{\mathbf{k},n}(\mathbf{r}_1', \mathbf{r}_2')\, d\mathbf{r}_1'\, d\mathbf{r}_2'. \quad \text{(L24)}$$

Of course $\Psi'_{\mathbf{k},n}(\mathbf{r}_1, \mathbf{r}_2)$ in (L24) is still not known, but has to be found by solving the integral equation (L20).

---

# (L20) can be put in the form

$$\Psi'_{\mathbf{k},n}(\mathbf{r}_1, \mathbf{r}_2) = \phi'_{\mathbf{k},n}(\mathbf{r}_1, \mathbf{r}_2) + \frac{1}{E_a' - H_0' + i\varepsilon}\, U'(\mathbf{r}_1, \mathbf{r}_2)\, \Psi'_{\mathbf{k},n}(\mathbf{r}_1, \mathbf{r}_2), \quad \text{(L20a)}$$

where $H_0'$, $U'$ are given in (L10). This can also be written

$$\Psi'_{\mathbf{k},n} = \phi'_{\mathbf{k},n} + \frac{1}{E_a' - H_0' + i\varepsilon - U'}\, U'\phi'_{\mathbf{k},n}, \quad \text{(L20b)}$$

as can be verified directly, or by using (M11) of the following section.

Next consider those states in (L20) with

$$k_m^2 < 0,$$

[such as correspond to highly excited discrete or continuous states $\xi_m(\mathbf{r}_2)$]. For these terms in the $\sum_m \int$, the asymptotic behavior of $\exp(ik_m|\mathbf{r}_1 - \mathbf{r}_1'|)/|\mathbf{r}_1 - \mathbf{r}_1'|$ is of the nature

$$\frac{\exp(-|\sqrt{-k_m^2}|r_1)}{r_1} \xi_m(\mathbf{r}_2) \int\int \cdots d\mathbf{r}_1' \, d\mathbf{r}_2'. \tag{L25}$$

These terms therefore do not correspond to outgoing spherical waves in particle 1 but rather to particle 1 being confined to a finite region around the nucleus, while $\xi_m(\mathbf{r}_2)$ may extend to large distances in $r_2$. Thus the solution $\Psi'_{\mathbf{k},n}(\mathbf{r}_1, \mathbf{r}_2)$ of (L20), while explicitly giving the "direct" scattering of particle 1 in both elastic and inelastic scatterings, also includes the possibilities of particle 1 being captured in some bound states and 2 being ejected. These possibilities correspond to what may be called the "exchange" or "rearrangement" collisions. The questions now are: (i) does the solution of (L20) also completely describe the "exchange" scatterings?, and (ii) to find an appropriate form of $\Psi_{\mathbf{k},n}(\mathbf{r}_1, \mathbf{r}_2)$ that describes these exchange scatterings explicitly.

## 2. "EXCHANGE" SCATTERING

The question (i) at the end of the preceding paragraph has been answered in the affirmative by Moses and by Lippmann. It can be shown by a direct transformation that the equation (L20) is completely equivalent to the following:#

---

# (L26) is equivalent to the following form

$$\Psi''_{\mathbf{k},n}(\mathbf{r}_1, \mathbf{r}_2) = \frac{i\varepsilon}{E_0' - H_0'' + i\varepsilon} \phi'_{\mathbf{k},n}(\mathbf{r}_1, \mathbf{r}_2) + \frac{1}{E_0' - H_0'' + i\varepsilon} U''(\mathbf{r}_1, \mathbf{r}_2)\Psi'_{\mathbf{k},n}(\mathbf{r}_1, \mathbf{r}_2) \tag{L26a}$$

where, instead of (L10), we regroup $H$ in (L8) as follows

$$\begin{aligned} H &= H_0'' + U'', \\ H_0'' &= -\Delta_2 + \mathcal{H}(r_1), \quad \mathcal{H}(r_1) = -\Delta_1 + U(r_1), \\ U'' &= U(r_2) + U(\mathbf{r}_1, \mathbf{r}_2), \end{aligned} \tag{L27}$$

$$\begin{aligned} &(H_0'' - E_a')\phi''_{k,n}(\mathbf{r}_1, \mathbf{r}_2) = 0, \\ &\phi''_{\mathbf{k},n}(\mathbf{r}_1, \mathbf{r}_2) = \exp(i\mathbf{k}\cdot\mathbf{r}_2)\, \xi_n(\mathbf{r}_1), \quad W_n < 0, \\ &\phi''_{\mathbf{k},\mathbf{p}}(\mathbf{r}_1, \mathbf{r}_2) = \exp(i\mathbf{k}\cdot\mathbf{r}_2)\, \xi_{\mathbf{p}}(\mathbf{r}_1), \quad W_p > 0. \end{aligned} \tag{L28}$$

$$(2\pi)^3\Psi''_{\mathbf{k},n}(\mathbf{r}_1,\mathbf{r}_2) = \lim_{\varepsilon\to 0}\left[\sum_m \int\int d\mathbf{k}'' \frac{i\varepsilon\phi''_{\mathbf{k}'',m}(\mathbf{r}_1,\mathbf{r}_2)}{k^2 + W_n - (k''^2 + W_m) + i\varepsilon} \times \int\int \phi''^{*}_{\mathbf{k}'',m}(\mathbf{r}'_1,\mathbf{r}'_2)\,\phi_{\mathbf{k},n}(\mathbf{r}'_1,\mathbf{r}'_2)\,d\mathbf{r}'_1\,d\mathbf{r}'_2\right]$$
$$-\frac{1}{4\pi}\sum_m \int\int\int \frac{\exp(ik_m|\mathbf{r}_2-\mathbf{r}'_2|)}{|\mathbf{r}_2-\mathbf{r}'_2|}[U(r'_2) + U(\mathbf{r}'_1,\mathbf{r}'_2)] \times \xi^*_m(\mathbf{r}'_1)\,\xi_m(\mathbf{r}_1)\,\Psi''_{\mathbf{k},n}(\mathbf{r}'_1,\mathbf{r}'_2)\,d\mathbf{r}'_1\,d\mathbf{r}'_2. \quad \text{(L26)}$$

We shall prove the equivalence of (L20) and (L26) in Sect. M below.

Now, for those terms in the second term in (L26) with

$$k_m^2 > 0$$

it is seen that the asymptotic behavior for large $r_2$ is

$$\frac{1}{4\pi}\sum_m \int \frac{\exp(ik_m r_2)}{r_2}\,\xi_m(\mathbf{r}_1)\int\int \exp(-i\mathbf{k}_m\cdot\mathbf{r}'_2)[U(r'_2) + U(\mathbf{r}'_1,\mathbf{r}'_2)] \times \xi^*_m(\mathbf{r}'_1)\,\Psi''_{\mathbf{k},n}(\mathbf{r}'_1,\mathbf{r}'_2)\,d\mathbf{r}'_1\,d\mathbf{r}'_2 \quad \text{(L29)}$$

which represents spherical outgoing waves in particle 2 and $\xi_m(\mathbf{r}_1)$ for particle 1, thus corresponding to "exchange" or "rearrangement" collisions. The scattering amplitude is

$$g_m(\vartheta_2,\varphi_2) = -\frac{1}{4\pi}\int\int \exp(-i\mathbf{k}_m\cdot\mathbf{r}'_2)\,\xi^*_m(\mathbf{r}'_1) \times \left[-\frac{2}{r'_2} + \frac{1}{|\mathbf{r}'_1-\mathbf{r}'_2|}\right]\Psi''_{\mathbf{k},n}(\mathbf{r}'_1,\mathbf{r}'_2)\,d\mathbf{r}'_1\,d\mathbf{r}'_2. \quad \text{(L30)}$$

Here $\Psi''_{\mathbf{k},n}(\mathbf{r}_1,\mathbf{r}_2)$ is still to be determined from (L26).

We shall now examine the first term in (L26). Here a fine distinction must be made between two cases, namely, when the initial state $\xi_n(\mathbf{r}_2)$ in $\phi_{\mathbf{k},n}(\mathbf{r}_1,\mathbf{r}_2)$ in (L20) is a discrete (or bound) state, and when $\xi_n(\mathbf{r}_2)$ is a state in the continuum as denoted by $\xi_{\mathbf{p}}(\mathbf{r}_2)$ in (L17a). We shall denote this case by $\Psi''_{\mathbf{k},\mathbf{p}}(\mathbf{r}_1,\mathbf{r}_2)$. Let the $\sum_m \int$ be separated into a summation over the discrete states $m$ and an integration over the continuum states $\mathbf{p}''$. Thus, for $\Psi''_{\mathbf{k},\mathbf{p}}(\mathbf{r}_1,\mathbf{r}_2)$, the first term in (L26) is

$$\lim_{\varepsilon\to 0}\sum_m \int d\mathbf{k}'' \frac{i\varepsilon}{k^2 + p^2 - (k''^2 + W_m) + i\varepsilon}\phi''_{\mathbf{k}'',m}(\mathbf{r}_1,\mathbf{r}_2) \times \int\int \phi''^{*}_{\mathbf{k}'',m}(\mathbf{r}'_1,\mathbf{r}'_2)\,\phi_{\mathbf{k},\mathbf{p}}(\mathbf{r}'_1,\mathbf{r}'_2)\,d\mathbf{r}'_1\,d\mathbf{r}'_2, \quad \text{(L31a)}$$

$$+\lim_{\varepsilon\to 0}\int d\mathbf{p}''\int d\mathbf{k}'' \frac{i\varepsilon}{k^2 + p^2 - (k''^2 + p''^2) + i\varepsilon}\phi''_{\mathbf{k}'',\mathbf{p}''}(\mathbf{r}_1,\mathbf{r}_2) \times \int\int \phi''^{*}_{\mathbf{k}'',\mathbf{p}''}(\mathbf{r}'_1,\mathbf{r}'_2)\,\phi_{\mathbf{k},\mathbf{p}}(\mathbf{r}'_1,\mathbf{r}'_2)\,d\mathbf{r}'_1\,d\mathbf{r}'_2. \quad \text{(L31b)}$$

Similarly the first term for $\Psi'_{\mathbf{k},n}(\mathbf{r}_1, \mathbf{r}_2)$ in (L26) is

$$\lim_{\varepsilon\to 0} \sum_m \int d\mathbf{k}'' \frac{i\varepsilon}{k^2 + W_n - (k''^2 + W_m) + i\varepsilon} \phi''_{\mathbf{k}'',m}(\mathbf{r}_1, \mathbf{r}_2) \times \iint \phi''^{*}_{\mathbf{k}'',m}(\mathbf{r}'_1, \mathbf{r}'_2)\, \phi'_{\mathbf{k},n}(\mathbf{r}'_1, \mathbf{r}'_2)\, d\mathbf{r}'_1\, d\mathbf{r}'_2, \quad \text{(L31c)}$$

$$+ \lim_{\varepsilon\to 0} \int d\mathbf{p}'' \int d\mathbf{k}'' \frac{i\varepsilon}{k^2 + W_n - (k''^2 + p''^2) + i\varepsilon} \phi''_{\mathbf{k}'',\mathbf{p}''}(\mathbf{r}_1, \mathbf{r}_2) \times \iint \phi''^{*}_{\mathbf{k}'',\mathbf{p}''}(\mathbf{r}'_1, \mathbf{r}'_2)\, \phi'_{\mathbf{k},n}(\mathbf{r}'_1, \mathbf{r}'_2)\, d\mathbf{r}'_1\, d\mathbf{r}'_2. \quad \text{(L31d)}$$

Of these four expressions, only (L31b) does not vanish. By (L18), (L28) and (L13), the double integral in (L31b) gives a factor $\delta(\vec{\mathbf{k}}'' - \vec{\mathbf{p}}'')\, \delta(\vec{\mathbf{k}} - \vec{\mathbf{p}}'')$, so that (L31b) becomes

$$\phi''_{\mathbf{p},\mathbf{k}}(\mathbf{r}_1, \mathbf{r}_2).$$

Hence one finally obtains from (L26), according as the particle 2 is initially in a bound or continuum state, the expression

$$\Psi'_{\mathbf{k},n}(\mathbf{r}_1, \mathbf{r}_2) = -\frac{1}{4\pi} \sum_m \iiint \frac{\exp(ik_m|\mathbf{r}_2 - \mathbf{r}'_2|)}{|\mathbf{r}_2 - \mathbf{r}'_2|} \times \xi_m(\mathbf{r}_1)\, \xi^*_m(\mathbf{r}'_1)\, U''(\mathbf{r}'_1, \mathbf{r}'_2) \Psi'_{\mathbf{k},n}(\mathbf{r}'_1, \mathbf{r}'_2) \cdot d\mathbf{r}'_1\, d\mathbf{r}'_2, \quad \text{(L32a)}$$

or

$$\Psi'_{\mathbf{k},\mathbf{p}}(\mathbf{r}_1, \mathbf{r}_2) = \phi''_{\mathbf{p},\mathbf{k}}(\mathbf{r}_1, \mathbf{r}_2) - \frac{1}{4\pi} \sum_m \iiint \frac{\exp(ik_m|\mathbf{r}_2 - \mathbf{r}'_2|)}{|\mathbf{r}_2 - \mathbf{r}'_2|} \times \xi_m(\mathbf{r}_1)\, \xi^*_m(\mathbf{r}'_1)\, U''(\mathbf{r}'_1, \mathbf{r}'_2) \Psi'_{\mathbf{k},\mathbf{p}}(\mathbf{r}'_1, \mathbf{r}'_2) \cdot d\mathbf{r}'_1,\, d\mathbf{r}'_2, \quad \text{(L32b)}$$

where

$$\phi''_{\mathbf{p},\mathbf{k}}(\mathbf{r}_1, \mathbf{r}_2) = \exp(i\mathbf{p}\cdot\mathbf{r}_2)\, \xi_{\mathbf{k}}(\mathbf{r}_1). \quad \text{(L33)}$$

We must emphasize that (L32a), (L32b) or (L26) have been obtained from (L20) by a direct transformation. Since (L20) has the asymptotic behavior (L22) which describes "direct" scatterings and since (L32a), (L32b) have the asymptotic behavior (L29) which describes "exchange" scatterings, we come to the result that the solution of (L20) can be made, by a transformation, to describe either "direct" or "exchange" scatterings. In either form, *all* scatterings are included. The solution of the Schrödinger equation in the form (L20) is in the appropriate form for describing the direct scatterings, but it also contains the totality of states (L26) that correspond to, and can be transformed to describe, the exchange scatterings. Similarly, the *same* solution of (L20) put in the form (L26) describes the exchange scatterings, but it also contains the totality of states that can be transformed back to describe the direct scatterings. It is not *a priori* evident that the solution of the Schrödinger equation subject to the asymptotic condition (L22) for direct scatterings *automatically* satisfies at the *same* time the asymptotic condition (L29) for exchange scatterings. That this is so is proved by the transformation from (L20) to (L26). (For proof, see Sect. M below.) The descriptions "direct" and "exchange" scatterings are two representations of the same wave function.#

In actual fact (L20) and (L26) are not the only forms of representation possible. For example, one might want to know the probability that both particles 1 and 2 come out in definite directions and energies (consistent with the energy and momentum laws). To obtain this knowledge, (L20), (L26) are not the appropriate forms, and one would analyze the scattered waves in another representation in which both particles 1 and 2 behave asymptotically as plane waves. Again one might wish to know the probability that the incident particle is captured in an inverse Auger transition, the final system being one of the quasi-stationary states (for example, the doubly excited, "discrete" states embedded in the continuum, i.e., above the ionization limit, so familiar in atomic spectra). For this information, one would analyze the wave function in a representation in which both particles are in "hydrogenic" states. All these show that the choice of the representation in describing a scattering problem is determined by the specific information one wishes to get, and that one uses one representation at a time. One must avoid the ambiguity of obtaining some direct and some exchange scattering at the same time by using a mixture of both representations, since each state of one representation contains many states of the other, the wave functions in the *two* representations being not orthogonal to each other (or, very simply we note that $H_0'$, $H_0''$ do not commute in general).

## 3. SCATTERING INVOLVING NONDISTINGUISHABLE PARTICLES

The results in the preceding sections 1, 2 can be immediately put together to describe the scattering in which the two particles 1 and 2 are indistinguishable.

---

# We may perhaps mention a historical point. It was conjectured (Mott and Massey, *Theory of Atomic Collisions*, 1st and 2nd eds.) that equation (L20) is equivalent to (L32a). This was questioned by Wu and the correct answer was given by Moses (ref. 5) and Lippmann (ref. 6) in the form given in (L26) or (L26a) above. The conjecture of Mott and Massey is correct whenever the initial state has one incident particle and the "atomic" particle in a *bound* state. When the "atomic" particle is in a continuous state, the first term in (L26) or (L26a) does not vanish. The presence of this term is essential for (L26) or (L26a) as an *exact* equation. It may also be noted that Altschuler (ref. 7) obtains (L26) or (L26a) without the first term in (L32b), so that his result is correct only when the "atomic" particle is initially in a bound state. The failure to obtain the first term in (L26) is probably due to the neglect of some contributions in the integrations corresponding essentially to (L31b).

From (L20) we have, if the initial state of 2 is a discrete state $n$,

$$\Psi''_{\mathbf{k},n}(\mathbf{r}_1, \mathbf{r}_2) = \phi'_{\mathbf{k},n}(\mathbf{r}_1, \mathbf{r}_2) - \frac{1}{4\pi}\sum_m \iint\int \frac{\exp(ik_m|\mathbf{r}_1 - \mathbf{r}'_1|)}{|\mathbf{r}_1 - \mathbf{r}'_1|} \times \xi_m(\mathbf{r}_2)\,\xi^*_m(\mathbf{r}'_2)[U(r'_1) + U(\mathbf{r}'_1, \mathbf{r}'_2)]\,\Psi''_{\mathbf{k},n}(\mathbf{r}'_1, \mathbf{r}'_2)\cdot d\mathbf{r}'_1\, d\mathbf{r}'_2. \quad \text{(L20)}$$

From (L32a), on interchanging 1 and 2,

$$\Psi''_{\mathbf{k},n}(\mathbf{r}_2, \mathbf{r}_1) = -\frac{1}{4\pi}\sum_m \iint\int \frac{\exp(ik_m|\mathbf{r}_1 - \mathbf{r}'_1|)}{|\mathbf{r}_1 - \mathbf{r}'_1|} \times \xi_m(\mathbf{r}_2)\,\xi^*_m(\mathbf{r}'_2)[U(r'_2) + U(\mathbf{r}'_1, \mathbf{r}'_2)]\,\Psi''_{\mathbf{k},n}(\mathbf{r}'_2, \mathbf{r}'_1)\cdot d\mathbf{r}'_1\, d\mathbf{r}'_2. \quad \text{(L34)}$$

From these, one obtains for the symmetric and antisymmetric wave functions the equations

$$\left.\begin{matrix}\Psi''^{s}_{\mathbf{k},n}\\ \Psi''^{a}_{\mathbf{k},n}\end{matrix}\right\} = \Psi''_{\mathbf{k},n}(\mathbf{r}_1, \mathbf{r}_2) \pm \Psi''_{\mathbf{k},n}(\mathbf{r}_2, \mathbf{r}_1)$$

$$= \phi'_{\mathbf{k},n}(\mathbf{r}_1, \mathbf{r}_2) - \frac{1}{4\pi}\sum_m \iint\int \frac{\exp(ik_m|\mathbf{r}_1 - \mathbf{r}'_1|)}{|\mathbf{r}_1 - \mathbf{r}'_1|} \times \xi_m(\mathbf{r}_2)\,\xi^*_m(\mathbf{r}'_2)[U(r'_1) + U(\mathbf{r}'_1, \mathbf{r}'_2)]\left\{\begin{matrix}\Psi''^{s}_{\mathbf{k},n}\\ \Psi''^{a}_{\mathbf{k},n}\end{matrix}\right\} d\mathbf{r}'_1\, d\mathbf{r}'_2. \quad \text{(L35)}$$

If 2 is initially in a continuous state, say $\mathbf{p}$, then, instead of (L32a), one must use (L32b), so that there is an additional term to the right of (L35), namely,

$$\pm\phi''_{\mathbf{p},\mathbf{k}}(\mathbf{r}_2, \mathbf{r}_1) = \pm\exp(i\mathbf{p}\cdot\mathbf{r}_1)\,\xi_{\mathbf{k}}(\mathbf{r}_2). \quad \text{(L36)}$$

### 4. BORN APPROXIMATION

If in the integral (L24) for $f_m(\vartheta, \varphi)$, one makes the approximation of replacing the wave function $\Psi''_{\mathbf{k},n}(\mathbf{r}_1, \mathbf{r}_2)$ in the integrand by the initial wave function, i.e.,

$$\phi'_{\mathbf{k},n}(\mathbf{r}_1, \mathbf{r}_2) = \exp(i\mathbf{k}\cdot\mathbf{r}_1)\,\xi_n(\mathbf{r}_2), \quad \text{(L37)}$$

one obtains

$$f_{mn}(\vartheta, \varphi) = -\frac{1}{4\pi}\iint \exp(-i\mathbf{k}_m\cdot\mathbf{r}'_1)\,\xi_m(\mathbf{r}'_2)\left[-\frac{2}{r'_1} + \frac{2}{|\mathbf{r}'_1 - \mathbf{r}'_2|}\right] \times \exp(i\mathbf{k}\cdot\mathbf{r}'_1)\,\xi_n(\mathbf{r}'_2)\,d\mathbf{r}'_1\,d\mathbf{r}'_2, \quad \text{(L38)}$$

where $\vartheta$ is the angle between $\mathbf{k}_m$ and $\mathbf{k}$. (L38) corresponds to the first Born approximation in (C19).

Similarly, if one makes the same approximation (L37) for $\Psi''_{\mathbf{k},n}(\mathbf{r}_1, \mathbf{r}_2)$ in (L30) for $g_m(\vartheta, \varphi)$, one obtains

$$g_{mn}(\vartheta, \varphi) = -\frac{1}{4\pi}\int\int \exp(-i\mathbf{k}_m\cdot\mathbf{r}'_2)\,\xi_m(\mathbf{r}'_1) \times \left[-\frac{2}{r'_2} + \frac{2}{|\mathbf{r}'_1 - \mathbf{r}'_2|}\right] \exp(i\mathbf{k}\cdot\mathbf{r}'_1)\,\xi_n(\mathbf{r}'_2)\,d\mathbf{r}_1\,d\mathbf{r}_2. \quad \text{(L39)}$$

Finally, if we replace $\Psi^s_{\mathbf{k},n}$ and $\Psi^a_{\mathbf{k},n}$ in the integrand in (L35) by the initial state wave function

$$\phi'_{\mathbf{k},n}(\mathbf{r}_1, \mathbf{r}_2) \pm \phi'_{\mathbf{k},n}(\mathbf{r}_2, \mathbf{r}_1) = \exp(i\mathbf{k}\cdot\mathbf{r}_1)\,\xi_n(\mathbf{r}_2) \pm \exp(i\mathbf{k}\cdot\mathbf{r}_2)\,\xi_n(\mathbf{r}_1) \quad \text{(L40)}$$

we obtain for the scattering amplitude in the singlet and the triplet state for the inelastic scattering from the initial state (L39) to the final state $(\mathbf{k}_m, m)$#

$$\left.\begin{matrix} {}^1f_{mn}(\vartheta) \\ {}^3f_{mn}(\vartheta) \end{matrix}\right\} = f_{mn}(\vartheta) \pm g_{mn}(\vartheta), \quad \text{(L41)}$$

where $f_{mn}(\vartheta)$, $g_{mn}(\vartheta)$ are given by (L37), (L39). This result was first obtained by Openheimer (ref. 1).

It is to be noted that although the same replacement (L37) of $\Psi'_{\mathbf{k},n}$ is made in the integrands of (L24) and (L30) for $f_{mn}(\vartheta)$ and $g_{mn}(\vartheta)$, the "degree" of approximation is really not quite the same in the two cases (L38) and (L39). In the case of $f_{mn}(\vartheta)$ in (L38), the interaction term $U'(\mathbf{r}_1, \mathbf{r}_2) = -2/r_1 + 2/\mathbf{r}_{12}$ may be regarded as a "small" perturbation to the $H'_0$ of (L10), of which $\exp(i\mathbf{k}_m\cdot\mathbf{r}_1)\,\xi_m(\mathbf{r}_2)$ and $\exp(-i\mathbf{k}\cdot\mathbf{r}_1)\,\xi_n(\mathbf{r}_2)$ are eigenfunctions. In (L39) for $g_{mn}(\vartheta)$, the $\exp(-i\mathbf{k}_m\cdot\mathbf{r}_2)\,\xi_m(\mathbf{r}_1)$ is an eigenfunction of $H''_0$ in (L27) while $\exp(i\mathbf{k}\cdot\mathbf{r}_1)\,\xi_m(\mathbf{r}_2)$ is an eigenfunction of $H'_0$ in (L10), and $U''(\mathbf{r}_1, \mathbf{r}_2) = -2/r_2 + 2/r_{12}$ is *not* "small" with respect to $H'_0$. Thus one may say that the formula (L39) for $g_{mn}(\vartheta)$ is a worse approximation than (L38) is for $f_{mn}(\vartheta)$. It may also be noted that such a difference is a general feature of all similar "Born approximations".

In this connection, we may emphasize a feature of the "Born approximation" for rearrangement collisions in general (see Sects. M.1, 2 below). In (L39), $g_{mn}(\vartheta)$ is given by the matrix element $(\phi_f, U''\phi_i)$ where $U''$ is the ("post") perturbation from the point of view of the final state of the scattering process. But it can be shown readily from (L10), (L15), (L28) that

$$(\phi_f, U''\phi_i) = (\phi_f, U'\phi_i),$$

where $U' = -2/r_1 + 2/r_{12}$ is the (prior) perturbation appropriate for the initial state. This equality disposes of any ambiguity as to which of the $U'$ or $U''$ to be used. There remains, however, an unsatisfactory point, namely, that since $\phi_i$ and $\phi_f$ are not orthogonal, changing $U'$ or $U''$ by an additive

# In the following we shall omit the $\varphi$ from $f(\vartheta, \varphi)$ and $g(\vartheta, \varphi)$ for brevity.

constant will yield a different result for the matrix element $(\phi_f, U\phi_i)$. It seems that we can resolve this ambiguity only by the requirement that $U'$ (or $U''$) vanish for large separations of the colliding (or resulting) parts of the system.

Since the electron–hydrogen atom is the simplest of three-body systems, it is desirable to have a comparison between some "exact" results and some Born approximation results on the basis of (L41), (L39), (L38), to test the validity of this approximation for various energies. Early calculations of $f_{mo}(\vartheta)$, $g_{mo}(\vartheta)$ involve the development of the integrands of (L38) and (L39) in infinite series of trigonometric Bessel functions and the approximation of the Bessel functions by their asymptotic expressions (Massey and Mohr, ref. 2). More recently, $f_{mo}(\vartheta)$, $g_{mo}(\vartheta)$ have been expressed in closed analytic forms for a few inelastic processes (Corinaldesi and Trainor, ref. 3), and the calculations carried out (Wu, ref. 3a).

i) Elastic scattering ($1s \rightarrow 1s$ for the H atom)

$$f(\vartheta) = \frac{1}{2P^2} - \frac{1}{2P^2(1 + P^2)^2}, \tag{L42}$$

where and in the following, $f(\vartheta)$, $g(\vartheta)$ are in units of the Bohr radius, and

$$P = k \sin \frac{\vartheta}{2}. \tag{L42a}$$

$\vartheta$ is the scattering angle, and $k^2$ is the energy of the incident electron in units of rydbergs.

$$g(\vartheta) = \frac{16}{(1 + k^2)^3} - \frac{4}{(1 + k^2)^3 P^2} \left\{ \frac{1}{(1 + P^2)^2} [P^4 + \tfrac{1}{2}(3 + k^4) P^2 - k^2] + \frac{1}{P}(k^2 + P^2) \sin^{-1} \frac{P}{\sqrt{1 + P^2}} \right\}. \tag{L43}$$

The first term in $f(\vartheta)$, $g(\vartheta)$ comes from $1/r_1$ and the second term from $1/r_{12}$.

It follows from (L42) and (L43) that the forward ($\vartheta = 0$) scattering amplitude (in units of the Bohr radius $a_B$) is

$$f(0) = 1, \qquad g(0) = \frac{1}{(1 + k^2)^3} [3 - \tfrac{10}{3}k^2 - k^4]. \tag{L44}$$

ii) Inelastic scattering ($1s \rightarrow 2s$ for the H atom)

Let

$$t^2 = 8k^2 - 3 - 4xk\sqrt{4k^2 - 3},$$

$$x = \cos \vartheta,$$

$$X = \sin^{-1} \left[ \frac{t^2 + 3}{(t^2 + 1)^{1/2}(t^2 + 9)^{1/2}} \right] + \sin^{-1} \left[ \frac{t^2 - 3}{(t^2 + 1)^{1/2}(t^2 + 9)^{1/2}} \right],$$

$$\kappa = 1 + 4k^2, \tag{L45}$$

$$B_1 = \frac{64}{(t^2 + 9)^2}, \qquad B_2 = -\frac{48}{t^2(t^2 + 9)} + \frac{8X}{t^3},$$

$$B_3 = \frac{6(t^2 + 3)}{t^4} + \frac{(t^4 - 10t^2 - 27)}{t^5} X,$$

$$C_1 = \frac{512}{3} \frac{(t^2 + 27)}{(t^2 + 9)^3}, \qquad C_2 = 128 \frac{1}{(t^2 + 9)^2}, \tag{L46}$$

$$C_3 = -\frac{8}{t^4} \frac{(t^2 - 27)}{(t^2 + 9)} + \frac{4}{t^5}(t^2 - 9) X,$$

$$C_4 = \frac{1}{3t^6}(9t^4 - 58t^2 - 135) + \frac{1}{2t^7}(t^6 - 7t^4 + 63t^2 + 135) X.$$

$$f(\vartheta) = -\frac{512\sqrt{2}}{(t^2 + 9^3)} \left(\text{from } \frac{1}{r_{12}} \text{ alone}\right), \tag{L47}$$

$$g(\vartheta) = \frac{256\sqrt{2}(4k^2 - 1)}{\kappa^4} - \frac{4\sqrt{2}}{\kappa^3}\Big[3\kappa^2 B_1 + 16\kappa B_2 + 32 B_3 - \frac{3}{8\kappa}(\kappa^3 C_1 + 8\kappa^2 C_2 + 128\kappa C_3 + 256 C_4)\Big]. \tag{L48}$$

iii) Inelastic collision ($1s \rightarrow 2p$, $m_l = 0$, for the atom)

$$D_1 = \frac{1024}{5} \frac{1}{(t^2 + 9)^3}, \qquad D_2 = -\frac{64}{t^4(t^2 + 9)^2}(5t^2 + 27) + \frac{32}{t^5} X,$$

$$D_3 = \frac{6}{t^6(t^2 + 9)}(t^4 + 40t^2 + 135) + \frac{1}{t^7}(t^4 - 30t^2 - 135) X, \tag{L49}$$

$$f(\vartheta) = -\frac{1536\sqrt{2}}{t^2(t^2 + 9)^3}(2k - x\sqrt{4k^2 - 3}), \tag{L50}$$

$$g(\vartheta) = \frac{1024\sqrt{2}}{\kappa^4} k - \frac{8\sqrt{2}}{\kappa^4}\Big[(\kappa^2 C_2 + 32\kappa C_3 + 96 C_4) + \frac{2k - x\sqrt{4k^2 - 3}}{16k} \kappa(15\kappa^2 D_1 + 48\kappa D_2 + 128 D_3)\Big], \tag{L51}$$

($1s \rightarrow 2p$, $m_l = \pm 1$, for the atom):

$$f(\vartheta) = 1536 \frac{\sqrt{1 - x^2}}{t^2(t^2 + 9)^3} \sqrt{4k^2 - 3}\, e^{im\phi}. \tag{L52}$$

$$g(\vartheta) = 0. \tag{L53}$$

The cross section is given by the weighted mean of the triplet and the singlet state scattering cross sections

$$d\sigma = \frac{k'}{k}\left[\tfrac{3}{4}|f(\vartheta) - g(\vartheta)|^2 + \tfrac{1}{4}|f(\vartheta) + g(\vartheta)|^2\right] d\cos\vartheta\, d\phi, \tag{L54}$$

where $k'^2$ is the energy, in units of rydbergs, of the scattered electron.

The total cross sections are given in Table 1 below for electron of energies 1, 4, 9, 16 rydbergs (1 rydberg = 13.6 eV). For the differential cross sections, reference is made to ref. 3a.[#]

TABLE 1. *Total cross sections of electron hydrogen atom scattering, in* $\pi a_B^2$. *The values in parentheses are the results of variational calculations of Massey and Moiseiwitsch, ref.* 11.

| Electron energy in rydbergs | $1s \to 1s$ | $1s \to 2s$ | $1s \to 2p$ $\Delta m = 0$ | $1s \to 2p$ $\Delta m = \pm 1$ |
|---|---|---|---|---|
| 1 | 2.104 (3.34) | 0.823 | 0.750 | 0.0415 |
| 4 | 0.574 (0.495) | 0.0862 | 0.3981 | 0.2449 |
| 9 | 0.256 | 0.0406 | 0.2039 | 0.1945 |
| 16 | 0.144 | 0.0236 | 0.1197 | 0.1437 |

## 5. VARIATIONAL METHOD: ELASTIC SCATTERING

### i) General formulation

The variational principles of Sect. D for the scattering of a particle by a static potential can be extended to the scattering of an electron by a hydrogen atom.

Consider the elastic scattering of low energy (below 10 eV) electrons by a H atom in the ground state. The Schrödinger equation for the system is (L9), Let $\psi^{(\pm)}$ denote the spatial wave functions which are symmetric (singlet) and antisymmetric (triplet) in the two electrons [$\psi^{(+)}$, $\psi^{(-)}$ are the $\psi^{(s)}$, $\psi^{(a)}$ in (L35)], and let $\psi_l^{(\pm)}$ be the wave functions of orbital angular momentum $l$ ($m_l = 0$). Thus

$$(H - E)\,\psi_l(1, 2) = 0, \tag{L55}$$

$$\psi_l^{(\pm)}(1, 2) = \pm\psi_l^{(\pm)}(2, 1). \tag{L56}$$

Asymptotically,[##]

$$\psi_l^{(\pm)} \to \frac{1}{r_2} Y_l(\vartheta_2)\left[\sin\left(kr_2 - \frac{l}{2}\pi\right) + \tan\eta_l^{(\pm)}\cos\left(kr_2 - \frac{l}{2}\pi\right)\right]\xi_0(1), \qquad (r_2 \to \infty)$$

$$\to \pm\frac{1}{r_1} Y_l(\vartheta_1)\left[\sin\left(kr_1 - \frac{l}{2}\pi\right) + \tan\eta_l^{(\pm)}\cos\left(kr_1 - \frac{l}{2}\pi\right)\right]\xi_0(2), \qquad (r_1 \to \infty), \tag{L57}$$

[#] Some errors in the numerical values there are corrected in Table 1 here.

[##] To avoid confusion with the variational sign $\delta$, the phase shift, true and trial, will be denoted by $\eta$ in the present subsection.

where $\xi_0(r)$ is the normalized ground state wave function of $H$

$$\xi_0(r) = \frac{1}{\sqrt{\pi}} e^{-r}$$

and $Y_l(\vartheta)$ is the normalized $Y_l(\vartheta) = \sqrt{(1/4\pi)(2l+1)}\, P_l(\cos\vartheta)$. The normalization in (L57) corresponds to that in (D17). [Analogous formulations corresponding to the choices of (D9), (D16) can be made.] The scattering amplitudes are given, analogously to (A17), by#

$$f^{\pm}(\vartheta) = \frac{1}{2ik} \sum_l (2l+1)\{\exp[2i\eta_l^{(\pm)}] - 1\} P_l(\cos\vartheta). \tag{L58}$$

The differential and total cross sections are, as in (L54),

$$\begin{aligned} d\sigma &= \tfrac{1}{4}\{|f^{+}(\vartheta)|^2 + 3|f^{-}(\vartheta)|^2\}\, d\cos\vartheta\, d\varphi, \\ \sigma &= \tfrac{1}{4}\sigma_{(+)} + \tfrac{3}{4}\sigma_{(-)}, \\ \sigma_{(\pm)} &= \frac{4\pi}{k^2} \sum (2l+1)\sin^2 \eta_l^{(\pm)}. \end{aligned} \tag{L59}$$

---

# (L58) is derived from (L35) in the following way. The unsymmetrized $\psi'_{\mathbf{k},0}$ (L20) has the asymptotic form (L22) and (L30)

$$\begin{aligned} \psi'_{\mathbf{k},0}(r_1, r_2) &\to \exp(i\mathbf{k}\cdot\mathbf{r}_1)\,\xi_0(r_2) + \frac{\exp(ikr_1)}{r_1} f_0(\vartheta_1)\,\xi_0(r_2), \quad (r_1 \to \infty) \\ &\to \frac{\exp(ikr_2)}{r_2} g_0(\theta_2)\xi_0(r_1), \quad (r_2 \to \infty) \end{aligned}$$

The properly symmetrized $\psi_{\mathbf{k},0}^{(\pm)}$, therefore, has the following form

$$\begin{aligned} \psi_{\mathbf{k},0}^{(\pm)}(r_1, r_2) &\to \exp(i\mathbf{k}\cdot\mathbf{r}_1)\,\xi_0(r_2) + [f_0(\vartheta_1) \pm g_0(\vartheta_1)] \\ &\qquad \times \exp(ikr_1)\,\xi_0(r_2)/r_1, \quad (r_1 \to \infty) \\ &\to \pm\exp(i\mathbf{k}\cdot\mathbf{r}_2)\,\xi_0(r_1) \pm [f_0(\vartheta_2) \pm g_0(\vartheta_2)] \\ &\qquad \times \exp(ikr_2)\,\xi_0(r_1)/r_2, \quad (r_2 \to \infty). \end{aligned}$$

The scattering amplitude $f^{\pm}(\vartheta) \equiv f_0(\vartheta) \pm g_0(\vartheta)$ as well as the wave function $\psi_{\mathbf{k},0}^{(\pm)}$ can be decomposed into partial waves see [(C42), (C45)].

$$f^{\pm}(\vartheta) = \sum_{l=0}^{\infty} \frac{(2l+1)}{k} f_l^{\pm} P_l(\cos\theta),$$

$$\begin{aligned} \psi_{\mathbf{k},0}^{(\pm)} \to \sum_{l=0}^{\infty} \frac{1}{kr_1}(2l+1)\, i^l \left\{\sin\left(kr_1 - \frac{l\pi}{2}\right) + \frac{1}{2i}[S_l^{(\pm)} - 1]\exp\left[i\left(kr_1 - \frac{l\pi}{2}\right)\right]\right\} \\ \times P_l(\cos\vartheta_1)\cdot\xi_0(r_2), \quad (r_1 \to \infty). \end{aligned}$$

Therefore [see (C35)]

$$f_l^{\pm} = \frac{1}{2i}(S_l^{\pm} - 1) = \frac{1}{2i}\{\exp[2i\eta_l^{(\pm)}] - 1\}.$$

To apply the variational method of Sect. D, we have

$$I_l^{(\pm)} = \int \psi_l^{(\pm)}(H - E)\,\psi_l^{(\pm)}\,d\tau_1\,d\tau_2 \tag{L60}$$

and analogous to (D19)

$$\delta[I_l^{(\pm)} - 2k\tan\eta_l^{(\pm)}] = 4\int \delta\psi_l^{(\pm)}(H - E)\,\psi_l^{(\pm)}\,d\tau_1\,d\tau_2 = 0. \tag{L61}$$

In an actual calculation, a trial wave function $\Phi_l^{(\pm)}$ [instead of $\psi_l$ in (L57)] containing a number of variational parameters $c_1, \ldots, c_n, \eta_l^{(\pm)}$ is used in (L60). The procedure described in Sect. D [following (D22)] is then used to determine first a "trial" $\eta_l^{(\pm)}$

$$\frac{\partial I_l^{(\pm)}}{\partial c_i} = 0, \quad i = 1, \ldots, n; \qquad \frac{\partial I_l^{(\pm)}}{\partial \tan\eta_l^{\pm}} = 2k, \tag{L62}$$

and finally a more accurate value of the tangent of the phase shift

$$\tan\eta_l^{(\pm)} - \frac{1}{2k} I_l^{(\pm)}. \tag{L63}$$

### *ii) Hartree–Fock approximation*

A special case of the trial function $\Phi_l^{(\pm)}$ is the Hartree–Fock function

$$\begin{aligned}\Phi_l^{(\pm)} &= \varphi(1)\,\xi_0(2) \pm \varphi(2)\,\xi_0(1)\\ &\equiv \frac{1}{r_1} F_l(r_1)\,Y_l(\vartheta_1)\,\xi_0(2) \pm \frac{1}{r_2} F_l(r_2)\,Y_l(\vartheta_2)\,\xi_0(1),\end{aligned} \tag{L64}$$

which must satisfy the asymptotic condition (L57). In units of rydbergs, the total energy $E = -1 + k^2$. Equation (L61) is[#]

$$\begin{aligned}\delta(I_l^{(\pm)} - 2k\tan\eta_l^{(\pm)}) = 4\int &\delta\varphi(1)\,\xi_0(2)[-\Delta_1 + U(1) + U(1,2) - k^2]\\ &\qquad\times \varphi(1)\,\xi_0(2)\,d\tau_1\,d\tau_2\\ \pm 2\int &\delta\varphi(1)\,\xi_0(2)[-\Delta_2 + U(2) + U(1,2) - k^2]\\ &\qquad\times \varphi(2)\,\xi_0(1)\,d\tau_1\,d\tau_2\\ \pm 2\int &\delta\varphi(1)\,\xi_0(2)[-\Delta_1 + U(1) + U(1,2) - k^2]\\ &\qquad\times \varphi(2)\,\xi_0(1)\,d\tau_1\,d\tau_2 = 0,\end{aligned}$$

or

$$\begin{aligned}&[-\Delta_1 + U(1) + \int U(1,2)\,\xi_0^2(2)\,d\tau_2 - k^2]\,\varphi(1)\\ &\quad \pm \int \xi_0(2)[-\tfrac{1}{2}\Delta_1 - \tfrac{1}{2}\Delta_2 + \tfrac{1}{2}U(1) + \tfrac{1}{2}U(2) + U(1,2) - k^2]\\ &\qquad\qquad \times \varphi(2)\,d\tau_2\cdot\xi_0(1) = 0.\end{aligned}$$

---

[#] The factors 2, 4 in the following expressions arise from the normalization (L64).

On expanding $V(1, 2) = 2/r_{12}$ by the well-known theorem

$$\frac{1}{r_{12}} = \sum_l \frac{r_<^l}{r_>^{l+1}} \sum_{-l}^{l} m \; Y_{lm}(\vartheta_1, \varphi_1) \; Y_{l,\,-m}(\vartheta_2, \varphi_2),$$

and hence

$$\int \xi_0(2) \, U(1, 2) \, \varphi(2) \, d\tau_2 = \frac{8\pi}{2l + 1} \, Y_{l0}(\vartheta_1) \int_0^\infty \xi_0(2) \, F_l(2) \, r_2 \, dr_2,$$

we obtain

$$\left[\frac{d^2}{dr_1^2} - \frac{l(l + 1)}{r_1^2} - W(r_1) + k^2\right] F_l(r_1)$$

$$= \pm\left[\frac{8\pi}{2l + 1} \int_0^\infty \frac{r_<^l}{r_>^{l+1}} F_l(r_2) \, \xi_0(r_2) \, r_2 \, dr_2 - 4\pi(1 + k^2) \, \delta_{0,l} \times \int_0^\infty F_l(r_2) \, \xi_0(r_2) \, r_2 \, dr_2\right] r_1 \xi_0(r_1), \quad \text{(L65)}$$

where

$$W(r) = U(1) + \int U(1, 2) \, \xi_0^2(2) \, d\tau_2 = -2\left(1 + \frac{1}{r}\right) e^{-2r}.$$

(L65) is the familiar equation of the Fock approximation.

If instead of (L64) one starts with the Hartree approximation

$$\Phi_l = \varphi(1) \, \xi_0(2) = \frac{1}{r_1} \, F_l(r_1) \, Y_l(\vartheta_1) \, \xi_0(2), \quad \text{(L66)}$$

the (L65) becomes

$$\left[\frac{d^2}{dr_1^2} - \frac{l(l + 1)}{r_1^2} - W(r_1) + k^2\right] F_l(r_1) = 0. \quad \text{(L67)}$$

The H atom is now represented by the screened field $W(r)$.

In the following table, the phase shifts $\eta_l$, for $l = 0, 1$, in the Fock (L64-5) and in the Hartree approximation (L66-7) are tabulated. It is seen that the effect of the Pauli principle is considerable.

For very small (zero) energies, the theory of scattering length and effective range of Sect. E can be used. Let the wave function at zero energy be normalized according to

$$\Psi_0 \to \left(\frac{1}{r_2} - \frac{1}{a_\pm}\right) \xi_0(1), \qquad (r_2 \to \infty)$$

$$\to \pm\left(\frac{1}{r_1} - \frac{1}{a_\pm}\right) \xi_0(2), \quad (r_1 \to \infty). \quad \text{(L68)}$$

The effective range is given by (ref. 10)

$$r_{0\pm} = \frac{1}{4\pi}\int_0^\infty (u_0^2 - \Psi_0'^2)\, d\tau_1\, d\tau_2 \mp 8\left(1 - \frac{2}{a_\pm}\right)^2, \tag{L69}$$

where

$$u_0 = \left(\frac{1}{r_2} - \frac{1}{a_\pm}\right)\xi_0(1) \pm \left(\frac{1}{r_1} - \frac{1}{a_\pm}\right)\xi_0(2).$$

The first term in (L69) corresponds to (E11a), and the second term comes from the exchange.

In the Hartree–Fock approximation, in units of Bohr radius (ref. 8a),

$$\begin{aligned} a_+ &= 8.06, & a_- &= 2.35, \\ r_{0+} &= 3.02, & r_{0-} &= 1.22. \end{aligned} \tag{L70}$$

The singlet system has one bound state $1s^{2\,1}S$ with a small binding energy, and $a_+$ is positive (see Sect. E). There does not seem to be a bound state in the triplet system. $a_-$ is positive. The phase shifts $\eta_0^\pm$ ($l = 0$) start from $\pi$ and decrease monotonically to zero at high energies.# For $l \geqslant 1$, in the Hartree–Fock approximation, $\eta_l^\pm \propto k^{2l+1}$ for small energies [see (A27)]. This relation is not valid when the electron–electron correlation is taken into account. The results of calculations are shown in Table 2.

---

# In this connection it may be remarked that Levinson's theorem [see (E22)] is not always valid when the Pauli principle must be taken into account. The triplet phase shift at $k = 0$ must be (or greater than) $\pi$. This can be shown in the following way. Let the antisymmetrized wave function of the triplet $S$ state be $\psi^{(-)}(1, 2)$. Constant the function $\varphi$,

$$\varphi(1) = \int \psi^{(-)}(1, 2)\, \xi_0(2)\, d\tau_2, \quad \varphi(1) \to \frac{\sin(kr_1 + \eta)}{r_1}.$$

The phase shift $\eta$ is defined as if $r_1\varphi(r_1)$ were the solution of a one-dimensional wave equation,

$$\eta = \lim_{m\to\infty} (m\pi - kr^{(m)}),$$

where $r^{(m)}$ is the $m$th zero in $\varphi(1)$. The following integral vanishes,

$$\int \varphi(1)\, \xi_0(1)\, d\tau_1 = \int \psi^{(-)}(1, 2)\, \xi_0(1)\, \xi_0(2)\, d\tau_1\, d\tau_2 = 0$$

because of the antisymmetry of $\psi^{(-)}$. Since $\xi_0(1)$ is positive definite, $\varphi(1)$ must have at least one node. Therefore the Pauli principle increases the phase shift at zero energy at least by $\pi$ even if a bound state does not exist. The zero energy scattering phase shift is given by $(n + m)\pi$ in general, where $n$ is the number of bound states and $m$ is the number of states from which the incident particle is excluded by the Pauli principle (see ref. 9).

TABLE 2. *The s and p phase shifts (in radians) of electron–hydrogen atom scattering in the Hartree and the Fock approximation*

| Electron energy in rydberg | $l = 0$ | | | $l = 1$ | | |
|---|---|---|---|---|---|---|
| | Hartree | Fock | | Hartree | Fock | |
| | | Singlet (+) | Triplet (−) | | Singlet (+) | Triplet (−) |
| 0.01 | 0.726 | 2.396 | 2.907 | 0.0003 | −0.0012 | 0.0021 |
| 0.04 | 0.973 | 1.870 | 2.679 | 0.0021 | −0.0084 | 0.0166 |
| 0.09 | 1.046 | 1.508 | 2.461 | 0.0066 | −0.0241 | 0.0511 |
| 0.25 | 1.045 | 1.031 | 2.070 | 0.026 | −0.0703 | 0.169 |
| 1.00 | 0.906 | 0.543 | 1.391 | 0.112 | −0.1059 | 0.358 |

### *iii) Electron–electron correlation*

To take into account the electron–electron correlation, one of the most effective ways is to introduce the electron–electron distance $r_{12}$ explicitly in the wave function, as was first done by Hylleraas in his classic treatment of the ground state of the two-electron atoms. Massey and Moiseiwitsch use the trial wave function

$$\psi^{(\pm)} = \Phi(1, 2) \pm \Phi(2, 1),$$

$$\Phi(1, 2) = \frac{1}{r_2}\{\sin kr_2 + [a + (b + cr_{12}) \exp(-r_2)] \times [1 - \exp(-r_2)] \cos kr_2\}\, \xi_0(1), \tag{L71}$$

where $a$, $b$, $c$ are variational parameters. The phase shifts $\eta_0^{(\pm)}$ so obtained are given in Table 3. The improvement of this method over the Hartree–Fock approximation is mainly due to the term $cr_{12}$ in (L71) which takes care of the

TABLE 3. *The s phase shifts (in radians) of electron–hydrogen atom scattering.* MM: *Massey and Moiseiwitsch* (*ref.* 11), TL: *Temkin and Lamkin* (*ref.* 11c).

| Electron energy in rydberg | Singlet | | | Triplet | | |
|---|---|---|---|---|---|---|
| | Fock | MM | TL | Fock | MM | TL |
| Scattering length | 8.10 | (7) | 5.8 | 2.35 | (2.3) | 1.9 |
| 0.01 | 2.396 | 2.48 | 2.58 | 2.907 | 2.91 | 2.945 |
| 0.04 | 1.870 | 2.00 | 2.11 | 2.679 | 2.68 | 2.73 |
| 0.09 | 1.508 | 1.65 | 1.75 | 2.461 | 2.45 | 2.52 |
| 0.25 | 1.031 | 1.25 | 1.25 | 2.070 | 2.04 | 2.13 |
| 1.00 | 0.543 | 0.71 | 0.76 | 1.391 | 1.40 | 1.46 |

"short-range correlation". In a recent work, Temkin and Lamkin (ref. 11c) attempt to include the long-range polarization# in a reasonable approximation (see next subsection). It is seen from Table 3 that the phase shifts are

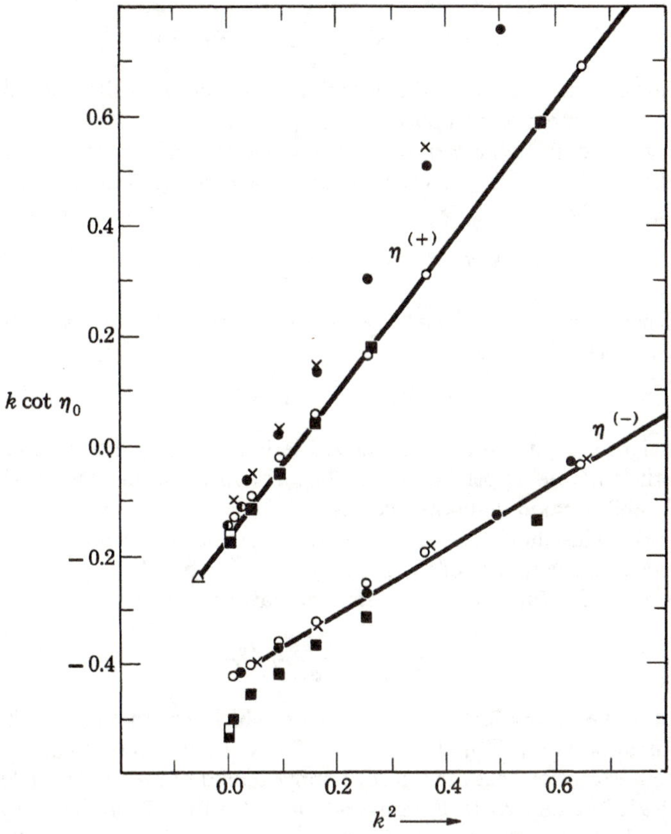


**Figure 1.** The values of $k$ cot $\eta$ of electron–hydrogen atom scattering obtained by various methods. Closed circles: the Fock approximation; Open circles: Massey and Moiseiwitsch, variational method; Half-closed circles: Bransden *et al.*; Crosses: Geltman; Open squares: Rosenberg, Spruch, and O'Malley; Closed squares: Temkin and Lamkin; Triangle: the bound state of $H^-$. [From Ohmura, T., and Ohmura, H., *Phys. Rev.* **121**, 513 (1961).]

improved appreciably by the short-range correlation for the singlet, and by the long-range correlation for the triplet state scattering. This is probably due

# This distinction between the "short-range" and the "long-range" correlation is made only for convenience and is strictly speaking not possible.

to the smaller probability of finding two electrons at very small separations in the triplet than in the singlet states on account of the Pauli exclusion principle. The best values of the scattering lengths for the singlet and the triplet state at the present time (ref. 11b) are

$$a_+ = 6.22(^1S), \qquad a_- = 1.91(^3S). \tag{L72}$$

These have been derived by the variational method with the normalization (E41b), and assured to be upper bounds (Sect. E.3).

The $s$ phase shift obtained by many workers will be summarized in Fig. 1. The solid line for the singlet system is drawn according to the effective range formula [see (D30)],

$$k \cot \eta_0^{(+)} = -\gamma + \frac{\rho}{2}(\gamma^2 + k^2).$$

$\gamma$ and $\rho$ can be calculated from the asymptotic form of the $H^-$ wave function [ref. 10; see (L75)] to be

$$\gamma = 0.2356 a_B^{-1}, \qquad \rho = 2.65 a_B.$$

The straight line gives $6.17a_B$ for the scattering length, which is close to and within the rigorous upper bound, $6.22a_B$. The line also fits Massey–Moiseiwitsch's and Temkin–Lamkin's values quite well up to rather high energies, say $k^2 = \frac{3}{4}$. This shows that the effective range approximation is better than expected from the nature of the approximation involved.

The solid line for the triplet system is drawn according to

$$k \cot \eta_0^{(-)} = -\frac{1}{a_-} + \frac{r_{0-}}{2} k^2,$$

with the values $a_- = 2.35$ and $r_{0-} = 1.22$, which are obtained in the Fock approximation [see (L70)]. The line also fits well a group of points, with the important exception that the rigorous upper bound of the scattering length is $a_- = 1.91$. The departure of Temkin–Lamkin's values from the straight line possibly shows the effect of the long range polarization, which will be discussed below.

### iv) Polarization potential

The effect of electron–electron correlation at small distances is most effectively represented by the use of $r_{12}$ in the wave function, as in (L71). At large distances, the same correlation gives rise to the polarization interaction which is not negligible especially for $l \geqslant 1$ and small energies.

In the adiabatic approximation, this interaction between an electron (coordinate $\mathbf{r}'$) and an H atom (electron coordinate $\mathbf{r}$) can be readily obtained from the perturbation theory. The perturbed wave function of the ground state is

$$\xi_0'(\mathbf{r}) = \xi_0(\mathbf{r}) - \sum_{m\neq 0} \frac{V_{0m}(r')}{E_m - E_0}\, \xi_m(\mathbf{r}),$$

$$V_{0m}(r') = \int -\frac{1}{|\mathbf{r} - \mathbf{r}'|}\, \xi_m^*(\mathbf{r})\, \xi_0(\mathbf{r})\, d\mathbf{r}. \tag{L73a}$$

Calculation gives

$$\xi_0'(\mathbf{r}) = \xi_0(\mathbf{r}) + \sum_{n\neq 0} \left(\frac{r}{r'}\right)^{n+1} \left(\frac{n}{n+1} + \frac{1}{nr}\right) P_n(\cos\vartheta), \quad (r' > r),$$

$\vartheta$ = angle between $\mathbf{r}$ and $\mathbf{r}'$. The interaction between the electron and the H atom is given by the second-order energy

$$V(r') = -\sum_{m\neq 0} \frac{|V_{m0}(r')|^2}{E_m - E_0}$$

$$= -\sum_{n\neq 0} \frac{1}{r'^{2n+2}}\, \frac{(2n+1)!\,(n+2)}{(2^{2n-1})\, n(n+1)} \tag{L73b}$$

$$\to -\frac{9}{2r'^4}. \tag{L73c}$$

This is the interaction between the electron and the induced dipole of the H atom in the ground state. About $\frac{2}{3}$ of this comes from the $2p$ state in (L73a). The expressions (L73a)–(L73c) are of course valid only for $r'$ large ($r' \to \infty$). For small distances $r'$, the adiabatic approximation itself is not valid and there is no meaning in holding one electron fixed at a distance $r'$ from the proton and calculating the energy of the other electron as a function of $r'$.

In the problem of the scattering of a proton (or even a $\mu$ or $\pi$ meson) by an H atom, the adiabatic approximation is justifiable on account of the large mass of the incident particle compared with that of the atomic electron.

The theory of scattering length and effective range of Sect. E needs some modification if the long-range correlation (L73c) is taken into account (ref. 14). For low energies at which only elastic scattering is possible, the radial wave function $\psi_0(\mathbf{r})$ (L57) satisfies the equation (ref. 13)

$$\left[\Delta + \frac{\beta^2}{r^4} + k^2\right] \psi_0(\mathbf{r}) = 0, \quad \text{for} \quad r \to \infty. \tag{L74}$$

The $s$-wave solution of (L74) in the limit of zero energy is

$$\phi(r) = \frac{r}{\beta} \sin\frac{\beta}{r} - \frac{r}{a} \cos\frac{\beta}{r}$$

$$= 1 - \left(\frac{r}{a}\right) + \left(\frac{\beta^2}{2ar}\right) - \left(\frac{\beta^2}{6r^2}\right) + \cdots.$$

The zero-energy $s$-wave $u_0(r)$ of (L74) approaches $\phi(r)$ as $r \to \infty$. The effective range $r_0$ defined in (E11a) for central-field, namely

$$r_0 = 2\int_0^\infty (v_0^2 - u_0^2)\,dr, \quad v_0 = 1 - \frac{r}{a},$$

is seen to diverge.# This suggests that there should be other terms between $-1/a$ and $\frac{1}{2}r_0k^2$ in the expansion of $k \cot \eta$. Actually, for potential with the tail $-\beta^2/r^4$, it can be shown (ref. 14) that

$$k \cot \eta = -\frac{1}{a} + \frac{\pi\beta^2}{3a^2}k + ck^2 \ln\left(\tfrac{1}{4}\beta k\right) + \tfrac{1}{2}r_0'k^2 + \cdots.$$

On the other hand, $k \cot \eta$ is an even function of $k$## and can be expanded in powers of $k^2$. The presence of the extra terms shows that $k \cot \eta$ is not analytic at $k = 0$.

We know that $k \cot \eta$ can be expanded accurately around the bound state if the binding energy is small. [See (E30) and (E31).] In the electron–hydrogen atom singlet system, such a state is the ground state of the negative hydrogen ion, with the small electron affinity (=binding energy) $\frac{1}{2}\gamma^2 = 0.0556$ rydberg or about 0.7 eV. The expansion has the form

$$k \cot \eta^{(+)} = -\gamma + \frac{\rho}{2}(\gamma^2 + k^2) + O[(\gamma^2 + k^2)^2]. \tag{L75}$$

The effective range $\rho$ in the ion state is obtainable from

$$\rho = \frac{1}{\gamma} - \frac{1}{4\pi^2c^2},$$

where $c$ is the coefficient appearing in the normalized asymptotic $H^-$ wave function $\psi(\mathbf{r}_1, \mathbf{r}_2)$,

$$\psi(r_1 = 0, |\mathbf{r}_2| = r) = \psi(|\mathbf{r}_1| = r, r_2 = 0) \to ce^{-\gamma r}/r, \quad (r \to \infty).$$

The value of $\rho$ derived from the Pekeris wave function of $H^-$ is $\rho = 2.65a_B$. The scattering length in the singlet state $a_+$ can be determined by setting $k = 0$ in (L75). The result is $a_+ = 6.17a_B$, in good agreement with (L72)

---

# In the case of electron–hydrogen atom scattering, $r_0$ is somewhat different [see (L69)], but essentially the same argument applies.

## From (G9a), $f(k) = f^*(-k)$,

$$\exp[2i\eta(k)] = S(k) = \frac{f(k)}{f(-k)} = \frac{f(k)}{f^*(k)}$$

$$\exp[2i\eta(-k)] = S(-k) = \frac{f(-k)}{f(k)} = \frac{f^*(k)}{f(k)}.$$

Therefore,

$$(-k)\cot \eta(-k) = k \cot \eta(k).$$

deduced through the use of a rigorous minimum principle. It is also noteworthy that, in spite of the singularity at $k = 0$, the formula (L75) with the values of $\gamma$ and $\rho$ determined by the $H^-$ wave function reproduces the phase shift values of MM and TL tabulated in Table 3 quite well up to $k^2 = 1$ (in rydberg) (refs. 10, 8a and 16). The long range polarization, which is the cause of the singularity, thus gives only a small correction to the singlet scattering but a moderate change for the triplet.

It has been shown (ref. 11a) that the phase shifts with $l \geqslant 1$ are proportional to $k^2$ for small energies rather than $k^{2l+1}$ as in the short-range potential scattering or in the Hartree–Fock approximation of electron–hydrogen scattering, if the long-range polarization is taken into account properly. This qualitative feature can also be seen from the Born approximation. The radial wave function $u_l(r)$ satisfies the equation

$$\left(\frac{d^2}{dr^2} + k^2 - \frac{l(l+1)}{r^2} + \frac{\beta^2}{r^4}\right) u_l(r) = 0, \quad r \to \infty.$$

The phase shift $\eta_l$ in the Born approximation is given by (C8a).

$$\eta_l = -\frac{\lambda\pi}{2}\int_0^\infty J^2_{l+\frac{1}{2}}(kr)\, U(r)\, r\, dr.$$

For small energies, $J^2_{l+\frac{1}{2}}(kr)$ is proportional to $(kr)^{2l+1}$, hence $\eta_l$ is proportional to $k^{2l+1}$ at low energies if $U(r)$ is short-range. However, if $\lambda U(r)$ is replaced by $-\beta^2/r^4$, we have

$$\eta_l \sim \frac{\pi\beta^2}{2}\int_0^\infty J^2_{l+\frac{1}{2}}(kr)\frac{dr}{r^3}, \quad kr = x$$

$$= \frac{\pi\beta^2 k^2}{2}\int_0^\infty J^2_{l+\frac{1}{2}}(x)\frac{dx}{x^3}.$$

Thus $\eta_l$ is proportional to $k^2$ (except for $l = 0$, where the potential at short distances is more important) at low energies on account of the long-range nature of the polarization potential. The correctness of this statement is demonstrated by the numerical values of Temkin (ref. 11c) in a systematic treatment of long-range polarization. His method is based on the fact that inclusion of $2p$ states in (L20) will cover about $\frac{2}{3}$ of the effect of the long-range polarization.

### v) *Résumé of variational calculations of scattering lengths*

We usually do not know how accurate the result is in an approximate calculation of an atomic scattering problem. Since the upper bound theorem on the scattering length has been established in the variational treatment of scattering problem (see Sect. E.3), we are now in a position to compare the results of various approximations on the $s$ wave scattering of electrons by hydrogen atoms at zero energy to see how the results are improved with the choice of the trial functions. Corresponding to (E41b) the total wave function $\psi_0$ with angular momentum zero has the asymptotic form [see (L68)],

$$\psi_0 \to \frac{1}{r_2}(a_\pm - r_2)\,\xi_0(1), \qquad (r_2 \to \infty)$$

$$\to \pm \frac{1}{r_1}(a_\pm - r_1)\,\xi_0(2), \quad (r_1 \to \infty).$$

The following is a list of results and trial wave functions employed by various authors.

1) Hartree–Fock approximation [Seaton, ref. 8a, also see (L70)],

$$\psi^{(\pm)} = \Phi(1, 2) \pm \Phi(2, 1), \qquad \Phi(1, 2) = u(2)\,\xi_0(1),$$

with the full flexibility of $u(r)$. The results are $a_+ = 8.058$, $a_- = 2.347$ (in atomic units).

2) Massey and Moiseiwitsch (1951, ref. 11).
The trial function is given by (L71) with the limit $k \to +0$. The number $n$ of adjustable parameters is three. The result is $a_- = 2.33$. ($a_+$ has not been computed at $k = 0$.)

3) Borowitz and Greenberg (1957, ref. 19).

$$\psi^{(+)} = \Phi(1, 2) + \Phi(2, 1),$$

$$\Phi(1, 2) = (a + br_{12})\exp(-\alpha r_1 - \beta r_2) + \xi_0(1)$$
$$\times \{1 - \exp(-\gamma r_2) + c[1 - \exp(-\gamma r_2)]^2\, r_2^{-1}\}.$$

$n = 3$ (namely $c$, $\gamma$ and $a$) with $b/a = 0.3121$, $\alpha = 1.075$ and $\beta = 0.4776$. The fixed values are chosen to give the best $H^-$ wave function for the first term (symmetrized version) of $\Phi$. The novel feature of this $\psi$ is that $\psi$ is reduced to an $H^-$ wave function at small electron separations. The result is $a_+ = 7.63$. [This is the revised value by Rosenberg *et al.*, see Phys. Rev. **118**, 184 (1960).]

4) Seaton (1957, ref. 8a).

$$\Phi(1, 2) = \xi_0(1)\{1 - ar_2^{-1}[1 - \exp(-2r_2)] + b\exp(-r_2) + c\exp(-r_{12})\}.$$

$a_+ = 7.03$, $n = 3$. (If $c = 0$, $a_+ = 8.11$, thus a good value for the Fock approximation.)

5) Martin, Seaton and Wallace (1958, ref. 18).

$$\Phi(1, 2) = \left[\xi_0(1) + \sum_m \frac{V_{2p}^m(2)\,\xi_{2p}^m(1)}{E_0 - E_{2p}}\right] u(2),$$

$$V_{2p}^m(r) = \int \xi_{2p}^{m*}(r')\,\frac{1}{|\mathbf{r} - \mathbf{r}'|}\,\xi_0(\mathbf{r}')\,d\mathbf{r},$$

where $E_0 = -1$, $E_{2p} = -\frac{1}{4}$ rydberg, and $u(r)$ is the same function obtained in the Fock approximation. The results are $a_+ = 6.372$ and $a_- = 1.993$. (The value of $a_+$ obtained has not been proved rigorously to be an upper bound.) The trial function here is seen to be improved over the Fock approximation in that the target hydrogen atom $\xi_0$ has an induced dipole term which is fixed according to the adiabatic approximation (L73a). The actual calculation includes the $2p$ states only. These results are the best among **1)–6)**, indicating the importance of the effect of the polarization of the target atom.

**6)** Geltman (1960, ref. 20).

$$\Phi(1, 2) = \xi_0(1)\{1 + ar_2^{-1}[1 - \exp(-r_2)]\} + \exp(-\gamma r_2) \sum_{n=0}^{2} C_n \xi_n(1),$$

where $\xi_0 = e^{-r}$, $\xi_1$ and $\xi_2$ are $1s$-, $2s$- and $3s$-wave functions, respectively. $n = 5$ (namely, $a$, $\gamma$, $C_0$, $C_1$ and $C_2$). $a_+ = 8.652$, $a_- = 2.351$. The results are even worse than the Fock approximation values indicating that the inclusion of $s$ states, unlike the $p$ states, is ineffective in allowing for the effect of polarization. Also note that Geltman's values are similar to the Fock approximation values for $k \neq 0$. See Fig. 1.

**7)** Rosenberg, Spruch and O'Malley (1960, ref. 11b).

$$\Phi(1, 2) = \xi_0(1)\{1 + A[1 - \exp(-br_2)]r_2^{-1} + B \exp(-cr_1 - dr_2) + C \exp(-fr_1 - gr_2 - hr_{12})\}.$$

$n = 9$. $a_+ = 6.23$, $a_- = 1.93$. If another term, $D \exp(-lr_1 - mr_2 - nr_{12})$, is added ($n = 12$) to $\Phi(1, 2)$, we have $a_- = 1.91$. They use a trial and error method to choose a number of sets of parameter values, and select the best one. If the full minimization (L62) is employed, still better results will be obtained. $a_- = 1.91$ is the best value hitherto obtained.

**8)** Hara, Ohmura and Yamanouchi (1961, ref. 11b).

$$\psi = \Phi(1, 2) \pm \Phi(2, 1) + f_\pm,$$

$$\Phi(1, 2) = \xi_0(1)\{1 + a_1(1 - e^{-r_2})\, r_2^{-1}\}$$

$$f_+ = e^{-s}(a_2 + a_3 s + a_4 u + a_5 s^2 + a_6 t^2 + a_7 u^2 + a_8 us),$$

$$f_- = e^{-s}(a_2 t + a_3 st + a_4 ut + a_5 u^2 t),$$

where $s = r_1 + r_2$, $t = r_2 - r_1$ and $u = r_{12}$. The successive improvement is shown in Table 4 as the number of adjustable parameters $n$ is increased. There is a positive jump of $a_+$ from $n = 0$ to $n = 1$

TABLE 4

| $n$ | $a_+$ | $a_-$ |
|---|---|---|
| 0 | −7.000 | 5.000 |
| 1 | 10.895 | 2.354 |
| 2 | 8.669 | 2.349 |
| 3 | 8.145 | 2.349 |
| 4 | 6.301 | 2.315 |
| 5 | 6.293 | 2.272 |
| 6 | 6.293 | |
| 7 | 6.265 | |
| 8 | 6.217 | |

(due to the existence of $H^-$ state), otherwise $a_\pm$ decreases monotonically and converges to the true value from above (see Sect. E.3). The convergence of $a_-$ is not good in view of $a_- = 1.93$ obtained in 7), while the convergence of $a_+$ is not bad. The inclusion of $a_4 e^{-s} u$ term reduces the value of $a_+$ from 8.15 to 6.30. The damping factor $e^{-s}$ in $f_\pm$ seems too strong to take into account the long range polarization effect sufficiently. Recently C. Schwartz has obtained by the variational method the much better values $a_+ = 5.97$ and $a_- = 1.77$, (ref. 11d).

## 6. EXPERIMENTAL WORK ON ELECTRON-HYDROGEN ATOM SCATTERING

The scattering of electrons by hydrogen atoms has been recently studied with modern experimental techniques. The elastic scattering ($1s \to 1s$) of low energy electrons (below 10 eV) has been measured by Bederson *et al.* (ref. 21) and by Fite *et al.* (ref. 22, 22a). The former group of authors found a (broad) maximum in the cross section at electron energies ~3 eV, which is not present in all the calculated cross sections and has not been found by Fite *et al.* It is believed that this maximum is spurious, but its exact "experimental origin" is not clear. The results of Fite *et al.* are in general agreement with the earlier measurements of Ramsauer and Kollath (ref. 22b), and with the theoretical results of Bransden *et al.* (ref. 11a).

For the inelastic scatterings at electron energies higher than one rydberg (13.6 eV), the analysis of the measured cross sections is complicated by such considerations as (i) the presence of cascade processes (such as the excitation $1s \to 3s$ followed by a transition $3s \to 2p$, say) and (ii) the degeneracy in the quantum number $l$ and the Stark transitions between the degenerate levels. The $1s \to 2s$ scattering has been studied by Lichten and Schultz (ref. 23) for electron energies below 45 eV and by Stebbings *et al.* (ref. 23a) for energies up to 600 eV. The $1s \to 2p$ scattering has been studied by Fite *et al.* (ref. 24) for electron energies 10–50 eV. In these cases, there are differences between the experimental results of different authors as well as between the experimental and the theoretical results. This is due to the fact that, for the "intermediate" energy range (for which higher partial waves become important and the simple Born approximation is not yet good enough), accurate theoretical calculations are difficult (ref. 25).

## REFERENCES


1. Oppenheimer, J. R., Phys. Rev. **32**, 361 (1928), treats the effect of Pauli principle in the scattering of electrons by hydrogen and helium atoms to the "Born approximation" (L39).

2. Massey, H. S. W. and Mohr, C. B. O., Proc. Roy. Soc. (London) **A132**, 605 (1931).

3. Corinaldesi, E. and Trainor, L., Nuovo Cimento **9**, 940 (1952).

3a. Wu, T. Y., Can. J. Phys. **38**, 1654 (1960).

4. Bates, D. R. and Miskelly, D., Proc. Phys. Soc. (London) **A70**, 539 (1957), calculate the Born cross sections for $1s \to 2s$, $1s \to 3s$ by partial wave method.

4a. Kingston, A. E., Moiseiwitsch, B. L. and Skinner, B. G., Proc. Roy. Soc. (London) **A258**, 237 (1960). $1s \to 2s$ excitation in second Born approximation.

5. Moses, H. E., Phys. Rev. **91**, 185 (1953), obtains the result (L26) in a different notation.

6. Lippmann, B. A., Phys. Rev. **102**, 264 (1956), obtains the result (L26) very simply by means of the Green's function method. See the following section.

7. Altschuler, S., Phys. Rev. **91**, 1167; **92**, 1157 (1953).

8. Morse, P. M. and Allis, W. P., Phys. Rev. **44**, 269 (1933); Feenberg, E., Phys. Rev. **42**, 17 (1932), Hartree–Fock approximation. The calculation for *s*, *p*, and *d* phase shifts in this approximation has been done by many workers, for example, Omidvar, K., New York Univ. Institute of Mathematical Sciences Report No. CX-37. (Unpublished) ; Treffetz, E. and John, T. L. (to be published).

8a. Seaton, M. J., Proc. Roy. Soc. (London) **A241**, 522 (1957); Ohmura, T. and Ohmura, H., Phys. Rev. **118**, 154 (1960).

9. Rosenberg, L. and Spruch, L., Phys. Rev. **121**, 1720 (1961); Temkin, A., J. Math. Phys., **2**, 336 (1961); Swan, P., Proc. Roy. Soc. (London) **A228**, 10 (1955).

10. Ohmura, T., Hara, Y. and Yamanouchi, T., Prog. Theor. Phys. (Kyoto) **20**, 82 (1958).

**11.** Massey, H. S. W. and Moiseiwitsch, B. L., Proc. Roy. Soc. (London) **A205**, 483 (1951). This is one of the earliest variational calculation on electron–hydrogen atom elastic scattering. There are many recent calculations, for example, see:

**11a.** Bransden, Dalgarno, John and Seaton, Proc. Phys. Soc. (London) **A71**, 877 (1958).

**11b.** Rosenberg, Spruch and O'Malley, Phys. Rev. **119**, 164 (1960); Hara, Ohmura and Yamanouchi, Prog. Theor. Phys. (Kyoto) **25**, 467 (1961).

**11c.** Temkin, A. and Lamkin, J. C., Phys. Rev. **121**, 788 (1961).

**11d.** Schwartz, C., Phys. Rev., **124**, 1468 (1961).

**12.** Dalgarno, A. and Lynn, N., Proc. Phys. Soc. (London) **A70**, 223 (1957). Evaluation of (L73).

**13.** Castillejo, Percival and Seaton, Proc. Roy. Soc. (London) **A254**, 259 (1960).

**14.** Spruch, L., O'Malley, T. F. and Rosenberg, L., Phys. Rev. Letters **5**, 375 (1960); J. Math. Phys., **2**, 491 (1961).

**15.** Temkin, A., Phys. Rev. Letters **4**, 566 (1960); **6**, 354 (1961). Other formulation of the electron–hydrogen atom scattering.

**16.** Gerjuoy, E., Rev. Modern Phys. **33**, 544 (1961).

**17.** Rotenberg, M. (to be published). Another formulation of the electron (and positron) hydrogen atom scattering.

**18.** Martin, Seaton and Wallace, Proc. Phys. Soc. (London) **72**, 701 (1958).

**19.** Borowitz, S. and Greenberg, H., Phys. Rev. **108**, 716 (1957).

**20.** Geltman, S. Phys. Rev. **119**, 1283 (1960).

**21.** Bederson, B., Hammer, J. M. and Malamud, N., Technical Report No. 2, College of Engineering, Phys. Dept., New York Univ. (1958).

**22.** Brackman, R. T., Fite, W. L. and Neynater, R. H., Phys. Rev. **112**, 1157 (1958).

**22a.** Gilbody, H. B., Stebbings, R. F. and Fite, W. L., Phys. Rev. **121**, 794 (1961). Angular distributions in elastic scattering.

**22b.** Ramsauer, C. and Kollath, R., Ann. Physik **12**, 529 (1932).

**23.** Lichten, W. and Schultz, S., Phys. Rev. **116**, 1132 (1959).

**23a.** Stebbings, R. F., Fite, W. L., Hammer, D. G. and Brackmann, R. T., Phys. Rev. **119**, 1939 (1960).

**24.** Fite, W. L., Stebbings, R. F. and Brackman, R. T., Phys. Rev. **116**, 356 (1959).

**25.** Hammer, D. G. and Seaton, M. J., Phys. Rev. Letters **6**, 471 (1961).

section

# M Scattering Involving Rearrangements

## 1. INTEGRAL EQUATION FORMALISM (LIPPMANN)

We shall obtain the Schrödinger equation in integral form for the description of scattering processes in general. Thus we may be interested in a process of the type

$$A + B \rightarrow C + D, \tag{M1}$$

where $A$, $B$, $C$, $D$ may be single or composite particles. In Sect. L, we have left the equivalence of (L20) and (L26) [or, of (L20a) and (L26a)] unproved. These rearrangement collisions can be most simply treated after the manner of Lippmann. For the system

$$H = H_0 + V, \tag{M2}$$

we introduce the operators (Green's functions) which are Hermitian and which correspond to outgoing waves (Sect. B).

$$\begin{aligned} G_0(\lambda) &\equiv G_0(E + i\varepsilon) = \frac{1}{E - H_0 + i\varepsilon} = \frac{1}{\lambda - H_0}, \\ G(\lambda) &\equiv G(E + i\varepsilon) = \frac{1}{E - H + i\varepsilon} = \frac{1}{\lambda - H}. \end{aligned} \tag{M3}$$

For these, one obtains, by direct operation with the "differential" operators $\lambda - H_0$, $\lambda - H$, and the $G_0(\lambda)$, $G(\lambda)$,

$$\begin{aligned} G(\lambda) &= G_0(\lambda)[1 + VG(\lambda)], \qquad G(\lambda) = [1 + G(\lambda)V]\, G_0(\lambda), \\ &\text{and} \\ G_0(\lambda) &= G(\lambda)[1 - VG_0(\lambda)], \qquad G_0(\lambda) = [1 - G_0(\lambda)V]\, G(\lambda). \end{aligned} \tag{M4}$$

On defining the operators $T$, $U$ by

$$T(\lambda) = V + VG(\lambda)V, \tag{M5}$$

$$U(\lambda) = 1 + G(\lambda)\, V, \tag{M6}$$

one obtains from (M4)

$$G = UG_0$$

and

$$G_0 = (1 - G_0V)\, G = (1 - G_0V)\, UG_0,$$

so that

$$(1 - G_0V)\, U(\lambda) = 1. \tag{M7}$$

Also

$$T = V(1 + GV) = VU. \tag{M8}$$

The Schrödinger equation (L20a)

$$\begin{aligned} \Psi_a &= \phi_c + \frac{1}{E_a - H_0 + i\varepsilon} V\Psi_a \\ &= \phi_a + G_0(\lambda)\, V\Psi_a \end{aligned} \tag{M9}$$

can therefore be written

$$[1 - G_0(\lambda)V]\,\Psi_a = \phi_a. \tag{M10}$$

Using the operator equation (M7) above, one has

$$[1 - G_0(\lambda)V]\,\Psi_a = [1 - G_0(\lambda)V]\, U(\lambda)\, \phi_a$$

so that Schrödinger equation (M10) becomes

$$\Psi_a = U(\lambda)\, \phi_a = [1 + G(\lambda)V]\, \phi_a. \tag{M11}$$

For describing the direct scatterings, we write as in (L10)

$$H = H_0' + V',$$

where $H_0'$ is the Hamiltonian of the system in the initial state, i.e., *before* the colliding particles come close to each other. Then equations (M3), (M7), (M9), (M11) above become

$$G_0'(\lambda) = \frac{1}{\lambda - H_0'}, \quad \lambda = E + i\varepsilon, \tag{M3'}$$

$$[1 - G_0'(\lambda)V']\, U'(\lambda) = 1, \tag{M7'}$$

$$\Psi_a'' = \phi_a' + G_0'(\lambda)\, V'\Psi_a'', \tag{M9'}$$

or

$$\Psi_a'' = U'(\lambda)\, \phi_a'. \tag{M11'}$$

For describing the "exchange" or rearrangement collisions, we write, as in (L27)

$$H = H_0'' + V''.$$

Here $H_0''$ is the Hamiltonian of the system in the final state, i.e., after the resulting particles have separated off to large distances at which their interaction $V''$ vanishes. On operating on (M7′) by $\lambda - H_0'$, one obtains

$$(\lambda - H_0')\, U' - V'U' = \lambda - H_0',$$

which becomes, on account of $H_0' + V' = H_0'' + V'' = H$,

$$(\lambda - H)\, U' = \lambda - H_0'' + V' - V''. \tag{M12}$$

Operating on (M12) by $G_0''(\lambda) = 1/(\lambda - H_0'')$, one obtains

$$\frac{1}{\lambda - H_0''}(\lambda - H_0'' - V'')\, U' = 1 + G_0''(\lambda)(V' - V''),$$

or

$$[1 - G_0''(\lambda)\, V'']\, U' = 1 + G_0''(\lambda)(V' - V''). \tag{M13}$$

Operating on (M11′) by $[1 - G_0''(\lambda) V'']$ and on using (M13), we have

$$[1 - G_0''(\lambda)\, V'']\Psi_a' = [1 + G_0''(\lambda)(V' - V'')]\, \phi_a'.$$

Now

$$[1 + G_0''(\lambda)(V' - V'')]\, \phi_a' = \frac{E - H_0'' + i\varepsilon + V' - V''}{E - H_0'' + i\varepsilon}\, \phi_a'$$
$$= \frac{E - H_0' + i\varepsilon}{E - H_0'' + i\varepsilon}\, \phi_a' = \frac{i\varepsilon}{E - H_0'' + i\varepsilon}\, \phi_a'.$$

Hence the Schrödinger equation describing the rearrangement collision is

$$\Psi_a' = \frac{i\varepsilon}{E - H_0'' + i\varepsilon}\, \phi_a' + \frac{1}{E - H_0'' + i\varepsilon}\, V''\Psi_a'. \tag{M14}$$

(M14) is (L26a), and leads to (L26) in the special case treated in Section L. This proves the equivalence of (M9′) and (M14), i.e., of (L20a) and (L26a). It is to be noted (i) that the function in (M14) describing the rearrangement process is the same function in (M9′) describing the direct scattering. (M14) has been obtained not independently, but by a transformation from (M9′); and (ii) that it is not correct to drop the first term in (M14) in general, as has also been shown in (L31)–(L32).

To describe the process (M1), let us write the Hamiltonian in two alternative forms

$$H = H_{AB}' + V_{AB}', \tag{M15}$$
$$= H_{CD}'' + V_{CD}''. \tag{M16}$$

The form (M15) is appropriate for the initial state in which $A$ and $B$ are far apart so that $V_{AB}'$ is small and $H_{AB}'$ is simply the sum of the Hamiltonians of the separate particles $A$ and $B$. (M16) is appropriate for the final state.

Equations (M9′) and (M14) now take the form

$$\Psi'_{AB} = \phi'_{AB} + \frac{1}{E - H'_{AB} + i\varepsilon} V'_{AB} \Psi'_{AB}, \tag{M17}$$

$$\Psi''_{AB} = \frac{i\varepsilon}{E - H''_{CD} + i\varepsilon} \phi'_{AB} + \frac{1}{E - H''_{CD} + i\varepsilon} V''_{CD} \Psi''_{AB}. \tag{M18}$$

(M17) describes the direct scattering

$$A + B \to A + B,$$

and (M18) the rearrangement process (M1). They are the Schrödinger equations, in integral form and containing the appropriate asymptotic form through the Green's function. To apply them to an actual process, it remains to formulate the $H'$, $H''$, $V'$, $V''$ in manageable forms and to solve these equations, and the remarks in the introduction in Sect. L apply. In almost all cases, the only approximate solution that can be carried out with ease is the Born approximation# in which $\psi'_{AB}$ on the right-hand side in (M17), (M18) is replaced by the initial wave function $\phi'_{AB}$. This approximation is illustrated by (L38), (L39) for the case of the scattering of an electron by a hydrogen atom.

Just as in the simple case of the scattering of a particle by a potential in Sect. B, the boundary (or asymptotic) conditions of a rearrangement collision are associated with the Green's function $G$ in (M17-8). For a detailed discussion of the boundary conditions see a paper by Gerjuoy (ref. 2).

## 2. GERJUOY'S THEORY

In a recent work, Gerjuoy (ref. 4) has suggested a formulation of the problem of rearrangement collisions in general, which starts from a somewhat different point of view from the standard theory of transition probabilities.

---

# In this connection, we may mention an important question recently raised about the application of the "Born approximations" to collisions involving rearrangements. [See paragraphs following (L41).] It will be shown in Sect. P that the transition amplitude for the transition from state $\phi_a$ to $\phi_b$ due to a perturbation $V$ is, (P41),

$$T_{ba} = (\phi_b, V\psi_a^{(+)}).$$

If we iterate (M9) and substitute into this, we get a series

$$T_{ba} = (\phi_b, V\phi_a) + (\phi_b, VG_0V\phi_a) + (\phi_b, VG_0VG_0V\phi_a) + \cdots$$

of which the first term is the (first) Born approximation [for example, (L39)]. Recently, Aaron, Amado and Lee (ref. 3) have found that for a class of problems the above series (in powers of $V$) diverges. The generality, however, of this result and its full understanding seem to call for further studies.

The basic ideas will be briefly described below. For the sake of clarity and without any loss of generality, let us consider the collisions

$$a + b \rightarrow a + b, \tag{M19}$$

$$a + b \rightarrow c + d, \tag{M20}$$

where $a, b, c, d$ are themselves composite particles (i.e., consisting of a number of particles, for example $a, b, c, d$ may be molecules). Let $u_a(\mathbf{s}_a)$, $u_b(\mathbf{s}_b)$, $u_c(\mathbf{s}_c)$, $u_d(\mathbf{s}_d)$ be the eigenstates of the $a, b, c, d$ (as separate systems), $\mathbf{s}_a, \mathbf{s}_b, \ldots$ their internal coordinates (vibration, rotation, electronic, for example). Let $\mathbf{r}_a$, $\mathbf{r}_b, \ldots$ be the radius vectors of the centers of mass of $a, b, \ldots$. If $a + b$ is the initial state (i.e., the system exists as two "aggregates" $a$ and $b$ at infinite separations, hence non-interacting), the Schrödinger equation and eigenfunction $\psi_i$ are given by

$$(H_i - E)\,\psi_i \equiv (H_a + H_b - E)\,\psi_i = 0, \tag{M21}$$

$$\psi_i = u_a(\mathbf{s}_a)\,u_b(\mathbf{s}_b)\exp(i\mathbf{k}_a\cdot\mathbf{r}_a + i\mathbf{k}_b\cdot\mathbf{r}_b). \tag{M22}$$

Similarly, a state $c + d$ ($c$, $d$ at infinite separations) is described by

$$(H_f - E)\,\psi_f \equiv (H_c + H_d - E)\,\psi_f = 0, \tag{M23}$$

$$\psi_f = u_c(\mathbf{s}_c)\,u_d(\mathbf{s}_d)\exp(i\mathbf{k}_c\cdot\mathbf{r}_c + i\mathbf{k}_d\cdot\mathbf{r}_d). \tag{M24}$$

Let the total number of particles in the system be $n$ (which may form $a + b$, $c + d$, or $e + f + g$, etc.), and let $H$, $\Psi$ be the total Hamiltonian and eigenstate of the system. Then

$$H = H_a + H_b + H_{ab} \equiv H_i + V_i, \tag{M25}$$

$$= H_c + H_d + H_{cd} \equiv H_f + V_f, \tag{M26}$$

and

$$(H - E)\,\Psi = 0. \tag{M27}$$

The wave function that represents outgoing wave in $a + b$ for the initial state $\psi_i$ in (M22) is given by (M17)

$$\Psi_i = \psi_i + \frac{1}{E + i\varepsilon - H_i} V_i \Psi_i \tag{M28}$$

which, by means of (M21) or (L20b), can be shown to have the solution

$$\Psi_i = \psi_i + \frac{1}{E + i\varepsilon - H} V_i \psi_i \equiv \psi_i + G^{(+)} V_i \psi_i \tag{M28a}$$

$$\equiv \psi_i + \varphi_i.$$

Here $V_i = H_{ab}$ in (M25) denotes the interaction between (the particles of) $a$ and (the particles of) $b$.

Let us rewrite the Hamiltonian $H$ in (M25) in the form

$$(H - E)\Psi \equiv (T + V - E)\Psi = 0, \tag{M29}$$

where $T$ is the kinetic energy operator of all the $n$ particles and $V$ the interaction between all pairs of particles

$$T = -\sum_j^n \frac{\hbar^2}{2M_j}\nabla_j^2, \qquad V = \sum_{i\neq j}^n V(\mathbf{r}_i, \mathbf{r}_j). \tag{M30}$$

If $\Psi_1$, $\Psi_2$ are two eigenstates of $H$ at the same energy ($\Psi_1$, $\Psi_2$ may be states $a + b$, $c + d$, $e + f + g$, etc.), then a generalized Green's theorem can readily be proved (from $\Psi_1 T\Psi_2 - \Psi_2 T\Psi_1 = 0$)

$$\int dSJ(\Psi_2, \Psi_1) \equiv \frac{1}{i\hbar}\int (d\mathbf{S}_\nu \cdot \mathbf{W}) = 0, \tag{M31}$$

where $\mathbf{W}$ is the probability current in the $3n$-dimensional space of $(\mathbf{r}_1, \ldots, \mathbf{r}_n)$, whose $j$th component (corresponding to the $j$th particle) is

$$\mathbf{W}_j = \frac{\hbar^2}{2M_j}(\Psi_2\nabla_j\Psi_1 - \Psi_1\nabla_j\Psi_2). \tag{M32}$$

$d\mathbf{S}_\nu$ is the (outward) normal component of the surface element, and the integration in (M31) is taken over an infinite sphere in the $3n$-space.

In the $3n$-configurational space, a surface element on the infinite sphere represents a configuration of the system when $|\mathbf{r}| = |\sum_j |\mathbf{r}_j|^2|^{1/2}$ becomes infinite. An aggregate, such as $a$, is represented by an element $dS_a$ such that the distances among the particles forming the aggregate $a$ remain finite while their distances from all other particles become infinite as $|\mathbf{r}| \to \infty$. The element $dS_{ab}$ will represent the state $a + b$, etc.

The plausible postulate is now made that the current operator $\mathbf{W}$ represents the scattered current, i.e., the probability of a particular rearrangement collision (or, a particular "channel" of the reaction) is given by the current through a particular surface element of the infinite sphere in the $3n$-space. In calculating this current, one needs the wave function $\Psi_i$ and $\Psi_f$ (for the final state, or channel), or rather their asymptotic forms. These asymptotic forms depend on the asymptotic form of the Green's functions for the various states ($\Psi_i$, $\Psi_f$). We shall not take up the details of the work to which the reader is referred, but only summarize the result that the theory leads to the same expression for the transition probability as given in the usual time-dependent theory [see Chapter 4, (P35-6), (Q4-6)].

A simple result of the theory is obtained by putting $\psi_1 = \psi_i$, $\psi_2 = \psi_i^*$, where $\psi_i$ is that given in (M22), into (M31). One obtains then

$$\int dSJ(\varphi_i^*, \varphi) = -\frac{2}{\hbar}\,\mathrm{Im}\int [d\mathbf{S}_\nu \cdot \mathbf{W}(\psi_i^*, \varphi_i)], \tag{M33}$$

where $\varphi_i = -G^{(+)}V_i\psi_i$ is the scattered wave. The integration is over the surface of the infinite sphere. On account of the individual aggregate wave functions $u(a)$, $u(b)$ in $\psi_i$ of (M22), the contributions to the integral on the right come only from those directions corresponding to the formation of the initial aggregates $a$ and $b$. Thus the relation (M33) is a sort of generalized optical theorem (A21), namely, a relation between the total cross section and the imaginary part of the forward scattered amplitude.

It is seen that the theory furnishes a novel point of view in obtaining essentially the same result as the theory of transitions. The claim that the theory has circumvented the difficulty associated with non-orthogonal states (discussed in Sect. L.2) should perhaps be taken in the following sense. In *any general, formal* formulation such as the theory of Lippmann in the preceding section, or of Lippmann and Schwinger (Sect. P), or Gell-Mann and Goldberger (Sect. Q), the *exact* transition matrix elements (P36) (Q48) etc. can always be written down also without any difficulty from the non-orthogonality of states. It is only in the actual calculations by approximate methods that the difficulty arises.

### 3. SCATTERING TREATED BY A SYSTEM OF DIFFERENTIAL–INTEGRAL EQUATIONS

The integral equations for scattering in Sect. M can be put in the form of an infinite system of differential–integral equations. The essence of the method can be sufficiently brought out by the following simple example.

Consider a simple particle $A$ (such as an electron, $\mu$ meson or proton) and a particle $X$ (such as an atom, a molecule or an atomic nucleus). Let $\mathbf{r}$ denote collectively all the coordinates in $X$ (in the case of a molecule, $\mathbf{r}$ denotes the electronic, vibrational and rotational coordinates), and let $\mathbf{R}$ denote the coordinate of $A$ with respect to the center of mass of $X$. The Hamiltonian of the system, after separating off the part due to the motion of the center of mass of $A$ and $X$, is then of the form

$$H(\mathbf{R}, \mathbf{r}) = K(\mathbf{R}) + \mathscr{H}_0(\mathbf{r}) + H_1(\mathbf{r}, \mathbf{R}) + V(\mathbf{r}, \mathbf{R}), \tag{M34}$$

where $K(\mathbf{R})$ is the kinetic energy of relative motion of $A$ and $X$, $\mathscr{H}_0(\mathbf{r})$ is the Hamiltonian of the $X$ alone and is itself in turn made up of terms of the type in (M34), $H_1(\mathbf{r}, \mathbf{R})$ is some cross terms in the $\nabla_r$, $\nabla_R$, and $V(\mathbf{r}, \mathbf{R})$ the interaction between $A$ and $X$. Let the eigenvalues and eigenfunctions of $\mathscr{H}_0(\mathbf{r})$ be denoted by $W_n$, $\phi_n(\mathbf{r})$ where $n$ denotes the totality of all the quantum numbers. We have

$$(\mathscr{H}_0 - W_n)\,\phi_n(\mathbf{r}) = 0, \tag{M35}$$

and assume that the $\phi_n(\mathbf{r})$ form a complete set. Let the wave function $\Psi(\mathbf{R}, \mathbf{r})$ of $H(\mathbf{R}, \mathbf{r})$

$$[H(\mathbf{R}, \mathbf{r}) - E]\,\Psi(\mathbf{R}, \mathbf{r}) = 0 \tag{M36}$$

be expanded in a series

$$\Psi(\mathbf{R}, \mathbf{r}) = \sum_m \!\!\!\!\!\!\int \psi_m(\mathbf{R})\, \phi_m(\mathbf{r}). \tag{M37}$$

On putting (M37) into (M36), and using (M35), there results the system of equations

$$(-K + E - W_n)\, \psi_n(\mathbf{R}) = \sum_m \!\!\!\!\!\!\int V_{nm}(\mathbf{R})\, \psi_m(\mathbf{R}), \tag{M38}$$

where

$$V_{nm}(\mathbf{R}) = \int \phi_n^*(\mathbf{r})[H_1(\mathbf{r}, \mathbf{R}) + V(\mathbf{r}, \mathbf{R})]\, \phi_m(\mathbf{r})\, d\mathbf{r}. \tag{M39}$$

The system (M38) is exact.

For the scattering problem with the initial condition

$$\Psi_0(\mathbf{R}, \mathbf{r}) = e^{ikZ}\phi_0(\mathbf{r}) \tag{M40}$$

and the asymptotic condition for an outgoing wave for $\mathbf{R}$,

$$\Psi(\mathbf{R}, \mathbf{r}) \to \Psi_0(\mathbf{R}, \mathbf{r}) + \frac{\exp(ik_nR)}{R} f_n(\vartheta)\, \phi_n(\mathbf{r}), \tag{M41}$$

$$\hbar^2k_n^2 = 2\mu(E - W_n),$$

one solves the system (M38) subject to these conditions. It is seen that (M38), together with (M40), (M41), is equivalent to the integral equation (M17).

In practice, one tries to obtain approximate solutions in various ways.

### i) Born approximation

The simplest approximation is to retain only one term in the sum in (M38), namely, $V_{n0}(\mathbf{R})\, \psi_0(\mathbf{R})$, and to replace $\psi_0(\mathbf{R})$ by the plane wave $e^{i\mathbf{k}\cdot\mathbf{R}}$. Then (M38) becomes a system of uncoupled inhomogeneous equations

$$(-K + E - W_n)\, \psi_n(\mathbf{R}) = V_{n0}(\mathbf{R})\, e^{i\mathbf{k}\cdot\mathbf{R}}, \tag{M42}$$

By the method of Sects. B, C, one obtains for the scattering amplitude of the inelastic collision (in which the particle $X$ has been excited or deexcited from state 0 to state $n$)

$$f_n(\vartheta) = -\frac{1}{4\pi}\int \exp(-i\mathbf{k}_n\cdot\mathbf{R}')\, U_{n0}(R')\exp(i\mathbf{k}\cdot\mathbf{R}')\, d\mathbf{R}', \tag{M43}$$

where

$$U_{n0}(R) = \frac{2\mu}{\hbar^2}V_{n0}(R), \quad \hbar^2k_n^2 = 2\mu(E - W_n),$$

$\mu$ being the reduced mass associated with the coordinate $R$ in $H_0(R)$. The differential cross section is given by

$$\sigma(\vartheta)\, d\Omega = \frac{k_n}{k}|f_n(\vartheta)|^2\, d\cos\vartheta\, d\varphi, \tag{M43a}$$

the factor $k_n/k$ coming from the ratio of the scattered and the incident flux.

### *ii) "Distorted wave" approximation*

In many cases (for example, when $A$ is a slow electron or a "heavy" particle such as an atom) the plane wave approximation for $\psi_0(\mathbf{R})$ is not valid. The following approximation has been used in the literature (ref. 11, Chapters VIII, 5; VI, 3). The underlying idea is to regard the $H_1 + V$ in (M39) as a perturbation so that the terms representing the scattered waves $\psi_m(R)$, $m \neq 0$, in (M37) are small compared with that representing the incident wave $\psi_0(\mathbf{R})$, and the terms $V_{0m}(R)\,\psi_m(\mathbf{R})$, $m \neq 0$, are negligible compared with $V_{m0}(R)\psi_0(\mathbf{R})$. The system (M38) now becomes

$$[-K + E - W_0 - V_{00}(R)]\,\psi_0(\mathbf{R}) = 0, \tag{M44a}$$

$$[-K + E - W_n - V_{nn}(R)]\,\psi_n(\mathbf{R}) = V_{n0}(R)\psi_0(R). \tag{M44b}$$

These are to be solved subject to the asymptotic condition

$$\begin{aligned} \psi_0(\mathbf{R}) &\to e^{i\mathbf{k}\cdot\mathbf{R}} + \frac{1}{R}\, e^{ikR} f_0(\vartheta), \\ \psi_n(\mathbf{R}) &\to \qquad\quad \frac{1}{R}\exp(ik_nR) f_n(\vartheta). \end{aligned} \tag{M45}$$

The solution $\psi_0(R)$ is that obtained in Sect. A, (A12)

$$\psi_0(\mathbf{R}) = \sum (2l + 1)\, i^l \exp[i\delta_l(0)]\, \frac{1}{kR}\, u_l(kR)\, P_l(\cos\vartheta)$$

where, asymptotically

$$u_l(kR) \to \sin[kR - (l\pi/2) + \delta_l(0)],$$

$\delta_0(0)$ being the phase shifts for elastic scattering by the field $V_{00}(R)$. The solution $\phi_n(\mathbf{R})$ is similarly expressible as

$$\psi_n(\mathbf{R}') = \sum (2l + 1)\, i^l \exp[i\delta_l(n)]\, \frac{1}{k_nR}\, L_l(k_nR')\, P_l[\cos(\pi - \Theta)],$$

where $\Theta$ is the angle between $\mathbf{R}'$ and the direction of the scattered wave $\mathbf{k}_n$, and $L_l(k_nR)$ is that solution of the homogeneous equation obtained by equating the left-hand side of (M44b) to zero, which is regular at $R = 0$ and behaves asymptotically as

$$\sin[k_nR - (l\pi/2) + \delta_l(n)],$$

where $\delta_l(n)$ is the phase shift in the field $V_{nn}(R)$. If $\vartheta$ is the scattering angle (between $\mathbf{k}$ and $\mathbf{k}_n$), $\vartheta'$, $\varphi'$ the polar angles of $\mathbf{R}'$ with $\mathbf{k}$ as the polar axis, then

$$\cos\Theta = \cos\vartheta\cos\vartheta' + \sin\vartheta\sin\vartheta'\cos(\varphi - \varphi').$$

The (inelastic) scattering amplitude is then given by

$$f_n(\vartheta) \equiv -\frac{1}{4\pi}\int \psi_n(R', \pi - \Theta)\, U_{n0}(R')\, \psi_0(R', \vartheta')\, d\mathbf{R}', \qquad \text{(M46)}$$

and the cross section again by (M43a) with this $f_n(\vartheta)$. This approximation is known as the "distorted wave" approximation because the $V_{00}(R)$, $V_{nn}(R)$ distort the $\psi_0$, $\psi_n$ from the plane wave states. (M46) reduces to the Born approximation (M43) if $\psi_0$, $\psi_n$ are replaced by plane waves.

### *iii) Two-state approximation*

As a better approximation for the scattering to the final state $n$, we may replace the infinite system (M38) by the coupled equations

$$\begin{aligned} [-K + E - W_0 - V_{00}(R)]\, \psi_0(\mathbf{R}) &= V_{0n}(R)\, \psi_n(\mathbf{R}), \\ [-K + E - W_n - V_{nn}(R)]\, \psi_n(\mathbf{R}) &= V_{n0}(R)\, \psi_0(\mathbf{R}), \end{aligned} \qquad \text{(M47)}$$

corresponding to replacing (M37) by

$$\psi(\mathbf{R}, \mathbf{r}) = \psi_0(\mathbf{R})\, \phi_0(\mathbf{r}) + \psi_n(\mathbf{R})\, \phi_n(\mathbf{r}). \qquad \text{(M48)}$$

If, instead of solving the pair (M47) exactly, $V_{0n}\,(=V_{n0}^*)$ is treated as a perturbation in this "two-state approximation", one gets the distorted wave approximation (M44).

This procedure of solving (M47) exactly is particularly necessary when $V_{0n}$ is large, as in the case of resonance. For further details of the treatment of the case of exact resonance, such as a charge transfer $He^+ + He \rightarrow He + He^+$, see ref. 11, Chapter VIII, § 6.

The simplest problem to which the approximate methods of these subsections can be applied is the capture of an electron from an H atom by a fast proton (velocity > electron orbital velocity in first Bohr orbit, or energy > 25 keV) (ref. 10),

$$H + H^+ \rightarrow H^+ + H.$$

It turns out that even for this simple system, the Born and the distorted wave approximation still do not give a completely satisfactory account of the experimental results. For a brief sketch of the many investigations of this problem, see refs. 87–96.

### *iv) Introduction of exchange*

The "exchange" scattering is often not negligible when identical particles are involved in collision phenomena. For example, the exchange effect is appreciable in the low-energy scattering of electrons by light atoms. To include such an effect the wave function (M37) must be properly symmetrized. We shall take up the scattering of electrons by hydrogen atoms as an example. The wave function (M48) is modified as follows,

$$\psi(\mathbf{R}, \mathbf{r}) = \psi_0(\mathbf{R})\,\phi_0(\mathbf{r}) + \psi_n(\mathbf{R})\,\phi_n(\mathbf{r}) \pm \psi_0(\mathbf{r})\,\phi_0(\mathbf{R}) \pm \psi_n(\mathbf{r})\,\phi_n(\mathbf{R}), \quad \text{(M49)}$$

where $+(-)$ corresponds to the singlet (triplet) state. On assuming infinitely heavy protons, we get the following coupled equations in the two-state approximation with exchange

$$[-K + E - W_0 - V_{00}(R)]\,\psi_0(\mathbf{R}) \pm \int G_{00}(\mathbf{R}, \mathbf{R}')\,\psi_0(\mathbf{R}')\,d\mathbf{R}'$$
$$= V_{0n}(R)\,\psi_n(\mathbf{R}) \mp \int G_{0n}(\mathbf{R}, \mathbf{R}')\,\psi_n(\mathbf{R}')\,d\mathbf{R}', \quad \text{(M50a)}$$

$$[-K + E - W_n - V_{nn}(R)]\,\psi_n(\mathbf{R}) \pm \int G_{nn}(\mathbf{R}, \mathbf{R}')\,\psi_n(\mathbf{R}')\,d\mathbf{R}'$$
$$= V_{n0}(R)\,\psi_0(\mathbf{R}) \mp \int G_{n0}(\mathbf{R}, \mathbf{R}')\,\psi_0(\mathbf{R}')\,d\mathbf{R}'. \quad \text{(M50b)}$$

$G_{00}$, $G_{nn}$, and $G_{0n} = G_{n0}^*$ are defined by

$$\begin{aligned} G_{00}(\mathbf{R}, \mathbf{R}') &= \phi_0^*(\mathbf{R})\,\phi_0(\mathbf{R}')\left(E - 2W_0 - \frac{e^2}{|\mathbf{R} - \mathbf{R}'|}\right), \\ G_{nn}(\mathbf{R}, \mathbf{R}') &= \phi_0^*(\mathbf{R})\,\phi_n(\mathbf{R}')\left(E - 2W_n - \frac{e^2}{|\mathbf{R} - \mathbf{R}'|}\right), \\ G_{0n}(\mathbf{R}, \mathbf{R}') &= \phi_0^*(\mathbf{R})\,\phi_0(\mathbf{R}')\left(E - W_0 - W_n - \frac{e^2}{|\mathbf{R} - \mathbf{R}'|}\right). \end{aligned} \quad \text{(M51)}$$

If $\psi_0(\mathbf{R})$ and $\psi_n(\mathbf{R})$ are solutions of (M50a) and (M50b) without the right-hand side terms, the improved version of the distorted wave approximation is given by

$$f_n(\vartheta, \phi) = -\frac{1}{4\pi}\int \psi_2^*(\mathbf{R})[V_{n0}(R)\psi_0(\mathbf{R}) \pm \int G_{n0}(\mathbf{R}, \mathbf{R}')\psi_0(\mathbf{R}')\,d\mathbf{R}']\,d\mathbf{R} \quad \text{(M52)}$$

with the boundary conditions

$$\psi_0(\mathbf{R}) \to \exp(i\mathbf{k}_0\cdot\mathbf{R}) + f_0'(\vartheta, \varphi)\exp(ik_0R)/R,$$
$$\psi_n(\mathbf{R}) \to \exp(i\mathbf{k}_n\cdot\mathbf{R}) + f_n'(\vartheta, \varphi)\exp(ik_nR)/R.$$

If $\psi_0$ and $\psi_n$ are replaced by plane waves in (M52), $f_n(\vartheta, \varphi)$ becomes $f_{n0}$ of (L41) (the Born approximation with exchange, i.e., the Oppenheimer expression), (M50a) is equivalent to the Hartree–Fock approximation, (L65), if the right-hand side term is omitted.

Let us examine the results of calculations, in various approximations, on the inelastic ($1s \to 2s$) scattering of electrons by hydrogen atoms. The cross sections shown in Table 1 have been obtained by taking into account $s$ electrons only. The effects of electron exchange and of distortion of the wave

TABLE 1. *The total cross section of* $H(1s) + e^- \to H(2s) + e^-$ *calculated in various approximations.* $a_B$ = *Bohr radius,* DW: *distorted wave approximation;* TS: *two-state approximation.* (DW *is taken from ref.* 5, TS *without exchange from ref.* 6, *and* TS *with exchange from ref.* 7.)

| Energy of incident electrons (atomic units) | (eV) | Cross sections in $\pi a_B^2$ | | | | | |
|---|---|---|---|---|---|---|---|
| | | Exchange neglected | | | Exchange included | | |
| | | Born | DW | TS | Born | DW | TS |
| 0.75 | 10.2 | 0 | 0 | 0 | 0 | 0 | 0 |
| 1.00 | 13.5 | 0.198 | 0.239 | 0.204 | 1.59 | 0.178 | 0.074 |
| 1.44 | 19.4 | 0.127 | 0.118 | 0.102 | 0.503 | 0.094 | 0.061 |
| 2.25 | 30.4 | $0.058_5$ | 0.045 | 0.045 | 0.104 | 0.035 | $0.031_5$ |
| 4.00 | 54.0 | 0.019 | 0.014 | $0.015_5$ | 0.020 | 0.011 | 0.012 |

function are quite appreciable for small incident energies. The last two columns show that even the distorted wave approximation with exchange is not very good for low energies. Furthermore, there is no assurance that the result of the last column is accurate unless we know that a "three-state approximation" does not alter the result significantly of the "two-state approximation". This is not obvious because the coupling of 2*s* and (added) 2*p*-states may not be small. A reliable calculation of the inelastic scattering is thus rather difficult even for this simplest example in atomic collisions.

For references to the experimental data of ($1s \to 2s$) excitation of hydrogen atom by electrons, see Sect. L.6.

Consider a somewhat different case where the incident electron excites the system to a state belonging to the same electron configuration as the initial state. The calculated cross sections by the distorted wave approximation (exchange included), for example, for $^3P \to {}^1S$ or $^3P \to {}^1D$ transition in atomic oxygen exceeds the theoretical upper limit by two orders of magnitude (ref. 9). This is due to the strong coupling among the $^3P$, $^1D$ and $^1S$ states, and therefore the three coupled equation should be solved accurately. Detailed accounts on atomic collisions are found in refs. 8, 11 and 13.

### v) "*Modified wave number*" *method*

A further approximation along the line of the distorted wave method has been used by Takayanagi (ref. 15) for dealing with the transfer of translational to rotational or vibrational energies in molecular collisions. This method consists in reducing the solution for higher angular momentum partial waves to that for *s* waves by replacing the terms $k^2 - l(l+1)/R^2$ (for $l$ not too large) by a mean $k^{*2}$, when the variation of $k^2 - l(l+1)/R^2$ over the range of relevant $R$ is small compared with that of the interaction $V_{0n}(R)$.

The need for such an approximation can be illustrated by the large class of processes involving collisions between atomic (and molecular) systems. Consider for example the elastic collision between two atoms such as potassium at thermal energies (of the order $10^{-2}$ eV). If we use the Born–Oppenheimer static-nuclei (usually called the adiabatic) approximation, we may represent the interaction between the two atoms by a potential $V(r)$, where $r$ is the interatomic distance, which has the following general characteristics: Starting from very large $r$, $V(r)$ is the van der Waals attraction (resulting from the interaction between the mutually induced dipoles in the two atoms). At smaller distances, $V(r)$ results from the "valence" or "exchange" interaction (which is essentially of the type treated by the Heitler–London theory). This "valence" attraction has a minimum if the order of $-1$ eV at a distance $r$ of the order 1 Å. At still smaller distances $r$, $V(r)$ becomes repulsive, rising rapidly with decreasing $r$. At very small $r$, this repulsive $V(r)$ is essentially the Coulomb interaction between the two nuclei.

Let us simplify the actual situation by simply considering the collision of two particles interacting with such a static central $V(r)$. The Schrödinger equation for the relative motion is

$$\left\{\frac{d^2}{dr^2} + \frac{2\mu}{\hbar^2}[E - V(r)] - \frac{l(l+1)}{r^2}\right\}\psi(\mathbf{r}) = 0.$$

It is seen that for a $V(r)$ as described above, partial waves of $l$ of the order $10^2$ contribute to the scattering. In the "exchange repulsion" region of $r$ [$r$ below the $r_e$ of minimum $V(r)$], $2\mu V(r)/\hbar^2 \gg l(l+1)/r^2$ for small $l$. Hence the modified-wave-number approximation.

## 4. SEMI-CLASSICAL (IMPACT PARAMETER AND ADIABATIC) METHODS

There is a large class of collision processes involving atoms, ions and molecules which can be treated by a semi-classical method. For the sake of definiteness, consider the excitation (or ionization) of an atom by a proton (or $\alpha$ particle)

$$\begin{aligned} A + H^+ &\rightarrow A' + H^+ \\ &\rightarrow A^+ + e + H^+. \end{aligned}$$

### i) *High velocity*

We shall first consider the case when the relative velocity $v$ between $H^+$ and $A$ is $\gg$ the orbital velocities of the electrons in $A$, and consequently the energy $\frac{1}{2}Mv^2$, where $M$ is the reduced mass of $H^+$ and $A$, is $\ggg$ the energy of excitation or ionization of $A$. In this case, the momentum and energy transfers due to the excitation or ionization of $A$ are negligible.

On account of the large mass of $H^+$ (compared with that of the electron) it is permissible to describe the motion of $H^+$ by a classical trajectory, which, under the conditions above, can be taken to be a straight line. The collision between $H^+$ and $A$ can be regarded as the action on $A$ by a moving center of force. Let us choose a cylindrical coordinate system with the $Z$-axis along the direction of motion, so that the distance $\mathbf{R}_{H^+} - \mathbf{R}_A \equiv \mathbf{R} = \mathbf{R}(z, \rho, \varphi)$ and $\rho$ is, under the condition of a straight trajectory, the impact parameter itself, and

$$R^2 = \rho^2 + z^2, \qquad z = vt.$$

The methods described below treat the electronic motions by quantum mechanics and the motions of the heavy particles by classical dynamics. For this reason they are called the semi-classical methods. The impact parameter $\rho$ obviously plays the role of the angular momentum through the relation $Mv\rho = l\hbar$.

Let $H_A$ be the Hamiltonian of the unperturbed $A$, and $V$ the interaction between the electrons of $A$ and $H^+$. Then

$$H = H_A + V(r, R) = H_A + \sum_i \frac{e^2}{|\mathbf{r}_i - \mathbf{R}|},$$

$$(H_A - E_n)\,\phi_n(r) = 0, \tag{M53}$$

where $r$ stands for all the electronic coordinates $\mathbf{r}_1, \ldots, \mathbf{r}_n$, and $\phi_n(\mathbf{r})$ the unperturbed atomic wave function. For the perturbed atom, the Schrödinger equation

$$\left(i\hbar\frac{\partial}{\partial t} - H_A - V\right)\Psi(\mathbf{r}, t; \mathbf{R}) = 0$$

can be solved by the method of Dirac (see Chapter 4, Sect. O.1) by expanding

$$\Psi(\mathbf{r}, t; \mathbf{R}) = \sum\!\!\!\!\!\!\int a_n(t)\,\phi_n(\mathbf{r})\exp(-iE_nt/\hbar).$$

This gives

$$i\hbar\dot{a}_n = \sum_m V_{nm}a_m(t)\exp[i(E_n - E_m)t/\hbar], \tag{M54}$$

where

$$V_{nm}(R) = \sum_i \int \phi_m^*(\mathbf{r})\frac{e^2}{|\mathbf{r}_i - \mathbf{R}|}\phi_n(\mathbf{r})\,d\mathbf{r}. \tag{M55}$$

Let the initial condition be the following:

$$a_0(t \to -\infty) = a_0(z \to -\infty) = 1,$$
$$\text{all other } a_m(t \to -\infty) = 0.$$

Remembering that $z = vt$, we obtain, on integrating to the first approximation,

$$i\hbar a_n(\rho) = \frac{1}{v}\int_{-\infty}^{\infty} dz\,V_{n0}(R)\exp(i\omega_{n0}z/v) \equiv \frac{1}{v}\,U(\rho), \tag{M56}$$

where

$$\hbar\omega_{n0} = E_n - E_0, \qquad \mathbf{R} = \mathbf{R}(\rho, z, \varphi).$$

The cross section is, upon integrating over all values of the impact parameter,

$$\sigma = 2\pi \int_0^\infty |a_n(\rho)|^2 \rho \, d\rho \tag{M57}$$

$$= \frac{2\pi}{\hbar^2 v^2} \int_0^\infty |U(\rho)|^2 \rho \, d\rho. \tag{M57a}$$

This is due to Gaunt (ref. 16).

It has been proved in a somewhat lengthy calculation by Frame (ref. 17) that the above result (M57a) is identical with the following

$$\sigma = \frac{k_n}{k_0} \int \left| \frac{M}{2\pi\hbar} \int V_{n0}(R) \exp\,[i(\mathbf{k} - \mathbf{k}_n)\cdot\mathbf{R}]\, d\mathbf{R} \right|^2 d\cos\vartheta\, d\varphi \tag{M58}$$

which is the Born approximation in which $H^+$ (its motion relative to $A$) is represented by a plane wave [see (M43)]. The proof of the equivalence of the impact parameter method and the Born approximation depends on the neglect of the coupling between the electronic and the nuclear motions (i.e., to the zeroth order Born–Oppenheimer approximation), but is otherwise rigorous. If we take the Born approximation as a good one in the high velocity limit, then the above equivalence establishes the semi-classical impact parameter method.#

---

# For velocities not very high, one may try to improve upon (M56) by using $i\hbar v(\partial/\partial z) \simeq a_0 V_{00}$ in (M54) thereby replacing $a_0 = 1$ by

$$a_0(z) = \exp\left(-\frac{i}{\hbar v}\int_0^z V_{00}\, dz\right),$$

and by retaining the term $a_n(z)V_{nn}$ in (M54). Then

$$i\hbar v \frac{\partial}{\partial z} a_n(z) = V_{n0} \exp\left(\frac{-i(E_n - E_0)z}{\hbar v} - \frac{i}{\hbar v}\int_0^z V_{00}\, dz\right),$$

and in place of (M56),

$$i\hbar a_n(\rho) = \frac{1}{v}\int_{-\infty}^{\infty} dz\, V_{n0}(R) \exp\left(\frac{i\omega_{n0}z}{v} + i\beta_{n0}\right), \tag{M59}$$

where

$$\beta_{n0} = \frac{1}{\hbar v}\int_0^z (V_{nn} - V_{00})\, dz. \tag{M60}$$

The factor $\exp\,(i\beta_{n0})$ is a correction for the "distortion" of the straight trajectory (plane wave) by a small deflection and a small change in the velocity $v$. This correction, introduced by Bates (ref. 18), has been applied to the excitation process

$$H^+ + H(1s) \rightarrow H^+ + H(2s),$$

and the correction is very large at low energies, reducing the Born approximation (M56) by a factor $\sim 20$ at energies $\sim 800$ eV. Such large corrections, however, throw some doubt on the whole method, since then one would have to investigate the corrections of higher orders.

### *ii)* $v \ll$ *orbital velocities, but* $\frac{1}{2}Mv^2 \gg E_n - E_0$

When the energy and momentum transfers to the excitation or ionization are very small compared with $\frac{1}{2}Mv^2$ and $Mv$, their effects on the trajectory may be neglected. But as $v$ is small compared with the orbital velocities of the electrons of $A$, it is permissible, in considering the electronic motions, to regard them as moving in the combined field of the nucleus of $A$ and $H^+$, both being spatially fixed. The system $A + H^+$ may be then regarded as forming a molecular ion $AH^+$.

The complete Hamiltonian of the system is given by (M34). In the "static-nuclei" or "adiabatic" approximation, we let $M \to \infty$. The Schrödinger equation for the electronic motions is, for a given $R$,

$$[H_A + V(r, R)]\, \phi_n(\mathbf{r};R) = \mathscr{E}_n(R)\, \phi_n(\mathbf{r}; R), \tag{M61}$$

where the eigenvalues $\mathscr{E}_n(R)$ are the electronic energies at a fixed $R$, and the electronic wave functions $\phi_n(\mathbf{r}; R)$ form a complete set of functions of $\mathbf{r}$ for any fixed $R$. In this semi-classical approximation, the motion in the interatomic coordinate $\mathbf{R}$ is not considered from the point of view of the Schrödinger equation. In the high velocity case, the $z$-component of $\mathbf{R}$ is represented by a uniform motion

$$z = vt \tag{M62}$$

and the coordinate $\rho$ = constant.

In this method, the zeroth order (static-nuclei) wave function is, for the initial state,

$$\Psi^{(0)}(\mathbf{r}, t; R) = \phi_0(\mathbf{r}; R) \exp(-i\mathscr{E}_0 t/\hbar). \tag{M63}$$

The motion (M62) will change the wave function, and hence causes transitions. If $v$ is very small, the condition of the system changes "adiabatically" and there are no transitions. Thus $v$ is the perturbation that causes transitions. To the first order (in the velocity $v$), we write

$$\Psi(\mathbf{r}, t; R) = \Psi^{(0)}(\mathbf{r}, t; R) + f(\mathbf{r}, t; R), \tag{M64}$$

where $f(\mathbf{r}, t; R)$ may be expanded

$$f = \sum a_n(t)\, \phi_n(\mathbf{r}; R) \exp(-i\mathscr{E}_n t/\hbar). \tag{M65}$$

From the equation

$$i\hbar \frac{\partial \Psi}{\partial t} = [H_A + V(\mathbf{r}, R)]\Psi,$$

with $\partial/\partial t = v(\partial/\partial z)$, and using (M61) for $\phi_n$, we get

$$v\frac{\partial \Psi^{(0)}}{\partial z} + \sum_n \left(\dot{a}_n\phi_n + a_n v \frac{\partial \phi_n}{\partial z}\right) \exp(-i\mathscr{E}_n t/\hbar) = 0.$$

For small $v$, we shall neglect the terms in $v$, and obtain

$$\dot{a}_n = -v \int \phi_0^*(\mathbf{r}, R) \frac{\partial}{\partial z} \phi_0(r, R)\, d\mathbf{r} \exp\left[\frac{i(\mathscr{E}_n - \mathscr{E}_0)t}{\hbar}\right], \tag{M66}$$

where $\mathscr{E}_0(R)$, $\mathscr{E}_n(R)$ are functions of $t$, or the component $z$. On integrating over the whole trajectory $z$ ($\rho$ being the fixed impact parameter), we get

$$a_n(\rho) = -\int_{-\infty}^{\infty} dz\, M_{n0}(R) \exp\left[i(\mathscr{E}_n - \mathscr{E}_0)z/\hbar v\right], \tag{M67}$$

where

$$M_{n0}(R) = \int \phi_n^*(\mathbf{r}, R) \frac{\partial z}{\partial} \phi_0(\mathbf{r}, R)\, d\mathbf{r}. \tag{M68}$$

The cross section is again given by

$$\sigma = 2\pi \int_0^{\infty} |a_n(\rho)|^2 \rho\, d\rho. \tag{M69}$$

This is called the "adiabatic approximation", in the sense that the heavy particles move slowly, compared with the electrons, and the electronic motions are described by molecular wave functions containing the interatomic distance $R$ as a parameter.

We shall now obtain an important result (Mott, ref. 19). Multiplying the equation (M61) for $\phi_0^*$ by $(\partial\phi_0/\partial z)\, d\mathbf{r}$ integrating over $\mathbf{r}$, subtracting this resulting equation from $\int d\mathbf{r}\phi_n^*$ [equation (M61) for $\phi_0$], we get, for the $M_{n0}(R)$ defined in (M68) above,

$$M_{n0}(R) = \frac{\int \phi_n^* \frac{\partial V}{\partial z} \phi_0(\mathbf{r}, R)\, d\mathbf{r}}{\mathscr{E}_n - \mathscr{E}_0}. \tag{M70}$$

Putting this into (M67), we get

$$a_n(\rho) = -\frac{1}{\hbar} \int_{-\infty}^{\infty} \frac{\partial}{\partial z} V_{n0}(R) \frac{e^{i\omega t}}{\omega}\, dz, \tag{M71}$$

where

$$\left.\begin{aligned} V_{n0}(R) &= \int \phi_n^*(\mathbf{r}; R)\, V(r, R)\, \phi_0(\mathbf{r}; R)\, d\mathbf{r},^{\#} \\ \hbar\omega(R) &= \mathscr{E}_n(R) - \mathscr{E}_0(R). \end{aligned}\right. \tag{M72}$$

It is seen that, on integrating (M71) by parts, the expression (M71) for $a_n(\rho)$

---

# For large values of the impact parameter $\rho$ (i.e. distant collisions), one may approximate the molecular functions $\phi_0(\mathbf{r}; R)$, $\phi_n(\mathbf{r}; R)$ in (M70), (M72) by the atomic wave functions $\phi_0(\mathbf{r})$, $\phi_n(\mathbf{r})$ of $A$ in (M53). In that case (M71) will reduce to the Born approximation (M58).

is identical with (M56) obtained for the high velocity case. Since one expects on general grounds that the Born approximation is good in the high velocity limit, one may conclude that the same formula (M71), or (M56), may be valid for slow as well as for fast collisions, although for slow collision, it does not depend on the Born approximation for its derivation.

*iii*) For still smaller velocities so that $\frac{1}{2}Mv^2$ is not much larger than the energy of excitation or ionization, the trajectory will no longer be a straight line, but will be bent, with a variable velocity along its path. Such a problem arises when we consider the excitation of $H$ from the $1s$ to the $2s$ state by a slow proton (just above the threshold, say). In pushing the adiabatic approximation further, we may adopt the following procedure: (1) Find the potential curves $\mathscr{E}_1(R)$, $\mathscr{E}_2(R)$ corresponding to the two asymptotic states $H(1s) + H^+$, $H(2s) + H^+$, for various angular momenta $l\hbar$. Also the electronic wave functions of $H_2^+$ in these states. (2) Find, for any fixed impact parameter $\rho$ or angular momentum $l\hbar$, the classical trajectory in two pieces, the incoming part corresponding to $\mathscr{E}_1(R)$, and the outgoing part to $\mathscr{E}_2(R)$, with the appropriate velocities $v_1(R)$, $v_2(R)$, respectively. In the region near the classical turning point, the two pieces are joined together on plausible arguments (Bauer, ref. 20). (3) The transition probability for a given $\rho$, $|a_2(\rho)|^2$, is calculated for each $R$ by using the Dirac perturbation theory (Sect. O.1) and integrating along the whole trajectory. (4) The total cross section is finally obtained by integrating over $\rho$ (M69). Such a semi-classical and semi-quantum mechanical method is plausible, but is open to many obvious objections (among them, the neglect of couplings of the two states in question and all the other states of the system $H + H^+$ or $H_2^+$). For such low velocity collisions, no really simple and yet reliable approximations are known.

The nature of the approximations made in the above "impact parameter" and the "adiabatic" methods will be brought out more clearly in the following subsection, where it will be seen how these methods can be obtained from a general, exact theory by making various approximations.

## 5. "PERTURBED STATIONARY STATE" OR "MOLECULAR WAVE FUNCTION" METHOD

For the large class of problems involving the slow collisions of an atom, ion or a molecule with another such particle, the following method can be formulated. It is in principle exact; but for the calculation of a specific problem to be practicable, approximations have to be made. We shall sketch the theory in its most general form, followed by an illustration by means of a simple process. It will then be shown how various approximations will lead to the approximate methods (Born, impact parameter and adiabatic) of the preceding subsection.

### *i) General formulation*

The guiding idea is that for slow [$v \ll$ orbital velocities of electrons, see 4, ii) above] collisions, the motions of the heavy particles are "adiabatic" so that, to solve the Schrödinger equation#

$$\left[-\frac{2\mu}{\hbar^2}\nabla_{\mathbf{R}}^2 + H_A + H_B + V(\mathbf{r}, \mathbf{R}) + \frac{e_A e_B}{R}\right]\Psi(\mathbf{r}, \mathbf{R}) = E\Psi \qquad \text{(M73)}$$

the most appropriate expansion will be in terms of the molecular wave functions $\phi_n(\mathbf{r}; R)$ which are the eigenfunctions, for fixed $R$, of

$$[H_A + H_B + V(\mathbf{r}, \mathbf{R})]\,\phi_n(\mathbf{r}; R) = \mathscr{E}_n(R)\,\phi_n(\mathbf{r}, R). \qquad \text{(M74)}$$

$\mathscr{E}_n(R)$ are the electronic energies at a given $R$, and the $\phi_n(\mathbf{r}; R)$ form a complete set of functions of $\mathbf{r}$ for any fixed $R$. Asymptotically for large $R$,

$$\begin{aligned} \mathscr{E}_n(R) &\to \mathscr{E}_n(\infty) = E_n(A) + E_n(B), \\ \phi_n(\mathbf{r}; R) &\to \phi_n(\mathbf{r}, \infty) = \varphi_n(A)\,\chi_n(B), \end{aligned} \qquad \text{(M75)}$$

where $E_n(A)$, $\varphi_n(A)$ are the energy and eigenfunction of the unperturbed atom $A$ in the state $\varphi_n$, etc. We shall assume that the $\mathscr{E}_n(R)$ and $\phi_n(\mathbf{r}; R)$ are already known for all $n$ and at all $R$ in the following discussion of the scattering problem. In actual fact, such a knowledge of $\mathscr{E}_n(R)$ and $\phi_n(\mathbf{r}; R)$ is not available even for the simplest of all molecular systems.

We shall now substitute the expansion##

$$\Psi(\mathbf{r}, R) = \sum\!\!\!\!\!\!\int \psi_n(R)\,\phi_n(\mathbf{r}; R) \qquad \text{(M76)}$$

into (M73), obtaining the (infinite) system of coupled equations

$$\left[\frac{\hbar^2}{2\mu}\nabla_{\mathbf{R}}^2 + E - \frac{e_A e_B}{R} - \mathscr{E}_n(R)\right]\psi_n(R) = \sum_m\!\!\!\!\!\!\int \frac{\hbar^2}{2\mu} C_{nm}(R)\,\psi_m(R), \qquad \text{(M77)}$$

---

# $H_A$, $H_B$ are the Hamiltonians of two non-interacting atomic systems. $\mathbf{r}$ stands for *all* the electronic coordinates, $\mathbf{R}$ the interatomic distance $A$–$B$, $\mu$ is the reduced mass, and $e_A$, $e_B$ are the charges of the nuclei of $A$ and $B$.

## We must note the mathematical similarity and the physical difference between the expansions (M76) and (M37). In (M37), the $\phi_m(\mathbf{r})$ are the wave functions of the unperturbed target system $X$, whereas in (M76), the $\phi_n(\mathbf{r}; R)$ are the "molecular wave functions" of all the electrons in the total (target + incident particles) system, in the static nuclei approximation. The expansion (M37) is the appropriate one for "fast" collisions, i.e., the relative velocity $v \gg$ orbital velocities of the electrons, while the expansion (M76) is the appropriate one for the "adiabatic collisions" (although in principle, both methods will give the same result in the limit of infinitely many terms in each expansion). The terminology "perturbed stationary state method" has perhaps been coined to mean that the "atomic" wave functions $\phi_n(\mathbf{r})$ in (M37) are now "perturbed" into the "molecular" wave functions $\phi_n(\mathbf{r}; R)$. Perhaps the name "method of molecular states" conveys the essence of the method more explicitly.

where the $C_{nm}$ are the operators

$$C_{nm}(R) = -\int \phi_n^* \nabla_{\mathbf{R}} \phi_m \, d\mathbf{r} \cdot \nabla_{\mathbf{R}} - \int \phi_n^* \nabla_{\mathbf{R}}^2 \phi_m \, d\mathbf{r}. \tag{M78}$$

We shall introduce the wave vectors $\mathbf{k}_0$, $\mathbf{k}_n$

$$\hbar^2 k_0^2 = 2\mu[E - \mathscr{E}_0(\infty)], \quad \hbar^2 k_n^2 = 2\mu[E - \mathscr{E}_n(\infty)]. \tag{M79}$$

and the "potential energy curve" for the state $\phi_n(\mathbf{r}; R)$ with the asymptotic state $\phi_n(\mathbf{r}; \infty)$ as zero energy,

$$U_n(R) = \frac{2\mu}{\hbar^2}\left[\frac{e_A e_B}{R} + \mathscr{E}_n(R) - \mathscr{E}_n(\infty)\right]. \tag{M80}$$

In our scattering problem, let the initial state be

$$\Psi \to \exp(i\mathbf{k}_0 \cdot \mathbf{R})\, \varphi_0(A)\, \chi_0(B). \tag{M81}$$

We now seek solutions of (M77) satisfying the asymptotic conditions

$$\begin{aligned} \psi_0(\mathbf{R})\,\phi_0(\mathbf{r}; R) &\to \left[\exp(i\mathbf{k}_0 \cdot \mathbf{R}) + \frac{\exp(ik_0 R)}{R} f_0(\Theta, \Phi)\right] \varphi_0(A)\, \chi_0(B), \\ \psi_n(\mathbf{R})\,\phi_n(\mathbf{r}; R) &\to \frac{\exp(ik_n R)}{R} f_n(\Theta, \Phi) \quad \varphi_n(A)\, \chi_n(B). \end{aligned} \tag{M82}$$

To obtain $f_0(\Theta, \Phi)$, $f_n(\Theta, \Phi)$, one solves the radial wave equation

$$\left[\frac{d^2}{dR^2} + k_n^2 - U_n(R) - C_{nn}(R) - \frac{l(l+1)}{R^2}\right] g_{n,l}(\mathbf{R}) = 0 \tag{M83}$$

for the solution which is regular at $R = 0$ and is asymptotically

$$g_{n,l}(\mathbf{R}) \to \frac{1}{k_n} \sin\left(k_n R - \frac{l}{2}\pi + \eta_{n,l}\right). \tag{M84}$$

The solution of the homogeneous equation of (M77),

$$[\nabla_{\mathbf{R}}^2 + k^2 - U_n(R) - C_{nn}(R)]\, \mathscr{F}_n(R, \Theta) = 0 \tag{M85}$$

subject to the asymptotic condition

$$\mathscr{F}_n(R, \Theta) \to \exp(ik_n R \cos\Theta) + \frac{\exp(ik_n R)}{R} f_n'(\Theta)$$

is then [see (A12a)]

$$\mathscr{F}_n(R, \Theta) = \frac{1}{R} \sum (2l + 1)\, i^l \exp(i\eta_{n,l})\, g_{n,l}(R)\, P_l(\cos\Theta). \tag{M86}$$

The scattering amplitude $f_n(\Theta, \Phi)$ in (M82) for the excitation from $\varphi_0(A)$ $\chi_0(B)$ to $\varphi_n(A)\, \chi_n(B)$ is then

$$f_n(\Theta) = -\frac{1}{4\pi} \int \mathscr{F}_n(R', \pi - \Theta) \sum_m C_{nm}(R')\, \mathscr{F}_m(R', \Theta')\, d\mathbf{R}'. \tag{M87}$$

From this the cross section for the process $0 \to n$ can be calculated.

The scheme sketched above is exact; but it is obvious that, without some simplifying approximations, the amount of computations will be prohibitive even for the simplest of processes. We shall discuss some usual approximations in the following.

### *ii) Two-state approximation*

An approximation similar to (M47) is now made, namely, that in considering the scattering from state 0 to state $n$, (M76) is replaced by

$$\Psi(r; R) = \psi_0(R)\,\phi_0(r; R) + \psi_n(R)\,\phi_n(r; R), \tag{M88}$$

and the system (M77) is replaced by the pair

$$[\nabla_{\mathbf{R}}^2 + k_0^2 - U_0(R) - C_{00}(R)]\,\psi_0(\mathbf{R}) = C_{0n}(R)\,\psi_n(\mathbf{R}), \tag{M88a}$$

$$[\nabla_{\mathbf{R}}^2 + k_n^2 - U_n(R) - C_{nn}(R)]\,\psi_n(\mathbf{R}) = C_{n0}(\mathbf{R})\,\psi_0(\mathbf{R}). \tag{M88b}$$

While this neglect of the coupling of infinitely many other states seems a reasonable approximation, it is difficult to assess its effect in general. We shall further simplify (M88a) by assuming that $C_{0n}\,\psi_n(\mathbf{R})$ is small compared with $C_{00}(\mathbf{R})\,\psi_0(\mathbf{R})$ and can be neglected. The two-state approximation (M88a, b) reduces the sum over $m$ in (M87) to one single term,

$$f_n(\Theta) = -\frac{1}{4\pi}\int \mathscr{F}_n(R', \pi - \Theta)\,C_{n0}(R')\,\mathscr{F}_0(R', \Theta')\,d\mathbf{R}'. \tag{M89}$$

To bring the expressions involving $C_{mn}(R)$ into more explicit forms, let us take up the simple system in which there is only one electron, so that the initial state (M81) consists of an electron bound to $A$ whose charge is $e_A$, and an incident particle $B$ of charge $e_B$. (If $A$ and $B$ are similar particles, then we have a molecular ion similar to $H_2^+$.) We may neglect $\nabla_{\mathbf{R}}\phi_n(\mathbf{r}; R)$ compared with $\nabla_{\mathbf{R}}\psi_n(\mathbf{R})$ so that

$$C_{n0}(R) \cong -2\int \phi_n^*(\mathbf{r}; R)\,\nabla_{\mathbf{R}}\phi_0(\mathbf{r}; R)\,d\mathbf{r}\cdot\nabla_{\mathbf{R}}.$$

Let us consider the case when both $\phi_n(\mathbf{r}; R)$, $\phi_0(\mathbf{r}; R)$ have zero angular momentum about the interatomic axis $R$, i.e., they are $\sigma$ states in the spectroscopic parlance. In this case, it can be shown (ref. 21) that

$$C_{n0}(R)\,\psi_0(\mathbf{R}) = M_{n0}(R)\,\frac{\partial}{\partial R}\,\psi_0(\mathbf{R}), \tag{M90}$$

where

$$M_{n0}(\mathbf{R}) = -2\int \phi_n^*(\mathbf{r}; R)\,\frac{\partial}{\partial R}\,\phi_0(\mathbf{r}; R)\,d\mathbf{r}, \tag{M91}$$

$$= \frac{2}{\mathscr{E}_n(R) - \mathscr{E}_0(R)}\int \phi_n^*\cdot(R - z)\,\frac{e_B}{|\mathbf{r} - \mathbf{R}|^3}\,\phi_0\,d\mathbf{r}. \tag{M91a}$$

The cross section of the excitation $0 \to n$ is then#

$$\sigma_{0n}(k_0) = 4\pi \frac{k_n}{k_0} \sum (2l+1)|q_{0n}(l)|^2, \tag{M92}$$

$$q_{0n}(l) = \exp\,[i(\eta_{0,l} + \eta_{n,l})] \times \int_0^\infty g_{n,l}(R')\, M_{n0}(R') \frac{\partial}{\partial R'} \left( \frac{1}{R'} g_{n,l}(R') \right) R'\, dR'. \tag{M93}$$

(M91–93) are, up to the two-state approximation (M88), the exact result, from which the Born approximation (M52) and the adiabatic approximation (M67–69) can be obtained by further approximations, as we shall show in *iii*) below.

The two-state approximation (M88) covers not only the processes in which the two particles $A$ and $B$ change their states of excitation [see (M75) and (M81)] but also those in which an electron has been captured from $A$ by $B$. For example, the capture of an electron from the H atom by an incident helium ion,

$$H + He^+ \to H^+ + He,$$

and the "resonance capture" processes such as

$$H^- + H \to H + H^-,$$

$$H + H^+ \to H^+ + H,$$

or

$$He + He^+ \to He^+ + He.$$

[See *iv*) below.]

For a bibliography of the applications of the theory to these and other processes and some experimental studies, see the references at the end of the present section.

### *iii) Semi-classical approximation*

From the "exact" theory (in the two-state approximation) of the preceding subsection, we shall make the following semi-classical approximation by solving (M83) for $g_{0,l}(R)$, $g_{n,l}(R)$ by the W-K-B method. This is a first-order approximation in an expansion in powers of $\hbar$ in solving the Schrödinger equation. Let

$$\kappa_{n,l}(R) = \left| k_n^2 - U_n(R) - \frac{1}{R^2}(l + \tfrac{1}{2})^2 \right|^{1/2}, \tag{M94}$$

and $R_0$ be the classical distance of closest approach (for $n$, $l$)

$$\kappa_{n,l}(R_0) = 0.$$

# If the $0 \to n$ is a $\sigma \to \pi$ transition (i.e., the state $n$ has an angular momentum $\hbar$ along the line **R**), (M93) will then be replaced by a similar formula (see ref. 21).

In the region $0 < R < R_0$, the solution $g_{n,l}(R)$ is

$$g_{n,l}(R) = \frac{1}{2\sqrt{k_n\kappa_{n,l}}} \exp\left[-\int_R^{R_0} \kappa_{n,l}(R)\, dR\right]. \tag{M95}$$

We shall neglect the contribution from this region of $R$. In $R_0 < R$, the solution that joins smoothly with (M95) is

$$g_{n,l}(R) = \frac{1}{\sqrt{k_n\kappa_{n,l}}} \sin\left[\int_{R_0}^{R} \kappa_{n,l}(R)\, dR + \frac{\pi}{4}\right]. \tag{M96}$$

Similarly for $g_{0,l}(R)$. Under conditions where the W-K-B method is valid, we may take $(R/\kappa_{0,l})(d\kappa_{0,l}/dR) \ll 1$, $R\kappa_{0,l}(R) \gg 1$ (except near $R_0$). Then from (M93), we find

$$q_{0n}(l) \simeq \tfrac{1}{2} \exp\left[i(\eta_{0,l} + \eta_{n,l})\right] \frac{1}{\sqrt{k_0 k_n}} \int_0^\infty dR \sqrt{\frac{\kappa_{0,l}}{\kappa_{n,l}}}\, M_{0n}(R)$$
$$\times \sin \int_{R_>}^{R} \left[\kappa_{n,l}(R') - \kappa_{0,l}(R')\right] dR', \tag{M97}$$

after neglecting a rapidly oscillating $\sin \int_{R_>}^{R} [\kappa_{n,l}(R') + \kappa_{0,l}(R')]\, dR'$. In (M97), $R_>$ is the larger one of the roots of $\kappa_{n,l}(R) = 0$ and $\kappa_{0,l}(R) = 0$.

Let $v_0(R)$, $v_n(R)$ be the relative velocities in the state $\phi_0(\mathbf{r}; R)$, $\phi_n(\mathbf{r}; R)$,

$$\mu^2 v_0^2(R) = \hbar^2[k_0^2 - U_0(R)], \qquad \mu^2 v_n^2(R) = \hbar^2[k_n^2 - U_n(R)]. \tag{M98}$$

and $\bar{v}$ be the "mean" velocity

$$2\bar{v}(r) = v_0(R) + v_n(R) \equiv v_0 + v_n.\text{\#}$$

If we approximate (M94) by

$$\kappa_{0,l} + \kappa_{n,l} \cong \frac{2\mu\bar{v}}{\hbar}\left[1 - \left(\frac{l\hbar}{\mu\bar{v}R}\right)^2\right]^{1/2}, \tag{M99}$$

then, from (M94), (M79) and (M80), it follows that

$$\kappa_{0,l} - \kappa_{n,l} = \frac{2\mu}{\hbar^2}(\mathscr{E}_n - \mathscr{E}_0) \cong \frac{\mathscr{E}_n - \mathscr{E}_0}{\hbar\bar{v}}\left[1 - \left(\frac{l\hbar}{\mu\bar{v}R}\right)^2\right]^{-1/2}. \tag{M99a}$$

Then (M97) becomes

$$q_{0n}(l) \cong \tfrac{1}{2} \exp\left[i(\eta_{0,l} + \eta_{n,l})\right] \frac{1}{\sqrt{k_0 k_n}} \int_0^\infty dR \sqrt{\frac{\kappa_{0,l}}{\kappa_{n,l}}}\, M_{0n}(R)$$
$$\times \sin\left[\int_{R_>}^{R} \frac{\mathscr{E}_n - \mathscr{E}_0}{\hbar\bar{v}}\left[1 - \left(\frac{l\hbar}{\mu\bar{v}R'}\right)^2\right]^{-1/2} dR', \tag{M100}$$

---

\# In the following, we shall, for simplicity, omit the argument $(R)$ from $\kappa(R)$, $v_n(R)$, $\bar{v}(R)$, $\varepsilon_n(R)$. Thus

$$\kappa_{0,l} \equiv \kappa_{0,l}(R) \quad \text{or} \quad \kappa_{0,l}(R'), \text{ etc.}$$

where $\mathscr{E}_n$, $\mathscr{E}_0$, $\bar{v}$ are functions of $R'$. Substitution of this into (M92) gives the cross section $\sigma_{0n}(k_0)$.

If the energy $\frac{1}{2}\mu v_0^2(\infty) = \hbar^2 k^2/2\mu$ is large compared with $(\hbar^2/2\mu)(k_0^2 - k_n^2)$, then $k_0/k_n \simeq 1$ and $\hbar\sqrt{k_0 k_n} \simeq \mu\bar{v}(\infty)$. If the summation over $l$ is replaced by an integration over the impact parameter $\rho = l\hbar/\mu\bar{v}(\infty)$, then

$$\sigma_{0n}(k_0) \cong 2\pi \int_0^\infty \rho\, d\rho \left| \int_{R>}^\infty dR\, M_{0n}(R) \right.$$
$$\left. \times \sin\left\{ \int_{R>}^R \frac{\mathscr{E}_n - \mathscr{E}_0}{\hbar\bar{v}} \left[ 1 - \left( \frac{\bar{v}(\infty)\rho}{\bar{v}R'} \right)^2 \right]^{-1/2} dR' \right\} \right|^2. \quad \text{(M101)}$$

Here $\mathscr{E}_n(R')$, $\mathscr{E}_0(R')$, $\bar{v}(R')$ are functions of $R'$.

If in (M101) we remember the relations

$$R^2 = \rho^2 + z^2, \qquad dR = \left( \frac{\partial R}{\partial z} \right) dz,$$

and make the approximation

$$(\alpha) \qquad \bar{v}(R) \cong \bar{v}(\infty), \quad \text{(M102)}$$

then (M101) becomes

$$\sigma_{0n}(k_0) \cong 2\pi \int_0^\infty \rho\, d\rho \left| \int_{z>}^\infty dz \left( \frac{\partial R}{\partial z} \right) M_{0n}(z) \sin\left( \frac{\mathscr{E}_n - \mathscr{E}_0}{\hbar\bar{v}} \right) \right|^2, \quad \text{(M103)}$$

or, on using (M91),

$$\sigma_{0n}(k_0) \cong 2\pi \int_0^\infty \rho\, d\rho \left| \int_{-\infty}^\infty \phi_n^*(\mathbf{r}; R) \frac{\partial}{\partial z} \phi_0(\mathbf{r}; R) \sin\left( \frac{\mathscr{E}_n - \mathscr{E}_0}{\hbar\bar{v}} z \right) \right|^2, \quad \text{(M104)}$$

which are the same as the formulas (M67–69).

If in (M101) we make the further approximation of neglecting the molecular potentials and put

$$(\beta) \qquad \mathscr{E}_n(R) - \mathscr{E}_0(R) \cong \mathscr{E}_n(\infty) - \mathscr{E}_0(\infty) = \text{excitation energy} \equiv \Delta E, \quad \text{(M105)}$$

this is equivalent to replacing $\mathscr{F}_n$, $\mathscr{F}_0$ in (M85, 89) by plane waves. Since $\sigma_{0n}$ in (M104) obviously depends on the sine function, this plane-wave approximation ($\beta$) of replacing $\sin\{[\mathscr{E}_n(R) - \mathscr{E}_0(R)]z/\hbar\bar{v}(R)\}$ by $\sin[\Delta E z/\hbar\bar{v}(\infty)]$ will in general introduce a serious error.

($\gamma$) If in addition to ($\alpha$), ($\beta$), we make the further approximation of replacing the molecular wave function $\phi_n(\mathbf{r}; R)$, $\phi_0(\mathbf{r}; R)$ by atomic functions $\varphi_n(A)$, $\varphi_0(A)$ [see (M75), (M81)], then (M104) reduces to the Born approximation (M58). For velocities $v(\infty) \ll$ electron orbital velocities, this approximation is a very poor one.

The formula (M100), or the approximate ones (M101), (M104), (M67), (M71), show that as $\bar{v}$ (or $k_0$) is varied, the cross section $\sigma$ should oscillate on account of the sine function in the integrand. This oscillatory variation has been beautifully demonstrated in the experiments of Ziemba, Everhart and Russek (ref. 22) on the process

$$\mathrm{He} + \mathrm{He}^+ \rightarrow \mathrm{He}^+ + \mathrm{He}.$$

The "perturbed stationary state method" in its various approximations has been applied to a variety of collision processes, notably by Massey, Bates and their associates. There have been also further studies of the methods of approximation themselves (see, for example, refs. 135–137). Unfortunately we shall not be able to go into these works and the detailed calculations. For these the reader must be referred to the published papers, of which an incomplete bibliography is appended to the end of Sect. M.

### *iv) Resonance, or exchange, scattering*

We shall consider the class of collisions typified by

$$\mathrm{H} + \mathrm{H}^+ \rightarrow \mathrm{H} + \mathrm{H}^+,$$

$$\mathrm{H} + \mathrm{H}^+ \rightarrow \mathrm{H}^+ + \mathrm{H},$$

involving the elastic scattering and the capture of the electron of the target H atom by the incident proton. Because of the identity of the two protons, it is necessary to formulate the problem by properly symmetrizing the wave function with respect to the two protons.

Let us again limit our discussion to the case of low energies ($\frac{1}{2}\mu v^2(\infty) < \frac{3}{4}$ rydberg) so that we shall consider only the two states $\phi_{1s\sigma g}(\mathbf{r}; R) \equiv \phi_g$, $\phi_{2p\sigma u}(\mathbf{r}; R) \equiv \phi_u$ of $H_2^+$ which asymptotically for large $R$ correspond to $H(1s) + \dot{H}^+$, or

$$\begin{aligned} \phi_g(\mathbf{r}; R) &\rightarrow \frac{1}{\sqrt{2}} [\varphi_0(\mathbf{r}_A) + \varphi_0(\mathbf{r}_B)], \\ \phi_u(\mathbf{r}; R) &\rightarrow \frac{1}{\sqrt{2}} [\varphi_0(\mathbf{r}_A) - \varphi_0(\mathbf{r}_B)]. \end{aligned} \tag{M106}$$

$\varphi_0(\mathbf{r}_A)$ being the hydrogenic wave function of the electron centered around the proton $A$, etc. $\phi_g$, $\phi_u$ are even and odd respectively with respect to the inversion about the center of the line $A$–$B$. The equations (M76), (M88a, b) are now of the form

$$\psi(\mathbf{r}, R) = \frac{1}{\sqrt{2}} [\phi_g(\mathbf{r}; R)\Psi_g(\mathbf{R}) + \phi_u(\mathbf{r}; R)\Psi_u(\mathbf{R})], \tag{M107}$$

$$\begin{aligned} [\nabla_{\mathbf{R}}^2 + k^2 - \mathscr{W}_u(R)]\Psi_u(\mathbf{R}) &= 0, \\ [\nabla_{\mathbf{R}}^2 + k^2 - \mathscr{W}_g(R)]\Psi_g(\mathbf{R}) &= 0, \end{aligned} \tag{M108}$$

since the coupling $C_{ug}(R)$ in (M78) vanishes by virtue of the difference in parity of $\Psi_u$, $\Psi_g$.

Let $\chi_s(\sigma_A, \sigma_B, \sigma_e)$, $\chi_a(\sigma_A, \sigma_B, \sigma_e)$ be the spin wave functions which are symmetric and antisymmetric in the spins of the two protons. The wave function of the system $p_A + p_B + e$ may have one of the following forms, on account of the Pauli principle (ref. 23),

$$\chi_s \times \begin{cases} \phi_g(\mathbf{r}; R)\,\Psi_{go}(\mathbf{R}), \\ \phi_u(\mathbf{r}; R)\,\Psi_{ue}(\mathbf{R}), \end{cases} \qquad \chi_a \times \begin{cases} \phi_g(\mathbf{r}; R)\,\Psi_{ge}(\mathbf{R}), \\ \phi_u(\mathbf{r}; R)\,\Psi_{uo}(\mathbf{R}), \end{cases} \tag{M109}$$

where $\Psi_{ge}(\mathbf{R})$, $\Psi_{go}(\mathbf{R})$ are the even and odd ($l$) waves of $\Psi_g$ in (M108), respectively; and similarly for $\Psi_{ue}$, $\Psi_{uo}$.

We seek solutions of (M108) which asymptotically behave as

$$\begin{aligned} \Psi_g(\mathbf{R}) &\to e^{i\mathbf{k}\cdot\mathbf{R}} + \frac{e^{ikR}}{R} f_g(\Theta), \\ \Psi_u(\mathbf{R}) &\to e^{i\mathbf{k}\cdot\mathbf{R}} + \frac{e^{ikR}}{R} f_u(\Theta), \end{aligned} \tag{M110}$$

where

$$\begin{aligned} f_g(\Theta) &= \frac{1}{2ik} \sum (2l+1)\,\{\exp[2i\delta_l(g)] - 1\}\, P_l(\cos\Theta), \\ f_u(\Theta) &= \frac{1}{2ik} \sum (2l+1)\,\{\exp[2i\delta_l(u)] - 1\}\, P_l(\cos\Theta), \end{aligned} \tag{M111}$$

$\delta_l(g)$, $\delta_l(u)$ being the phase shifts of the radial parts of (M108). Asymptotically, $\psi(\mathbf{r}; R)$ in (M107) is

$$\psi(\mathbf{r}; \mathbf{R}) \to e^{i\mathbf{k}\cdot\mathbf{R}}\,\varphi_0(\mathbf{r}_A) + \frac{e^{ikR}}{2R}\,[(f_g + f_u)\,\varphi_0(\mathbf{r}_A) + (f_g - f_u)\,\varphi_0(\mathbf{r}_B)], \tag{M112}$$

which describes a scattering process in which the electron is initially bound with the proton $A$, and is, after the collision, associated with $A$ and $B$ with the amplitude $\frac{1}{2}(f_g + f_u)$, $\frac{1}{2}(f_g - f_u)$ respectively.

According to (M109),

$$\text{for the spin state } \chi_s, \quad \begin{matrix} \phi_g \\ \phi_u \end{matrix}\Bigg\} \text{ can contain only } \quad l = \begin{cases} \text{odd} \\ \text{even} \end{cases} \text{ waves.}$$

The differential cross sections for "direct scattering" and "exchange capture" are then

$$\chi_s: \quad \begin{aligned} d\sigma_{\mathrm{D}} &= \frac{2\pi}{4}\,|f_g(l\text{ odd}) + f_u(l\text{ even})|^2\, d\cos\vartheta, \\ d\sigma_{\mathrm{Ex}} &= \frac{2\pi}{4}\,|f_g(l\text{ odd}) - f_u(l\text{ even})|^2\, d\cos\vartheta. \end{aligned} \tag{M113}$$

Their contributions to the total cross section will be the same since the interference terms vanish on integrating over $\vartheta$. Similarly, for the spin state $\chi_a$ of (M109),

$$\chi_a: \quad \begin{aligned} d\sigma_{\mathrm{D}} &= \frac{2\pi}{4}\,|f_g(l \text{ even}) + f_u(l \text{ odd})|^2\, d\cos\vartheta, \\ d\sigma_{\mathrm{Ex}} &= \frac{2\pi}{4}\,|f_g(l \text{ even}) - f_u(l \text{ odd})|^2\, d\cos\vartheta. \end{aligned} \tag{M114}$$

For unpolarized protons, the $\chi_s$, $\chi_a$ contributions are in the ratio 3:1, and in terms of phase shifts, the total cross section is given by

$$\sigma = \frac{4\pi}{8k^2}\Big[3\sum_{\text{odd}}(2l+1)\sin^2\delta_l(g) + 3\sum_{\text{even}}(2l+1)\sin^2\delta_l(u) + \sum_{\text{even}}(2l+1)\sin^2\delta_l(g) + \sum_{\text{odd}}(2l+1)\sin^2\delta_l(u)\Big]. \tag{M115}$$

The above theory is exact [up to the two-state approximation (M88)]. Failure to take into account the effect of the spins of the protons on the symmetry of the total wave functions (M109) will have led to (M113-4) without the restriction to even or odd $l$, and to a totally different expression from (M115) for $\sigma$.

The resonance in (M105) disappears for the processes

$$\mathrm{H} + d \rightarrow \begin{cases} \mathrm{H} + d \\ p + \mathrm{D} \end{cases} \tag{M116}$$

where $d$ is a deuteron and D the deuterium atom. The case in which the H atom in (M116) is replaced by a mu-mesic atom is of particular interest, in this case, on account of the difference in the reduced masses of $p - \mu$ and $d - \mu$, the ground state of $\mathrm{D}_\mu$ is lower than that of $\mathrm{H}_\mu$ by about 135 eV. For a treatment of (M105) and (M116), see ref. 23.

## 6. RÉSUMÉ OF COLLISIONS BETWEEN PARTICLES

At the beginning of the present chapter, we have mentioned the various types of collision processes. In Sects. L and M, we have presented the exact, but formal, formulation of the theory for any scattering process, and various approximate methods appropriate under different conditions. We shall in the following list a number of typical processes of physical interest and indicate the approximate methods best applicable to them.

In spite of the considerable amount of calculations that have been carried out for various types of processes by different approximate methods, it seems still difficult in general to assess accurately the errors introduced by the various approximations. This is especially so when the conditions of the processes under consideration are "intermediary ones" such that none of the approximate methods is very suitable; for example, when the collision between two atoms is neither "fast" enough for the Born approximation (with distorted waves), nor "slow" enough for the "method of molecular state", to be a very good approximation. In such cases, one may have to carry out the calculation by two different approximate methods to see how reliable or otherwise the results might be. The suggested approximations for (M118)–(M130) are merely general guides and deviations from them are possible in specific processes. Familiarity with the findings of the existing literature will be helpful. For these the works of Massey (refs. 8, 11–14) will prove most useful.

**1)** Collision between slow electrons and atoms or molecules

$$\begin{aligned} e + X &\to X + e, && \text{elastic,} \\ &\to X' + e, && \text{electronic excitation,} \\ &\to X^+ + e + e, && \text{ionization.} \end{aligned} \tag{M117}$$

Examples are the excitation of the $^1D$, $^1S$ of the oxygen atom from the ground state $^3P$ by electron impact. The forbidden transitions $^1D \to {}^3P$, $^1S \to {}^1D$ are the prominent radiations of the aurora and the night sky. The process "detachment"

$$e + \mathrm{O}^- \to \mathrm{O} + e + e \tag{M118}$$

may play a role in the ionized layers of the Earth's upper atmosphere (ref. 24).

**2)** Excitation of molecular vibrations or rotations by electron impact (refs. 25, 26)

$$e + X_2 \to e + X_2'. \tag{M119}$$

**3)** Dissociative recombinations between electrons and molecular ions, such as

$$e + \mathrm{O}_2^+ \to \mathrm{O}' + \mathrm{O}''. \tag{M120}$$

This process may play an important role in the ionized layers of the Earth's upper atmosphere (ref. 27).

The inverse process of (M120) is

$$\mathrm{O}' + \mathrm{O}'' \to \mathrm{O}_2^+ + e. \tag{M121}$$

**4)** Three-body recombination processes are typified by

$$e + \mathrm{O}^+ + X \to \mathrm{O}' + X, \quad \sigma_3, \tag{M122}$$

where $X$ may be an electron, an atom or a molecule. Such processes are important in many laboratory experiments, and (M122) may actually play an important role in those layers of the Earth's atmosphere where the pressure is not too low.

The cross section $\sigma_3$ of the three-body process (M122) can be more conveniently found in terms of the cross section $\sigma_2$ of the inverse process

$$O' + X \rightarrow O^+ + X + e, \quad \sigma_2 \qquad \text{(M123)}$$

$\sigma_3$ and $\sigma_2$ being related together by a relation that can be obtained on statistical mechanical considerations alone (ref. 28).

5) Collisions between a positive ion and an atom or molecule
The simplest process of this kind is

$$\begin{aligned} H + H^+ &\rightarrow H + H^+, \text{ elastic,} \\ &\rightarrow H^+ + H, \text{ resonance capture.} \end{aligned} \qquad \text{(M124)}$$

A particular case of (M124) is when H is a mu-mesonic hydrogen atom and $H^+$ is a deuteron

$$\begin{aligned} H_\mu + d &\rightarrow H_\mu + d, \text{ elastic} \\ &\rightarrow D_\mu + p, \text{ exchange.} \end{aligned} \qquad \text{(M125)}$$

These processes play an important role in experiments involving a stopped muon in a liquid hydrogen bubble chamber, for example, in the phenomenon of nuclear fusion by catalysis, whose overall result is

$$p + d + \mu \rightarrow {}^3He + \mu + \text{energy} \qquad \text{(M126)}$$

[see ref. 23 for (M125)].

6) Excitation transfer: de-excitation of atoms by atoms or molecules

$$A' + X \rightarrow A + X', \qquad \text{(M127)}$$

where the excitation of $X$ after collision may be electronic (M127a), vibrational (M127b) or translational (M127c).

7) Excitation transfer: de-excitation of molecular vibration by atoms or molecules

$$X_2' + Y \rightarrow X_2 + Y'. \qquad \text{(M128)}$$

Such processes are of basic importance in the phenomenon of supersonic dispersion (refs. 14, 15 and 29).

8) Charge transfer [other than the cases of degeneracy resonance (M124)]
The simplest example is

$$H + He^+ \rightarrow H^+ + He. \qquad \text{(M129)}$$

9) Chemical reactions
The simplest is of the type

$$A + BC \rightarrow AB + C. \qquad \text{(M130)}$$

Such processes form the basic problem in gaseous chemical kinetics (ref. 30).

| Processes | Methods applicable<br>Better approximation | <br>Next choice |
|---|---|---|
| (M117) "fast" | "Distorted wave"<br>Sect. M.3, *ii)* | Born<br>Sect. M.3, *i)* |
| (M117) "slow" | "Two-state"<br>Sect. M.3, *iii)* | Variational |
| (M118)<br>(M119) | " | "Distorted wave" |
| (M121) "slow" | "Perturbed Stationary state"<br>Sect. M.5, *ii)* | "Semi-classical" impact parameter<br>Sect. M.4, *iii)* |
| (M123) "slow" | " | " |
| (M123) "fast" | "Semi-classical" impact parameter<br>Sect. M.4, *ii)* | Born |
| (M124) "slow" | "Pert. Stat. State" Resonance<br>Sect. M.5, *iv)* | |
| (M124) "fast" | "Distorted Wave"<br>Sect. M.3, *ii)* | Born<br>Sect. L |
| (M125) "slow" | "Pert. Stat. State":<br>"Two-state"<br>Sect. M.5, *ii)* | |
| (M127) | " | |
| (M128) | " | "Modified wave number<br>Sect. M.3, *v)* |
| (M129) | " | Landau–Zener |
| (M130) | " | |

## REFERENCES

1. Lippman, B. A., Phys. Rev. **102,** 264 (1956).

2. Gerjuoy, E., Phys. Rev. **109,** 1806 (1958). This paper discusses the boundary conditions for rearrangement collisions.

3. Aaron, R., Amado, R. D. and Lee, B. W., Phys. Rev. **121,** 319 (1961).

4. Gerjuoy, E., Annals of Phys. **5,** 58 (1958).

5. Erskine, G. A. and Massey, H. S. W., Proc. Roy. Soc. (London) **A212,** 521 (1952). Distorted wave treatment of $1s \rightarrow 2s$ excitation of H atom by electron impact.

6. Bransden, B. H. and McKie, J. S., Proc. Phys. Soc. (London) **A69**, 422 (1956); **A70**, 398 (1957). $1s \rightarrow 2s$ excitation of H atom by electron impact, using the two-state approximation (Sect. M.3, iii) without exchange.

7. Marriott, R., Proc. Phys. Soc. (London) **72,** 121 (1958). $1s \rightarrow 2s$ excitation of H by electron impact, two-state approximation with exchange.

8. Massey, H. S. W., Rev. Modern Phys. **28**, 199 (1956). A review of the various approximate methods (Sect. M.3, i, ii, iii, iv), "close coupling and the effect of intermediate states", as applied to the slow collisions between electrons and atoms.

9. Yamanouchi, Inui and Amemiya, Proc. Phys. Math. Soc. Japan, **22**, 847 (1940). Excitation of $^1D$, $^1S$ of 0 atom from the ground state $^3P$, calculated with distorted wave approximation with exchange.

10. For a brief review of the literature on this problem, see refs. 87–96 in the following.

11. Mott, N. F. and Massey, H. S. W., *Theory of Atomic Collisions*. Oxford, Clarendon Press (1949).

12. Massey, H. S. W., article in Handbuch der Physik Vol. **36**, 232, Springer (1956). This article treats the scattering by a central potential and the various approximations of the "perturbed stationary state" (Chap. VIII of ref. 11).

13. Massey, H. S. W., article in Handbuch der Physik Vol. **36**, 307, Springer (1956). This article treats the various approximation methods for the collision between (slow) electrons and atoms, and also the experimental methods of measurements.

14. Craggs, J. D. and Massey, H. S. W., article in Handbuch der Physik Vol. **37/1**, 314 (1959). This article treats the general theory and its approximations for the collision between electrons and molecules.

References 8, 11, 12, 13, 14 contain many other topics and references not given in the present work. The reader is referred to them for these and more detailed treatments.

15. Takayanagi, K., (a) Progr. Theor. Phys. **8**, 497 (1952); (b) **9**, 578; **10**, 369 (1953). The modified-wave-number method. This method has been applied to a number of problems; (c) Takayanagi, K. and Kaneko, S., Sc. Rep. Saitama University **AI**, 111 (1954); (d) Takayanagi, K., *ibid.* **AII**, 1 (1958). Vibrational transitions in molecular collisions; (e) Takayanagi, K., J. Phys. Soc. Japan **14**, 75 (1959). Excitation of molecular vibrations by collisions; (f) Takayanagi, K., Sc. Rep. Saitama Univ. **AIII**, 65, 87, 101 (1959); (g) J. Phys. Soc. Japan **14**, 1458 (1959). Rotational transitions in molecular collisions.

16. Gaunt, J. A., Proc. Camb. Phil. Soc. **23**, 732 (1927).

17. Frame, J. W., Proc. Camb. Phil. Soc. **27**, 511 (1931).

18. Bates, D. R., Proc. Phys. Soc. (London) **73**, 227 (1959).

19. Mott, N. F., Proc. Camb. Phil. Soc. **27**, 553 (1931).

20. Bauer, E., Jour. Chem. Phys. **23**, 1087 (1955).

21. Bates, D. R., Massey, H. S. W. and Stewart, A. L., Proc. Roy. Soc. (London) **A216**, 437 (1953).

22. Ziemba, E. P. and Everhart, E., Phys. Rev. Letters **2**, 299 (1959); Ziemba and Russek, A., Phys. Rev. **115**, 922 (1959).

23. Wu, T. Y., Rosenberg, R. L. and Sandstrom, H., Nuclear Phys. **16**, 432 (1960).

24. For the application of atomic collision theory to the problems of the Earth's upper atmosphere, see Massey, H. S. W. and Boyd, R. L. F.: *The Upper Atmosphere*, Hutchinson, London (1958).

25. For the excitation of molecular ($H_2$) vibrations by electrons (a) Theoretical: Massey, H. S. W., Trans. Faraday Soc. **31**, 556 (1935); Wu, T. Y., Phys. Rev. **71**, 111 (1947); Morse, P. M., Phys. Rev. **90**, 51 (1953); Carson, T. R., PPS **A67**, 908 (1954); J. C. Y. Chen and J. L. Magee, to be published.
(b) Experimental: Ramien, H., Z. Physik **70**, 353 (1931); Chao, Wang and Shen, Sc. Record (China) **2**, 358 (1949). See ref. 14 above.

26. For the excitation of molecular rotations by electron—Morse, P. M., ref. 25 above; Gerjuoy, E. and Stein, S., Phys. Rev. **97**, 1671; **98**, 1848 (1954). See ref. 14 above.

27. Dissociative recombinations: Bates, D. R. and Massey, H. S. W., Proc. Roy. Soc. (London) **A192**, 1 (1947); Bates, Phys. Rev. **78**, 492 (1950).

28. For the application of the principle of detailed balancing to the three-body recombination coefficient, see Fowler, R. H., *Statistical Mechanics*. Cambridge Univ. Press (1936), Sect. 17.6; Moses, H. E. and Wu, T. Y., Phys. Rev. **83**, 109 (1951), Appendix 5.

29. Salkoff, M. and Bauer, E., J. Chem. Phys. **29**, 26 (1958); Bauer, E. and Cummings, F. W., *De-excitation of Molecular Vibration by Collision* Aeronutronic Res. Labs. Newport Beach, Calif., Publ. No. U-1207, (1961); Schwartz, R. N. and Herzfeld, K. F., J. Chem. Phys. **22**, 767 (1954); Takayanagi, ref. 15(f) above; Nikitin, E. E., Optika i Spektrosk **6**, 141 (1959). Calculation of vibrational excitation of molecules by collisions (two diatomic molecules).

30. Bauer, E. and Wu, T. Y., J. Chem. Phys. **21**, 726, 2072 (1953). See remarks concerning the intermolecular interactions in the two paragraphs preceding Sect. L.

### ADDITIONAL REFERENCES

There are many interesting topics that have not been covered in the text above. The following bibliography, arranged according to the nature of the collision processes and to chronological order, is intended to make some references to some topics and to the many applications of the various approximate methods discussed in the text to specific collision processes.

In the following, we shall abbreviate "Proc. Phys. Soc. (London)" to PPS, and "Proc. Roy. Soc. (London)" to PRS.

(1) *Elastic scattering of electrons and positrons by* H *atoms. See refs.* 8–15 *in Sect. L*

**31.** Swan, P., PPS **A67,** 1086 (1954). Elastic scattering by H atoms in $2s$, $2p$ state.

**32.** Moiseiwitsch, B. L., PPS **72,** 139 (1958). Elastic scattering of slow positrons by H atoms.

**33.** Moiseiwitsch and Williams, A., PRS **A250,** 337 (1959). Elastic scattering of fast electrons and positrons by H and He atoms.

**34.** Moussa, A. H., PPS **74,** 101 (1959). Effect of polarization on the elastic scattering of positrons by H atoms.

**35.** John, T. L., PPS **76,** 532 (1960). Numerical solution of the exchange equation for slow electron collisions with H atoms.

**36.** Smith, K., Miller, W. F. and Mumford, A. J. P., PPS **76,** 559 (1960). The elastic and inelastic scattering of electrons and positrons from $S$ states of H atoms.

(2) *Excitation of* H *atoms by electrons. See refs.* 5, 6, 7, 11–13

**37.** Moiseiwitsch, B. L., Phys. Rev. **82,** 753 (1951). Extension of the variational methods of Hulthén and Kohn (see Sect. D) to inelastic scattering.

**38.** Massey and Moiseiwitsch, PPS **A66,** 406 (1953). $1s \to 2s$, variational method.

**39.** McCarroll, R., PPS **A70,** 460 (1957). Excitation of discrete states of H atoms by fast electrons.

**40.** Bates, D. R. and Miskelley, D., PPS **A70,** 539 (1957). $1s \to 2s$; $1s \to 3s$.

**41.** Boyd, T. J. M., PPS **72,** 523 (1958).

**42.** Lynn, N., PPS **73,** 515 (1959). Variational calculation $1s \to 2s$.

**43.** Kingston, A. E., Moiseiwitsch and Skinner, B. G., PRS **A258,** 245 (1960). ($1s \to 2s$).

**44.** Khashaba, S. and Massey, PPS **71,** 579 (1958). Distorted wave method. ($1s \to 2p$).

**45.** McCrea, D. and McKirgan, T. V. M., PPS **75,** 235 (1960). Excitation of H atom in $2p$ state.

**46.** Swan, P., PPS **A68,** 1157 (1955). Ionization of H atom in $2s$, $2p$ states.

(3) *Elastic scattering of electrons by* He *atoms. See ref.* 13

**47.** Bransden and Dalgarno, A., PPS **A66,** 268 (1953). Variational method, with asymptotic Coulomb field wave functions.

**48.** Moiseiwitsch, PRS **A219**, 102 (1953). Distorted wave method.

**49.** Stewart, D. T. and Gabathuler, E., PPS **74**, 473 (1959).

(4) *Inelastic collision of electron and* He *atoms and* $H^-$ *ions. See ref.* 13

**50.** Abdelnabi, I. and Massey, PPS **A66**, 288 (1953). Townsend's ionization coefficient.

**51.** Bransden, Dalgarno and King, N. M., PPS **A66**, 1097 (1953). Distorted wave approximation, $1s \rightarrow 2s$.

**52.** Massey and Moiseiwitsch, PRS **A227**, 38 (1954). Distorted wave method, $1\,^1S \rightarrow 2\,^1S$, $2\,^3S$.

**53.** Massey and Moiseiwitsch, PRS **A258**, 147 (1960). ($1\,^1S \rightarrow 2\,^3P$).

**54.** Miller, W. F. and Platzman, R. L., PPS **A70**, 299 (1957). Theory of inelastic scattering.

**55.** Corrigan, S. J. B. and Von Engel, A., PPS **72**, 786 (1958). Excitation of He by low energy electrons.

**56.** Phelps, A. V., Pack, J. L. and Frost, L. S., Phys. Rev. **117**, 470 (1960). Measurement of cross section at zero energies ($\sim$ a few hundredths of an eV).

**57.** Geltman, S., PPS **75**, 67 (1960). Calculates $H^- + e \rightarrow H + e + e$.

(5) *Collision of electrons and* $H_2$ *molecules. See refs.* 14, 24, 25

**58.** Massey and Ridley, R. O., PPS **A69**, 659 (1956). Variational method, scattering of slow electron by $H_2$.

**59.** Carter, C., March, N. H. and Vincent, D., PPS **71**, 2 (1958). X-ray and electron scattering by $H_2$.

**60.** Arthurs, A. M. and Dalgarno, A., PRS **A256**, 540 (1960). Theory of scattering by a rigid rotator.

(6) *Collisions of electrons with other atoms. See ref.* 13

**61.** Layzer, D., Phys. Rev. **84**, 1221 (1951). A new approximate method for treating the scattering of electrons by atoms. See ref. 67.

**62.** Massey and Mohr, C. B. O. PPS **A65**, 845 (1952). Inelastic scattering.

**63.** Seaton, M. J., PRS **A218**, 400 (1953). Excitation of forbidden lines occurring in gaseous nebulae.

**64.** Seaton, M. J., Phil. Trans. Roy. Soc. **A245**, 469 (1953).

**65.** Seaton, M. J., PRS **A231**, 37 (1955). Perturbation calculation of electron excitation of forbidden lines.

**66.** Seaton, M. J., PPS **A68**, 457 (1955). Excitation by $e$ and $H^+$ of $2s \to 2p$ in H and $3s \to 3p$ in Na atoms.

**67.** Öpik, U., PPS **A68**, 377 (1955). On Layzer's approximation.

**68.** Percival, I. C., PPS **A70**, 241 (1957). Excitation of $^3P$ of OI.

**69.** Seaton, M. J., PPS **A70**, 620 (1957). The use of extrapolated quantum defects as a check on the calculated phases for the scattering of electrons by positive ions.

**70.** Stewart, D. J. and Gabathuler, E., PPS **72**, 287 (1958). Cross sections for N and O atoms.

**71.** Arthurs, A. M. and Moiseiwitsch, PRS **A247**, 550 (1958). $K$-shell ionization of atoms by high energy electrons.

**72.** Massey and Mousoa, A. H. A., PPS **71**, 38 (1958). Elastic scattering of positrons by atoms and molecules.

**73.** Martin, V. M., Seaton and Wallace, J. B. G., PPS **72**, 701 (1958). Adiabatic approximation for electron-atom collisions.

**74.** Klein, M. M. and Brueckner, K. A., Phys. Rev. **111**, 1115 (1958). Effective range calculation for electron–O atom scattering.

(7) *Excitation and ionization of atoms by electrons near the threshold. See ref.* 13, *Sect.* 30)

**75.** Wigner, E. P., Phys. Rev. **73**, 1002 (1948). From the theory of the $R$ matrix (see Sect. X), very general laws have been obtained for the energy dependence near the threshold for various types of reactions, atomic and ionic. For neutral atoms, the excitation cross section is proportional to the square root of the energy above the threshold. For a positive ion, it is independent of the energy above the threshold. *See Sect. T*, 8, *and ref.* 10 *there.*

**75a.** Wannier, G. H., Phys. Rev. **90**, 817 (1953). The ionization cross section of a neutral atom by an electron is proportional to the 1.127th power of the energy excess above the threshold. For positive ions, the exponent lies between 1 and 1.127. The derivation of these laws is not entirely general.

**76.** Gerjuoy, E., Jour. App. Phys. **30**, 28 (1959).

**77.** Geltman, S., Phys. Rev. **102**, 171 (1956). The cross section for the simultaneous ejection of $n$ electrons is proportional to the $n$th power of the energy excess above the threshold.

**78.** Fite, W. L., Stebbings, R. F. and Brackmann, R. T., Phys. Rev. **116**, 356 (1959). $\sigma_{ex} \propto \sqrt{\Delta E}$ relation verified for $1s \to 2p$ excitation of H atom.

**79.** Schulz, G. J., Phys. Rev. **112**, 150 (1958); **116**, 1141 (1959); Schulz and Fox, R. E., Phys. Rev. **106**, 1179 (1957). Find $\sigma_{ex} \alpha \Delta E$ near threshold for excitation of He.

**80.** Fox, R. E., Article "Study of multiple ionization in He and Xe by electron impact" in *Advances in Mass Spectrometry*, Pergamon Press (1959). $\sigma_n \alpha (\Delta E)^n$ relation obeyed in the double ionization of He, but not in the case of Xe.

**81.** Krauss, M., Reese, R. M. and Dibeler, V. H., Jour. Res. Natl. Bur. Standards **63A**, 201 (1959).

**82.** Morrison, J. D. and Nicholson, A. J. C., J. Chem. Phys. **31**, 1320 (1959). These two groups of investigators find the $\sigma_n \alpha (\Delta E)^n$ law satisfied in the case of Xe and other rare gases.

(8) *Elastic collisions between heavy particles*

**83.** McDowell, M. R. C., PPS **72**, 1087 (1958). Elastic scattering of slow ions in their parent gases.

**84.** Haywood, C. A., PPS **A68**, 932 (1955). Approximation for the high order phase shifts in the elastic scattering of slow protons in inert gases.

(9) *Collisions between heavy particles involving charge transfer* (*or capture*)

**85.** For the basic theories of Stueckelberg, Landau and Zener, see the treatments in refs. 11 and 12.

**86.** Hasted, J. B., PRS **A205**, 421 (1951). Exchange of charge between ions and atoms. The problem of the capture of an electron from a H atom by a "fast" proton (energy $>25$ keV) has a long history of development and is still not completely settled, as is seen from the following references.

**87.** Brinkman, H. C. and Kramers, H. A., Proc. Acad. Sci. Amsterdam **33**, 973 (1930). Calculated the electron capture cross section by the impact parameter method (Sect. M.4, or ref. 19). Only the electron–proton interaction appears as the perturbation.

**88.** Bates, D. R. and Dalgarno, A., PPS **A65**, 919 (1952). Resonance capture of electrons from H atoms by fast protons; PPS **A66**, 972 (1953), capture of electrons from H atoms by protons into excited states. The result is similar to that of ref. 90 below.

**89.** Dalgarno, A. and Yadav, H. N., PPS, **A66**, 173 (1953). Resonance capture of electrons from H atoms by slow protons, up to 100 keV, calculated by the perturbed stationary state method (ref. 21).

**90.** Jackson, J. D. and Shiff, H., Phys. Rev. **89**, 359 (1953). Capture of electron from H atom by proton, calculated with the proton–proton and the proton–electron interactions $(e^2/|R_A - R_B|) - (e^2/|R_A - r_e|)$ as the perturbation in the first Born approximation.

**91.** Jackson, J. D., PPS **A70**, 26 (1957). This paper discusses the justification for using the "full interaction" as the perturbation, and shows that the contribution from $e^2/|R_A - R_B|$ up to the second Born approximation vanishes. This is also found by

Drisko, R. M., Unpublished thesis, Carnegie Inst. of Technology, Pittsburgh (1955). This may be taken to support the argument for not including the proton–proton interaction in the perturbation rather than including it.

**92.** Pradhan, T., Phys. Rev. **105**, 1250 (1957). The problem of electron capture from an H atom by a proton is formulated by following the formal theory of scattering (Gell-Mann and Goldberger, see Sect. Q below) and making the impulse approximation such that the initial and the final wave functions are orthogonal. In this way, the proton–proton interaction has no contribution to the transition matrix element if the proton mass is regarded as $\infty$. The resulting cross sections for the charge transfer agree with the existing experimental data (references given in this paper) from 25 to 150 keV, somewhat better than the results of ref. 90.

**93.** Tuan, T. F. and Gerjuoy, E., Phys. Rev. **117**, 756 (1960). Most previous electron-capture by proton cross sections have been measured in $H_2$ gas, and the comparison of these data with the calculated $H + H^+ \rightarrow H^+ + H\sigma$ is made on the supposition that for high energy protons, the $H_2$ molecule may be regarded as two H atoms. This paper investigates the electron capture from $H_2$ and shows that the above supposition is not justified.

**94.** Bessel, R. H. and Gerjuoy, E., Phys. Rev. **117**, 749 (1960). By correcting for the distortion of the trajectory of the proton by the proton–proton interaction, this paper finds for the cross section the Brinkman–Kramers result in the limit of high energies. An unjustified approximation is found in Pradhan's work (ref. 92).

**95.** See discussions in Sect. Q. 4.

The most recent measurements of the capture cross section of $H + H^+ \rightarrow H^+ + H$ are those of

**96.** Fite, W. L., Stebbings, R. F., Hummer, D. G. and Brackmann, R. T., Phys. Rev. **119**, 663 (1960). These measurements show that the $\sigma$ for energies above $\sim 30$ keV do seem to come closer to the value calculated according to Brinkman and Kramers (ref. 87), i.e., without the proton–proton interaction. There remain, however, discrepancies among the observed $\sigma$, that of Bessel and Gerjuoy (ref. 94) and that of Brinkman and Kramers, at energies of 30–40 keV. Thus the problem of the charge transfer $H + H^+ \rightarrow H^+ + H$ is yet not completely satisfactory.

**97.** Bransden, Dalgarno and King, PPS **A67**, 1075 (1954). Capture of electron from He by fast protons.

**98.** Bates, D. R., PPS **A68**, 344 (1955). Charge transfer and ion-atom interchange collision $X^+ + YZ \rightarrow X + YX^+$.

**99.** Bates, D. R. and Moiseiwitsch, B. L., PPS **A67**, 805 (1954). $H + Be^{++} \rightarrow H^{+} + Be^{+}$ (also capture by $Si^{++}$, $Mg^{++}$). The Landau–Zener formula for transition probability when the potential energy curves "cross".

**100.** Dalgarno, A., PPS **A67**, 1010 (1954). Charge transfer from H to $Al^{3+}$, $B^{2+}$, $Li^{2+}$, $Al^{2+}$.

**101.** Bates and Lewis, J. T., PPS **A68**, 173 (1955). $H^{-} + H^{+} \rightarrow H + H$. The Landau–Zener formula is used.

**102.** Bates and Boyd, T. J. M., PPS **A69**, 901 (1956). Ionic recombinations.

**103.** Boyd and Moiseiwitsch, PPS **A70**, 809 (1957). Charge transfer from neutral atoms to doubly and triply charged ions.

**104.** Hasted, J. B., PRS **A222**, 84 (1954). Detachment of electron from negative ions.

**105.** Hasted, J. B., PRS **A227**, 466 (1955). Charge exchange and electron detachment.

**106.** Hasted and Smith, R. A., PRS **A233**, 349 (1956). Detachment of electrons from negative ions.

**107.** Hasted and Smith, PRS **A235**, 354 (1956). Partial charge transfer of doubly charged ions.

**108.** Sida, D. W., PPS **A68**, 240 (1955). Detachment of $e$ from $H^{-}$ by impact with neutral atoms.

**109.** Dalgarno, A. and McDowell, M. R. C., PPS **A69**, 615 (1955). Perturbed stationary state method for charge transfer $H^{-} + H \rightarrow H + H^{-}$.

**110.** Moiseiwitsch, B. L., PPS **A69**, 653 (1956). Interaction energy and charge exchange between the atoms and ions.

**111.** Arthurs, A. M. and Hyslop, J., PPS **A70**, 849 (1957). Radiative charge transfer from H atoms to $He^{++}$ ions.

**112.** Boyd, J. M. and Dalgarno, A., PPS **72**, 694 (1958). Charge transfer of proton in excited atomic H.

**113.** Bates and McCarroll, R., PRS **A245**, 175 (1957) Electron capture in slow collisions. Refinement of ref. 21.

**114.** Bates, D. R., PRS **A247**, 294 (1958). Electron capture in fast collisions.

**115.** Bates and Lynn, PRS **A253**, 141 (1959). Electron capture of the accidental resonance type.

**116.** McDowell, M. R. C. and Peach, G., PPS **74**, 463 (1959). Electron loss from fast $H^{-}$ passing through H.

**117.** Haywood, C. A., PPS **73**, 201 (1959). Electron capture cross sections for protons in He calculated by the impact parameter formulation of the perturbed stationary state method.

**118.** Ferguson, A. F. and Moiseiwitsch, PPS **74**, 457 (1959). Double charge transfer $He^{++}$ in He.

**119.** Sil, N. C., PPS **75**, 194 (1960). Electron capture by $H^+$ through H.

**120.** Karmohapatro, S. B., PPS **76**, 416 (1960). Charge exchange between Ar ions and atoms.

**121.** Bates, D. R., PRS **A257**, 22 (1960). Collisions involving the crossing of potential energy curves.

**122.** Stebbings, R. F., Fite, W. L. and Hummer, D. G., J. Chem. Phys. **33**, 1226 (1960). Measure $H + O^+ \rightarrow H^+ + O$.

**123.** Hummer, D. G., Stebbings, R. F., Fite, W. L., and Branscomb, L. M., Phys. Rev. **119**, 668 (1960). Measure the $\sigma$ of $H^- + H \rightarrow H + H^-$ for energies of $H^-$ from 40 to 10,000 eV.


(10) *Excitation and ionization collisions between heavy particles*


**124.** Hasted, J. B., PRS **A212**, 235 (1952). Inelastic collisions between ions and atoms.

**125.** Buckingham, R. A. and Dalgarno, PRS **A213**, 406 (1952). Excitation transfer of metastable He in gaseous He

**126.** Bates, D. R. and Griffing, G. W., PPS **A66**, 961 (1953). Excitation and ionization of H atoms by fast protons.

**127.** Bates and Griffing, PPS **A67**, 663 (1954); **A68**, 90 (1955). Double transitions in collisions of fast H and H atoms.

**128.** Moiseiwitsch and Stewart, PPS **A67**, 1069 (1954). Excitation of H atom by fast H atoms, $H^+$ and $He^+$.

**129.** Seaton, M. J., PPS **A68**, 457 (1955).

**130.** Dalgarno, A. and Griffing, G. W., PRS **A232**, 423 (1955). Energy loss of proton through hydrogen.

**131.** Boyd, Moiseiwitsch and Stewart, PPS **A70**, 110 (1957). Ionization of H by fast $He^+$.

**132.** Adler, J. and Moiseiwitsch, PPS **A70**, 117 (1957). Excitation of $2s\,^3S$, $2p\,^3P$, $3p\,^3P$ of fast He atoms by H and Ne atoms.

**133.** Bates and Williams, A., PPS **A70**, 306 (1957).

**134.** Gilbody, H. B. and Hasted, J. B., PRS **A238**, 334 (1956); **A240**, 382 (1957). Ionization by positive ions $H^+ + A \rightarrow H^+ + A^+ + e$.

**135.** Bates, D. R., PRS **A240**, 437 (1957). Approximations for inelastic atomic collisions. The paper introduces the "perturbed rotating atom" form of the "perturbed stationary state method".

**136.** Bates, D. R., PRS **A243**, 15 (1957). Slow collisions between heavy particles. The paper considers the effect of some usually neglected coupling terms in the "perturbed stationary state method" on the passage to the Born approximation.

**137.** Bates, D. R., PRS **A245**, 299 (1958). Impact parameter treatment of certain proton–H and H–H excitation collisions. The paper considers the Born approximation and the "perturbed rotating atom" approximation.

**138.** Kingston, Moiseiwitsch and Skinner, PRS **A258**, 237 (1960). Excitation $1s \rightarrow 2s$ of H atoms by protons.

**139.** Frost, L. S. and Phelps, A. V., Westinghouse Research Lab., Pittsburgh (U.S.A.) Research Report (1957). These authors find excitation transfer cross sections of

$$\begin{aligned} \mathrm{He}(5\,^1P) + \mathrm{He}(1\,^1S) &\rightarrow \mathrm{He}(1\,^1S) + \mathrm{H}(5\,^1D) \\ &\rightarrow \mathrm{He}(1\,^1S) + \mathrm{H}(5\,^3D) \end{aligned}$$

to be $\sim 9 \times 10^{-14}$, $7 \times 10^{-14}$ cm$^2$ respectively.

**140.** Gabriel, A. H. and Heddle, W. O., PRS **A258**, 123 (1960), find similarly large singlet $\rightarrow$ singlet transfer but much smaller singlet $\rightarrow$ triplet transfer cross sections.

**141.** St. John, R. M. and Fowler, R. G., Phys. Rev. **122**, 1813 (1961). Electronic excitation energy transfer between two helium atoms.

section

# N Scattering of a Particle by a System of Particles

## 1. COMPLEX POTENTIAL AND OPTICAL MODEL

In the problems of the scattering of an electron by an atom and of that of a nucleon by a nucleus, an exact treatment involving the interactions between the incident particle and all the particles in the "target particle" can be seen to be a complicated, and difficult, one. It is hence desirable to devise reasonable approximations which are valid for certain purposes. One such approximation, known as the optical model (or as the cloudy crystal ball model) has been proposed and extensively employed for the collision between a nucleon and a nucleus. It seems that the model has proved most fruitful, and a brief description of it will be given below.

The basic ideas of this "optical model" are as follows: (i) Consider first a high energy particle $A$ incident on a composite particle $P$. The particle $A$ will be scattered, either elastically or inelastically. Instead of treating the interaction of $A$ with the individual particles in $P$, one seeks an "effective" potential $V(r)$ between $A$ and the whole system $P$, such that the scattering by $V(r)$ describes the observed scattering. If there is only elastic scattering, the theory is then either exactly that treated in Chapter 1, Sect. G, where the problem of determining $V(r)$ from the observed scattering data is discussed, or that of determining $V(r)$ to give the observed differential cross section at a given energy. (ii) If there are inelastic scatterings, i.e., the particle $A$ is scattered after exciting $P$ to some state, one seeks an effective complex potential whose imaginary part will represent the inelastic scatterings. In constructing such a phenomenological complex potential, the choice of the strength and the shape of both the real and the imaginary parts is at our disposal. (iii) By properly introducing isobaric spin, etc., it is possible to include the exchange scattering such as the capture of an incident neutron with the ejection of a proton, etc.

### *i) Complex potential*

That the scattering of a particle by a composite particle (i.e., consisting of a system of particles), under certain conditions, can be described by a complex potential (i.e., consisting of a real and an imaginary part) can be demonstrated in an elementary way (Glauber, ref. 1).

For simplicity, let us start with the expression (C78b) for the scattering amplitude

$$f(\vartheta) = \frac{k}{2\pi i}\int d^2\boldsymbol{\rho}\, \exp\,[i(\mathbf{k} - \mathbf{k}')\cdot\boldsymbol{\rho}]$$

$$\times\left\{\exp\left[-\frac{i}{\hbar v}\int_{-\infty}^{\infty} V(|\boldsymbol{\rho} + \mathbf{z}|)\, dz\right] - 1\right\}, \quad d^2\boldsymbol{\rho} = d\varphi\, \rho\, d\rho, \quad \text{(N1)}$$

where, unlike (C20), we have chosen the direction of $\mathbf{k}$ to be $z$-axis of the cylindrical coordinate system $(z, \rho, \varphi)$ so that $\rho$ is the impact parameter of the classical trajectory.# Obviously, $V(|\boldsymbol{\rho} + \mathbf{z}|) = V(\sqrt{\rho^2 + z^2})$ as before.

It is possible to generalize the above expression (N1) for the scattering of a (high energy) particle by a potential field $V(r) = V(|\boldsymbol{\rho} + \mathbf{z}|)$ to the case of the scattering of a particle by a system of particles. Let the coordinates of these particles be $q_1 \ldots q_n$ and the wave function of the system in state $a$ be denoted by

$$\Psi_a(\mathbf{q}_1 \ldots \mathbf{q}_n). \quad \text{(N2)}$$

Let the interaction of the incident particle with these particles be

$$\sum_j V_j(|\mathbf{r} - \mathbf{q}_j|). \quad \text{(N3)}$$

We now make the plausible steps of replacing $V(|\boldsymbol{\rho} + \mathbf{z}|)$ in (N1) by the sum

$$\sum V_j(|\mathbf{r} - \mathbf{q}_j|) = \sum V_j(|\boldsymbol{\rho} - \mathbf{s}_j - \mathbf{p}_j + \mathbf{z}|), \quad \text{(N4)}$$

where $\mathbf{q}_j = \mathbf{s}_j + \mathbf{p}_j$, $\mathbf{s}_j$, $\mathbf{p}_j$ being the component $\perp$ and $\parallel$ to the $\mathbf{z}$-axis ($\mathbf{k}$) respectively, and of averaging over all values of $\mathbf{q}_1 \ldots \mathbf{q}_n$ by means of the

---

# To arrive at this form, we have only to use, instead of (C22),

$$P_l(\cos\vartheta) \approx J_0[(l + \tfrac{1}{2})\sin\vartheta] = J_0[2\rho k \sin(\vartheta/2)\cos(\vartheta/2)]$$

$$= \frac{1}{2\pi}\int_0^{2\pi} \exp\left(i|\mathbf{k} - \mathbf{k}'|\rho\cos\frac{\vartheta}{2}\cos\varphi\right) d\varphi, \text{ etc.}$$

$\Psi_a(\mathbf{q}_1 \ldots \mathbf{q}_n)$ in (N2). Thus, if the target system has not changed its state $a$ by the collision, we have for the elastic scattering

$$f(\vartheta) = \frac{k}{2\pi i}\int d^2\boldsymbol{\rho} \exp\,[i(\mathbf{k} - \mathbf{k}')\cdot\boldsymbol{\rho}] \int\cdots\int \{\exp\,[2i\delta(\rho)] - 1\} \times |\Psi_a(\mathbf{q}_1 \ldots \mathbf{q}_n)|^2\, d\mathbf{q}^1 \ldots d\mathbf{q}_n, \quad \text{(N5)}$$

and for the inelastic scattering in which the target system changes from the initial state $\Psi_a$ to the final state $\Psi_f$,

$$f_{fa}(\vartheta) = \frac{k}{2\pi i}\int d^2\boldsymbol{\rho} \exp\,[i(\mathbf{k} - \mathbf{k}')\cdot\boldsymbol{\rho}] \int\cdots\int \{\exp\,[2i\delta(\rho)] - 1\} \times \Psi_f^*\Psi_a\, d\mathbf{q}_1 \ldots d\mathbf{q}_n. \quad \text{(N6)}$$

Here

$$\delta(\rho) \equiv \delta(\rho;\mathbf{s}_1 \ldots \mathbf{s}_n) = -\frac{1}{2\hbar v}\sum_j \int_{-\infty}^{\infty} V_j(|\boldsymbol{\rho} - \mathbf{s}_j + \mathbf{z}|)\, dz, \quad \text{(N7)}$$

where we have combined $\mathbf{p}_j$ with $\mathbf{z}$ since the range of integration in $\mathbf{z}$ extends from $-\infty$ to $\infty$.

Before proceeding to show the "natural appearance" of a complex potential, let us note that the approximation (N7) of adding the phase shifts due to the various pair interactions (N3) is *not* the same as adding the scattered amplitudes from the individual particles of the target system. The latter procedure will not take into account the possibility of multiple scatterings, whereas the former to a certain extent does, provided the energy of the particle, the range $a$ of the interactions $V_j$ in (N3) and the size of the target system $R$ satisfy the relations

$$ka \gg 1, \qquad ka^2 \gg R. \quad \text{(N8)}$$

The expression (N5) could have been obtained in a somewhat different manner, by starting from the integral Schrödinger equation (M9) and putting for the exact wave function of the integrand an appropriate wave function. For more detailed discussions, the reader is referred to the article of Glauber.

Let us seek an effective potential $V_{op}(r)$ that produces the same $f(\vartheta)$ as (N5). By (N1), as applied to this $V_{op}(r)$ and by (N5) with (N7), we have, by hypothesis,

$$\exp\,[2i\delta_{op}(\rho)] = \int\cdots\int |\Psi_a(\mathbf{q}_1 \ldots \mathbf{q}_n)|^2 \exp\,[2i\delta(\rho;\mathbf{s}_1 \ldots \mathbf{s}_n)]\, d\mathbf{q}_1 \ldots d\mathbf{q}_n, \quad \text{(N9)}$$

where

$$\delta_{op}(\rho) = -\frac{1}{2\hbar v}\int_{-\infty}^{\infty} V_{op}(|\boldsymbol{\rho} + \mathbf{z}|)\, dz. \quad \text{(N10)}$$

As $\exp [2i\delta(\rho; \mathbf{s}_1 \ldots \mathbf{s}_n)]$ is a unitary operator, we have, writing $\delta$ for $\delta(\rho; \mathbf{s}_1 \ldots \mathbf{s}_n)$ for the moment,

$$(e^{2i\delta})^\dagger (e^{2i\delta}) = 1,$$

or

$$\sum_m (e^{2i\delta} \Psi_a, \Psi_m)(\Psi_m, e^{2i\delta} \Psi_a) = 1,$$

which can be written

$$\left| \int \Psi_a^* e^{2i\delta} \Psi_a \, d\mathbf{q}_1 \ldots d\mathbf{q}_n \right|^2 + \sum_{m \neq a} \left| \int \Psi_m^* e^{2i\delta} \Psi_a \, d\mathbf{q}_1 \ldots d\mathbf{q}_n \right|^2 = 1. \quad \text{(N11)}$$

By (N9), we have

$$|\exp [2i\delta_{op}(\rho)]|^2 = 1 - \sum_{m \neq a} \left| \int \Psi_m^* e^{2i\delta} \Psi_a \, d\mathbf{q}_1 \ldots d\mathbf{q}_n \right|^2,$$
$$\leqslant 1. \quad \text{(N12)}$$

From this it follows that $\delta_{op}(\rho)$ has a (in general) nonvanishing *positive* imaginary part, and from (N10), it follows that the effective potential $V_{op}(r)$ has a (in general) nonvanishing *negative* imaginary part, i.e.,

$$V_{op}(r) = V_{op \text{ real}} - iV_{op \text{ imag}}, \quad \text{(N13)}$$

where the real part may be attractive or repulsive but the imaginary part $V_{op \text{ im}}(r) \geqslant 0$.

The potential $V_{op}$ in (N13) is called the optical potential, and the model of a target particle represented by such a potential is called the optical model (or the cloudy crystal ball model) by analogy with classical optics. In physical optics, the diffraction of light by a spherical medium of refractive index $n(1 - i\kappa)$ is described by an equation similar to (N1). The real part $n$ gives the refraction while the imaginary part $i\kappa n$ accounts for the absorption by the medium.

Since the original suggestion of the basic ideas of this optical model by Fernbach, Serber and Taylor, and Feshbach, Porter and Weisskopf, a considerable amount of work has been done, both on calculations of a phenomenological potential from the empirical data, and on refinements of the theory. Thus attempts have been made to develop an optical model on more basic and general ground than the rather intuitively clear idea of a complex potential. There are such questions as resonance levels, rearrangement collisions etc. It is clear that when refinements are pushed further and further, one ultimately comes to face the basic problem of the structure of a finite nucleus, which problem has not yet been completely solved either in theory or in practice.

In the phenomenological approach, the potential between a nucleon and a nucleus is represented by a sum of a central $V^c$ and an $\mathbf{L\cdot S}$ term, each of which is complex. Thus

$$V^c = V^c_{\rm re}(r) + iV^c_{\rm im}(r), \tag{N14}$$

$$V^{\mathbf{L\cdot S}} = [V^{\mathbf{L\cdot S}}_{\rm re}(r) + iV^{\mathbf{L\cdot S}}_{\rm im}(r)](\mathbf{L}\cdot\boldsymbol{\sigma}). \tag{N14a}$$

The $\mathbf{L\cdot S}$ potential is needed to account for the polarization data (such as the polarization of an unpolarized beam by scattering. See Chapter 2, Sect. J. 1–5). Of the four functions $V^c_{\rm re}$, $V^c_{\rm im}$, $V^{\mathbf{L\cdot S}}_{\rm re}$, $V^{\mathbf{L\cdot S}}_{\rm im}$, each has a different depth and radial dependence (form factor, or thickness of diffuse boundary). They are fitted to various nuclei for various scattering energies. Although there is always the question of uniqueness in the potential so obtained, there is little doubt that the optical model has been very successful in representing the nucleon–nucleus scattering data on the whole. See Chapter 5, Sect. U.

For a calculation of the effect of $V^{\mathbf{L\cdot S}}$ of (N25) on the polarization in the Born approximation, see Chapter 2, Sect. J.5.

### *ii) Inelastic scattering and generalized optical theorem*

The complex potential $V_{op}(r)$ has an immediate consequence. Firstly it is possible to obtain a generalized, but very important, theorem of the relation (A21). From the Schrödinger equation

$$[\nabla^2 + k^2 - (U_r - iU_{\rm im})]\,\Psi(\mathbf{r}) = 0,$$

and proceeding as in Chapter 1, Sect. G (G2–G6), one obtains, for two wave vectors $\mathbf{k}'$, $\mathbf{k}$,

$$\nabla\cdot[\Psi^*_{\mathbf{k}'}\nabla\Psi_{\mathbf{k}} - \Psi_{\mathbf{k}}\nabla\Psi^*_{\mathbf{k}'}] + 2iU_{\rm im}\,\Psi^*_{\mathbf{k}'}\,\Psi_{\mathbf{k}} = 0. \tag{N15}$$

Using Green's theorem and the asymptotic form (G4) for $\Psi_{\mathbf{k}}$, $\Psi_{\mathbf{k}'}$, etc., one obtains, as in (G6),

$$\frac{1}{2i}[f(\mathbf{k}',\mathbf{k}) - f^*(\mathbf{k},\mathbf{k}')] = \frac{k}{4\pi}\int f^*(\mathbf{k}'',\mathbf{k})f(\mathbf{k}'',\mathbf{k})\,d\Omega_{\mathbf{k}''} + \frac{1}{4\pi}\int U_{\rm im}(r)\,\Psi_{\mathbf{k}'}\,\Psi_{\mathbf{k}}\,d\mathbf{r}. \tag{N16}$$

For $\mathbf{k}' = \mathbf{k}$, i.e., in the forward direction,

$$\operatorname{Im} f(\mathbf{k},\mathbf{k}) = \frac{k}{4\pi}\,\sigma_{sc} + \frac{1}{4\pi}\int U_{\rm im}(r)\,\Psi^*_{\mathbf{k}}\,\Psi_{\mathbf{k}}\,d\mathbf{r}, \tag{N17}$$

where $\sigma_{sc}$ is the scattering cross section. Now, from (N15), one obtains for the current $\mathbf{J}$ the equation

$$\nabla\cdot\mathbf{J} = -\frac{\hbar}{m}\,U_{\rm im}(r)\,\Psi^*_{\mathbf{k}}\,\Psi_{\mathbf{k}}$$

which shows that $\mathbf{J}$ is not divergenceless at points where $U_{\rm im}(r) \neq 0$. Thus $U_{\rm im}(r)\,d\mathbf{r}$ is an element of a "sink". This disappearance of the particles of momentum $|\mathbf{k}|$ may be ascribed to an apparent absorption (which is in fact due to the inelastic scatterings). As the incident beam $e^{i\mathbf{k}\cdot\mathbf{r}}$ is normalized to $\frac{k\hbar}{m} = v$ particles per unit area per second, we may define an "absorption cross section" $\sigma_{ab}$ by

$$v\sigma_{ab} = -\int \nabla\cdot\mathbf{j}\, d\mathbf{r} = \frac{\hbar}{m}\int U_{\text{im}}(r)|\Psi_{\mathbf{k}}|^2\, d\mathbf{r}. \tag{N18}$$

Then (N17) becomes, on writing $f(\mathbf{k}, \mathbf{k}) \equiv f(\mathbf{k}, \vartheta) = f(\mathbf{k}, 0)$,

$$\text{Im}\, f(\mathbf{k}, 0) = \frac{k}{4\pi}(\sigma_{sc} + \sigma_{ab}) = \frac{k}{4\pi}\sigma(k), \tag{N19}$$

where $\sigma(k)$ is the total cross section for elastic + inelastic collisions. This relation is a generalized theorem of (A21).#

To show that $\sigma_{ab}$ in (N18), (N19) is due to inelastic scatterings, one has from (N19), (N5), (N9)

$$\sigma(k) = \frac{4\pi}{k}\,\text{Im}\, f(\mathbf{k}, 0) = 2\int [1 - \text{Re}\exp(2i\delta_{op})]\, d^2\boldsymbol{\rho}. \tag{N20}$$

The (elastic) scattering cross section is given by (C82) which in the present case is ($\sigma_{el} \equiv \sigma_{sc}$)

$$\sigma_{el} = \int |f(\mathbf{k}, \vartheta)|^2\, d\Omega = \int |1 - \exp(2i\delta_{op})|^2\, d^2\boldsymbol{\rho}. \tag{N21}$$

From (N20), (N21), it is seen that there is a difference $\sigma - \sigma_{el}$

$$\sigma_{ab} = \sigma - \sigma_{el} = \int [1 - |\exp(2i\delta_{op})|^2]\, d^2\boldsymbol{\rho}. \tag{N22}$$

---

# A formal but exact derivation of (N19) without the explicit use of the optical potential is also possible. Using the notation of (T33) and (T34), we have the total cross section $\sigma$

$$\begin{aligned}\sigma &= \sigma_{sc} + \sigma_r \\ &= \frac{\pi}{k^2}\sum(2l+1)\left(|1-\eta_l|^2 + 1 - |\eta_l|^2\right) \\ &= \frac{\pi}{k^2}\sum(2l+1)(2 - \eta_l - \eta_l^*).\end{aligned} \tag{N24}$$

The elastic scattering amplitude $f(\vartheta)$ is obtained from the expression $\psi_{sc}$ just before the equation (T33).

$$f(\vartheta) = \frac{1}{2ik}\sum_{l=0}^{\infty}(2l+1)(\eta_l - 1)P_l(\cos\vartheta). \tag{N25}$$

Using (N24), (N25) and $P_l(1) = 1$ we get equation (N19).

That this difference is due to *inelastic* collisions is seen from (N12), which on comparison with (N22) shows

$$\sigma_{ab} = \sum_{m \neq a} \left| \int \Psi_m^* \Psi_a \exp\,[2i\delta(\rho; s_1 \ldots s_n)]\, d\mathbf{q}_1 \ldots d\mathbf{q}_n \right|^2 \tag{N23}$$

which clearly shows that $\sigma_{ab}$ of (N22) is the inelastic scattering cross sections in which the target has changed from the initial state $\Psi_a$ to any of all the other states $\Psi_m$, $m \neq a$.

## 2. THEORY OF MULTIPLE SCATTERING OF WATSON

We shall treat the problem of the scattering of a particle by a system of particles [such as a charged particle (electron or $\mu$-meson) by an atom, or a nucleon by a nucleus] in a general, and therefore formal, manner. This problem is seen to be closely related to the theory of optical potential introduced in the last section. Historically, it is from the work of Watson and his coworkers on this problem that Brueckner and his coworkers start their important theory of many-body problem (nuclear matter and finite nuclei). A simplified account of the work of Watson *et al.* will be given below.

### *i) General theory of multiple scattering*

Let the Hamiltonian of the composite particle (consisting of $N$ particles) be $H_N$, the coordinates collectively denoted by $\boldsymbol{\xi}$, the eigenfunctions $\phi_{\gamma,m}(\boldsymbol{\xi})$, where $m$ is the quantum number of the $z$-component of the angular momentum, and $\gamma$ stands for all the other quantum numbers of the system. Thus

$$(H_N - W_\gamma)\,\phi_{\gamma m} = 0. \tag{N26}$$

We shall for simplicity write $\gamma$ for $\gamma m$ in the following.

Let $K$ be the kinetic energy operator, $\varepsilon$ the energy, of the incident particle (coordinate denoted by $\mathbf{x}$),

$$(K - \varepsilon_q)\,\varphi_q = 0, \qquad \varphi_q = \frac{1}{(2\pi)^{3/2}}\, e^{i\mathbf{q}\cdot\mathbf{x}}. \tag{N27}$$

For the non-interacting system (i.e., $\boldsymbol{\xi}$, $\mathbf{x}$ at infinite separations), in the initial state $\phi_{\gamma m}$ $(= \phi_n)$, $\varphi_q$, one has

$$(H_N + K - E_n)\,\phi_n\,\varphi_q = 0, \qquad E_n = W_n + \varepsilon_q, \tag{N28}$$

or

$$(H_0 - E_n)\,\chi_n = 0, \qquad H_0 = H_N + K, \qquad \chi_n = \phi_n\,\varphi_q. \tag{N28a}$$

Let $V_\alpha = V_\alpha(\boldsymbol{\xi}_\alpha, \mathbf{x})$ be the interaction between the incident particle and the $\alpha$th particle in the target, and

$$V = \sum_{\alpha=1}^{N} V_\alpha. \tag{N29}$$

For the collision, the Schrödinger equation is

$$(H_0 + V - E_n)\Psi = 0, \tag{N30}$$

which can be written in the integral form [see (M17)]

$$\Psi = \chi_n + \frac{1}{E_n - H_0 + i\eta} V\Psi \equiv \chi_n + \frac{1}{a} V\Psi \tag{N31}$$

which contains the condition $(+i\eta)$ for outgoing waves and the incident wave $e^{i\mathbf{q}\cdot\mathbf{x}}$ through $\chi_n$ in (N27), (N28a).

In Sect. M, equation (N28) is solved by expanding $\Psi(\boldsymbol{\xi}, \mathbf{x})$ in the complete set $\phi_\sigma(\boldsymbol{\xi})$ of (N26),

$$\Psi = \sum \phi_\sigma(\boldsymbol{\xi})\, \psi_{\sigma k}(\mathbf{x}) \tag{N32}$$

$$= \phi_\gamma(\boldsymbol{\xi})\, \pi_{\gamma q}(\mathbf{x}) + \sum_\sigma{}' \phi_\sigma(\boldsymbol{\xi})\, \psi_{\sigma q}(\mathbf{x}), \tag{N32a}$$

where $\phi_\gamma$ stands for the $\phi_{\gamma m}$ of (N26) for the state $\gamma$, $m$ and the prime denotes the exclusion of $\sigma = \gamma$ in the sum. From (N28), (N30), one gets

$$(K - \varepsilon_{\gamma q})\, \psi_{\gamma q}(\mathbf{x}) = -\sum_\sigma < \gamma | V(\boldsymbol{\xi}, \mathbf{x}) | \sigma > \psi_{\sigma q}(\mathbf{x}), \tag{N33}$$

where

$$\langle \gamma | V | \sigma \rangle = \int \phi_\gamma^*(\boldsymbol{\xi})\, V(\boldsymbol{\xi}, \mathbf{x})\, \phi_\sigma(\boldsymbol{\xi})\, d\boldsymbol{\xi}, \tag{N33a}$$

$V(\boldsymbol{\xi}, \mathbf{x})$ being given by (N29) and $d\boldsymbol{\xi} = d\boldsymbol{\xi}_1 \ldots d\boldsymbol{\xi}_N$. (N33) together with similar equations for $\psi_{\sigma q}(\mathbf{x})$, $\sigma \neq \gamma$, form an infinite system of coupled differential–integral equations. One then seeks the solution $\psi_{\gamma q}(\mathbf{x})$ that describes an incident plane wave (N27) and an outgoing wave.

i) Coherent and incoherent scattering

In the method of Watson, instead of using (N30), (N33), one proceeds as follows. Let

$$\Psi_n(\boldsymbol{\xi}, \mathbf{x}) = \phi_n(\boldsymbol{\xi})\, \psi_{nq}(\mathbf{x}), \tag{N34}$$

where $\phi_n(\boldsymbol{\xi})$ is the initial state of the target, and $\psi_{nq}(\mathbf{x})$ is a function that describes the *elastic* scattering by the target in the state $\phi_n(\boldsymbol{\xi})$. For this, an "effective" potential $U_n(\mathbf{x})$ is sought such that the "exact" solution $\psi_n \equiv \psi_{nq}$ of

$$[K - \varepsilon_n + U_n(\mathbf{x})]\, \psi_n(\mathbf{x}) = 0 \tag{N35}$$

describes the elastic scattering. In general, $U_n(\mathbf{x})$ will be different for various initial states $\phi_n(\boldsymbol{\xi})$ and incident momenta $\mathbf{q}$; it is not a static potential. But in some approximations, $U_n(\mathbf{x})$ becomes static, as we shall see later on. A part of the theory is to determine $U_n(\mathbf{x})$.

(N35) can be combined with (N26) into one equation, in view of (N34),

$$(K + H_N - W_n + U_n(\mathbf{x}) - \varepsilon_n)\,\Psi_n(\boldsymbol{\xi}, \mathbf{x}) = 0, \tag{N36}$$

which can be put in the integral form

$$\Psi_n(\boldsymbol{\xi}, \mathbf{x}) = \chi_n(\boldsymbol{\xi}, \mathbf{x}) + \frac{1}{a}\,U_n(\mathbf{x})\,\Psi_n, \tag{N37}$$

where $\chi_n$ is given in (N28a) and $a$ by (N31),

$$a = \varepsilon_n + W_n - K - H_N + i\eta = E_n - H_0 + i\eta.$$

Now we seek an operator $F$ such that the solution $\Psi$ of (N30) or (N31) can be obtained by operating by $F$ on $\Psi_n$.

$$\Psi(\boldsymbol{\xi}, \mathbf{x}) = F\Psi_n(\boldsymbol{\xi}, \mathbf{x}).^{\#} \tag{N38}$$

Since $\Psi_n$ is by hypothesis to describe the elastic (and hence coherent) scattering, $(F - 1)\Psi_n$ will describe the inelastic (and hence incoherent, i.e., no interference between the scattered and the incident waves) scattering. For this, we require that $(F - 1)\Psi_n$ be orthogonal to $\phi_n$,

$$[\phi_n(\xi), (F - 1)\,\Psi_n] = 0, \tag{N39a}$$

i.e.,

$$\langle n|F|n\rangle = 1. \tag{N39}$$

[We shall see in (N56) below how this is satisfied.] It will be noted that $(F - 1)\Psi_n$ does not vanish even when inelastic scattering is energetically not possible.

For $F\Psi_n$ to be a solution of (N31), we impose on $F\Psi_n$ the integral equation

$$F\Psi_n = \Psi_n + \frac{1}{a - U_n}(V - U_n)\,F\Psi_n. \tag{N40}$$

The sufficiency of this can be verified by operating on the left by (the differential operator) $a - U_n$, and using (N36), thereby leading to (after allowing $\eta \to 0$)

$$(\varepsilon_n + W_n - K - H_N - V)\,F\Psi_n = 0,$$

---

# The solution $\Psi$ of (N31) is expressible in terms of the wave operator $\Omega$ of Møller as

$$\Psi = \Omega\chi_n. \tag{N38A}$$

On comparing this with (N35), one obtains for $\Omega$ the operator equation

$$\Omega = 1 + \frac{1}{a}\,V\Omega. \tag{N40A}$$

The operator $F$ is an operator in a similar sense [see (N41) below], although it is not exactly the Møller operator $\Omega$ since it operates on $\Psi_n$ of (N37) containing not the "plane wave" but the "outgoing wave".

showing that $F\Psi_n$ satisfies the same equation (N30) as $\Psi$. One can write (N40) in the form of

$$F = 1 + \frac{1}{a - U_n}(V - U_n)\,F, \tag{N40a}$$

if one remembers that this is an operator equation to operate on $\Psi_n$ (not an operator equation valid for any wave function).

**ii)** The optical potential $U(\mathbf{x})$

It must be noted that $F$ and the state $\phi_n$ [giving the elastic scattering $\psi_n(\mathbf{x})$] are interrelated by (N38) and (N37), or, explicitly, by (N40). For each state $\phi_n$ [and momentum $\mathbf{q}$, see $\chi_n$ in (N37) and (N28a)], there is a $U_n$ and an associated $F$. A more explicit relation between $U$ and $F$ can be found as follows.

Let us consider the elastic scattering by the target in the state $\phi_\gamma(\boldsymbol{\xi}) \equiv \phi_{\gamma m}(\boldsymbol{\xi})$. From (N36) (on changing $n$ into $\gamma$), or from (N35) directly,

$$(K - \varepsilon_q)\,\psi_{\gamma q}(\mathbf{x}) = -\langle\gamma|U_\gamma(\mathbf{x})|\gamma\rangle\,\psi_{\gamma q}(\mathbf{x}), \tag{N35a}$$

where

$$\langle\gamma|U_\gamma(\mathbf{x})|\gamma\rangle = U_\gamma(\mathbf{x}). \tag{N41}$$

Since by hypothesis (N39a), $(F - 1)\,\phi_\gamma(\boldsymbol{\xi})\,\psi_{\gamma q}(\mathbf{x})$ is orthogonal to $\phi_\gamma(\boldsymbol{\xi})$, it follows that $\psi_{\gamma q}(\mathbf{x})$ in (N35a) (namely, the coefficient of $\phi_\gamma(\boldsymbol{\xi})$ in the expansion of $F\Psi_n$) is *identical* with the $\psi_{\gamma q}(\mathbf{x})$ in (N32a), which is given by (N33),

$$(K - \varepsilon_q)\,\psi_{\gamma q}(\mathbf{x}) = -\sum_\sigma \langle\gamma|V(\boldsymbol{\xi}, \mathbf{x})|\sigma\rangle\,\psi_{\sigma q}(\mathbf{x}). \tag{N33}$$

Now, from (N32) and (N38), on replacing $n$ by the general state $\gamma$ as in (N40–41), we have

$$F\phi_\gamma(\boldsymbol{\xi})\,\psi_{\gamma q}(\mathbf{x}) = \sum_\sigma \phi_\sigma(\boldsymbol{\xi})\,\psi_{\sigma q}(\mathbf{x})$$

or

$$F|\gamma\rangle\,\psi_{\gamma q}(\mathbf{x}) = \sum_\sigma |\sigma\rangle\,\psi_{\sigma q}(\mathbf{x}).$$

(N33) above can therefore be written

$$(K - \varepsilon_q)\,\psi_{\gamma q}(\mathbf{x}) = -\langle\gamma|V(\boldsymbol{\xi}, \mathbf{x})\,F|\gamma\rangle\,\psi_{\gamma q}(\mathbf{x}). \tag{N42}$$

On comparing this with (N35a), one gets#

$$U_\gamma(\mathbf{x}) = \langle\gamma|U_\gamma(\mathbf{x})|\gamma\rangle = \langle\gamma|VF|\gamma\rangle \tag{N43}$$

which, we must remember, is an operator to operate on $\psi_{\gamma q}(\mathbf{x})$. This $F$ satisfies (N40), (N40a) with $n$ replaced by $\gamma$. $U_\gamma(\mathbf{x})$ is defined only for the diagonal

---

# Note that the derivation of the exact relation (N43) here is general and independent of the operator $P$. Compare refs. 2, 3.

elements, i.e., $\langle\gamma'|U_\gamma(\mathbf{x})|\gamma\rangle = 0$ for $\gamma' \neq \gamma$, and is not related to $\langle\gamma'|VF|\gamma\rangle$ for $\gamma' \neq \gamma$. This may be expressed by

$$U_\gamma(\mathbf{x}) = \delta_{\gamma'\gamma}\langle\gamma'|VF|\gamma\rangle. \tag{N43a}$$

In the following, we shall, for simplicity, drop the subscript $n$ or $\gamma$ from $U_n$, $U_\gamma$, it being understood that the $U$ and the $F$ are always related to each other by (N40) or (N43), and that $F$ is an operator in the sense of (N40).

Instead of working with (N40), we may seek an operator $P$ (operating on $V$, $F$, but not on $\Psi$) such that (N40) is equivalent to

$$F\Psi_\gamma = \Psi_\gamma + \frac{1}{a - U} PVF\Psi_\gamma, \tag{N44}$$

or, from (N40) and (N43),

$$[(1 - P)V - U]\, F\Psi_\gamma = 0. \tag{N44a}$$

This relation connecting $P$, $U$, etc., insures that the solution of (N44) is a solution of (N40).

From (N43) and (N44a), one can obtain a condition on $P$. Writing $\Psi_\gamma = \phi_\gamma(\xi)\, \psi_{\gamma q}(\mathbf{x}) = |\gamma\rangle\, \psi_{\gamma q}(\mathbf{x})$, and since $U$, $F$, $\Psi_\gamma$ are such that $F$ and $U$ are only defined for $\phi_\gamma$ $\psi_{\gamma q}$, (N44a) implies

$$\langle\gamma|UF|\gamma\rangle\, \psi_{\gamma q} = \langle\gamma|(1 - P)VF|\gamma\rangle\, \psi_{\gamma q}.$$

Since $U$ is diagonal, by (N39), one obtains

$$U_\gamma = \langle\gamma|(1 - P)\, VF|\gamma\rangle \tag{N44b}$$

which will be consistent with (N43) if

$$\langle\gamma|PVF|\gamma\rangle = 0. \tag{N45}$$

**iii)** The operator $t$

Let us for a moment consider the scattering of one particle by another free (i.e., unbound to another system) particle. Let $\chi_n$ be the incident wave and $\Psi$ the solution of the Schrödinger equation for outgoing wave. Then, by (N38A), (N40A) in the preceding footnote, in terms of the Møller wave operator $\Omega$,

$$\psi = \Omega\chi_n,$$

$$\Omega = 1 + \frac{1}{E - H_0 + i\eta} V\Omega,$$

where $V$ is the interaction between the two particles. Let us introduce the two-particle scattering operator $t$ by

$$t = V\Omega = V\left(1 + \frac{1}{E - H_0 + i\eta} V\Omega\right) \tag{N46}$$

$$= V + V \frac{1}{E - H_0 + i\eta} t. \tag{N46a}$$

When this is allowed to operate on $\chi_n$, one has an integral equation determining $t$ from the potential $V$.

Next consider the case of the collision between two particles, where now the incident particle not only interacts with a potential $V_{12}$ with the other, but also with a field $U$. The equation (N44) in operator form [to operate on $\Psi_n$ as in (N40)] is

$$F = 1 + \frac{1}{a - U} P V_{12} F. \tag{N44c}$$

Again we define the scattering operator $t$ (for two particles) by

$$\begin{aligned} t &= VF \\ &= V + V\frac{1}{a - U}Pt. \end{aligned} \tag{N47}$$

With $t = VF$, (N44c) can be put in the form

$$F = 1 + \frac{1}{a - U}Pt. \tag{N48}$$

This describes the scattering of a particle in terms of an "effective" interaction operator $t$. The $U$ in the denominator of Green's function represents the effect of the field $U$ on the incident particle. This is usually expressed by saying that the particle is in a "dispersive medium".

Coming back to the case of the $N$-particle target, one introduces the two-body operator $t_\alpha$ (between the incident and the $\alpha$th particle) by

$$t_\alpha = V_\alpha + V_\alpha \frac{1}{a - U} P t_\alpha. \tag{N49}$$

(The operator $P$ is made necessary by the presence of $U$.) It can be easily verified that (N44) is satisfied if we put

$$F = 1 + \frac{1}{a - U} \sum_{\alpha=1}^{N} P t_\alpha F_\alpha, \tag{N50}$$

$$F_\alpha = 1 + \frac{1}{a - U} \sum_{\beta \neq \alpha}^{N} P t_\beta F_\beta. \tag{N51}$$

(N50) can be iterated, by means of (N51) into the series

$$F = 1 + \frac{1}{a - U} \sum \left\{ P t_\alpha + P t_\alpha \frac{1}{a - U} P t_\beta + P t_\alpha \frac{1}{a - U} P t_\beta \frac{1}{a - U} P t_\gamma + \cdots \right\}, \tag{N50a}$$

where the summation is over $\alpha, \beta, \gamma, \ldots$ independently, with the restriction that no two successive indices be the same. This simply means that any particle does not scatter the incident particle twice in *immediate* succession. (N50a) now expresses the scattering of the incident particle by the $\alpha$th; by $\beta$th after $\alpha$th; by $\gamma$th after $\beta$ and $\alpha$; etc., and thus contains the theory of multiple scattering.

From (N49) and (N51), we have

$$\sum_{\alpha=1}^{N} t_\alpha F_\alpha = \sum_{\alpha=1}^{N} V_\alpha \left(1 + \frac{1}{a - U} \sum_{\beta \neq \alpha}^{N} P t_\beta F_\beta\right) + \sum_\alpha V_\alpha \frac{1}{a - U} P t_\alpha F_\alpha$$

$$= V + V \frac{1}{a - U} \sum_{\alpha=1}^{N} P t_\alpha F_\alpha = VF. \tag{N52}$$

**iv)** Choice of operator $P$, and the optical potential $U$

In (N44), we have an operator $P$ which we still have to determine. In (N45), for example, $P$ operates on $VF$, or by (N52), on $t_\alpha F_\alpha$. In taking the matrix elements of $VF$, we deal with the matrix elements of $t_\alpha$, $F_\alpha$, etc. between various (end, and intermediate) states. Therefore we may make the choice of $P$ from the point of view of its effect on $t_\alpha$.

Let us choose a $P_0$ such that in the succession of (multiple) scatterings by the target, all the initial, intermediate and final states are different, i.e., all matrix elements of $\langle \gamma' | PVF | \gamma \rangle$ vanish unless $\gamma'$, $\gamma$ and all the intermediate states are different from one another. To see if such a $P_0$ satisfy the condition (N44a), it is immediately seen that by hypothesis,

$$\langle \gamma | P_0 VF | \gamma \rangle = 0, \tag{N53}$$

i.e., it does satisfy (N45), which by (N44b), leads to

$$U_\gamma = \langle \gamma | U_\gamma | \gamma \rangle = \langle \gamma | VF | \gamma \rangle, \tag{N54}$$

which is of course the result (N43) obtained earlier.

Other choices of $P$ satisfying the condition (P19a) or (N45) may be possible; but the $P_0$ defined above seems to be the simplest.#

Concerning the potential $U_\gamma(\mathbf{x})$ given by (N43) or (N44), we have from (N35)

$$\left[-\frac{\hbar^2}{2\mu} \nabla^2 + U_\gamma(\mathbf{x})\right] \psi_\gamma(\mathbf{x}) = \varepsilon_\gamma \psi_\gamma(\mathbf{x}), \tag{N55}$$

---

# In ref. 2, Watson gives another choice $P_1$ which is so defined that $\langle \gamma' | P t_\alpha | \gamma \rangle = \langle \gamma' | t_\alpha | \gamma \rangle$ for $\gamma' \neq \gamma$, and $\langle \gamma | P t_\alpha | \gamma \rangle = 0$. This means that the matrix elements $\langle \gamma | PVF | \gamma \rangle$ vanish whenever two successive states are the same in the expansion of $F_\alpha$, according to (N51) in $\sum_\alpha \langle \gamma | P t_\alpha | \gamma' \rangle \langle \gamma' | F_\alpha | \gamma \rangle$. We have not been able to show that his result for this $P_1$,

$$U_\gamma = \langle \gamma | \sum{}' t_\alpha | \gamma \rangle$$

satisfies the necessary exact condition (N43).

which is the equation for a particle in an effective field $U_\gamma(\mathbf{x})$ such that $\psi_\gamma(\mathbf{x})$ correctly describes the elastic scattering of the original system defined by

$$H_0 + V = H_N - \frac{\hbar^2}{2\mu}\nabla^2 + \sum_{\alpha=1}^{N} V_\alpha(\mathbf{x}, \boldsymbol{\xi}_\alpha)$$

in (N30) or (N31). This potential $U_\gamma(\mathbf{x})$ is called the "optical potential" and has been used recently in describing the nucleon–nucleus scattering. It may be real and may be complex, as calculated from (N54). Of course the operator $F$ still has to be determined from the equation (N40a) which can be iterated to

$$F = 1 + \frac{1}{a - U}(V - U) + \frac{1}{a - U}(V - U)\frac{1}{a - U}(V - U) + \cdots, \tag{N56}$$

but which still contains the unknown $U$. As a first approximation, one may set $F_\alpha \cong 1$ thereby getting

$$\begin{aligned} U_\gamma &= \langle\gamma|U_\gamma|\gamma\rangle = \langle\gamma|VF|\gamma\rangle \\ &= \langle\gamma\left|\sum_{\alpha=1}^{N} t_\alpha F_\alpha\right|\gamma\rangle \fallingdotseq \langle\gamma\left|\sum_{\alpha=1}^{N} t_\alpha\right|\gamma\rangle. \end{aligned}$$

On using (N49), (N51)

$$U_\gamma = \langle\gamma\left|\sum_{\alpha=1}^{N} V_\alpha\right|\gamma\rangle + \langle\gamma\left|\sum_{\alpha=1}^{N} V_\alpha \frac{1}{a - U} PV_\alpha\right|\gamma\rangle + \cdots. \tag{N57}$$

The first term corresponds to the Hartree approximation

$$U_{\gamma H} = \langle\gamma\left|\sum_{\alpha=1}^{N} V_\alpha\right|\gamma\rangle. \tag{N58}$$

The approximation (N57) has been used and discussed by Watson *et al.* For details, the reader is referred to the original papers.

v) Rearrangement collisions

The theory outlined above can be extended to the case of rearrangement collisions in which the incident particle (say, a neutron colliding with a nucleus) is captured and a new particle emerges (say, a proton). For any specific process, it is only necessary to express the $\Psi$ in the form having the appropriate asymptotic condition for the correct outgoing particle. This is done by choosing the appropriate Green's function in (N40) for $F$. The essence of this method of choosing the Green's function for direct and rearrangement collisions has been described in Sects. L (L20a), (26a) and M. For the details of the expansion of the theory of the present section to such cases as pick-up (the incoming particle picking up a particle and emerging as a new unit) and stripping (an incoming deuteron being stripped and emerging as a proton) processes, the reader will be referred to Chapter V, Sect. V and the references.

**vi)** Remarks on the relation between the theory of Watson and Brueckner's theory of many-body systems.

Perhaps it is of some historical interest to make a few remarks in passing concerning the relation between Watson's theory of multiple scattering and Brueckner's theory of many-body systems. In the latter, one deals with such questions as the energy of the ground state $\phi_0$ of a system of a large number $N$ of interacting particles [through two-body interactions $V = \sum_{i \neq j} V(ij)$]. When the energy $E$ of the system is calculated according to the familiar Schrödinger perturbation theory but with the use of the formalism of the reaction matrix $R$ (see below) similar to that of the scattering matrix $t$ in (N46), it is found that the energy due to the interaction $V$ is given by an infinite series

$$\begin{aligned} E_1 = \langle V \rangle &+ \langle V \frac{1}{b} V \rangle + \langle V \frac{1}{b} V \frac{1}{b} \rangle - \langle V \rangle \langle V \frac{1}{b^2} V \rangle \\ &+ \langle V \frac{1}{b} V \frac{1}{b} V \frac{1}{b} \rangle - \langle V \rangle \langle V \frac{1}{b} V \frac{1}{b^2} V \rangle - \langle V \rangle \langle V \frac{1}{b^2} V \frac{1}{b} V \rangle \\ &+ \langle V \rangle^2 \langle V \frac{1}{b^3} V \rangle - \langle V \frac{1}{b} V \rangle \langle V \frac{1}{b^2} V \rangle + \cdots, \end{aligned} \tag{N59}$$

where

$$\frac{1}{b} \equiv \frac{Q}{E_0 - H_0} \tag{N60}$$

is the propagator of free particles whose kinetic energy operator is $H_0$. [Infra (O74)]. $Q$ is the projection operator of the state $\phi_0$ such that

$$Q\phi_0 = 0, \qquad Q\phi_a = \phi_a \quad \text{for} \quad a \neq 0,$$

and

$$\langle V \rangle \equiv \langle \phi_0 | V | \phi_0 \rangle.$$

$\phi_0$ is the zeroth-order ground state wave function of the system (of non-interacting particles). As the number $N \to \infty$ (the corresponding volume $\Omega$ also becomes $\infty$ in such a way that the number density $N/\Omega$ remains finite), many of the matrix elements (N59) diverge. The essence of the great contributions of Brueckner and others is that (i) the divergent terms (with respect to the volume $\Omega$, or number $N$) in (N59) can be shown to cancel among themselves exactly, (ii) the great varieties of the infinitely many remaining terms can be summed up in closed forms. This was achieved by the introduction of new interactions $K_{ij}$ (called "vertex operators") to replace the original interaction $V_{ij}$, and of a new Green's function $1/e$ (called the "propagators") to replace the $1/b$ in (N60). These are given by

$$\frac{1}{e} = \frac{Q}{E + K_{ii}^0 + K_{jj}^0 - H_0 - K_{ii} - K_{ij}}, \tag{N61}$$

and the non-linear integral equation

$$K_{ij} = V_{ij} + V_{ij} \frac{1}{e} K_{ij}. \tag{N62}$$

Without making any attempt to explain anything in detail, we may still hope to make the following remarks intelligible. The $K_{ij}$ is an improved form of the reaction matrix $R_{ij}$

$$R_{ij} = V_{ij} + V_{ij} \frac{1}{b} R_{ij}$$

representing the interaction between the two particles $i$, $j$, but not with the other particles of the system,

$$\langle R_{ij} \rangle = \langle V_{ij} \rangle + \langle V_{ij} \frac{1}{b} V_{ij} \rangle + \langle V_{ij} \frac{1}{b} V_{ij} \frac{1}{b} V_{ij} \rangle + \cdots.$$

Hence $K_{ij}$ now includes the interaction between a particle $i$ and all the (unexcited) particles of the system to all orders. The propagator $1/e$ (which is, like $1/b$, a generalized Green's function) containing the operators $K$, in the denominator, now takes into account that the particles $i$, $j$ are not freely moving in an empty medium [as $1/b = Q/(E_0 - H_0)$ represents], but interact with the other particles of the system. One may say, in the language of optics, that the particles are in a "dispersive" medium.

On coming back to the theory of Watson of this section, it is clear that many similar features are obvious. Thus the "propagator" $1/(a - U)$ in (N40), (N44) etc. is a simple expression of the fact that the incident particle travels in a dispersive medium (the target system) on account of the interaction with the target particles. The integral equation (N49) for example, when iterated, is a simple expression of the situation that the incident particle interacts in any (and all) number of collisions, and the operator $P$ replaces the projection operator $Q$ in (N60) and (N61).

### *ii) Scattering of a charged particle by an atom*

The general theory of the preceding section has been applied to the scattering of a charged particle (electron, $\mu$-meson, proton, etc.) by a neutral atom by Mittleman and Watson.

For the atom, (N26) now is written

$$(H_A - W_n)\, \phi_n(\xi) = 0, \tag{N63}$$

where $\xi$ stands for all the electron coordinates, The Schrödinger equation of the system: atom + incident particle is

$$[H_A + K + V_N(\mathbf{x}) + V(\mathbf{x}, \xi) - \varepsilon_n - W_n]\, \Psi(\mathbf{x}, \xi) = 0, \tag{N64}$$

where $\varepsilon_n$ is the initial kinetic energy of the incident particle of coordinate $\mathbf{x}$, and $W_n$ the energy of the atom in the initial state $\phi_n(\xi)$ (which may be the ground or an excited state). $V_N(\mathbf{x})$ is the (Coulomb) interaction of $\mathbf{x}$ with the nucleus and $V(\mathbf{x}, \xi)$ the (Coulomb) interaction of $\mathbf{x}$ with the atomic electrons. (N64), together with the asymptotic condition for outgoing waves in $\mathbf{x}$, is equivalent to the integral equation

$$\Psi = \chi_n + \frac{1}{\varepsilon_n + W_n - K - H_A + i\eta}(V_N + V)\Psi, \tag{N65}$$

$$\equiv \chi_n + \frac{1}{a}(V_N + V)\Psi,$$

where

$$\chi_n = \phi_n(\xi)\frac{1}{(2\pi)^{3/2}}\exp(i\mathbf{k}_n\cdot\mathbf{x}). \tag{N66}$$

Similarly to (N34), (N35), (N37), we write

$$\Psi_n(\xi, \mathbf{x}) = \phi_n(\xi)\,\psi_n(\mathbf{x}), \tag{N67}$$

and seek an "optical" potential $U_n(\mathbf{x})$ for $\psi_n(\mathbf{x})$

$$[K - \varepsilon_n + U_n(\mathbf{x})]\,\psi_n(\mathbf{x}) = 0 \tag{N68}$$

such that $\psi_n(\mathbf{x})$ describes the elastic scattering, i.e.,

$$\psi_n(\mathbf{x}) = (2\pi)^{-3/2}\exp(i\mathbf{k}_n\cdot\mathbf{x}) + \frac{1}{\varepsilon_n - K + i\eta}U_n(\mathbf{x})\,\psi_n(\mathbf{x}), \tag{N69}$$

or, on combining (N68) with (N63), we can write (N69) in the form

$$\Psi_n = \chi_n + \frac{1}{a}U_n\Psi_n. \tag{N70}$$

We shall from now on drop the subscript $n$ from $U_n$ and remember that $U$ is an operator such that $U\psi_n = U_n\psi_n$ for any $n$, and that $\langle\psi_m|U|\psi_n\rangle = 0$ if $m \neq n$. Analogously to (N38), we introduce an operator $F$ by

$$\Psi(\xi, \mathbf{x}) = F\Psi_n(\xi, \mathbf{x}) \tag{N71}$$

such that (i) $F\Psi_n$ is a solution of (N70), (ii) $\Psi_n$ describes the elastic (coherent) and $(F - 1)\Psi_n$ the inelastic (incoherent) scattering, i.e., $F$ satisfies the integral equation [see (N40)]

$$F\Psi_n = \Psi_n + \frac{1}{a - U}(V_N + V - U)\,F\Psi_n, \tag{N72}$$

or, the operator equation (operating on $\Psi_n$)

$$F = 1 + \frac{1}{a - U}(V_N + V - U)F, \tag{N72a}$$

and, similarly to (N39),

$$\langle n|F|n\rangle = \int \phi_2^*(\xi)\,F\phi_n(\xi)\,d\xi = 1 \tag{N73}$$

as an operator equation operating on $\psi_n(x)$.

Again, as in (N44), (N44a), we seek an operator $P$ such that

$$F = 1 + \frac{1}{a - U} P(V_N + V)\, F \tag{N74}$$

is equivalent to (N72a). We shall choose for $P$ the same operator $P_0$ as in Subsect. 2 (iv), (N53), namely, the initial, all the intermediate and the final states in the matrix element of $P(V_N + V)\, F$ are different from one another. For this $P_0$, we have, as in (N53), (N54)

$$\langle n|P_0(V_N + V)F|n\rangle = 0, \tag{N75}$$

and

$$\begin{aligned} U_n \equiv \langle n|U|n\rangle &= \langle n|(V_N + V)F|n\rangle \\ &= V_N + \langle n|V(\mathbf{x}, \xi)F|n\rangle. \end{aligned} \tag{N76}$$

since $V_N = V_N(\mathbf{x})$ is not a function of the $\xi$'s.

All the results above, in particular the expression (N76) for $U_n$, are exact. But the theory is still only formal since $F$ and hence $U_n(\mathbf{x})$ are still not known. One may formally obtain, by iterating $F$ according to (N74),

$$\begin{aligned} U_n(\mathbf{x}) &= \langle n|U|n\rangle = \langle n|(V_N + V)F|n\rangle \\ &= V_N + \langle n|V|n\rangle + \left\langle n\middle|(V_N + V)\frac{1}{a - U}P(V_N + V)\middle|n\right\rangle \\ &\quad + \left\langle n\middle|(V_N + V)\frac{1}{a - U}P(V_N + V)\frac{1}{a - U}P(V_N + V)\middle|n\right\rangle + \cdots. \end{aligned} \tag{N77}$$

Alternatively, similarly to (N49), (N51), one may introduce

$$F_\alpha = 1 + \frac{1}{a - U}\sum_{\beta \neq \alpha} Pt_\beta F_\beta, \tag{N78}$$

$$t_\alpha = (V_N + V)_\alpha + (V_N + V)_\alpha \frac{1}{a - U} Pt_\alpha, \tag{N79}$$

so that

$$\begin{aligned} U_n(x) &= \langle n|\sum_\alpha t_\alpha F_\alpha|n\rangle^{\#} \\ &= \langle n|\sum_\alpha t_\alpha|n\rangle + \langle n|\sum_\alpha t_\alpha \frac{1}{a - U}\sum_{\beta \neq \alpha} Pt_\beta|n\rangle + \cdots. \end{aligned} \tag{N80}$$

The first-order terms (linear in $V_N$, $V$) are of course the same in (N77) and (N80), namely the nuclear Coulomb field $V_N$ and the (Hartree) screening potential.

$$\langle n|V|n\rangle_{\text{Hartree}} = \langle\phi_0(\xi)|V(\xi, \mathbf{x})|\phi_0(\xi)\rangle. \tag{N81}$$

---

\# Setting $F_\alpha = 1$ does not form a consistent development in higher powers of $t_\alpha$ or $(V + V_N)$.

In (N77), the higher order terms are in powers of $V_N$ and $V$, while in (N80), the expansion is in the scattering operator $t_\alpha$. In principle, (N77) and (N80) are equivalent; in practice, the use of the $t$ operator may have an essential advantage over $V_N + V$ in case the potentials $V_N$ and $V$ have infinitely large values for small distances so that the expansion (N77) becomes divergent and meaningless.# On the other hand, the $t_\alpha$ matrix given by (N79) may be finite and the series (N80) convergent. The higher order terms either in (N77) or in (N80) represent the effect of the incoming particle on the states (wave functions) of the target atom, i.e., the usual polarization (in the "adiabatic" limit) and the fast recoil of the atomic nucleus (in the case of high-energy collisions).

To proceed further in an actual application of the theory, it is possible, under certain conditions, to make simplifying approximations, such as the "adiabatic" and the "high energy" approximations. Some approximation can be made by modifying the denominator $a - U$ in the Green's function (N72)

$$a - U = \varepsilon_n + W_n - K - H_A - U + i\eta. \tag{N82}$$

Let us now consider the case of scattering of a particle, of initial kinetic energy $\varepsilon_0$, by an atom in the (ground) state $\phi_0$ so that $a - U$ now becomes

$$a - U = \varepsilon_0 - K - U + W_0 - H_A + i\eta. \tag{N83}$$

We shall introduce the total energy operator

$$\varepsilon = K + U \tag{N84}$$

such that

$$d \equiv a - U = \varepsilon_0 - \varepsilon + W_0 - H_A + i\eta. \tag{N85}$$

When we take the matrix elements of $F$ in (N77), $1/d$ operates on an intermediate state, say $\phi_m$ [and by (N63) $H_A\phi_A = W_m\phi_m$], and we shall hence be dealing with

$$d = \varepsilon_0 - \varepsilon + W_0 - W_m + i\eta. \tag{N85a}$$

i) High-energy approximation.

For high energies of the incident particles, we shall compare the magnitudes of $W_0 - W_m$ and $\varepsilon_0 - \varepsilon$ in (N85a). The difference $\varepsilon_0 - \varepsilon$ is of the order of the $(p/M)\,\delta p$ where $p$, $M$ are the momentum and mass of the incident particle and $\delta p$ the momentum transfer to the atomic electrons. The difference

# This is the case with the nucleon–nucleon interaction for which the scattering and polarization data indicate a strongly repulsive core. The use of the scattering operator $t$ by Brueckner is an essential step in his theory of nuclear matter, not only because by means of it one can handle such singular potentials, but also it leads to a successful method of summing over infinite series for the energy. See end of last section.

$W_m - W_0$ is of the order $(\delta p)^2/\mu$ where $\mu$ is the electronic mass. The average excitation energy $\langle W_m - W_0\rangle_{\text{Av}}$ is

$$\frac{(\delta p)^2}{\mu} \simeq \langle W_m - W_0\rangle \simeq \overline{Z^2}\,\frac{e^2}{a_B}, \tag{N86}$$

where $\overline{Z^2}$ is the mean square of the nuclear charge (varying from $Z = N$, the atomic number, to $Z = 1$, corresponding to excitations of electrons in the various shells). If

$$\varepsilon_0 - \varepsilon \simeq \frac{p}{M}\,\delta p \gg \frac{(\delta p)^2}{\mu} \simeq \langle W_m - W_0\rangle, \tag{N87}$$

we have

$$\frac{p^2}{M} \gg \frac{M}{\mu}\,\frac{(\delta p)^2}{\mu},$$

or

$$\varepsilon_0 \gg \frac{M}{\mu}\,\frac{\overline{Z^2}e^2}{a_B}. \tag{N88}$$

If (N88) is valid, then (N87) shows that $\varepsilon_0 - \varepsilon \gg \langle W_m - W_0\rangle$ and the Green's function may be replaced by

$$\frac{1}{d_H} \doteqdot \frac{1}{\varepsilon_0 - \varepsilon + i\eta}. \tag{N89}$$

ii) Adiabatic approximation.

When the incident particle moves slowly enough in traversing the atom so that the electrons of the atom are able to adjust their state at each moment as if the incident particle were at rest at the position **x**, we talk of an "adiabatic" approximation. One condition is that the time of traversing the atom be long compared with the periodic times of the atomic electrons, or the velocity $v_{in}$ of the incident particle in the atom by $\ll v_{el}$ of the atomic electrons. The other condition is that the incident particle can be described by a trajectory, i.e., the de Broglie wave length $\lambda_{in} = (v_{in}\hbar)/T_{in} \ll \lambda_{el} = (v_{el}\hbar)/T_{el}$, where $T_{in}$, $T_{el}$ are the kinetic energy of the incident particle (inside the atom) and the electrons, respectively. When this is true, we may make the approximation of the commutability of the kinetic energy operator and $F$,

$$(FK - KF)\,\psi_0(\mathbf{x}) \doteqdot 0 \tag{N90}$$

which implies that $K$ can be treated as a function of **x** for $F$ (not for $\psi_0$). From (N64), we have

$$[H_A + V_N(\mathbf{x}) + V(\mathbf{x}, \boldsymbol{\xi}) - W_0]\,F\phi_0(\boldsymbol{\xi})\,\psi_0(\mathbf{x}) = -(K - \varepsilon_0)\,F\phi_0(\boldsymbol{\xi})\,\psi_0(\mathbf{x}), \tag{N91}$$

and from (N35),

$$[K - \varepsilon_0 + U_0(\mathbf{x})]\,\phi_0(\boldsymbol{\xi})\,\psi_0(\mathbf{x}) = 0.$$

By making the approximation (N90), this last equation may be written

$$F[K - \epsilon_0 + U_0(\mathbf{x})]\,\phi_0(\boldsymbol{\xi})\,\psi_0(\mathbf{x}) \fallingdotseq (K - \epsilon_0 + U_0)\,F\phi_0(\boldsymbol{\xi})\,\psi_0(\mathbf{x}),$$

so that (N91) becomes

$$[H_A + V_N(\mathbf{x}) + V(\mathbf{x}, \boldsymbol{\xi}) - W_0]\,F\phi_0(\boldsymbol{\xi})\,\psi_0(\mathbf{x}) \doteqdot U_0(\mathbf{x})\,F\phi_0(\boldsymbol{\xi})\,\psi_0(\mathbf{x}). \quad \text{(N92)}$$

$U_0(\mathbf{x})$ is to be considered to be the energy eigenvalue of the modified atomic state $F\phi_0(\boldsymbol{\xi})$ for each fixed $\mathbf{x}$. We may say that by the adiabatic approximation (N90), the exact equation (N91) is reduced to (N92) in which the "static-nuclei" or "adiabatic potential" (in the sense of the potential energy curves of diatomic molecules) is the optical potential $U_0(\mathbf{x})$. This $U_0(\mathbf{x})$ naturally contains the polarization of the atomic electrons

For the adiabatic approximation to be valid, we visualize the incident particle to be heavy (such as a proton or a meson) compared with an electron. The kinetic energy of such a particle changes very little so that the energy operator $\varepsilon$ in (N84) is such that

$$K = \varepsilon_0 - U_0 = \varepsilon - U_m, \quad \text{(N93)}$$

and the operator $1/d$ (N85a) when operating on a state $\phi_m$ of the atom is

$$d_A \fallingdotseq U_0(\mathbf{x}) - U_m(\mathbf{x}) + W_0 - W_m + i\eta. \quad \text{(N94)}$$

It can readily be shown that using this $d_A$ in (N72a),

$$F = 1 + \frac{1}{d_A}(V_N + V - U)\,F$$

[operating on $\phi_0(\boldsymbol{\xi})\,\psi_0(\mathbf{x})$], and applying $d_A$ on the left, one gets the equation (N92), i.e., the adiabatic approximation discussed from (N90) to (N92).

If, in some cases, $U(\mathbf{x})$ is not much dependent on the state of the atom,

$$U_0 - U_m \ll W_0 - W_m,$$

one may further approximate $d_A$ by

$$d_A \fallingdotseq W_0 - W_m + i\eta. \quad \text{(N95)}$$

Under conditions in which the adiabatic approximation is valid, namely, the periodic times of the atomic electrons are very short compared with the time taken by the incident particle to traverse the atom, the electronic wave functions of the atom for each "static" position $\mathbf{x}$ of the incident particle form an orthogonal set so that all non-diagonal elements of $V_N$ vanish, i.e., for $n \neq n'$

$$\langle n'|V_n(\mathbf{x})|n\rangle \equiv \langle \phi_{n'}(\boldsymbol{\xi})|V_n(\mathbf{x})|\phi_n(\boldsymbol{\xi})\rangle = 0.$$

Since with the choice (N75) for the operator $P$, the initial, intermediate and the final states are all different, we have only to do with non-diagonal elements of $P_0 V$, and they vanish for the reason just mentioned,

$$\langle n'|P_0 V_n|n\rangle = 0.$$

In this case, we may replace (N74) by

$$F = 1 + \frac{1}{a - U} P_0 VF. \tag{N96}$$

Now, on iterating (N96) in (N76), one obtains, as in (N57),

$$U_0(\mathbf{x}) = V_N(\mathbf{x}) + \langle 0|V(\boldsymbol{\xi}, \mathbf{x})|0\rangle + \left\langle 0\left|V\frac{1}{d}P_0 V\right|0\right\rangle + \left\langle 0\left|V\frac{1}{d}P_0 V\frac{1}{d}P_0 V\right|0\right\rangle + \cdots, \tag{N97}$$

where $d$ is given by (N94). The first terms are the same as those in (N77). The higher order terms give the effect of polarization of the atom by the incident particle.

When the incident (or scattered) particle is at large distances from the target atom, the first two terms in the series (N97) for $U_0(\mathbf{x})$ almost cancel each other (resulting in the familiar exponentially decreasing function of $x$). The third term shows a Van der Waals force. If we expand $V(\boldsymbol{\xi}, \mathbf{x})$ in inverse powers of $x$,

$$V = \frac{ZeQ}{x} + \sum_{k=1}^{Z} \frac{eQ}{x^3} (\mathbf{x}\cdot\boldsymbol{\xi}_j) + \cdots, \tag{N98}$$

where $Q$ is the charge of the incident particle, $Z$ the atomic number of the target atom. The $U(\mathbf{x})$ in the denominator in (N94) tends to zero as $x \to \infty$, and we have

$$\left\langle 0\left|V\frac{1}{d_A}PV\right|0\right\rangle = -\frac{Q^2\alpha}{2x^4}, \tag{N99}$$

where $\alpha$ is the polarizability of the atom in state $\phi_0$ given by

$$\alpha = 2e^2 \sum_{n\neq 0} \frac{\left|\left\langle n\left|\frac{1}{x}\sum_j^Z (\mathbf{x}\cdot\boldsymbol{\xi}_j)\right|0\right\rangle\right|^2}{W_n - W_0}. \tag{N100}$$

This is in accord with the result obtained by the usual perturbation method.

It is of interest to formulate a variational expression for the optical potential $U(\mathbf{x})$. Let us denote $F\phi_0(\boldsymbol{\xi})$ by $\psi^{(+)}$ so that (N96)# becomes

$$\psi^{(+)} = \phi_0 + \frac{1}{d_+} PV\psi^{(+)}, \tag{N101}$$

---

# Note that the use of (N96) implies adiabatic approximation, contrary to the statement in ref. 3, Sect. 6.B.

where + denotes $+i\eta$ in the Green's function. Similarly let $\psi^{(-)}$ be the function given by (N101) with $d_-$, i.e., $-i\eta$. Hence $\psi^{(-)}$ is the "complex conjugate" of $\psi^{(+)}$. From (N76)

$$\begin{aligned}\langle 0|U|0\rangle - V_n &= (\phi_0, VF\phi_0)\\ &= (\phi_0, V\psi^{(+)}) = (\psi^{(-)}, V\phi_0)\\ &= \left(\psi^{(-)}, \left[V - V\frac{1}{d_+}PV\right]\psi^{(+)}\right).\end{aligned} \tag{N102}$$

It can readily be verified that the variational equation

$$\delta J = 0, \qquad J = \frac{(\phi_0, V\psi^{(+)})(\psi^{(-)}, V\phi_0)}{\left(\psi^{(-)}, \left[V - V\frac{1}{d_+}PV\right]\psi^{(+)}\right)}, \tag{N103}$$

for arbitrary $\delta\psi^{(+)}$, $\delta\psi^{(-)}$ leads to (N101) and the corresponding equation for $\psi^{(-)}$. It is also seen that $J$ above gives a stationary expression for $\langle 0|U|0\rangle - V_n$.

To apply the theory to the problem where the "high energy" approximation in **i)** or the adiabatic approximation in **ii)** is not valid and the successive approximation (N77) is difficult, some other device for each individual case will be useful. Let us take the elastic scattering of an electron by a hydrogen atom in the triplet state as an example. The anti-symmetry spatial wave function $\psi^{(-)}(1, 2)$ may be expanded:

$$\psi^{(-)}(1, 2) = \sum_{n=0} \!\!\!\!\!\!\!\int \phi_n(1)\, \psi_n(2),$$

where $\phi_n$ are the hydrogen atom wave functions.

$$\psi_0(2) = \int \psi^{(-)}(1, 2)\phi_0(1)\, d\tau_1$$

satisfies the equation with the optical potential $U$

$$[\Delta_2 + U(2) + k^2]\, \psi_0(2) = 0.$$

One of the tests for the approximate wave function obtained by an approximate optical potential is provided by the requirement that $\psi_0(2)$ be orthogonal to $\phi_0(2)$, since $\int \psi^{(-)}(1, 2)\, \phi_0(1)\, \phi_0(2)\, d\tau_1\, d\tau_2$ vanishes for symmetry reasons. The optical model method has recently been applied by Lippmann and Schey (ref. 4) to the low-energy electron–hydrogen scattering.

## 3. THEORY OF MULTIPLE SCATTERING (OF A PARTICLE THROUGH A FOIL)

In the analysis of such data as the angular distribution of high-energy electrons or $\mu$ mesons through a foil of matter, we have the problem of finding the probability of a total deflection angle $\vartheta$ after $n$ successive scatterings of the particle in traversing the foil. The problem may be considered as made up of two parts, namely, i) the angular distribution or differential cross section of a single encounter between the incident particle and an atom of the foil, and ii) the statistical–geometrical problem of compounding the (small) deflections of successive collisions. The first part may be simplified, for many cases of experimental observations, by treating the scattering as that of a central field—the screened field of the nucleus by the "atomic" electrons. To this problem, the theory of Chapter 1, in particular, Sects. A and C, applies.

The second statistical–geometrical part has been treated by Goudsmit and Saunderson (ref. 6). Their theory involves only the assumptions that the path lengths of all particles in the foil are the same, that all scatterings are elastic and that there are no backward scatterings. Its validity is hence not restricted to small angles of scattering and to some specific forms for the law governing the single scattering. Many approximate treatments of the problem of multiple scattering have been given by various authors (refs. 10–13). In the following, we shall follow Goudsmit and Saunderson in deriving the basic formula for the intensity distribution in multiple scattering, and sketch the work of Nigam, Sundaresan and Wu (ref. 7) in developing the general formula to the second Born approximation for a charged fermion of relativistic energies.

### i) *Theory of Goudsmit and Saunderson*

We shall restrict our treatment to cases in which the individual scattering has axial symmetry. The probability that a particle in a single collision is scattered into the solid angle $d\Omega = \sin\vartheta\, d\vartheta\, d\varphi$ is

$$w_1(\vartheta, \varphi)\, d\Omega = \frac{\sigma(\vartheta)}{\sigma}\, d\Omega, \tag{N104}$$

where $\sigma(\vartheta)\, d\Omega$, $\sigma$ are the differential and the total cross section respectively. Let

$$g_l \equiv \langle P_l(\cos\vartheta)\rangle = \frac{1}{\sigma}\int P_l(\cos\vartheta)\, \sigma(\vartheta)\, d\Omega \tag{N105}$$

be the average value of $P_l(\cos\vartheta)$ in one single collision. Then it follows that

$$w_1(\vartheta, \varphi)\, d\Omega = \frac{1}{4\pi}\sum (2l+1)\, g_l P_l(\cos\vartheta)\, d\Omega. \tag{N106}$$

Let the polar angles of the direction of the particle after a second collision (with another atom) be $(\vartheta_2, \varphi_2)$ referred to the direction $(\vartheta_1, \varphi_1)$ after the first scattering, and $(\vartheta, \varphi)$ referred to the direction of the incident beam, so that

$$\cos\vartheta_2 = \cos\vartheta_1 \cos\vartheta + \sin\vartheta_1 \sin\vartheta \cos(\varphi_1 - \varphi).$$

The probability that, after two successive, independent collisions, the particle will be scattered into $d\Omega = \sin\vartheta\, d\vartheta\, d\varphi$ is given by integrating the product $w_1(\vartheta_1, \varphi_1)\, w_1(\vartheta_2, \varphi_2)$ over all values of $\vartheta_1, \varphi_1$ for a fixed $\vartheta, \varphi$,

$$w_2(\vartheta, \varphi)\, d\Omega = \int d\Omega_1\, w_1(\vartheta_1, \varphi_1)\, w_1(\vartheta_2, \varphi_2)\cdot d\Omega. \tag{N107}$$

This, on using the expansion of $P_l(\cos\vartheta_2)$ in terms of the $P_l^m(\cos\vartheta_1) \exp(im\varphi_1)$, $P_l^m(\cos\vartheta)\exp(im\varphi)$, leads to

$$w_2(\vartheta, \varphi) = \frac{1}{4\pi}\sum (2l + 1)(g_l)^2 P_l(\cos\vartheta). \tag{N108}$$

By continuing on in a similar manner, one gets for the probability that, after $n$ collisions, the particle will be scattered into $d\Omega = \sin\vartheta\, d\vartheta\, d\varphi$

$$w_n(\vartheta, \varphi)\, d\Omega = \frac{1}{4\pi}\sum (2l + 1)(g_l)^n P_l(\cos\vartheta)\, d\Omega. \tag{N109}$$

Let $W(n)$ denote the probability that the particle makes $n$ collisions in traversing the foil. Then the probability that the particle, in traversing the foil, be scattered by an angle between $\vartheta$ and $\vartheta + d\vartheta$ is[#]

$$\begin{aligned} f(\vartheta)\sin\vartheta\, d\vartheta &= 2\pi\sum_n W(n)\, w_n(\vartheta)\sin\vartheta\, d\vartheta \\ &= \tfrac{1}{2}\sum_l (2l + 1)\left[\sum_n W(n)(g_l)^n\right] P_l(\cos\vartheta)\sin\vartheta\, d\vartheta. \end{aligned} \tag{N110}$$

Let $\nu$ be the average number of collisions of the particle in traversing a foil of thickness $t$ and with $N$ atoms per unit volume, i.e.,

$$\nu = Nt\int \sigma(\vartheta)\, d\Omega. \tag{N111}$$

For large $\nu$ ($\nu \gg 1$), we may take $W(n)$ as given by the Poisson distribution

$$W(n) = \frac{e^{-\nu}\nu^n}{n!}.$$

Then

$$\sum_{n=0}^{\infty} W(n)(g_l)^n \simeq \exp\left[-\nu(1 - g_l)\right].$$

The intensity distribution for multiple scattering is then

$$f(\vartheta) = \tfrac{1}{2}\sum_l (2l + 1)\exp(-Q_l)\, P_l(\cos\vartheta), \tag{N112}$$

where

$$Q_l = 2\pi Nt\int_0^{\pi} \sigma(\chi)[1 - P_l(\cos\chi)]\sin\chi\, d\chi. \tag{N113}$$

---

[#] Note that $f(\vartheta)$ is here the intensity distribution function, not the scattering amplitude as in the rest of the book.

Let us express $Q_l$ in a form usually used in the literature. Let $Ze$, $ze$ be the charge of the atomic nucleus and of the incident particle respectively. Let $q(\chi)$ be the ratio of the $\sigma(\chi)$ of the particle in the screened field and the Rutherford expression (A47)

$$\sigma_R(\chi) = \left(\frac{Zze^2}{2pv}\right)^2 \frac{1}{\sin^4(\chi/2)}, \tag{N114}$$

i.e.,

$$\sigma(\chi) = q(\chi)\,\sigma_R(\chi). \tag{N115}$$

We shall introduce a $\chi_c$ defined by#

$$\chi_c^2 = 4\pi Nt\left(\frac{Zze^2}{pv}\right)^2 \tag{N116}$$

(N113) becomes

$$Q_l = \tfrac{1}{4}\chi_c^2 \int_0^1 q(y)\frac{1}{y^3}\,[1 - P_l(1 - 2y^2)]\,dy, \quad y = \sin\frac{\chi}{2}. \tag{N118}$$

On expanding $P_l(1 - 2y^2)$ in powers of $y^2$, we get

$$Q_l = \tfrac{1}{2}\chi_c^2\left[l(l+1)\int_0^1 q(y)\frac{dy}{y} - \sum_2^l (-1)^k \frac{(l+k)!}{(l-1)!(k!)^2}\int_0^1 q(y)\,y^{2k-3}\,dy\right]. \tag{N119}$$

So far we have only treated the statistical–geometrical problem of finding the distribution of the scattered particles after going through a foil of thickness $t$ and number density of atom $N$, in terms of the elastic-scattering cross section $\sigma(y) = q(y)\,\sigma_R(y)$ of single-collisions.

Now $q(y)$ has the obvious properties:

$$q(0) = 0, \qquad q(y) \simeq 1 \quad \text{for} \quad y > y_0, \tag{N120}$$

i.e., for scattering angles $\chi$ larger than a certain $\chi_0$ (or impact parameter < a certain value), the screening effect of the atomic electrons is small. If one approximates $q(y)$ in the integral $\int_0^1 q(y)\,y^{2k-3}\,dy$, $k \geqslant 2$, by

$$q(y) = 1 - \exp(-y/y_0), \tag{N121}$$

---

# The meaning of $\chi_c$ is given by

$$2\pi Nt \int_{\chi_c}^{\pi} \sigma(\chi)\sin\chi\,d\chi = 1, \tag{N117}$$

i.e., the probability of the particle being scattered in a single collision by an angle $\chi > \chi_c$ in going through the foil is unity. If $\chi_c \ll 1$, and $\sigma(\chi)$ behaves as $\sigma_R$ for small $\chi$, then (N116) results from (N117). It is considered better to replace $Z^2$ by $Z(Z + 1)$ in the definitions (N116), on semi-theoretical ground.

then

$$Q_l = \tfrac{1}{2}\chi_c^2 l(l+1)\left\{\int_0^1 q(y)\,\frac{dy}{y} + 1 - \Psi(l) - C + O[(l+1)ly_0^2]\right\}, \quad \text{(N122)}$$

where $\Psi(l) = (d/dl)\ln\Gamma(l)$, $C$ = Euler's constant = 0.5772...

For small $l$, the last term is unimportant. For large $l$ such that $ly_0$ is $\sim 1$, it is necessary to take it into account to avoid a spurious divergence in the summation over $l$ (see refs. 9 and 7 about this point).

### *ii) Evaluation of $Q_l$ and $f(\vartheta)$*

Let us calculate $Q_l$ and $f(\vartheta)$ for the scattering of a relativistic fermion of charge $ze$ and momentum $p$ by a screened Coulomb field

$$V(r) = \frac{Zze^2}{r}e^{-\lambda r}. \quad \text{(N123)}$$

Let

$$\chi_0 \equiv \frac{\hbar\lambda}{p}, \quad \text{(N124)}$$

$p$ being the momentum of the particle. For this case, the cross section, up to the second Born approximation, has been calculated by Dalitz (ref. 8). We have $[y = \sin(\chi/2)]$,

$$q(y) = \left(\frac{4y^2}{\chi_0^2 + 4y^2}\right)^2\left\{1 - \beta^2y^2 + 2\frac{\alpha}{\beta}\left(4 + \frac{\chi_0^2}{y^2}\right)X\tan^{-1}(\chi_0 X) + \alpha\beta(\chi_0^2 + 4y^2)\right.$$

$$\left.\times\left[\left(\frac{1}{2}\frac{\chi_0^2}{y^2} - 1\right)X\tan^{-1}(\chi_0 X) - \frac{1}{2}\tan^{-1}\left(\frac{2}{\chi_0}\right) + \frac{1}{2y}\tan^{-1}\left(\frac{y}{\chi_0}\right)\right]\right\}, \quad \text{(N125)}$$

where

$$X = y(\chi_0^4 + 4\chi_0^2 + 4y^2)^{-1/2}, \qquad \alpha = \frac{Zze^2}{\hbar c}, \qquad \beta = v/c.$$

The evaluation of $Q_l$ is somewhat lengthy (ref. 7). Let us define

$$\chi_\alpha^2 = \chi_0^2\left[1 + 4\alpha\chi_0\left(\frac{1-\beta^2}{\beta}\ln\chi_0 + \frac{0.231}{\beta} + 1.448\beta\right)\right], \quad \text{(N126)}$$

$$\ln\frac{2}{\chi_\alpha'} = \ln\frac{2}{\chi_\alpha} + \frac{1}{2} - C + (\xi - 1)(1 - C), \quad \text{(N127)}$$

$$\xi = 1 + \frac{2\alpha\chi_0}{\beta}(1 - \beta^2), \quad \text{(N128)}$$

then#

$$Q_l = \tfrac{1}{2}\chi_c^2\left\{l(l+1)\left[\ln\frac{2}{\chi_\alpha} - \xi\Psi(l)\right] + 2\pi\alpha\beta l - (\beta^2 + \pi\alpha\beta)[\Psi(l) + C]\right\}. \tag{N129}$$

We shall further introduce $z$, $b$, $B$, $u$ defined by

$$\begin{aligned} z &\equiv (l + \tfrac{1}{2})\,\chi_c, \\ b &= \xi \ln(\chi_c^2/4) - \ln(\chi_\alpha'^2/4), \\ b &= B - \xi \ln B, \\ u &= \sqrt{B}z. \end{aligned} \tag{N130}$$

We shall replace $P_l(\cos\vartheta)$ by

$$P_l(\cos\vartheta) \simeq J_0[(l+\tfrac{1}{2})\vartheta] = J_0\left(\frac{\vartheta z}{\chi_c}\right). \tag{N131}$$

and the summation over $l$ by an integration

$$\sum_{l=0}^{\infty}(l+\tfrac{1}{2})\,g(l) \simeq \frac{1}{\chi_c}\left[\int_0^\infty dy\,g(y) + \frac{1}{24}g'(0) + \cdots\right]. \tag{N132}$$

Then it can be shown that

$$f(\vartheta) = \frac{K}{\chi_c^2 B}\int_0^\infty g(u)\,du, \tag{N133}$$

where

$$\begin{aligned} g(u) = {} & K u^{[1 + \frac{1}{2}\chi_c^2(\beta^2 + \pi\alpha\beta - \frac{1}{4}\xi)]} J_0\left(\frac{\vartheta u}{\chi_c\sqrt{B}}\right)\exp\left(-\tfrac{1}{4}u^2\right) \\ & \times \left[1 + \frac{1}{B}\left\{-\pi\alpha\beta\chi_c\sqrt{B}u + \xi\frac{u^2}{4}\ln\frac{u^2}{4}\right\}\right. \\ & \left. + \frac{1}{2!B^2}\left\{-2\pi\alpha\beta\chi_c\sqrt{B}u\xi\frac{u^2}{4}\ln\frac{u^2}{4} + \left(\xi\frac{u^2}{4}\ln\frac{u^2}{4}\right)^2\right\} + \cdots\right] \end{aligned} \tag{N134}$$

$$K = \exp\left\{\frac{B\chi_c^2}{16}\left[1 + \frac{8\pi\alpha\beta}{B} + 2\xi\frac{\ln 2}{B} + \frac{8}{B}(\beta^2 + \pi\alpha\beta)[C - \ln(\chi_c'\sqrt{B})]\right]\right\}.$$

---

# Had we worked only up to the first Born approximation for $q(y)$, (N129) would have been

$$Q_l = \tfrac{1}{2}\chi_c^2\left\{l(l+1)\left[\ln\frac{2}{\chi_\alpha} - \frac{1}{2} + 1 - \Psi(l) - C\right] - \beta^2\left[\Psi(l) + C\right]\right\}. \tag{N129a}$$

In the notation of ref. 6, $S_l \equiv 1 - \psi(l) - C$, and

$$\int_0^1 q(y)\frac{dy}{y} \equiv \ln\xi \quad \text{which is} \quad \ln(2/\chi_\alpha) - \tfrac{1}{2} \text{ here.}$$

The term in $\beta^2$, coming from the spin in the relativistic theory in (N125), is absent in ref. 6.

The $\beta^2$ in the exponent of $u$ comes from the spin of the particle. The terms in $1/B$, $1/B^2$, with $\xi = 1$, come from the first Born approximation. The terms in $1/\sqrt{B}$, $1/B^{3/2}$, as well as the deviation of $\xi$ from 1, represent the effect of the second Born approximation. Let (N133) and (N134) be written in the form

$$f(\vartheta) = \frac{K}{\chi_c^2 B}\left[f^{(0)}\left(\frac{\vartheta}{\chi_c\sqrt{B}}\right) + \frac{1}{B}(f^{(1)'} + f^{(1)}) + \frac{1}{2B^2}(f^{(2)'} + f^{(2)}) + \cdots\right], \tag{N135}$$

where the argument of $f^{(1)'}$, $f^{(1)}, \ldots$ is also $\vartheta/\chi_c\sqrt{B}$. The terms $f^{(0)}, f^{(1)}$ (without spin) have been tabulated by Bethe (ref. 9). They give $f(\vartheta)$ up to the first Born approximation. The terms $f^{(1)'}$, $f^{(2)'}$ are due to the second Born approximation. $f^{(1)'}$ has been calculated in ref. 7, but $f^{(2)'}$ remains to be evaluated.

The term $f^{(0)}$ alone is, for small $\chi_c$ ($\chi_c \ll 1$), a simple Gaussian distribution

$$\begin{aligned} f^{(0)}(\vartheta/\chi_c\sqrt{B}) &\cong \int_0^\infty du\, u \exp\left(-\tfrac{1}{4}u^2\right) J_0(\vartheta u/\chi_c\sqrt{B}\,) \\ &= 2\exp\left[-(\vartheta/\chi_c\sqrt{B}\,)^2\right]. \end{aligned} \tag{N136}$$

From this it follows that the mean square $\vartheta$ is

$$\overline{\vartheta^2} = \chi_c^2 B, \tag{N137}$$

where $B$ is given in terms of the scattering angle $\chi_c$ and the screening angle $\chi_{\alpha'}$ in (N130). For $\chi \simeq 1$,

$$B = \ln B\left(\frac{\chi_c}{\chi_\alpha}\right)^2. \tag{N138}$$

For scattering angles somewhat beyond the region represented by $f^{(0)}$ alone, it is necessary to calculate $f(\vartheta)$ in (N135) with $f^{(0)}, f^{(1)'}, f^{(1)}, f^{(2)}$. It is found that for $\vartheta/\chi_c\sqrt{B} \lesssim 3$, these terms together still can be approximately described by a Gaussian distribution, although the "$1/e$ width" is then somewhat different from $\chi_c\sqrt{B}$ of $f^{(0)}$ in (N137), i.e.,

$$f(\vartheta) \propto \exp\left(-\vartheta^2/\overline{\vartheta^2}\right), \tag{N139}$$

where

$$\overline{\vartheta^2} = A\chi_c^2 B. \tag{N137a}$$

$A$ being a number of order 1 (ref. 7).

For larger angles ($\vartheta/\chi_c\sqrt{B} \geqslant 4$), $f(\vartheta)$ decreases with $\vartheta$ much less rapidly than the Gaussian law, and $f^{(2)'}$ and even higher order terms in (N134) or (N135) arising from the effect of the second and higher order Born approximations become important. Such calculations have not yet been carried out.

The theory represented by (N135), but without $f^{(2)'}$, has been applied to the experimental data on the scattering of 15 MeV electrons by gold and beryllium foils, and the agreement between the theoretical and the observed

$f(\vartheta)$ for the Gaussian region is good. For the details of the calculations of $Q_l$ and $f(\vartheta)$, and the discussion of the choice of $\chi_0$, reference must be made to the literature (ref. 7).

From (N116), (N124) and (N126-7), it is seen that, for relativistic energies ($\beta \simeq 1$)

$$\left(\frac{\chi_c}{\chi_c'}\right)^2 = 4\pi Nt\left(\frac{\alpha}{\beta\lambda}\right)^2 e^{-0.154}\left(1 + 4 \times 1.68\alpha\frac{\hbar\lambda}{p}\right)$$

which is almost independent of $p$ since $\chi_0 = \hbar\lambda/p$ is $\ll 1$. Equation (N137a)

$$(p\beta c)^2 = A\,\frac{4\pi NtZ^2z^2e^4B}{\overline{\vartheta^2}} \tag{N137b}$$

then furnishes a relation for the determination of the (relativistic) momentum of charged particles from the mean square scattering angle, $\overline{\vartheta^2}$. This is used in the photographic emulsion technique for the measurement of the energies of fast cosmic ray particles.

## 4. IMPULSE APPROXIMATION

For the collision between a fast nucleon and a deuteron, an approximate method has been introduced by Chew (ref. 14). It is known as the "impulse approximation" and has been applied to other collisions, elastic and inelastic, between a fast nucleon and a nucleus. The basic ideas and the assumptions made in this approximation are the following. In order to reduce the complexity of the many-body problem, it is assumed that the incident nucleon interacts with one nucleon of the nucleus at a time. For fast incident nucleons, the time $\tau$ taken in traversing over the range of the nucleon–nucleon interaction is short compared with the characteristic time period of a nucleus. The collision then may be regarded as an "impulse" during which the binding forces on the struck nucleon from the rest of the target system play no part. Thus during the interaction time $\tau$, the struck nucleon may be regarded as "free", the binding forces serving only to determine the "momentum distribution" of the nucleon wave function. It is also assumed that the amplitude of the incident nucleon wave on each nucleon of the target nucleus is unaffected by the presence of the other nucleons.

It is seen that, unlike the Born approximation, the assumption is not made that the interaction between the incident nucleon and the target nucleus is small. To illustrate the formulation of this method and its relation to the Born approximation, we shall for simplicity first consider the scattering of a particle 1 by another particle 2 which is bound by a potential $V(r_2)$.#

---

# This is similar to the problem, considered by Fermi, ref. 4 of Sect. E, of the scattering of neutrons by chemically bound protons in a molecule.

The Schrödinger equation of the system is

$$\left[-\frac{\hbar^2}{2m_1}\Delta_1 - \frac{\hbar^2}{2m_2}\Delta_2 + V_0(r_2) + V(r_{12})\right]\Psi(1, 2) = E\Psi(1, 2). \quad \text{(N140)}$$

Let the wave functions and energies of the target particle be $\phi_n(2)$, $W_n$. They satisfy the equation

$$\left[-\frac{\hbar^2}{2m_2}\Delta_2 + V_0(r_2)\right]\phi_n(2) = W_n\phi_n(2). \quad \text{(N141)}$$

Let the initial and the final state of the system be $(\phi_0, \mathbf{k}_0)$, $(\phi_n, \mathbf{k}_n)$, where

$$\frac{1}{2m_1}\hbar^2 k_0^2 + W_0 = \frac{1}{2m_1}\hbar^2 k_n^2 + W_n = E. \quad \text{(N142)}$$

We seek a solution $\Psi(1, 2)$ which contains the incident wave $\exp(i\mathbf{k}_0\cdot\mathbf{r}_1)\phi_0(\mathbf{r}_2)$ and scattered waves. The scattering amplitude is proportional to the matrix element

$$M = \int \exp(-i\mathbf{k}_n\cdot\mathbf{r}_1)\,\phi_n^*(r_2)\,V(|\mathbf{r}_1 - \mathbf{r}_2|)\,\Psi(\mathbf{r}_1, \mathbf{r}_2)\,d\mathbf{r}_1\,d\mathbf{r}_2. \quad \text{(N143)}$$

We shall now consider the usual Born approximation for $M$ in order to understand the nature of the impulse approximation. In the Born approximation, $\Psi$ is replaced by $\exp(i\mathbf{k}_0\cdot\mathbf{r}_1)\,\phi_0(\mathbf{r}_2)$ so that

$$M_B = \int g_n^*(\mathbf{K}')\,g_0(\mathbf{K})\,d\mathbf{K}'\,d\mathbf{K}\int \exp[-i(\mathbf{k}_n\cdot\mathbf{r}_1 + \mathbf{K}'\cdot\mathbf{r}_2)]\,V(|\mathbf{r}_1 - \mathbf{r}_2|)$$
$$\times \exp[i(\mathbf{k}_0\cdot\mathbf{r}_1 + \mathbf{K}\cdot\mathbf{r}_2)]\,d\mathbf{r}_1\,d\mathbf{r}_2, \quad \text{(N144)}$$

where $g_0$, $g_n$ are the Fourier transforms of $\phi_0(\mathbf{r}_2)$, $\phi_n(\mathbf{r}_2)$, i.e.,

$$g_0(\mathbf{K}) = \frac{1}{(2\pi)^3}\int e^{-i\mathbf{K}\cdot\mathbf{r}}\,\phi_0(\mathbf{r})\,d\mathbf{r},$$
$$g_n(\mathbf{K}') = \frac{1}{(2\pi)^3}\int e^{-i\mathbf{K}'\cdot\mathbf{r}}\,\phi_n(\mathbf{r})\,d\mathbf{r}. \quad \text{(N145)}$$

Now in the impulse approximation, the last factor $\exp[i(\mathbf{k}_0\cdot\mathbf{r}_1 + \mathbf{K}\cdot\mathbf{r}_2)]$ in (N144) is replaced by the "exact" wave function

$$\psi_{\mathbf{k}_0,\mathbf{K}}(\mathbf{r}_1, \mathbf{r}_2)$$

which represents the scattering of particle 1 with momentum $\mathbf{k}_0$ by a free particle 2 with momentum $\mathbf{K}$, i.e., it is a solution of

$$\left[-\frac{\hbar^2}{2m_1}\Delta_1 - \frac{\hbar^2}{2m_2}\Delta_2 + V(|\mathbf{r}_1 - \mathbf{r}_2|)\right]\psi_{\mathbf{k}_0,\mathbf{K}}(\mathbf{r}_1, \mathbf{r}_2)$$
$$= \left(\frac{1}{2m_1}\hbar^2 k_0^2 + \frac{1}{2m_2}\hbar^2 K^2\right)\psi_{\mathbf{k}_0,\mathbf{K}}(\mathbf{r}_1, \mathbf{r}_2). \quad \text{(N146)}$$

Then

$$M_I = \int g_n^*(\mathbf{K}')\, g_0(\mathbf{K})(\mathbf{k}_n, \mathbf{K}'|V|\mathbf{k}_0, \mathbf{K})\, d\mathbf{K}'\, d\mathbf{K}, \qquad \text{(N147)}$$

where

$$(\mathbf{k}_n, \mathbf{K}'|V|\mathbf{k}_n, \mathbf{K}) = \int \exp\,[-i(\mathbf{k}_n\cdot\mathbf{r}_1 + \mathbf{K}'\mathbf{r}_2)] \times V(|\mathbf{r}_1 - \mathbf{r}_2|)\, \psi_{\mathbf{k}_0,\mathbf{K}}(\mathbf{r}_1, \mathbf{r}_2)\, d\mathbf{r}_1\, d\mathbf{r}_2. \qquad \text{(N148)}$$

Thus the impulse approximation consists in replacing the exact $\Psi(\mathbf{r}_1, \mathbf{r}_2)$ in (N143) by

$$\Psi(\mathbf{r}_1, \mathbf{r}_2) \to \int d\mathbf{K} g_0(\mathbf{K})\, \psi_{\mathbf{k}_0,\mathbf{K}}(\mathbf{r}_1, \mathbf{r}_2)$$

which will go over into the Born approximation $\Psi(\mathbf{r}_1, \mathbf{r}_2) = \exp\,(i\mathbf{k}_0\cdot\mathbf{r}_1)\phi_0(\mathbf{r}_2)$ in the limit of $V(|\mathbf{r}_1 - \mathbf{r}_2|) \to 0$ in (N146). $M_I$ can be expected to be a better approximation than $M_B$ since $\psi_{\mathbf{k}_0,\mathbf{K}}(\mathbf{r}_1, \mathbf{r}_2)$ takes $V(|\mathbf{r}_1 - \mathbf{r}_2|)$ into account whereas the Born $\exp\,(i\mathbf{k}_0\cdot\mathbf{r}_1)\, \phi_0(\mathbf{r}_2)$ does not.

Let us separate from $\psi_{\mathbf{k}_0,\mathbf{K}}$ the center of mass motion by introducing (assuming $m_1 = m_2$ for simplicity here)

$$\mathbf{R} = \tfrac{1}{2}(\mathbf{r}_1 + \mathbf{r}_2), \qquad \mathbf{r} = \mathbf{r}_1 - \mathbf{r}_2, \qquad \text{(N149)}$$

and writing

$$\psi_{\mathbf{k}_0,\mathbf{K}}(\mathbf{r}_1, \mathbf{r}_2) = \exp\,[i(\mathbf{k}_0 + \mathbf{K})\cdot\mathbf{R}]\, \varphi_{\mathbf{k}_0-\mathbf{K}}(\mathbf{r}). \qquad \text{(N150)}$$

Substituting (N150) into (N148) and integrating over $\mathbf{R}$, we obtain

$$(\mathbf{k}_n, \mathbf{K}'|V|\mathbf{k}_0, \mathbf{K}) = (2\pi)^3\, \delta(\mathbf{k}_n + \mathbf{K}' - \mathbf{k}_0 - \mathbf{K}) \times \int \exp\,[i(\mathbf{K}' - \mathbf{k}_n)\cdot\mathbf{r}/2]\, V(r)\, \varphi_{\mathbf{k}_0-\mathbf{K}}(\mathbf{r})\, d\mathbf{r}, \qquad \text{(N151)}$$

and

$$M_I = (2\pi)^3 \int g_n^*(\mathbf{k}_0 - \mathbf{k}_n + \mathbf{K})\, g_0(\mathbf{K})\, d\mathbf{K} \times \int \exp\,[i(\mathbf{K}' - \mathbf{k}_n)\cdot\mathbf{r}/2]\, V(r)\, \varphi_{\mathbf{k}_0-\mathbf{K}}(\mathbf{r})\, d\mathbf{r}, \qquad \text{(N152)}$$

where

$$\mathbf{k}_n + \mathbf{K}' = \mathbf{k}_0 + \mathbf{K}.$$

The cross section is proportional to $|M_I|^2$.

If one makes the assumption (approximation) that the integrand of the second integral in (N152) is a slowly varying function of $\mathbf{K}$, and if $\phi_n$ is an unbound state, one may put for $\mathbf{K}$ in $\phi_{\mathbf{k}_0-\mathbf{K}}$ the value at which $g_n^*(\mathbf{k}_0 - \mathbf{k}_n + \mathbf{K})$ [see (N145)] is singular, thereby approximating $M_I$ by the product of two factors. The factor

$$(2\pi)^3 \int g_n^*(\mathbf{k}_0 - \mathbf{k}_n + \mathbf{K})\, g_0(\mathbf{K})\, d\mathbf{K} = f_{n0}(\mathbf{k}_n - \mathbf{k}_0), \qquad \text{(N153)}$$

where

$$f_{n0}(\mathbf{k}_n - \mathbf{k}_0) = \int \exp\,[i(\mathbf{k}_n - \mathbf{k}_0)\cdot\mathbf{r}]\,\phi_n^*(\mathbf{r})\,\phi_0(\mathbf{r})\,d\mathbf{r} \tag{N154}$$

is the Fourier transform of $\phi_n^*(\mathbf{r})\,\phi_0(\mathbf{r})$ and is thus connected with the momentum distribution of the target wave functions.

The impulse approximation was originally suggested for treating the scattering of a neutron by a deuteron (one has simply to let 1, 2, denote the incident neutron and the proton in the deuteron in the formulas above). It has been generalized and its assumptions have been discussed by Chew, Wick and Ashkin (ref. 15). For further details, the reader will be referred to the literature.

We have considered the target with only one particle. To see the nature of the Impulse Approximation for the general case, a more formal formulation is convenient. Corresponding to (N140) and (N146) we have, respectively,

$$H\Psi_a = E_a\Psi_a, \qquad H = K + V + V_0, \tag{N155}$$

$$(K + V)\,\psi_n = E_n\psi_n. \tag{N156}$$

The solution of (N155) can be written

$$\Psi_a = \Phi_a + \frac{1}{E_a + i\eta - K - V_0}\,V\Psi_a, \quad (\eta > 0, \eta \to 0), \tag{N157}$$

where $\Phi$ is a product wave function of the incident particle and the target system satisfying

$$(K + V_0)\,\Phi_a = E_a\Phi_a. \tag{N158}$$

Similarly $\psi_n$ in (N156) can be expressed by

$$\psi_n = \chi_n + \frac{1}{E_n + i\eta - K}\,V\psi_n, \tag{N159}$$

where $K\chi_n = E_n\chi_n$. The transition matrix element from state $a$ to $b$ is

$$M_{ba} = (\Phi_b, V\Psi_a) = (\Phi_b, T\Phi_a). \tag{N160}$$

The $T$-operator is defined from (N157) by

$$T = V + V\frac{1}{E_a + i\eta - K - V_0}\,T. \tag{N161}$$

Similarly (N159) is converted to

$$t = V + V\frac{1}{E_n + i\eta - K}\,t. \tag{N162}$$

Now it will be seen in the following that the replacement of $T$ in $M$, (N160), by $t$ consists in the impulse approximation

$$M_{ba}^{(I)} = (\Phi_b, t\Phi_a). \tag{N163}$$

$t$ is also expressed by the following matrix element in the $\chi_n$ representation [compare (N159) and (N162)],

$$t_{mn} = (\chi_m, V\psi_n).$$

Therefore $t\Phi_a$ in (N163) can be written as

$$\begin{aligned} t\Phi_a &= \sum_{m,n} \chi_m t_{mn}(\chi_n, \Phi_a) \\ &= V\sum_n \psi_n(\chi_n, \Phi_a) \equiv V\Psi''_a. \end{aligned}$$

It will be seen that $\Psi''_a$ is equivalent to $\int d\mathbf{K} g_0(\mathbf{K})\, \psi_{\mathbf{k}_0,\mathbf{K}}(\mathbf{r}_1, \mathbf{r}_2)$ [see the formula following (N148)], if we write down the various quantities explicitly,

$$\begin{aligned} &\Phi_a \to \exp(i\mathbf{k}_0 \cdot \mathbf{r}_1)\, \phi_0(2), \\ &\chi_n \to \exp[-i(\mathbf{k}_1 \cdot \mathbf{r}_1 + \mathbf{K}\cdot\mathbf{r}_2)], \\ &\psi_n \to \psi_{\mathbf{k}_0,\mathbf{K}}(1, 2), \\ &(\quad) \to \int d\mathbf{r}_1\, d\mathbf{r}_2, \\ &\sum_n \to \int d\mathbf{k}_1\, d\mathbf{K}. \end{aligned}$$

Thus the replacement: $T \to t$ is equivalent to the impulse approximation. A straightforward operational calculation leads to the relation:

$$T = t + T' \frac{1}{E_a + i\eta - K - V_0} [V_0, u], \quad t \equiv Vu,$$

where

$$T' = V + V \frac{1}{E_a - i\eta - (K + V + V_0)} V.$$

(The proof will not be given here.)

The factor $1/(E_a + i\eta - K - V_0)$ represents in a sense the reciprocal of energy conservation violation ($\Delta E$), during the collision, and the commutator $[V_0, u]$ is a quantity of the order of the inverse period of the target system. Hence if the collision time ($\sim\hbar/\Delta E$) is short enough compared with the characteristic time period of the target system, the approximation is good. This fact can be also confirmed from the time-dependent formulation [see Epstein, S. T., Phys. Rev. **119**, 458 (1960)].

If the target contains many particles, $V_0$ stands for all the binding potentials of the particles in the target, and $V$ is the sum of the interactions between the incident particle and the target particles. The impulse approximation assumes that the total $T$ operator is just the sum of the two-particle (namely, the incident particle and one of the target particles) $t_i$ operators.

$$T \doteq t_1 + t_2 + \cdots + t_n, \qquad t_i = V_i + V_i \frac{1}{E_n + i\eta - K} t_i. \quad \text{(N164)}$$

If the interactions $V_i$ are "weak", this is again reduced to the Born approximation. The approximation (N164) corresponds to the lowest approximation in the multiple scattering theory of Watson except that the $t_i$ defined in (N49) are further approximated by a two-particle operator $t_i$ here. [See (N50a).] A many-particle Green's function $(a-U)^{-1}$ appears in (N49). Therefore the multiple scattering correction is one of the errors which does not appear when the target has only one particle. Thus the approximation is good if the scattering by each target particle is almost the same as if that particle were alone in the target. Hence, if the mean free path of the incident particle in the target is short compared with the dimension of the target, the approximation is not good. Besides this "transparency" assumption we must also impose the condition that the incident particle never interacts strongly with more than one particle in the target at the same time, otherwise the total amplitude can not be a simple sum of independent scattering amplitudes. This condition will require a small ratio of the range of the interaction of the incident particle with a target particle to the average interparticle distances in the target.

## REFERENCES

The treatment of the (complex) optical potential here is based on an article by

1. Glauber, R. J. in *Lectures in Theoretical Physics*, Vol. I. Edited by W. E. Brittin and L. G. Dunham. Interscience Publishers, New York (1959).

The treatment of Watson's theory of multiple scattering here is based on

2. Watson, K. M., Phys. Rev. **105**, 1388 (1957), and

3. Mittleman, M. H. and Watson, K. M., Phys. Rev. **113**, 198 (1959). Ref. 2 considerably simplifies the treatment given in a few earlier papers: Watson, Phys. Rev. **89**, 575 (1953); Francis, N. C. and Watson, Phys. Rev. **92**, 291 (1953); Takeda, G. and Watson, Phys. Rev. **97**, 1336 (1955).

4. Lippmann, B., Mittleman, M. H. and Watson, K. M., Phys. Rev. **116**, 920 (1959); Mittleman and Watson, Annals of Phys. **10**, 268 (1960); Mittleman, Annals of Phys. **14**, 94 (1961); Lippmann and Schey, H. M., Phys. Rev. **121**, 1112 (1961). These papers apply the method to atomic scattering including the effect of the Pauli principle.

In a paper "Remarks on the multiple scattering theory in nuclei", by Yang Li-Ming, in the Science Reports, Peking University, No. 3, 277 (1958) (in Chinese), Watson's theory is treated in a slightly different formulation and an iteration procedure is given for obtaining the optical potential to higher approximations.

Brueckner's theory of many-body system is most clearly treated in an article in

5. The Many Body Problem, Cours donnés a l'école d'été de physique theorique, les Houches—Session 1958, John Wiley and Sons, New York (1959).

6. Goudsmit, S. A. and Saunderson, J. L., Phys. Rev. **57**, 24 (1940); 58, 36 (1940).

7. Nigam, B. P., Sundaresan, M. K. and Wu, T. Y., Phys. Rev. **115**, 491 (1959).

8. Dalitz, R. H., Proc. Roy. Soc. (London) **A206**, 509 (1951).

9. Bethe, H. A., Phys. Rev. **89**, 1256 (1953).

10. Williams, E. J., Proc. Roy. Soc. (London) **A169**, 531 (1938); Phys. Rev. **58**, 292 (1940); Rev. Mod. Phys. **17**, 217 (1945).

11. Molière, G., Z. Naturforsch, **2a**, 133 (1947); **3a**, 78 (1948). The first of these two papers treats the single scattering and gives a relation corresponding to (N126) in the present section, which however does not correspond to any consistent approximation (see ref. 7 above). The second paper gives a treatment of the multiple scattering which has been shown by Bethe (ref. 9) to be obtainable from that of Goudsmit and Saunderson (ref. 6) by making a few approximations.

12. Snyder, H. and Scott, W. T., Phys. Rev. **76**, 220 (1949); Scott, W. T., Phys. Rev. **85**, 245 (1952). These papers give the theory for small angles and for the screened Coulomb law (N123) only.

13. Lewis, H. W., Phys. Rev. **78**, 526 (1950). This paper treats the theory for the screened Coulomb law (N123) and for arbitrary angles of scattering.

14. Chew, G. F., Phys. Rev. **80**, 196 (1950).

15. Chew, G. F. and Wick, G. C., Phys. Rev. **85**, 108, 636 (1952); Ashkin, J. and Wick, G. C., Phys. Rev. **85**, 686 (1952): Chew, G. F. and Goldberger, M. L., Phys. Rev. **87**, 778 (1952).

chapter

4

# Time-Dependent Theory of Scattering

In the preceding chapters, we have treated the problem of scattering by the stationary state method, i.e., the time-independent Schrödinger equation. In the simplest case, namely, the scattering of a particle by a potential field $V(r)$ (Chapters 1, 2), the theory gives the scattering amplitudes $f(k, \vartheta)$ and cross section $\sigma(k, \vartheta)$ in terms of the phase shifts or the scattering matrix $S$ [(A34) and Chapter 6]. In the more general case of rearrangement collisions (Chapter 3), the theory describes the scattering by an integral equation which embodies the initial and the asymptotic conditions, and the cross section can again be expressed in terms of the $S$ matrix (Chapter 6). In all these cases, the stationary-state method is reduced to a boundary-value problem. The total Hamiltonian $H$ is assumed to approach an unperturbed Hamiltonian $H_0$ when the "colliding" parts of the system are far apart. The eigenstates of $H$ which we denote by $\Psi$ are assumed to approach eigenstates of $H_0$ which we denote by $\Phi$.# However, the eigenstates $\Psi$ are not unique and one has to give boundary conditions on them in order that they describe the scattering. For the stationary-state method to be applicable, a necessary condition is that the Hamiltonian of the system be time-independent.

There are, however, problems in which the Hamiltonian depends explicitly on time, for example, the interaction of a system with an external time-dependent field. In such cases, the system does not remain in any stationary state, and the behavior of the system is governed by the Schrödinger equation

$$i\hbar \frac{\partial \Psi}{\partial t} = H(\mathbf{r}, t)\Psi(\mathbf{r}, t). \tag{O1}$$

The usual problem will now be an "initial value" problem of solving (O1) for a prescribed initial condition, i.e., $\Psi$ at a certain $t = t_0$.

The importance, however, of the time-dependent theory of scattering lies in the following situation. Even if the Hamiltonian $H$ of a system does not depend explicitly on time, it is still possible to treat the scattering problem on the basis of the time-dependent Schrödinger equation. The point of view is as follows: Let us suppose that *initially* (we shall denote this "initial" state by $t \to -\infty$) the colliding parts of the system are at infinite spatial separations and the system is in a state $\Psi_a(-\infty)$ which is an eigenstate of $H_0$ (to which the Hamiltonian $H$ of the system approaches as $t \to -\infty$ and the interaction $V$ between the colliding parts vanishes). As $t$ increases from $-\infty$ to a finite time $t$ (corresponding to the approach of the colliding parts in space), the interaction $V$ (which may or may not depend explicitly on time) becomes operative, and the state of the system will evolve from the initial $\Psi_a(-\infty)$ to $\Psi(t)$, in accordance with Equation (O1). As $t \to +\infty$ (corresponding to the recession of the colliding parts), the interaction $V$ again vanishes, and the system will be in one of many possible states $\Psi_b(+\infty)$. The problem is then one of finding the relation between the final state and the initial state.

For this purpose, the following artifice of "switching on and off" an interaction $V$ has been introduced. Even when $V$ is time-dependent (such as that between an atom and an incident light wave), one may "turn on" the interaction "abruptly" at definite instances, so that

$$\begin{aligned} H &= H_0, & -\infty < t < 0, \\ H &= H_0 + V, & 0 \leqslant t \leqslant \tau, \\ H &= H_0, & \tau < t. \end{aligned} \tag{O2}$$

# If, instead of the Schrödinger representation, we wish to picture the colliding components of the system as spatially separated in the initial and the final states, then we must use wave packets and the initial and the final states will not be eigenstates of $H_0$ but are superpositions of them.

In the case of time-independent $V$ (such as the scattering of a particle by a static potential), one may still switch on and off the interaction $V$ by replacing $V$ by $e^{-\varepsilon|t|}V$ where $\tau = 1/\varepsilon$ is a very long time constant (very long compared with the characteristic times of the systems). Thus the Hamiltonian $H = H_0 + e^{-\varepsilon|t|}V$ approaches the $H_0$ of the "non-interacting" system as $t \to -\infty$ and $t \to +\infty$, and is $H = H_0 + V$ at $t = 0$. The scattering problem is then one of solving (O1) with this $H = H_0 + e^{-\varepsilon|t|}V$ with the initial state $\Psi_a(-\infty)$ (which is a prescribed eigenstate of $H_0$). The parameter $\varepsilon$ introduced is eventually allowed to approach zero so that the "switching" is adiabatic. The relation between the initial state $\Psi_a(-\infty)$ and the final state $\Psi_b(+\infty)$ is of course given by the solution of (O1), and is expressible in terms of an operator $S$ such that $\Psi_b(+\infty) = S_{ab}\Psi_a(-\infty)$.

It is clear that, in order that the stationary state method and the time-dependent treatment give identical results for the scattering problem, the $S$ matrices introduced in these two treatments be identical with each other. For this to be true, it is necessary that the adiabatic switching introduced in the time-dependent theory have no relevant consequences.

The earliest treatment of the problem of transitions is the perturbation method of Dirac (ref. 1) in which the switching process (O2) is used. A more general and elegant method is that based on a linear unitary "time translation" operator. A still more recent method based on the use of the Green's function has been developed by Feynman. It has proved especially useful when applied to the quantum field theories. In the following section, we shall treat these methods briefly. The theory based on the unitary operator has been further developed by many authors, and an account of the work of Lippmann and Schwinger, Gell-Mann and Goldberger, of Friedrichs, Moses will be given in Sects. P. Q. R. A brief introduction to the more mathematical aspects of the scattering theory will be given in Sect. S.

section

# O Methods of Unitary Operator and of Green's Function

## 1. DIRAC'S METHOD OF "VARIATION OF CONSTANTS"

i) This theory is so well known that only a brief sketch will be given here. We consider the following problem. A system has the Hamiltonian $H_0(\mathbf{r})$ which is independent of time. The Schrödinger equation

$$i\hbar \frac{\partial \Psi^{(0)}(\mathbf{r}, t)}{\partial t} = H_0 \Psi^{(0)}$$

can be solved, with

$$\Psi^{(0)}(\mathbf{r}, t) = \sum_m \!\!\!\!\!\!\int a_m^{(0)} u_m(\mathbf{r}) \exp (iE_m^0 t/\hbar), \tag{O3}$$

where the probability amplitudes $a_m^{(0)}$ are constants. $|a_m^{(0)}|^2$ gives the probability of finding the system in the stationary state $u_m$. Let a perturbation $V(\mathbf{r}, t)$ be introduced so that the system becomes

$$i\hbar \frac{\partial \Psi}{\partial t} = (H_0 + V)\Psi. \tag{O4}$$

This equation is solved by Dirac by the method of "variation of constants" which is familiar in the solution of non-linear equations, for example. Let $\Psi(\mathbf{r}, t)$ be expanded in terms of the complete set $u_m$

$$\Psi(\mathbf{r}, t) = \sum \!\!\!\!\!\!\int a_m(t)\, u_m(\mathbf{r}) \exp (-iE_m^0 t/\hbar), \tag{O5}$$

where the probability amplitudes $a_m(t)$ of finding the system in various states

$u_m$ are functions of time. On putting (O5) into (O4) we obtain the (infinite) system of equations

$$\frac{da_k}{dt} = -\frac{i}{\hbar}\sum_m \int a_m(t) \int d\mathbf{r} u_k^* V u_m \exp\,[i(E_k^0 - E_m^0)t/\hbar] \tag{O6}$$

which are equivalent to (O4) and are hence *exact*.

If $V$ is "small", we replace $V$ by $\lambda V$, where $\lambda$ is a parameter having any arbitrary value of $0 \leqslant \lambda \leqslant 1$, and expand

$$a_m(t) = a_m^{(0)} + \lambda a_m^{(1)} + \lambda^2 a_m^{(2)} + \cdots. \tag{O7}$$

On putting this into (O6) and equating the coefficients of each power of $\lambda$ on both sides, we obtain the system

$$\frac{da_k^{(0)}}{dt} = 0,$$

$$\frac{da_k^{(s+1)}}{dt} = -\frac{i}{\hbar}\sum_m \int a_m^{(s)}(t)\, V_{km}(t) \exp\,[i(E_k^0 - E_m^0)t/\hbar], \quad s = 1, 2, \ldots, \tag{O8}$$

where

$$V_{km} = \int d\,\mathbf{r}\, u_k^*(\mathbf{r})\, V(\mathbf{r}, t)\, u_m(\mathbf{r}). \tag{O9}$$

Let us take as an example the case of a periodic perturbation

$$V = V'(r)(e^{i\omega t} + e^{-i\omega t}), \tag{O10}$$

and assume that $V$ is switched on at $t = 0$ and off at $t = \tau$ in the manner of (O2), and that before $t = 0$, the system is in a definite state $u_n$, i.e.,

$$a_n = 1, \qquad a_m = 0, \qquad m \neq n, \quad \text{at} \quad t < 0. \tag{O11}$$

Then, up to the order $\lambda$ in (O7), we obtain on integrating (O8) from $t = 0$ to $t = \tau$,

$$a_k^{(1)}(\tau) = -\langle k|V'|n\rangle$$
$$\times \left\{\frac{\exp\,[i(E_k^0 - E_n^0 + \hbar\omega)\tau/\hbar] - 1}{E_k^0 - E_n^0 + \hbar\omega} + \frac{\exp\,[i(E_k^0 - E_n^0 - \hbar\omega)\tau/\hbar] - 1}{E_k^0 - E_n^0 - \hbar\omega}\right\}.$$

If $E_k^0 - E_n^0 > 0$, the important contributions come from the second term, so that

$$|a_k^{(1)}(\tau)|^2 = |\langle k|V'|n\rangle|^2 \frac{\sin^2 x}{x^2}\cdot\left(\frac{\tau}{\hbar}\right)^2, \tag{O12}$$

where

$$x = \frac{(E_k^0 - E_n^0 - \hbar\omega)}{2\hbar}\,\tau.$$

(O12) gives the probability that at the time $t = \tau$ the system is found in the state $k$. Now the function $\sin^2 x/x^2$ has an oscillatory character in the argument, with a central maximum at $x = 0$, which is the narrower in $E_k^0 - E_n^0 - \hbar\omega$ the longer the interval of $\tau$ and vice versa. This is the consequence of the uncertainty principle according to which the shorter the time interval $\tau$ over which $V$ acts, the greater the uncertainty with which the energy of the system can be specified.

Let us consider the case in which $\omega$ in $V$ is fixed and that $u_n^{(0)}$ is a discrete state, but $H_0$ has a continuous distribution of states $u_k^{(0)}$ in the neighborhood of

$$E_k^0 = E_n^0 + \hbar\omega. \tag{O13}$$

Let $\rho(E_k^0)\, dE_k^0$ be the number of states in the energy range $dE_k^0$ at $E_k^0$. For $\tau \gg$ periods of $H_0$, the maximum of $(\sin^2 x)/x^2$ is very narrow in $E_k^0 - E_n^0 - \hbar\omega$ so that the sum of the probability of the state $n$ going into any state $k$ near $E_k^0$ is given very closely by

$$\begin{aligned} w_{n,k}(\tau) &= \int |a_k^{(1)}(\tau)|^2\, \rho(E_k^0)\, dE_k^0 \\ &\cong |\langle k|V'|n\rangle|^2\, \rho(E_k^0) \int_{-\infty}^{\infty} \frac{\sin^2 x}{x^2} \frac{2\tau}{\hbar}\, dx \\ &= \frac{2\pi}{\hbar} |\langle k|V'|n\rangle|^2\, \rho(E_k^0)\, \tau. \end{aligned}$$

If $\tau$ is not too long [i.e., $\tau$ is such that $w_{n,k}(\tau)$ is still $\ll 1$], one obtains the probability per unit time of the system making the transition from state $E_n^0$ to state $E_k^0$ [or, more accurately, to a state in the neighborhood of $E_k^0$ given by (O13)]

$$w_{n,k} = \frac{w_{n,k}(\tau)}{\tau} = \frac{2\pi}{\hbar} |\langle k|V'|n\rangle|^2\, \rho(E_k^0). \tag{O14}$$

Second-order calculations can be carried out. We shall not go into these here.

We may remark that the method given above is applicable to two physically different classes of problems, namely, those in which the Hamiltonian $H = H_0 + V(\mathbf{r}, t)$ is actually time dependent (such as an atom in a radiation field), and those in which $V$ is not explicitly dependent on time (such as the scattering of a particle by a field $V(r)$ or by a system of particles). For the latter class of problem, it is necessary to introduce some process of switching on and off the interaction $V(\mathbf{r})$ at certain instants of time. This switching on and off may be done in the abrupt manner of (O2) and the problem treated in the manner above; or it may be done in a continuous or adiabatic manner which is more conveniently treated by means of the integral form of the Schrödinger equation, described briefly in the following section. Since the switching process is equivalent to replacing $V(\mathbf{r})$ by $Vf(t)$, the solution of (O4) in general depends on $f(t)$. It is hence necessary, for the cases of time-independent $V(\mathbf{r})$, that the physical result be independent of the parameters appearing in the function $f(t)$. See Sect. R.5, 6 below.

**ii)** Transition probability and cross section:

As a simple illustration of the result (O14), consider the case of the scattering of a particle by a static field $V'(r)$. Let us inquire about the probability per second that the particle originally in the state

$$u_n^{(0)}(\mathbf{r}) = \frac{1}{\sqrt{\Omega}} e^{i\mathbf{k}\cdot\mathbf{r}}, \tag{O15}$$

makes a transition to the state

$$u_k^{(0)}(\mathbf{r}) = \frac{1}{\sqrt{\Omega}} e^{i\mathbf{k}'\cdot\mathbf{r}}, \quad |\mathbf{k}'| = |\mathbf{k}|. \tag{O16}$$

The functions $u_n^{(0)}$, $u_k^{(0)}$ are normalized to one particle in the volume $\Omega = L^3$. The density $\rho(E_k^0)$, namely the number of states per unit energy range, is

$$\rho(E_k^0) = \frac{\Omega}{(2\pi\hbar)^3} mp\, d\cos\vartheta\, d\varphi, \tag{O17}$$

where $\vartheta$, $\varphi$ are the polar angles of the momentum $\mathbf{p} = \hbar\mathbf{k}'$. The transition probability per unit time is, according to (O14),

$$w = \frac{2\pi}{\hbar}\frac{mp}{(2\pi\hbar)^3\Omega}\left|\int e^{-i\mathbf{k}'\cdot\mathbf{r}}\, V'(r)\, e^{i\mathbf{k}\cdot\mathbf{r}}\, d\mathbf{r}\right|^2 d\cos\vartheta\, d\varphi. \tag{O18}$$

This $w$ involves the incidence of $v/L = p/mL$ particles per second. One may define a differential cross section $d\sigma$ by

$$d\sigma = \frac{w}{\text{no. of particle crossing unit area per second}}$$

$$= \frac{\frac{w}{v}}{L\cdot L^2} = \frac{w}{v}\Omega \tag{O19}$$

so that from (O17),

$$d\sigma = \left(\frac{m}{2\pi\hbar^2}\right)^2\left|\int e^{-i\mathbf{k}'\cdot\mathbf{r}}\, V'(r)\, e^{i\mathbf{k}\cdot\mathbf{r}}\, d\mathbf{r}\right|^2 \sin\vartheta\, d\vartheta\, d\varphi. \tag{O20}$$

This expression is the same as (C36) obtained in the first Born approximation in the stationary state treatment.

**iii)** Rearrangement collisions

We shall treat the problem of rearrangement collisions (M1) (which include exchange scattering as a special case) by the time-dependent method.

The system goes from the non-interacting state $A + B$ to the final non-interacting state $C + D$. Let the Hamiltonian of the system be grouped in two alternative forms

$$\begin{aligned} H &= H_0' + V', \\ &= H_0'' + V'', \end{aligned} \tag{O21}$$

such that $H_0'$ is the Hamiltonian of the non-interacting $A$ and $B$, $H_0''$ is that of the non-interacting $C$ and $D$. The Schrödinger equation (O4) can be put in two forms

$$\left(i\hbar \frac{\partial}{\partial t} - H_0'\right)\Psi = V'\Psi, \tag{O22}$$

$$\left(i\hbar \frac{\partial}{\partial t} - H_0''\right)\Psi = V''\Psi. \tag{O22a}$$

Let $u_n$, $v_n$ be the eigenfunctions of $H_0'$, $H_0''$ respectively

$$(H_0' - E_n')\, u_n = 0, \tag{O23}$$

$$(H_0'' - E_n'')\, v_n = 0. \tag{O23a}$$

For direct collisions (i.e., $A + B \rightarrow A + B$), we expand $\Psi$ in (O22) in terms of the complete set $u_n$ $(\mathbf{r})$

$$\Psi = \sum \!\!\!\!\int a_n(t)\, u_n \exp(-iE_n't/\hbar). \tag{O24}$$

Let the initial state of the system be $u_0$, i.e.,

$$a_0(0) = 1, \qquad a_n(0) = 0 \quad \text{for} \quad n \neq 0.$$

To the first approximation, i.e., on assuming

$$\int d\mathbf{r}\, u_n^* V'[\Psi - u_0 \exp(iE_0't/\hbar)] \ll \int d\mathbf{r}\, u_n^* V' u_0 \exp(iE_0't/\hbar), \tag{O25}$$

one obtains for the transition probability

$$w_{di} = \frac{2\pi}{\hbar} |(u_n, V'u_0)|^2 \rho(E_n'). \tag{O26}$$

For the "rearrangement" collision $(A + B \rightarrow C + D)$, we expand the $\Psi$ on the left in Equation (O22a)

$$\Psi = \sum_n \!\!\!\!\int b_n(t)\, v_n \exp(-iE_n''t/\hbar), \tag{O27}$$

but use the expansion (O24) on the right. If one assumes

$$\int d\mathbf{r}\, V''v_n^*[\Psi - u_0 \exp(-iE_0't/\hbar)] \ll \int d\mathbf{r}\, V''v_n^* u_0 \exp(-iE_0't/\hbar), \tag{O28}$$

one obtains in an entirely similar manner the transition probability for rearrangement collision

$$w_{re} = \frac{2\pi}{\hbar} |(V''v_n, u_0)|^2 \rho(E_n''). \tag{O29}$$

From the hermiticity of the Hamiltonian in (O21), it can be shown that (ref. 3)

$$(V''v_n, u_0) = (v_n, V'u_0), \tag{O30}$$

so that one can put (O29) in the form

$$w_{re} = \frac{2\pi}{\hbar} |(v_n, V'u_0)|^2 \rho(E_n''). \tag{O31}$$

In view of the relation (O20), the results (O26), (O29), (O31) are in agreement of the result (L38), (L39) in the case of electron scattering by the stationary state method.

In $w_{di}$, $V'$ is by hypothesis small compared with $H_0'$ in (O21), of which $u_0$, $u_n$ are eigenfunctions. In (O29) for $w_{re}$, $V''$ is not small relative to $H_0'$ of which $u_0$ is an eigenfunction [or, in (O31), $V'$ is not "small" with respect to $H_0''$]. The same remarks following (L41) hold.

In our discussion of the direct and rearrangement collision in Sect. L.2 by the stationary state method, we have emphasized that depending on the information we wish to obtain (i.e., the cross section $|f(\vartheta)|^2$ for direct scattering, *or* the cross-section $|g(\vartheta)|^2$ for rearrangement scattering), we choose the appropriate representation accordingly. We wish to note that we have the same situation in the time-dependent treatment. Thus for the "direct" scattering, we use the expansion (O24) so that

$$\sum_n \int |a_n(t)|^2 = 1, \tag{O32}$$

i.e., the total probability of all "direct" scatterings is unity. If we are interested in the rearrangement collisions, we use the expansion (O27) from which we get

$$\sum_n \int |b_n(t)|^2 = 1, \tag{O33}$$

i.e., the total probability of all "rearrangement" collisions is unity. These statements do not imply any contradiction in the sense that (O32) would exclude the possibility of rearrangement collision or *vice versa*. It is just that in general $H_0'$ and $H_0''$ do not commute and the states $u_n$, $v_n$ form two distinct complete sets so that a state $u_n$ of $H_0'$ is in general a superposition of the states $v_m$ of $H_0''$ and *vice versa*. The Schrödinger equation (O1), together with a given initial condition, completely determines the $\Psi$ of the system at subsequent times. This can be analyzed either with respect to $H_0'$ as in (O24), or with respect to $H_0''$ as in (O27), yielding $w_{di}$ or $w_{re}$ respectively.

## 2. METHOD OF UNITARY (TIME-TRANSLATION) OPERATOR

A more general method of treating the transitions of a system due to a perturbation by the time-dependent Schrödinger equation is to introduce a unitary operator $U(t, t_0)$ that effects the transformation from a state $\Psi(\mathbf{r}_0, t_0)$ at the space-time $\mathbf{r}_0$, $t_0$ to a state $\Psi(\mathbf{r}, t)$ at $\mathbf{r}$, $t$ in accordance with the Schrödinger equation (O4)

$$i\hbar \frac{\partial \Psi}{\partial t} = (H_0 + V)\Psi. \tag{O4}$$

This method, fully treated by Dirac in his *Quantum Mechanics* (ref. 2) and briefly described below, has been further developed by other authors whose work will be given in the following sections of the present chapter.

Let us assume that the Hamiltonian $H$ of the system is

$$H = H_0 + V, \tag{O34}$$

where $H_0$ is time-independent and has stationary states, and $V$ may or may not depend on time. Let us make a unitary transformation

$$\mathscr{H}(t) = \exp(iH_0 t/\hbar)\, V \exp(-iH_0 t/\hbar), \tag{O35}$$

$$\Psi'(t) = \exp(iH_0 t/\hbar)\, \Psi(t). \tag{O36}$$

Then (O4) becomes

$$i\hbar \frac{\partial \Psi'}{\partial t} = \mathscr{H}\Psi', \tag{O37}$$

which now contains only the "perturbation" or "interaction" $V$ through $\mathscr{H}$. (O37) has come to be referred to as the Schrödinger equation in the "interaction picture". #

Let us rewrite (O37) in terms of the eigenkets $|\alpha, t\rangle$ instead of the $\Psi'(t)$ (which are the representatives of $|\alpha, t\rangle$ in the Schrödinger, or coordinate, representation)

$$i\hbar \frac{\partial}{\partial t} |\alpha, t\rangle = \mathscr{H}(t)|\alpha, t\rangle. \tag{O37a}$$

Let $U(t, t_0)$ be a unitary (linear) operator

$$U^+(t, t_0)\, U(t, t_0) = 1 \tag{O38}$$

such that, for any $\alpha$, the following relation holds:

$$|\alpha, t\rangle = U(t, t_0)|\alpha, t_0\rangle. \tag{O39}$$

---

# The terminology "interaction picture" is preferable to "interaction representation" sometimes used in the literature, since the word "representation" has a different meaning in the established usage of "coordinate representation", "momentum representation" etc.

From (O39) and (O38), the following properties follow:

$$U(t, t) = 1, \tag{O40a}$$

$$U(t, t_1)\, U(t_1, t_0) = U(t, t_0), \tag{O40b}$$

$$U^+(t, t_0) = U(t_0, t). \tag{O40c}$$

On putting $|\alpha, t\rangle$ in (O39) into (O37a), we obtain the operator equation for $U(t, t_0)$

$$i\hbar \frac{\partial U(t, t_0)}{\partial t} = \mathscr{H}(t)\, U(t, t_0). \tag{O41}$$

This equation and the "boundary condition" (O40a) can be replaced by the integral equation

$$U(t, t_0) = 1 - \frac{i}{\hbar}\int_{t_0}^{t} dt' \mathscr{H}(t')\, U(t', t_0).^{\#} \tag{O42}$$

The operator $U(t, t_0)$ can be represented by the unitary (projection) operator

$$U(t, t_0) = \int |\alpha'',t\rangle\, d\alpha'' \langle\alpha'',t_0|, \tag{O46}$$

for then

$$U(t, t_0)|\alpha,t_0\rangle = \int |\alpha'',t\rangle\, d\alpha'' \langle\alpha'',t_0|\alpha,t_0\rangle$$

$$= \int |\alpha'',t\rangle\, d\alpha'' \delta(\alpha'' - \alpha) = |\alpha,t\rangle \tag{O47}$$

which satisfies (O39). We can show that the form (O46) for $U(t, t_0)$ also satisfies the equation (O42). Thus

$$-\frac{i}{\hbar}\int_{t_0}^{t} \mathscr{H}(t')\, U(t', t_0)\, dt' = \int_{t_0}^{t} \frac{\partial}{\partial t'} U(t', t_0)\, dt',$$

$$= \int_{t_0}^{t} dt' \frac{\partial}{\partial t'} \int |\alpha'', t'\rangle\, d\alpha'' \langle\alpha'', t_0|$$

$$= \int \{|\alpha'', t\rangle - |\alpha'', t_0\rangle\}\, d\alpha'' \langle\alpha'', t_0|$$

$$= U(t, t_0) - 1. \qquad \text{Q.E.D.}$$

---

# On taking the hermitian conjugate of (O41) and using (O40c), we get

$$-i\hbar \frac{\partial U(t_0, t)}{\partial t} = U(t_0, t)\, \mathscr{H}(t). \tag{O43}$$

On interchanging $t$ and $t_0$, this becomes

$$-i\hbar \frac{\partial U(t, t_0)}{\partial t_0} = U(t, t_0)\, \mathscr{H}(t_0) \tag{O44}$$

which is equivalent to the equation

$$U(t, t_0) = 1 + \frac{i}{\hbar}\int_{t}^{t_0} dt' U(t, t')\, \mathscr{H}(t'), \quad t_0 < t. \tag{O45}$$

To obtain the transition probabilities, let us form the matrix element of $U(t, t_0)$:

$$\langle \alpha', t_0 | U(t, t_0) | \alpha'', t_0 \rangle \overset{(O46-7)}{=\!=} \langle \alpha', t_0 | \alpha'', t \rangle, \tag{O48}$$

the right-hand side being the probability amplitude of the transition from the state $|\alpha', t_0\rangle$ at time $t_0$ to the state $|\alpha'', t\rangle$ at time $t$. It is seen that the transition probability $|\langle \alpha', t_0 | \alpha'', t\rangle|^2$ is independent of the time order of $t_0$ and $t$. The total transition probability from $\langle \alpha', t_0|$ to all states $\alpha''$ at $t$ is unity, as

$$\sum \int |\langle \alpha', t_0 | \alpha'', t\rangle|^2 \, d\alpha'' = \sum \int \langle \alpha', t_0 | \alpha'', t\rangle \, d\alpha'' \langle \alpha'', t | \alpha', t_0\rangle$$
$$= \langle \alpha', t_0 | \alpha', t_0 \rangle = 1. \tag{O49}$$

We shall now specify the $|\alpha, t\rangle$ above to be the eigenstates of $H_0$ of (O34), namely,

$$i\hbar \frac{\partial}{\partial t} |E_n, t\rangle = H_0 |E_n, t\rangle, \tag{O50}$$

where $E_n$ are the eigenvalues of $H_0$. We shall further assume that $V$ in (O34) is small, so that Equation (O42) can be solved by iteration (in powers of $\mathscr{H}$), yielding [on calling $t_0 = 0$, and writing $U(t) \equiv U(t, 0)$]

$$U(t) = 1 - \frac{i}{\hbar} \int_0^t dt' \mathscr{H}(t')\, U(t')$$
$$= 1 + U_1(t) + U_2(t) + \cdots + U_n(t) + \cdots, \tag{O51}$$

where

$$\begin{aligned} U_1(t) &= -\frac{i}{\hbar} \int_0^t \mathscr{H}(t')\, dt', \\ U_n(t) &= \left(-\frac{i}{\hbar}\right)^n \int_0^t \mathscr{H}(t_1)\, dt_1 \int_0^{t_1} \mathscr{H}(t_2)\, dt_2 \ldots \int_0^{t_{n-2}} \mathscr{H}(t_n)\, dt_n. \end{aligned} \tag{O52}$$

For the transition $|E_0, t_0\rangle \to |E_n, t\rangle$, to the first order, the transition amplitude is, by (O48), (O35),

$$\langle E_0, t_0 | U_1(t) | E_n, t_0\rangle = -\frac{i}{\hbar} \int \langle E_0, t_0 | V | E_n, t_0\rangle \exp\,[i(E_0 - E_n)t'/\hbar]\, dt'. \tag{O53}$$

If $V$ is independent of $t$, this gives the transition probability

$$|\langle E_0, t_0 | U_1(t) | E_n, t_0\rangle|^2 = \frac{|\langle E_0 | V | E_n\rangle|^2}{\chi^2} \sin^2 \chi \cdot \left(\frac{t}{\hbar}\right)^2, \quad \chi = \frac{(E_n - E_0)t}{2\hbar}$$

which is seen to be the same as (O12) obtained before. If $\langle E_0|V|E_n\rangle$ vanishes or is very small, it is then necessary to go to the next order.

$$\langle E_n, t_0|U_2(t)|E_0, t_0\rangle = -\frac{1}{\hbar^2}\int_0^t dt_1 \int_0^{t_1} dt_2 \sum_m \langle E_n|V|E_m\rangle\langle E_m|V|E_0\rangle$$
$$\times \exp[i(E_n - E_m)t_1/\hbar] \exp[i(E_m - E_0)t/\hbar]$$
$$= \sum_m \frac{\langle E_n|V|E_m\rangle\langle E_m|V|E_0\rangle}{E_m - E_0}$$
$$\times \left\{\frac{\exp[i(E_n - E_0)t/\hbar] - 1}{E_n - E_0} - \frac{\exp[i(E_n - E_m)t/\hbar] - 1}{E_n - E_m}\right\}. \qquad (O54)$$

Unless the first order amplitude (O53) vanishes, it is necessary to add the amplitudes in (O53) and (O54) *before* obtaining the transition probability,

$$|\langle E_n, t_0|U_1|E_0, t_0\rangle + \langle E_n, t_0|U_2|E_0, t_0\rangle|^2, \qquad (O55)$$

since there is interference between the amplitudes. (O54) also shows the role played by the intermediate states $E_m$.

To obtain the probability per unit time of transitions from an initial state $|E_0, t_0\rangle$ to any of the states having energy in the neighborhood of $E_n$, the expression (O55) is substituted for $|\langle k|V|n\rangle|^2$ in (O12) and the same prodceure as described in (O12)–(O14) is followed. Similar calculations to higher orders can be made.

The unitary operator $U(t, t_0)$, as $t_0 \to -\infty$, $t \to \infty$, is related to the scattering operator $S$ [see Sect. A (A34)–(A36)] which is in turn related to other operators $K$, $T$, $R$ (see Sect. A.4) used in the theory of scattering. Further developments of the theory involving these operators will be made in Sects. P, Q, R, W, X.

### 3. METHOD OF GREEN'S FUNCTION

In Chapter 1, Sect. B, and also Chapter 3, Sect. M.1, we have transformed the time-independent Schrödinger equation into an integral form and have shown how the boundary conditions for the scattering problem can be satisfied by an appropriate choice of the Green's function. It is to be expected that the time-dependent Schrödinger equation can be similarly expressed in an integral form. It turns out that the theory based on the integral equation is far more useful and important than the mere mathematical equivalence with the other methods would have suggested. This usefulness and importance are due to the new point of view and the interpretation given to the Green's functions by Feynman (refs. 4 and 5). When applied to the quantum theory of fields, the theory of Feynman provides the familiar, but most advantageous, method of "Feynman diagrams". We shall not be able to treat these field theoretical developments, but shall confine ourselves to a brief account of the main ideas of the method of Green's functions in the theory of scattering.

Let the Hamiltonian of the system $H = H_0 + V$, where $H_0$ is, for simplicity, that of a free-particle. The Schrödinger equation is (O4), which, when expressed in terms of the $|\alpha t\rangle$ as in (O37a), is

$$\left(i\hbar\frac{\partial}{\partial t} - H_0 - V\right)|\alpha,t\rangle = 0. \qquad \text{(O56)}$$

We wish to transform (O56) into an integral equation, and to find the Green's function $G(\mathbf{r}, t)$ of (O56),

$$\left(i\hbar\frac{\partial}{\partial t} - H_0 - V\right) G(\mathbf{r}, t) = \delta(\mathbf{r})\,\delta(t). \qquad \text{(O57)}$$

*i) Unperturbed system $H_0$*

Consider

$$\left(i\hbar\frac{\partial}{\partial t} - H_0\right) G_0(\mathbf{r}, t) = \delta(\mathbf{r} - \mathbf{r}_1)\,\delta(t - t_1). \qquad \text{(O58)}$$

It can be seen that this equation is satisfied by

$$G_0(\mathbf{r} - \mathbf{r}_1, t - t_1) = \frac{1}{(2\pi)^4}\int_\Gamma \frac{\exp\left[i\mathbf{k}\cdot(\mathbf{r} - \mathbf{r}_1) - i\omega(t - t_1)\right]}{\hbar\omega - E_k}\,d\mathbf{k}\,d\omega, \qquad \text{(O59)}$$

where $E_k = \hbar^2k^2/2m \equiv \hbar\omega_k$ is the energy of the particle. $\Gamma$ denotes the path of integration in the complex $\omega$-plane. We shall introduce the physical condition of causality in the theory by properly specifying the path $\Gamma$. In view of the physical meaning of $G_0(\mathbf{r} - \mathbf{r}_1, t - t_1)$ that we shall presently give below [following (O62)], this causality condition will be expressed by requiring $G_0(\mathbf{r} - \mathbf{r}_1, t - t_1)$ to vanish for $t - t_1 < 0$. This condition on $G_0(\mathbf{r} - \mathbf{r}_1, t - t_1)$ is met by writing

$$G_0(\mathbf{r} - \mathbf{r}_1, t - t_1) = \frac{1}{(2\pi)^4}\lim_{\varepsilon\to 0}\int_\Gamma \frac{\exp\left[i\mathbf{k}\cdot(\mathbf{r} - \mathbf{r}_1) - i\omega(t - t_1)\right]}{\hbar\omega - E_k + i\varepsilon}\,d\mathbf{k}\,d\omega$$

and by choosing $\Gamma$ to be, for $t - t_1 < 0$, the contour from $-\infty$ to $+\infty$ along the real axis and along a large semi-circle in the upper $\omega$-plane (in the counter-clockwise sense); and for $t - t_1 > 0$, along the real axis and a semi-circle in the lower $\omega$-plane in the clockwise sense. This defines $G_0(\mathbf{r} - \mathbf{r}_1, t - t_1)$ as

$$G_0(\mathbf{r} - \mathbf{r}_1, t - t_1) = \begin{cases} -\dfrac{i}{(2\pi)^3\hbar}\displaystyle\int \exp(i\mathbf{k}\cdot\mathbf{r})\exp\left[-i\omega_k(t - t_1)\right] \\ \qquad\times\exp(-i\mathbf{k}\cdot\mathbf{r}_1)\,d\mathbf{k}, & \text{for } t - t_1 > 0, \quad \text{(O59a)} \\ 0, & \text{for } t - t_1 < 0. \end{cases}$$

On carrying out the integration over **k**, one obtains

$$G_0(\mathbf{r} - \mathbf{r}_1, t - t_1) = \left[\frac{m}{2\pi i\hbar(t - t_1)}\right]^{3/2} \exp\left\{\frac{im(r - r_1)^2}{2\hbar(t - t_1)}\right\}.$$

This $G_0(\mathbf{r} - \mathbf{r}_0, t - t_0)$ may be regarded as the representative, in the coordinate representation, of the operator $G_0(t - t_1)$

$$\left(i\hbar\frac{\partial}{\partial t} - H_0\right) G_0(t - t_1) = \delta(t - t_1), \tag{O60}$$

i.e.,

$$G_0(t - t_1) = -\frac{i}{2\pi} \lim_{\varepsilon \to 0} \int_\Gamma \frac{\exp[-i\omega(t - t_1)]}{\hbar\omega - E_k + i\varepsilon} d\omega, \quad \varepsilon > 0,$$

$$= -\frac{i}{2\pi} \lim_{\varepsilon \to 0} \int_0^\infty d\xi \int_\Gamma \exp[i\xi(\hbar\omega - H_0 + i\varepsilon) - i\omega(t - t_1)]\, d\omega$$

$$= -i \int_0^\infty d\xi \exp(-i\xi H_0)\, \delta[\xi\hbar - (t - t_1)]$$

$$= \begin{cases} -i\exp[-iH_0(t - t_1)/\hbar], & t - t_1 > 0, \\ 0, & t - {}_1 < 0. \end{cases} \tag{O61}$$

The $\langle\mathbf{r}|, |\mathbf{r}_1\rangle$ matrix element of $G_0(t - t_1)$ is

$$\langle\mathbf{r}|G_0(t - t_1)|\mathbf{r}_1\rangle = -i \int \langle\mathbf{r}|E_k\rangle \exp[-iE_k(t - t_1)/\hbar] \langle E_k|\mathbf{r}_1\rangle\, d\mathbf{k}$$

$$= \frac{-i}{(2\pi)^3} \int \exp(i\mathbf{k}\cdot\mathbf{r}) \exp[-i\omega_k(t - t_1)] \times \exp[-i\mathbf{k}\cdot\mathbf{r}_1]\, d\mathbf{k}, \tag{O62}$$

which is (O59). In the form (O61) or (O62), we have found a meaning for $G_0(\mathbf{r} - \mathbf{r}_1, t - t_1)$ as follows. Let $\Psi_n(\mathbf{r}_1, t_1)$ be a wave function $\langle\mathbf{r}_0|E_n\rangle$, and let us form

$$i\hbar \int \langle\mathbf{r}|G_0(t - t_1)|\mathbf{r}_1\rangle\, d\mathbf{r}_1 \langle\mathbf{r}_1|E_n,t_1\rangle. \tag{O63}$$

Now the formal solution of

$$\left(i\hbar\frac{\partial}{\partial t} - H_0\right)|\alpha,t\rangle = 0, \tag{O64}$$

is

$$|\alpha,t\rangle = \exp[-iH_0(t - t_1)/\hbar]|\alpha,t_1\rangle. \tag{O65}$$

Taking the representative in the coordinate representation, we have

$$\langle\mathbf{r}|\alpha,t\rangle = \int \langle\mathbf{r}|\exp[-iH_0(t - t_1)/\hbar]|\mathbf{r}_1\rangle\, d\mathbf{r}_1 \langle\mathbf{r}_1|\alpha,t_1\rangle. \tag{O66}$$

(O66) states that $\langle\mathbf{r}|\exp[iH_0(t - t_1)/\hbar]|\mathbf{r}_1\rangle$ is the transition amplitude for a state $\langle \mathbf{r}_1|\alpha,t_1\rangle$ at $t_1$ to go, in accordance with the Schrödinger equation (O64), into $\langle \mathbf{r}|\alpha,t\rangle$ at time $t(>t_1)$. On comparing (O63) and (O66), it is seen that

$$i\hbar\langle \mathbf{r}_2|G_0(t_2 - t_1|\mathbf{r}_1\rangle = i\hbar G_0(\mathbf{r}_2 - \mathbf{r}_1, t_2 - t_1) \equiv i\hbar G_0(2, 1)$$

is the transition amplitude, of the system with Hamiltonian $H_0$, from the state $\langle \mathbf{r}_1|\alpha,t_1\rangle = \Psi_\alpha(\mathbf{r}_1, t_1)$ to $\langle \mathbf{r}_2|\alpha,t_2\rangle = \Psi_\alpha(\mathbf{r}_2, t_2)$. This meaning of the Green's function $G_0(2, 1)$ leads to the following important result. The probability amplitude that a state $\Psi(\mathbf{r}_1, t_2)$ at time $t_1$ goes over to the state $\Psi(\mathbf{r}_2, t_2)$ at time $t_2$ is given by

$$\Psi(\mathbf{r}_2, t_2) = i\hbar \int G_0(\mathbf{r}_2 - \mathbf{r}_1, t_2 - t_1)\, \Psi(\mathbf{r}_1, t_1)\, d\mathbf{r}_1. \tag{O67}$$

With this interpretation of $G_0(2, 1)$, the reason for the causality condition $G_0(\mathbf{r}_2 - \mathbf{r}_1, t_2 - t_1) = 0$ for $t_2 - t_1 < 0$ is clear.

*ii) Perturbed system $H = H_0 + V$*

In terms of $G_0(\mathbf{r}_2 - \mathbf{r}_1, t_2 - t_1)$, the solution of (O56) can be written

$$|\alpha,t\rangle = |H_0,t\rangle + \int_{-\infty}^{t} G_0(t - t')\, V(t')|\alpha,t'\rangle\, dt', \tag{O68}$$

where $|H_0,t\rangle$ is an eigenket of $[i\hbar(\partial/\partial t) - H_0]|H_0,t\rangle = 0$. On taking the representatives in the coordinate representation, (O68) becomes, as in (O62),

$$\Psi(\mathbf{r}, t) = \Psi_0(\mathbf{r}, t) + \int_{-\infty}^{t} dt' \int d\mathbf{r}' G_0(\mathbf{r} - \mathbf{r}', t - t')\, V(\mathbf{r}', t')\, \Psi(\mathbf{r}', t'), \tag{O69}$$

which is the Schrödinger equation (O56) in integral form.

Before proceeding further, let us consider the case in which $H = H_0 + V$ is independent of time. In this case,

$$\Psi(\mathbf{r}, t) = \exp(-iEt/\hbar)\, \psi(\mathbf{r}), \quad \Psi_0(\mathbf{r}, t) = \exp(-iEt/\hbar)\, \psi_0(\mathbf{r}).$$

By using the method of integration in Sect. B, it can easily be shown that (O69) reduces to

$$\psi(\mathbf{r}) = \psi_0(\mathbf{r}) - \frac{1}{4\pi}\frac{2m}{\hbar^2} \int \frac{\exp(ik|\mathbf{r} - \mathbf{r}'|)}{|\mathbf{r} - \mathbf{r}'|}\, V(r')\, \psi(\mathbf{r}')\, d\mathbf{r}' \tag{O70}$$

which is (B14) or (M9), namely, the Schrödinger equation for scattering of a particle by a field $V(r)$.

Let us next find the Green's function $G(\mathbf{r} - \mathbf{r}_1, t - t_1)$ of the Hamiltonian $H = H_0 + V$ in (O57). By a calculation similar to that used in obtaining (M4), we have the operator equation

$$G = G_0 + G_0 V G, \tag{O71}$$

where $G \equiv G(\mathbf{r} - \mathbf{r}_1, t - t_1)$ of (O57) and $G_0 \equiv G_0(\mathbf{r} - \mathbf{r}_1, t - t_1)$ of (O58), or (O59). If $V$ is "small", we may iterate (O71), obtaining, in powers of $V$,

$$G = G_0 + G_0VG_0 + G_0VG_0VG_0 + \cdots. \tag{O72}$$

The $n$th-order term is

$$G_0VG_0 \ldots VG_0 = \int \cdots \int G_0(n+2, n+1)\, V(\mathbf{r}_{n+1}, t_{n+1})\, G_0(n+1, n) \ldots V(\mathbf{r}_2, t_2)\, G_0(2, 1)\, d\mathbf{r}_{n+1}\, dt_{n+1}\, d\mathbf{r}_n\, dt_n \ldots d\mathbf{r}_2\, dt_2$$

where, on account of the condition (O59), the sequence of times is

$$t_1 < t_2 < t_3 < \cdots < t_{n+2} \tag{O73}$$

for the $G_0$ not to vanish.

We shall consider $G$ up to the first order in $V$ in (O72). The zeroth order term $G_0(3, 1)$ comes from $H_0$ and is independent of $V$. This amplitude thus comes from the "propagation" (in space-time) from $\mathbf{r}_1, t_1$ to $\mathbf{r}_3, t_3$ of a free particle. The first-order term

$$\int G_0(3, 2)\, V(\mathbf{r}_2, t_2)\, G_0(2, 1)\, d\mathbf{r}_2\, dt_2 \tag{O74}$$

represents the propagation from $\mathbf{r}_1, t_1$ to $\mathbf{r}_2, t_2$ (as a free particle), the scattering by $V$ "at" $\mathbf{r}_2, t_2$, and the propagation from $\mathbf{r}_2, t_2$ to $\mathbf{r}_3, t_3$, the time sequence being, as (O73),#

$$t_1 < t_2 < t_3.$$

Similarly for the higher order terms in (O72). The series for $G$ is then the *total* transition amplitude for the propagation from $\mathbf{r}_1, t_1$ to the final $\mathbf{r}, t$ after any and all numbers of scatterings. It is for this meaning that the Green's function has been known as the "propagator". $G_0(\mathbf{r} - \mathbf{r}_1, t - t_1)$ is the propagator of a free particle (according to our assumption about $H_0$).

Let us calculate the transition probability amplitude from a state $\Psi_n(\mathbf{r}_1, t_1)$ to $\Psi_m(\mathbf{r}_3, t_3)$ up to the first order in $V$. By (O74), the amplitude is

$$a_{mn}^{(1)} = i\hbar \int \cdots \int \Psi_m^*(\mathbf{r}_3, t_3)\, G_0(\mathbf{r}_3 - \mathbf{r}_2, t_3 - t_2)\, d\mathbf{r}_3\, V(\mathbf{r}_2)\, G(\mathbf{r}_2 - \mathbf{r}_1, t_2 - t_1) \times \Psi_n(\mathbf{r}_1, t_1)\, d\mathbf{r}_1\, d\mathbf{r}_2\, dt_2.$$

By using (O67) and its complex conjugate [with $G_0^*(2, 1) = G_0(1, 2)$ from (O59)], this becomes

$$a_{mn}^{(1)} = \frac{1}{i\hbar} \int \Psi_m^*(\mathbf{r}, t)\, V(\mathbf{r}')\, \Psi_i(\mathbf{r}', t')\, d\mathbf{r}'\, dt'.$$

---

# Since the "scattering" at the intermediate state (near $\mathbf{r}_2, t_2$) is confined to very short time intervals ($< t_3 - t_1$), the energy of the system in the intermediate state is not known with unlimited precision on account of the uncertainty principle.

Let the $\Psi'$ be normalized as in (O15, 6). The probability of the transition in the interval $t$ is

$$w_{mn} = \int |a_{mn}^{(1)}|^2 \rho_m(\omega)\, d\omega,$$

where $\rho(\omega)$ is the density of state near the final state $\Psi'_m$. Entirely identical calculations as in (O12)–(O20) give the same results (O18), (O20) for the probability per unit time and the cross section respectively.

The greatly simplified system (scattering of a particle by a potential) treated above merely illustrates the ideas and method of the theory. The theory is much more general and is applicable to the Dirac equation as well as to the Schrödinger equation. For further expositions of the physical ideas of the theory and its applications to the quantized fields, the reader must be referred to the original papers of Feynman and others.

## REFERENCES

1. Dirac, P. A. M., Proc. Roy. Soc. (London) **A112**, 661 (1926); **A114**, 243 (1927).

2. Dirac, P. A. M., *The Principles of Quantum Mechanics.* Clarendon Press, Oxford, 3rd ed. (1947), §§ 27, 44. The unitary matrix $U(t, t_0)$ in the present book is denoted by $T$ in Dirac's book.

3. Schiff, L. I., *Quantum Mechanics.* McGraw-Hill, New York, 2nd ed. (1955).

4. Feynman, R. P., Phys. Rev. **76**, 749 (1949); § 2, Green's function treatment of the Schrödinger equation.

5. Feynman, R. P., Rev. Mod. Phys. **20**, 367 (1948). Space-Time Approach to Non-Relativistic Quantum Mechanics.

section

# P Time-Dependent Theory of Scattering: Variational Principles of Lippmann and Schwinger

In Chapter 1, Sect. D, we have treated the variational methods for the scattering problem on the basis of the time-independent Schrödinger equation. Lippmann and Schwinger have developed more general variational principles for the operators $S$, $K$, $T$ which are relevant for the scattering problem and which we have only briefly mentioned in Chapter 1, Sect. A [*infra* (A33)]. The starting point is the time-dependent Schrödinger equation (O4)

$$i\hbar \frac{\partial \Psi}{\partial t} = (H_0 + V)\Psi. \tag{P1}$$

We assume that $H_0$ is time-independent and has the eigenfunctions $\Phi_a$

$$(H_0 - E_a)\Phi_a = 0. \tag{P2}$$

The same unitary transformation (O35, 36) is made to the interaction picture, in which (P1) becomes

$$i\hbar \frac{\partial \Psi'}{\partial t} = \mathscr{H}(t)\Psi(t)'. \tag{P3}$$

We shall in the following drop the prime from $\Psi'$ for simplicity.

Let $\Psi(-\infty)$, $\Psi(\infty)$ be the initial and final state function of the system when the interaction $V$ is zero, and let $U_+(t)$ be the unitary operator that yields $\Psi(t)$ from $\Psi(-\infty)$

$$\Psi(t) = U_+(t)\Psi(-\infty), \tag{P4}$$

It is seen that the $U_+(t)$ is the $U(t, -\infty)$ of (O39), and in terms of eigenkets instead of $\Psi$, (P4) reads

$$|\alpha, t\rangle = U_+(t)|\alpha, -\infty\rangle. \tag{P5}$$

The unitarity of $U_+$ means#

$$U_+^\dagger(t)\, U_+(t) = 1, \tag{P6}$$

and (O40a) is now

$$U_+(-\infty) = 1. \tag{P7}$$

The equations corresponding to (O41), (O42) are

$$i\hbar \frac{\partial U_+(t)}{\partial t} = \mathscr{H}(t)\, U_+(t), \tag{P8}$$

$$U_+(t) = 1 - \frac{i}{\hbar}\int_{-\infty}^{t} \mathscr{H}(t')\, U_+(t')\, dt', \tag{P9}$$

or

$$= 1 - \frac{i}{\hbar}\int_{-\infty}^{\infty} \eta(t - t')\mathscr{H}(t')\, U_+(t')\, dt', \tag{P9a}$$

where

$$\eta(t - t') = \begin{cases} 1 & \text{for} \quad t > t' \\ 0 & \text{for} \quad t < t'. \end{cases} \tag{P10}$$

Similarly, if $U_-(t)$ [which is $U(t, \infty)$ of (O39)] represents the unitary operator that generates $\Psi(t)$ from the final state $\Psi(\infty)$

$$\Psi(t) = U_-(t)\, \Psi(\infty), \tag{P11}$$

with

$$U_-^\dagger(t)\, U_-(t) = 1, \tag{P12}$$

and

$$U_-(\infty) = 1, \tag{P13}$$

then, corresponding to (P8), (P9), (P9a), one has

$$i\hbar \frac{\partial U_-(t)}{\partial t} = \mathscr{H}(t)\, U_-(t) \tag{P14}$$

$$U_-(t) = 1 + \frac{i}{\hbar}\int_{t}^{\infty} \mathscr{H}(t')\, U_-(t')\, dt' \tag{P15}$$

$$= 1 + \frac{i}{\hbar}\int_{-\infty}^{\infty} dt'\mathscr{H}(t')\, U_-(t')\, \eta(t' - t). \tag{P15a}$$

Thus, formally, the problem of solving the Schrödinger equation (P1) is replaced by one of solving the integral equation (P9a) or (P15a).

---

# For the unitarity of $U(t, t_0)$ for infinite times, see (Q22) below.

## 1. SCATTERING OPERATOR S AND VARIATIONAL PRINCIPLE

The two differential equations (P8), (P14), or their respective equivalent integral equations (P9a) and (P15a) are equivalent to a variational equation. Thus, if we denote by

$$S = 1 - \frac{i}{\hbar}\int_{-\infty}^{\infty} [U_-^\dagger(t)\mathscr{H}(t) + \mathscr{H}(t)\,U_+(t)]\,dt + \frac{i}{\hbar}\int_{-\infty}^{\infty} U_-^\dagger(t)\mathscr{H}(t)\,U_+(t)\,dt$$
$$+ \left(\frac{i}{\hbar}\right)^2 \int_{-\infty}^{\infty} dt \int_{-\infty}^{\infty} dt'\,U_-^\dagger(t)\mathscr{H}(t)\,\eta(t-t')\,\mathscr{H}(t')\,U_+(t'), \quad \text{(P16)}$$

then the variation of $\delta S$ due to arbitrary variations in $U_-^\dagger$ and $U_+$ is

$$\delta S = \frac{i}{\hbar}\int_{-\infty}^{\infty} dt\,\delta U_-^\dagger(t)\mathscr{H}(t)\left[U_+(t) - 1 + \frac{i}{\hbar}\int_{-\infty}^{\infty}\eta(t-t')\,\mathscr{H}(t')\,U_+(t')\,dt'\right]$$
$$+ \frac{i}{\hbar}\int_{-\infty}^{\infty} dt\left[U_-(t) - 1 - \frac{i}{\hbar}\int_{-\infty}^{\infty} dt'\mathscr{H}(t')\,U_-(t')\,\eta(t'-t)\right]^\dagger \mathscr{H}(t)\,\delta U_+(t). \quad \text{(P17)}$$

It is seen that on requiring $\delta S = 0$ for arbitrary $\delta U_-^\dagger$ and $\delta U_+$, one obtains exactly (P9a) and the adjoint of (P15a).

The stationary value of $S$ obtained by putting in the third term in (P16) the expression (P9a) is

$$S = 1 - \frac{i}{\hbar}\int_{-\infty}^{\infty}\mathscr{H}(t)\,U_+(t)\,dt, \quad \text{(P18)}$$

which, by (P9) is

$$S = U_+(\infty). \quad \text{(P19)}$$

Alternatively, if one puts in (P16) the adjoint of $U_-(t)$ from (P15a), one obtains for the stationary values of $S$

$$S = 1 - \frac{i}{\hbar}\int_{-\infty}^{\infty} U_-^\dagger(t)\mathscr{H}(t)\,dt, \quad \text{(P18a)}$$

or

$$S^\dagger = 1 + \frac{i}{\hbar}\int_{-\infty}^{\infty}\mathscr{H}(t)\,U_-(t)\,dt, \quad \text{(P20)}$$

which by (P15) is

$$S^\dagger = U_-(-\infty). \quad \text{(P21)}$$

The operator $S$ given by (P19) and (P21) has the following important property: From (P5) and (P11), we have

$$\Psi(\infty) = U_+(\infty)\Psi(-\infty) = S\Psi(-\infty), \quad \text{(P22)}$$
$$\Psi(-\infty) = U_-(-\infty)\Psi(\infty) = S^\dagger\Psi(\infty), \quad \text{(P22a)}$$

hence

$$\Psi(\infty) = SS^\dagger\Psi(\infty),$$

and hence

$$SS^\dagger = 1, \quad \text{or} \quad S^\dagger = S^{-1}, \quad \text{(P23)}$$

i.e., $S$ is a unitary operator. From (P5), (P11) and (P22), it immediately follows that the time-independent $S$ satisfies the relation

$$U_+(t) = U_-(t)\, S. \tag{P24}$$

The meaning of $S$ is clear from (P22). If $\Psi(-\infty)$ is regarded as an initial state, then $S\Psi(-\infty)$ gives the resulting state when the system has evolved in the course of (infinite) time in accordance with the Schrödinger equation (P1). Similarly, $S^\dagger$ generates, through $S^\dagger\Psi(\infty)$, the "initial" state $\Psi(-\infty)$ from the "final" state $\Psi(\infty)$. $S$ is called the scattering operator.#

The expression (P16) for $S$ suggests the procedure of obtaining an approximate $S$ with an approximate $U(t)$. Thus if one simply puts $U_-^\dagger(t) = U_+(t) = 1$ in (P16), one gets

$$S = 1 - \frac{i}{\hbar}\int_{-\infty}^{\infty} \mathscr{H}(t)\, dt + \left(\frac{i}{\hbar}\right)^2 \iint_{-\infty}^{\infty} dt\, dt'\, \mathscr{H}(t)\, \eta(t - t')\, \mathscr{H}(t) \tag{P24}$$

corresponding to the second Born approximation (to the second order in the "perturbation" $\mathscr{H}$). This $S$, however, does not satisfy the requirement of unitarity (P23). See Sect. 4 below.

## 2. TRANSITION PROBABILITY

Let $\Phi_a$ be the eigenstate of $H_0$ at $t \to -\infty$. During the time $-\infty < t < \infty$, there is an interaction $V$ on the system. This $V$ may arise from the close approach, followed by separation, of the component parts forming the system which is described by $H_0$ at infinite separation; or $V$ may arise from an external time-dependent interaction. In any case, we shall use the artifice of introducing a fictitious switching on and off by an adiabatic process, i.e., by accompanying $V$ with a factor

$$V \exp(-\varepsilon|t|/\hbar) \tag{P25}$$

so that the Hamiltonian $H = H_0 + V$ approaches $H_0$ as $t \to -\infty$ and reverts to $H_0$ as $t \to \infty$.## $\varepsilon$ is allowed to approach zero eventually. We want to find the probability that the system will be found in an eigenstate $\Phi_b$ ($\neq \Phi_a$).

According to (P22), an initial state $\Phi_a$ at $t \to -\infty$ evolves to $S\Phi_a$ as

---

# In Chapter 1, (A34), and also in Chapter 6, Sect. W, the scattering operator $S$ is introduced in the time-independent theory. In this and the following Sects. P, Q, R, the scattering operator is introduced in the time-dependent theory. See Chapter 6, Sect. W for further discussions of the properties of $S$, and Chapter 7 for the dispersion relations for $S$.

## See the introductory paragraphs of the present Chapter, also Sect. Q.1 and Sect. R.5, for a discussion of this "adiabatic switching".

$t \to +\infty$. The probability of finding the system in any state $\Phi_b$ ($\neq \Phi_a$) is hence given by

$$w_{ba} = |(\Phi_b, S\Phi_a)|^2 \tag{P26}$$

or, on introducing the "transition operator"$T$#

$$T = S - 1, \tag{P27}$$

$$w_{ba} = |(\Phi_b, T\Phi_a)|^2 = |T_{ba}|^2 \tag{P28}$$

since $(\Phi_b, \Phi_a) = \delta_{ba}$. Here the two states may or may not be on the same energy shell (i.e., $E_a$, $E_b$ may or may not be equal). On using (P18), (O35) and (P25), we obtain##

$$T_{ba} = -\frac{i}{\hbar}\int_{-\infty}^{\infty} dt(\Phi_b, \exp(iH_0t)\,V\exp(-\varepsilon|t|)\exp(-iH_0t)\,U_+(t)\,\Phi_a)$$

$$= -\frac{i}{\hbar}(\Phi_b, V\Psi_a^{(+)}(E_b)), \tag{P29}$$

where

$$\Psi_a^{(+)}(E_b) = \int_{-\infty}^{\infty} dt \exp[i(E_b - H_0)t - \varepsilon|t|]\,U_+(t)\,\Phi_a. \tag{P30}$$

Now, according to (P5), $U_+(t)\,\Phi_a$ is the eigenstate of $H$ at time $t$ in the interaction picture, which has evolved from $\Psi(-\infty) = \Phi_a$. According to (O36), $\exp(-iH_0t)\,U_+(t)\,\Phi_a$ is then the eigenstate of $H$ at time $t$ in the Schrödinger picture. If we denote by $\Psi_a^{(+)}$ the stationary eigenstate of $H$ with the eigenvalue $E_a$ at $t_0$, then $\exp(-iHt)\,\Psi_a^{(+)} = \exp(-iE_at)\,\Psi_a^{(+)}$ is the eigenstate of $H$, at time $t$, in the Schrödinger picture. Hence

$$\exp(-iH_0t)\,U_+(t)\,\Phi_a = \exp(-iE_at)\,\Psi_a^{(+)}. \tag{P31}$$

(P30) then becomes, as $\varepsilon \to 0$,

$$\Psi_a^{(+)}(E_b) = 2\pi\hbar\delta(E_b - E_a)\,\Psi_a^{(+)}. \tag{P32}$$

$\Psi_a^{(+)}$ is defined for the energy shell $E_a$. As we shall presently see, $\Psi_a^{(+)}$ is the eigenstate describing the outgoing wave, starting with the initial state $\Phi_a$. From (P31), using (P9) for $U_+(t)$, we get

$$\exp(-iH_0t)\,\Phi_a - \frac{i}{\hbar}\int_{-\infty}^{t} \exp(-iH_0t)\exp(iH_0t')\,V\exp(-\varepsilon|t'|)$$
$$\times \exp(-iH_0t')\,U_+(t')\,dt'\Phi_a = \exp(-iE_at)\,\Psi_a^{(+)}.$$

---

# For $T$ in the case of potential scattering, see Sect. A.5.

## Here and in the following, we omit the factor $1/\hbar$ in all the exponents, for simplicity in printing.

On using (P31) again, introducing $t - t' = \tau$, and carrying out the integration over $\tau$, we get

$$\Psi_a^{(+)} = \Phi_a + \frac{1}{E_a - H_0 + i\varepsilon} V\Psi_a^{(+)}, \tag{P33}$$

where the operator $1/(E_a - H_0 + i\varepsilon)$ has the meaning (B22). It is seen that (P33) is the Schrödinger integral equation (B20), (M9) that has been obtained before by the stationary state method.

If, instead of (P27), we start with the adjoint

$$T^\dagger = S^\dagger - 1 = S^{-1} - 1, \tag{P27a}$$

and use (P20), (O35), (P25), we would have obtained in an entirely analogous manner the following relations:

$$T_{ba}^\dagger = \frac{i}{\hbar} (\Phi_b, V\Psi_b^{(-)}(E_b)), \tag{P29A}$$

$$\Psi_a^{(-)}(E_b) = \int_{-\infty}^{\infty} dt \exp\left[i(E_b - H_0)t - \varepsilon|t|\right] U_-(t)\,\Phi_a, \tag{P30A}$$

$$\exp(-iH_0t)\,U_-(t)\,\Phi_a = \exp(-iE_at)\,\Psi_a^{(-)}, \tag{P31A}$$

$$\Psi_a^{(-)}(E_b) = 2\pi\hbar\delta(E_b - E_a)\,\Psi_a^{(-)}, \tag{P32A}$$

$$\Psi_a^{(-)} = \Phi_a + \frac{1}{E_a - H_0 - i\varepsilon} V\Psi_a^{(-)}. \tag{P33a}$$

To obtain the transition probability per unit time from the initial state $\Phi_a$ at $t \to -\infty$ to a state $\Phi_b \neq \Phi_a$, we have, from (P4),

$$w_{ba} = \frac{\partial}{\partial t} |(\Phi_b, U_+(t)\Phi_a)|^2 \tag{P34}$$

and by (P8)

$$w_{ba} = \frac{i}{\hbar} (\mathscr{H}(t)\,U_+(t)\,\Phi_a, \Phi_b)(\Phi_b, U_+(t)\Phi_a) + \text{comp. conj.}$$

Using (P9) for $U_+(t)$ in the second matrix element, with (O35) and (P25), we find

$$w_{ba} = \frac{1}{\hbar^2} \int_{-\infty}^{t} dt'(\exp\left[i(E_b - H_0)t\right] e^{-\varepsilon|t|} U_+(t)\Phi_a, V\Phi_b)$$
$$\times\, (\Phi_b, V\exp\left[i(E_b - H_0)t'\right] e^{-\varepsilon|t|} U_+(t')\Phi_a) + \text{comp. conj.}$$

Using (P31) and

$$\int_{-\infty}^{\infty} \exp\left[i(E_b - E_a)t\right] e^{-\varepsilon|t|}\,dt = 2\pi\hbar\delta(E_b - E_a), \quad \varepsilon \to 0,$$

we obtain

$$w_{ba} = \frac{2\pi}{\hbar} |\mathbf{T}_{ba}|^2 \,\delta(E_b - E_a), \tag{P35}$$

where

$$\mathbf{T}_{ba} = (\Phi_b, V\Psi_a^{(+)}). \tag{P36}$$

By working with

$$w_{ba} = \frac{\partial}{\partial t} |(U_-(t)\Phi_b, \Phi_a)|^2$$

instead of (P34), we obtain in an entirely similar manner,

$$\mathbf{T}_{ba} = (\Psi_a^{(-)}, V\Phi_a)^{\#} \tag{P36a}$$

The expressions (P36), (P36a) are exact. It is seen from (P35) that the transition probability between two states $\Phi_a$, $\Phi_b$ does not vanish only when the energy is conserved ($E_a = E_b$). The $\mathbf{T}_{ba}$ in (P35), (P36), (P36a) is defined only for the same "energy shell" $E_a = E_b = E$.

### 3. TRANSITION OPERATOR $T$ AND VARIATIONAL PRINCIPLE

It can be shown that the variational principle for $S$ in (P16)–(P18), through the relation (P27), leads (i) to a stationary expression for $T_{ba}$ of (P36), and (ii) to the equations (B20). Let us form the matrix element $(\Phi_b, T\Phi_a)$. By (P27) and (P16),

$$T_{ba} = (\Phi_b, S\Phi_a)$$

$$\begin{aligned} T_{ba} = {} & -\frac{i}{\hbar}\int_{-\infty}^{\infty} dt\Big[(\exp\,[i(E_a - H_0)t]\, U_-(t)\Phi_b, V\Phi_a) \\ & \qquad\qquad + (\Phi_b, V\exp\,[i(E_b - H_0)t]\, U_+(t)\Phi_a)\Big] \\ & + \frac{i}{\hbar}\int_{-\infty}^{\infty} dt(\exp\,(-iH_0t)\, U_-(t)\Phi_b, V\exp\,(-iH_0t)\, U_+(t)\Phi_a) \\ & + \left(\frac{i}{\hbar}\right)^2 \int_{-\infty}^{\infty} dt \int_{-\infty}^{t} dt'\,(\exp\,(-iH_0t)\, U_-(t)\Phi_b, \\ & \qquad V\exp\,[-iH_0(t - t')]\, V\exp\,(-iH_0t')\, U_+(t')\Phi_a), \end{aligned} \tag{P37}$$

where $V$ contains the factor $e^{-\varepsilon|t|}$ of (P25). On using (P31), (P31a), we have, on integrating over $t$,

$$\begin{aligned} T_{ba} = -2\pi i\delta(E_b - E_a)\Big\{ & (\Psi_b^{(-)}, V\Phi_a) + (\Phi_b, V\Psi_a^{(+)}) - (\Psi_b^{(-)}, V\Psi_a^{(+)}) \\ & + \left(\Psi_b^{(-)}, V\frac{1}{E + i\varepsilon - H_0} V\Psi_a^{(+)}\right)\Big\}. \end{aligned} \tag{P38}$$

---

# For a discussion of the ingoing wave $\Psi_a^{(-)}$ for the final state, see the following section, Q.3, ii), below.

For arbitrary variations $\delta\Psi_b^{(-)}$, $\delta\Psi_a^{(+)}$,

$$\delta T_{ba} = \left(\delta\Psi_b^{(-)}, V\left[\Phi_a + \frac{1}{E + i\varepsilon - H_0} V\Psi_a^{(+)} - \Psi_a^{(+)}\right]\right)$$

$$+ \left(\left[\Psi_b + \frac{1}{E - i\varepsilon - H_0} V\Psi_b^{(-)} - \Psi_b^{(-)}\right], V\delta\Psi_a^{(+)}\right) \quad \text{(P39)}$$

which leads, on requiring $\delta T_{ba} = 0$ for arbitrary $\delta\Psi_b^{(-)}$, $\delta\Psi_a^{(+)}$, to the following equations

$$\Psi_a^{(+)} = \Phi_a + \frac{1}{E + i\varepsilon - H_0} V\Psi_a^{(+)},$$

$$\Psi_a^{(-)} = \Phi_a + \frac{1}{E - i\varepsilon - H_0} V\Psi_a^{(-)},$$

which are again the integral equations (P33), (P33A). That these equations have now been obtained from the time-dependent theory (contrasted with the stationary state theory of Sect. B) shows the equivalence of the two methods. (See the introductory paragraphs of the present Chapter and also Sect. R, especially R.5 and R.6.)

The stationary value of $T_{ba}$ is, on using (P33), (P33A) in succession, in (P38), given by

$$T_{ba} = -2\pi i\delta(E_b - E_a)\,\mathbf{T}_{ba}, \quad \text{(P40)}$$

where

$$\mathbf{T}_{ba} = (\Phi_b, V\Psi_a^{(+)}) = (\Psi_b^{(-)}, V\Phi_a), \quad \text{(P41)}$$

which are the $\mathbf{T}_{ba}$ in (P36), (P36A) that appear in (P35) for the transition probability $w_{ba}$.

On iterating $\Psi_a^{(+)}$ in (P40)

$$\Psi_a^{(+)} = \Phi_a + \frac{1}{E - H_0 + i\varepsilon} V\Phi_a + \frac{1}{E - H_0 + i\varepsilon} V \frac{1}{E - H_0 + i\varepsilon} V\Phi_a + \cdots \quad \text{(P42)}$$

and putting this into (P35), one has, up to the second order in $V$,

$$T_{ba} = (\Phi_b, V\Phi_a) + \sum_n (\Phi_b, V\Phi_n)\left(\Phi_n, \frac{1}{E - H_0 + i\varepsilon} V\Phi_a\right)$$

$$= (\Phi_b, V\Phi_a) + \sum_n \frac{(\Phi_b, V\Phi_n)(\Phi_n, V\Phi_a)}{E - E_n}. \quad \text{(P43)}$$

Substitution of this into (P35) gives the transition probability up to the second order. In problems of physical interest, one seeks the transition probability from a definite initial state $\Phi_a$ to all the continuously distributed states $\Phi_b$ in the neighborhood of $E_b = E_a$. This is obtained by multiplying $w_{ba}$ in (P35) by the density $\rho(E_b)$ of states and integrating over $E_b$. [See (O12), (O53)–(O55)].

### 4. REACTION OPERATOR $K$ AND VARIATIONAL PRINCIPLE

It is mentioned at the end of Sect. 1 that the approximate $S$ in (P24) does not satisfy the condition of unitarity (P23). It will therefore be convenient to express $S$ in terms, instead of $U_+(t)$, of another operator which will ensure the unitarity of $S$ in all approximations.

From (P24) and (P23), we have

$$\frac{2U_+(t)}{1+S} = \frac{2U_-(t)S}{1+S} = \frac{2U_-(t)}{1+S^{-1}}.$$

Let a $W(t)$ be defined by

$$W(t) \equiv \frac{2U_+(t)}{1+S} = \frac{2U_-(t)}{1+S^{-1}}. \tag{P44}$$

From (P7) and (P13),

$$W(-\infty) = \frac{2}{1+S}; \qquad W(\infty) = \frac{2S}{1+S}, \tag{P45}$$

and hence

$$W(\infty) + W(-\infty) = 2, \qquad W^+(-\infty) = \frac{2}{1+S^{-1}} = W(\infty). \tag{P46}$$

The properties (P45), (P46) of $W(\infty)$, $W(-\infty)$ can be ensured by means of a Hermitian operator $K$

$$W(\infty) = 1 - \tfrac{1}{2}iK, \quad W(-\infty) = 1 + \tfrac{1}{2}iK. \tag{P47}$$

From these and (P45), we have

$$S = \frac{1 - \frac{1}{2}iK}{1 + \frac{1}{2}iK}. \tag{P48}$$

This $K$ is variously called the reaction matrix, the reactance matrix or Heitler's matrix. An approximate, but Hermitian, $K$ will now give a unitary $S$.# The problem is next to formulate for $K$ a variational principle which ($i$) is

---

# From (P48), one obtains $(1 + \frac{1}{2}iK)\,S = 1 - \frac{1}{2}iK$ so that for any Hermitian $K$, one has $S^\dagger = \dfrac{1 + \frac{1}{2}iK}{1 - \frac{1}{2}iK}$. Hence $S^\dagger S = 1$.

equivalent to that for $S$ which determines $U_+(t)$ etc., (*ii*) ensures the Hermitian property of $K$, and (*iii*) from which approximate $K$ can be obtained which gives, according to (P48), approximate but unitary $S$.

From (P9), (P15), (P44), (P46), one obtains the integral equations

$$W(t) = W(-\infty) - \frac{i}{\hbar}\int_{-\infty}^{t} \mathscr{H}(t')\, W(t')\, dt',$$
$$W(t) = W(\infty) + \frac{i}{\hbar}\int_{t}^{\infty} \mathscr{H}(t')\, W(t')\, dt', \tag{P49}$$

and on adding,

$$W(t) = 1 - \frac{i}{2\hbar}\int_{-\infty}^{\infty} \varepsilon(t - t')\, \mathscr{H}(t')\, W(t')\, dt', \tag{P50}$$

where

$$\varepsilon(t - t') = \begin{cases} 1, & t' < t, \\ -1, & r < t'. \end{cases} \tag{P51}$$

Let us define a Hermitian operator $K$ by

$$K = \frac{1}{\hbar}\int_{-\infty}^{\infty} [\mathscr{H}(t)W(t) + W^\dagger(t)\mathscr{H}(t)]\, dt - \frac{1}{\hbar}\int_{-\infty}^{\infty} W^\dagger(t)\mathscr{H}(t)W(t)\, dt$$
$$- \frac{i}{2\hbar^2}\iint_{-\infty}^{\infty} W^\dagger(t)\mathscr{H}(t)\varepsilon(t - t')\mathscr{H}(t')W(t')\, dt\, dt'. \tag{P52}$$

The variation $\delta K$ due to variations $\delta W^\dagger$ and $\delta W$ is

$$\delta K = -\frac{1}{\hbar}\int_{-\infty}^{\infty} dt[\delta W^\dagger(t)\mathscr{H}(t)I(t) + I(t)\mathscr{H}(t)\delta W(t)], \tag{P53}$$

where

$$I(t) = W(t) - 1 + \frac{i}{2\hbar}\int_{-\infty}^{\infty} \varepsilon(t - t')\, \mathscr{H}(t')\, W(t')\, dt'. \tag{P54}$$

Thus the requirement that $\delta K = 0$ for arbitrary $\delta W^\dagger$, $\delta W$ leads to $I(t) = 0$, which is (P50). The stationary value of $K$ is obtained by putting in the second term in (P52) the expression (P50) for $W(t)$,

$$K = \frac{1}{\hbar}\int_{-\infty}^{\infty} \mathscr{H}(t)\, W(t)\, dt, \tag{P55}$$

which is identical with (P47) and (P49). It is seen from (P52) that $K$ is Hermitian for arbitrary $W(t)$, and so is $K$ in (P55). With any approximate $K$ from any approximate $W(t)$ in (P55), one gets an approximate but unitary $S$ by means of (P48).

Quite analogous to the development (P37)–(P43), a variational principle can be formulated for $K$. Let us form the matrix element $(\Phi_b, K\Phi_a)$ of (P52)

$$K_{ba} = \frac{1}{\hbar}\int_{-\infty}^{\infty} dt\Big[(\exp\,[i(E_a - H_0)t]\, W(t)\,\Phi_b, V\Phi_a) + (\Phi_b, V\exp\,[i(E_b - H_0)t]\, W(t)\,\Phi_a)\Big]$$
$$- \frac{1}{\hbar}\int_{-\infty}^{\infty} dt\,(\exp\,(-iH_0t)\, W(t)\,\Phi_b, V\exp\,(-iH_0t)\, W(t)\,\Phi_a)$$
$$- \frac{i}{2\hbar^2}\int_{-\infty}^{\infty} dt\int_{-\infty}^{t} dt'\,(\exp\,(-iH_0t)\, W(t)\,\Phi_b, V\exp\,[-iH_0(t-t')]\, V\exp\,(-iH_0t')\, W(t')\,\Phi_a)$$
$$+ \frac{i}{2\hbar^2}\int_{-\infty}^{\infty} dt\int_{t}^{\infty} dt'\,(\exp\,(-iH_0t)\, W(t)\,\Phi_b, V\exp\,[-iH_0(t-t')]\, V\exp\,(-iH_0t')\, W(t')\,\Phi_a), \quad \text{(P56)}$$

where, as in (P37), $V$ contains the $e^{-\varepsilon|t|}$ of (P25). On writing

$$\exp\,(-iH_0t)\, W(t)\,\Phi_a = \exp\,(-iE_at)\,\Psi_a^{(1)},$$

and integrating over time, we obtain

$$K_{ba} = 2\pi\delta(E_b - E_a)\Big\{(\Psi_a^{(1)}, V\Phi_a) + (\Phi_b, V\Psi_a^{(1)}) - (\Psi_b^{(1)}, V\Psi_a^{(1)}) + \left(\Psi_b^{(1)}, VP\left(\frac{1}{E - H_0}\right) V\Psi_a^{(1)}\right)\Big\} \quad \text{(P57)}$$

By analogous calculations as in (P38), (P39), the requirement $\delta K_{ba} = 0$ for arbitrary $\delta\Psi_b^{(1)}$, $\delta\Psi_a^{(1)}$ leads to the equation

$$\Psi_b^{(1)} = \Phi_a + P\left(\frac{1}{E_a - H_0}\right) V\Psi_a^{(1)}. \quad \text{(P58)}$$

Here the meaning of the operator $P[1/(E_a - H_0)]$ is "taking the principal part of the integral", and written out explicitly as in (B21), we have

$$P\left(\frac{1}{E_a - H_0}\right)\chi(\mathbf{r}) = \lim_{\varepsilon\to 0}\frac{1}{2}\int dE'\left[\frac{1}{E_a - E' + i\varepsilon} + \frac{1}{E_a - E' - i\varepsilon}\right] \times \Phi_{E'}(\mathbf{r})\int \Phi_{E'}^{*}(\mathbf{r}')\,\chi(\mathbf{r}')\,d\mathbf{r}'. \quad \text{(P59)}$$

Thus $\Psi_a^{(1)}$ in (P58), instead of an outgoing wave $\Psi_a^{(+)}$ or an incoming wave $\Psi_a^{(-)}$, corresponds to a "standing" wave.

The stationary value of $K_{ba}$ is, from (P59) and (P58),

$$K_{ba} = 2\pi\delta(E_b - E_a)\,\mathbf{K}_{ba}, \quad \text{(P60)}$$

where

$$\mathbf{K}_{ba} = (\Phi_b, V\Psi_a^{(1)}) = (\Psi_b^{(1)}, V\Phi_a). \quad \text{(P61)}$$

Here, the $\mathbf{K}_{ba}$ is defined only for the same energy shell $E_b = E_a = E$, just as the $\mathbf{T}_{ba}$ in (P35) and (P36).

## 5. CROSS SECTION FOR THE SCATTERING BY A CENTRAL FIELD

The general, formal theory can be illustrated by applying it to the case of the scattering by a central field.

Still quite generally, from (P27) and (P48), we have

$$T + \tfrac{1}{2}iKT = -iK. \tag{P62}$$

Taking the $(\Phi_b, \Phi_a)$ matrix element and using (P40) and (P60), we obtain#

$$\mathbf{T}_{ba} + i\pi \sum_c \mathbf{K}_{bc}\delta(E_c - E)\,\mathbf{T}_{ca} = \mathbf{K}_{ba}, \tag{P63}$$

where $E = E_a = E_b$. Let $f_{bA}$ be the eigenfunctions of $\mathbf{K}$ and $\mathbf{K}_A$ the eigenvalues, so that they are given by

$$\sum_a \mathbf{K}_{ba}\delta(E_a - E)\,f_{aA} = \mathbf{K}_A f_{bA}, \tag{P64}$$

with

$$\sum_a f^*_{aA}\delta(E_a - E)\,f_{aB} = \delta_{AB}. \tag{P65}$$

From (P64),

$$\mathbf{K}_{ba} = \sum_A f_{bA}\mathbf{K}_A f^*_{aA}. \tag{P66}$$

From (P62), it is seen that $\mathbf{TK} = \mathbf{KT}$ so that $f_{bA}$ are also the eigenfunctions

---

# This equation is known as Heitler's equation. See the following subsection 6. (P63) can also be obtained as follows. From (P41),

$$\mathbf{T}_{ba} = (\Phi_b, V\Psi_a^{(+)}),$$

and from (P33) and (P58), one obtains by (B22) and (P41),

$$\begin{aligned}\Psi_a^{(+)} - \Psi_a^{(1)} &= \frac{P}{E_a - H_0} V(\Psi_a^{(+)} - \Psi_a^{(1)}) + \left[\frac{1}{E_a - H_0 + i\varepsilon} - \frac{P}{E_a - H_0}\right] V\Psi_a^{(+)} \\ &= \frac{P}{E_a - H_0} V(\Psi_a^{(+)} - \Psi_a^{(1)}) - i\pi \sum \Phi_c\delta(E_a - E_c)\,T_{ca}\end{aligned}$$

and by (P58) again for $\Phi_c$,

$$\left(1 - \frac{P}{E_a - H_0} V\right)[\Psi_a^{(+)} - \Psi_a^{(1)} + i\pi \sum_c \Psi_c^{(1)}\delta(E_a - E_c)\,T_{ca}] = 0.$$

Unless the operator $[P/(E_a - H_0)]\,V$ has the eigenvalue 1, the expression within the square brackets must vanish. On putting the $\Psi_a^{(+)}$ given by this equation into (P41), one obtains (P63).

of $T$. The eigenvalues $\mathbf{T}_A$ are related to $\mathbf{K}_A$ by (P63) in the representation in which $\mathbf{K}$, $\mathbf{T}$ are diagonal

$$\mathbf{T}_A = \frac{\mathbf{K}_A}{1 + i\pi\mathbf{K}_A}. \tag{P67}$$

From (P63) and (P66), we obtain

$$\mathbf{T}_{ba} = \sum_A f_{bA}\mathbf{T}_A f^*_{aA}. \tag{P68}$$

Since $\mathbf{K}$ is a Hermitian matrix, the eigenvalues $\mathbf{K}_A$ are real and we may write#

$$\mathbf{K}_A = -\frac{1}{\pi}\tan\delta_A, \tag{P69}$$

where $\delta_A$ is real. From (P67), we then have

$$\mathbf{T}_A = -\frac{1}{\pi}\sin\delta_A\exp(i\delta_A). \tag{P70}$$

The transition probability $w_{ba}$ in (P35) is, on using (P68) and (P70)

$$w_{ba} = \frac{2}{\pi\hbar}\left|\sum_A \sin\delta_A\exp(i\delta_A)f_{bA}f^*_{aA}\right|^2\delta(E_a - E_b). \tag{P71}$$

The total transition probability from a given state $\Phi_a$ to all states $\Phi_b$ in the neighborhood of $E_b = E_a$ is, in view of (P65),

$$\sum_b w_{ba} = \frac{2}{\pi\hbar}\sum_A \sin^2\delta_A|f_{bA}|^2. \tag{P72}$$

Now for the special case of the scattering of a particle by a central field, we may take $f_{aA}$ to be spherical harmonics $Y_{lm}(\vartheta_a, \varphi_a)$ which are the eigenfunctions of a central field. Here $\vartheta_a$, $\varphi_a$ are the polar angles of the wave vector $\mathbf{k}$ in the state $\Phi_a$, and $l$, $m$ are the indices $A$ so that $\delta_A = \delta_{lm}$ ($=\delta_l$, on account of spherical symmetry). Thus we put

$$f_{aA} = CY_{lm}(\vartheta_a, \varphi_a). \tag{P73}$$

If $\rho_E$ is the number of states per unit energy range in (O17),

$$\rho_E = \frac{L^3}{(2\pi\hbar)^3}mp = \frac{L^3k^2}{8\pi^3\hbar v}, \tag{P74}$$

the normalization of $f_{aA}$ by (P65), i.e.,

$$|C|^2\int \rho_E Y^*_{lm}(\vartheta, \varphi)\,Y_{l'm'}(\vartheta, \varphi)\sin\vartheta\,d\vartheta\,d\varphi = \delta_{ll'}\delta_{lm'},$$

gives

$$|C|^2\rho = 1. \tag{P75}$$

---

# With (P69), one obtains for the scattering matrix $S$ (P48)

$$S = \exp(2i\delta_A). \tag{P69a}$$

From (P71), we obtain for the probability that a particle originally in the direction $\mathbf{k}_a(\vartheta_a, \varphi_a)$ is scattered into $\mathbf{k}_b(\vartheta_b, \varphi_b)$,

$$w_{ba} = \frac{2}{\pi\hbar}\left|\sum_{lm} \sin\delta \exp(-\delta_l)|C|^2\, Y_{lm}(\vartheta_b, \varphi_b)\, Y^*_{lm}(\vartheta_a, \varphi_a)\right|^2 \rho_E \sin\vartheta_b\, d\vartheta_b\, d\varphi_b.$$

Using the relation

$$\sum_{m=-l}^{l} Y_{lm}(\vartheta_1, \varphi_1)\, Y^*_{lm}(\vartheta_2, \varphi_2) = \frac{2l+1}{4\pi} P_l(\cos\vartheta), \tag{P76}$$

where $\vartheta$ is now the angle between $\mathbf{r}_1(\vartheta_1, \varphi_1)$ and $\mathbf{r}_2(\vartheta_2, \varphi_2)$, and (P74), (P75), we have

$$w_{ba} = \frac{v}{L^3k^2}\left|\sum_l (2l+1)\sin\delta_l \exp(i\delta_l)\, P_l(\cos\vartheta)\right|^2 d\cos\vartheta\, d\varphi.$$

The differential cross section $d\sigma$ is, by (O19),

$$d\sigma = \frac{w_{ba}}{v} L^3 = \left|\frac{1}{2ik}\sum (2l+1)[\exp(2i\delta_l) - 1]\, P_l(\cos\vartheta)\right|^2 d\cos\vartheta\, d\varphi \tag{P77}$$

and the total cross section, either from (P72) or from (P77), is

$$\sigma = \frac{4\pi}{k^2}\sum (2l+1)\sin^2\delta_l \tag{P78}$$

which are (A17), (A18) of Chapter 1.

## 6. *K* MATRIX AND "EQUATION OF RADIATION DAMPING" OF HEITLER

In (P60), the matrix $K$, or Heitler's matrix, is given by

$$K_{ba} = 2\pi\delta(E_b - E_a)\,\mathbf{K}_{ba}, \tag{P60}$$

where

$$\mathbf{K}_{ba} = (\Phi_b, V\Psi'^{(1)}_a) = (\Psi'^{(1)}_b, V\Phi_a), \tag{P61}$$

where $\Phi_a$, $\Phi_b$ are the eigenstates of $H_0$ and $\Psi'^{(1)}_a$, $\Psi'^{(1)}_b$ the standing-wave eigenstates of $H$ given in (P58). The $\mathbf{K}_{ba}$ are related to the $\mathbf{T}_{ba}$ of (P42) by the relation (P63)

$$\mathbf{T}_{ba} = \mathbf{K}_{ba} - i\pi\sum_c \mathbf{K}_{bc}\delta(E_c - E)\,\mathbf{T}_{ca}. \tag{P63}$$

$\mathbf{T}_{ba}$, $\mathbf{K}_{ba}$ are defined only for states $b$, $a$ of equal energies, or, "on the same energy shell" $E_a = E_b = E$. (P63) is an exact equation; it involves the transition matrix elements between states of equal energy, which correspond to real, observable scattering processes.

One may introduce the off-energy shell matrix $K$ by the definition#

$$K_{\beta\alpha} = (\Phi_\beta, V\Psi_\alpha^{(1)}), \quad E_\beta \neq E_\alpha. \tag{P79}$$

On using (P58) and denoting by $B_{\beta\alpha}$ the first-order approximation to $\mathbf{K}_{\beta\alpha}$

$$B_{\beta\alpha} = (\Phi_\beta, V\Phi_\alpha), \quad E_\alpha = E_\beta, \tag{P80}$$

(P79) becomes

$$\begin{aligned} K_{\beta\alpha} &= B_{\beta\alpha} + \left(\Phi_\beta, V\frac{P}{E_a - H_0} V\Psi_a^{(1)}\right), \\ &= B_{\beta a} + P\sum_\alpha \frac{B_{\beta\alpha}K_{\alpha a}}{E_a - E_\alpha}, \quad E_\beta \neq E_a. \end{aligned} \tag{P81}$$

As shown in the footnote following (P38), any approximate but Hermitian $K$ will lead to a unitary $S$-matrix. In Heitler's theory of radiation damping (in an attempt to avoid the divergence difficulties in quantum electrodynamics before the development of the covariant formulation and the renormalization technique), $\mathbf{K}_{ba}$ is replaced by its first (non-vanishing) approximation, such as $\mathbf{B}_{ba}$ of (P80), leading by (P63) to the equation

$$\mathbf{T}_{ba} = B_{ba} - i\pi \sum_c B_{bc}\delta(E_c - E)\,\mathbf{T}_{ca}, \tag{P82}$$

where $E_b = E_a = E$. The solution of this equation leads to finite results and contains *some* correct higher order effects. But it neglects the higher order terms of the *off-energy shell K* matrix elements (which appear in the second term in (P81) and represent the "virtual" processes), and this neglect is unjustifiable in some cases.

It has been suggested that, in place of the integral equation (P33, 33a), one might obtain good results by solving (P81) by approximation methods (iteration) and solving (P63) by exact methods, if possible. The advantage of

---

# The off-energy shell matrix $K_{\beta a}$ is not Hermitian, as can be seen as follows. From (P61) and (P58), we have

$$\begin{aligned} K_{\beta a} &= (\Psi_\beta^{(1)}, V\Psi_a^{(1)}) - \left(\frac{P}{E_\beta - H_0} V\Psi_\beta^{(1)}, V\Psi_a^{(1)}\right) \\ &= \left(V\Psi_\beta^{(1)}, \left[\Psi_a^{(1)} - \frac{P}{E_a - H_0} V\Psi_a^{(1)}\right]\right) \\ &\qquad + \left(V\Psi_\beta^{(1)}, \left[\frac{P}{E_a - H_0} - \frac{P}{E_\beta - H_0}\right] V\Psi_a^{(1)}\right) \\ &= K_{\beta a}^{+} + (E_\beta - E_a)\left(\Psi_\beta^{(1)}, V\frac{P}{(E_a - H_0)(E_\beta - H_0)} V\Psi_a^{(1)}\right), \end{aligned}$$

i.e., $K_{\beta a} \neq K_{\beta a}^{+}$. The on-the-shell $\mathbf{K}_{ba}$ matrix is Hermitian, however. We shall use the notation $E_a = E_b = E_c \neq E_\alpha = E_\beta$.

replacing (P33, 33a) by the pair of equations (P63), (P81) is that, apart from a possible better convergence in the iteration of (P81) than of (P33, 33a), the pair (P63), (P81) brings out the separation of the real and the virtual processes clearly. On this view of taking (P63), (P81) as the basic equations, it is of interest to have a derivation of these equations "independent" of the results of the preceding sections. It has been shown by Goldberger that they can be obtained from variational principles (in the time-independent formalism). Thus the requirement that

$$\delta \mathbf{I}_{ba} = \delta \frac{(\mathbf{KT})_{ba}(\mathbf{TK})_{ba}}{(\mathbf{TT})_{ba} + i\pi(\mathbf{TKT})_{ba}} = 0 \tag{P83}$$

for arbitrary $\delta\mathbf{T}$ leads to (R62), with the stationary value of $\mathbf{I}_{ba}$

$$\mathbf{I}_{ba} = \frac{i}{\pi}(\mathbf{T}_{ba} - \mathbf{K}_{ba}). \tag{P84}$$

In (P83), the matrix multiplication for matrices defined only on the same energy shell is given by

$$(\mathbf{KT})_{ba} = \sum_c \mathbf{K}_{bc}\delta(E - E_c)\,\mathbf{T}_{ca}, \quad E_a = E_b = E. \tag{P85}$$

The equation (P81) is given by the variational principle

$$d\mathbf{J}_{ba} = 0, \quad \text{for arbitrary } \delta K,\ \delta K^+, \tag{P86}$$

$$\mathbf{J}_{ba} = -\frac{(K^+B)_{ba}(BK)_{ba}}{(K^+K)_{ba} - (K^+BK)_{ba}}, \tag{P86a}$$

with the stationary value

$$\mathbf{J}_{ba} = -\sum_\alpha \frac{B_{b\alpha}K_{\alpha a}}{E - E_\alpha} \overset{\text{(P81)}}{=\!=} \mathbf{B}_{ba} - \mathbf{K}_{ba}. \tag{P87}$$

In (P86a), the multiplication is defined by

$$(K^+B)_{ba} = P\sum_\alpha \frac{K^+_{b\alpha}B_{\alpha a}}{E - E_\alpha}, \text{ etc.} \tag{P88}$$

From (P87), it is seen that the variational principle (P86), (P86a) leads to the stationary value of $\mathbf{K}_{ba}$ which enters in (P63). Thus the variational formulation above has attractive features for applications to actual problems. For further discussions of the theory and applications to examples, the reader is referred to the work of Goldberger (ref. 2).

## REFERENCES

1. Lippmann, B. A. and Schwinger, J., Phys. Rev. 79, 469 (1950); The subsections 1–5 are based on this work which also contains an application of the variational principle [in terms of $T_{ab}$ in (P38)] to the scattering of a

neutron by a proton bound in a molecule, leading to Fermi's result [Ricerca Scient. VII–II, 13 (1936)] and that of Breit *et al.* [Breit, G., Phys. Rev. **71**, 215 (1947); Breit and Zilsel, P. R., Phys. Rev. **71**, 232 (1947); Breit, Zilsel and Darling, Phys. Rev. **72**, 576 (1947)] in successive approximations.

**2.** Goldberger, M. L., Phys. Rev. **82**, 757 (1951); **84**, 929 (1951), on which subsection 6 is based. An earlier work on variational method in this connection is Ma, S. T. and Hsueh, C. F., Phys. Rev. **67**, 303 (1945).

For Heitler's theory of radiation damping, see:

**3.** Heitler, W., Proc. Camb. Phil. Soc. **37**, 291 (1941); Wilson, A. H., Proc. Camb. Phil. Soc. **37**, 301 (1941); Ma, S. T. and Hsueh, C. F., Proc. Camb. Phil. Soc. **40**, 167 (1944); also a review article by Wentzel, G., Rev. Mod. Phys. **19**, 1 (1947).

**4.** Yang, Tse-sen, refs. 5 and 6, Sect. Q.

section

# Q Time-Dependent Theory of Scattering: Treatment of Gell-Mann and Goldberger

The theory of scattering on the basis of the time-dependent Schrödinger equation has been further developed by Gell-Mann and Goldberger (ref. 1). To present their different points of view, the following account is given in spite of some repetitions with the treatments given in the preceding two sections. To simplify writing, we shall put $\hbar = 1$ here.

## 1. TRANSITION PROBABILITY AND ADIABATIC SWITCHING

The Schrödinger equation (P1)

$$i\frac{\partial \Psi}{\partial t} = (H_0 + V)\Psi, \tag{Q1}$$

is transformed, as in (O35, 36, 37), to the interaction picture

$$\Psi'(t) = \exp(iH_0 t)\,\Psi(t), \tag{Q2}$$

$$i\frac{\partial \Psi'(t)}{\partial t} = \mathscr{H}(t)\,\Psi'(t), \tag{Q3}$$

$$\mathscr{H}(t) = \exp(iH_0 t)\, V \exp(-iH_0 t). \tag{Q4}$$

Let the equation (Q1) when $V = 0$ be

$$i\frac{\partial \Phi(t)}{\partial t} = H_0 \Phi(t) \tag{Q5}$$

and its solution

$$\Phi_j(t) = \exp(-iE_j t)\,\phi_j, \tag{Q5a}$$

where on account of $H = H_0 + V$,

$$(H - E_j)\,\phi_j = V\phi_j. \tag{Q6}$$

$\Phi_j(t)$ describes the stationary state when the two colliding components of the whole system $H$ are at infinitely great spatial separations, such as at an initial instant $t_0$ in the distant past ($t_0$ will eventually be made to become $-\infty$).

The scattering problem now is the following: starting with the system in the state $\Phi_a(t_0)$ at time $t_0$ in the distant past, to find the probability $w_{ba}(t)$ that, at time $t$, the system is in state $\Phi_b(t)$. To find this probability, it is noted that the formal solution of (Q1) is

$$\Psi_j(t) = \exp\,[-iH(t - t_0)]\,\Phi_a(t_0). \tag{Q7}$$

The transition amplitude is the inner product

$$f_{ba}(t) = \langle\Phi_b(t)|\Psi_a(t)\rangle,$$

where the $\Phi_b(t) = \exp\,[-iE_b t]\,\phi_b$ is normalized to unity in a (large) volume $L^3$, and $\Psi_a(t)$ according to

$$\langle\Psi_a(t)|\Psi_a(t)\rangle = N_a = \text{constant}.$$

The probability $w_{ba}(t)$ is then

$$w_{ba}(t) = |f_{ba}(t)|^2\,N_a^{-1}. \tag{Q8}$$

Now the equation (Q7) above implies that at $t_0$ [when the system is in the state $\Phi_a(t_0)$ of $H_0$], the system is suddenly subjected to the interaction $V$. A more physical picture and a better mathematical formulation of the collision problem is to require that, as $t_0 \to -\infty$, the Hamiltonian $H$ approaches $H_0$, and $\Psi_a(t)$ approaches the stationary state $\phi_a$, asymptotically. In this view, it is convenient to introduce the artifice of assuming the interaction $V$ to be "switched on" adiabatically from $V \to 0$ as $t_0 \to -\infty$ to the full strength $V$ at $t = 0$, and to $V \to 0$ as $t \to +\infty$. The $\Psi_a(t)$ at $t$ is then the result of the $\exp\,[-iH(t - t_0)]\,\Phi_a(t_0)$ of (Q7) for all $t_0$ from $-\infty$ to 0. This is most conveniently represented by the average

$$\Psi_a(t) = \varepsilon \int_{-\infty}^{0} dt_0 \exp(\varepsilon t_0) \exp\,[-iH(t - t_0)]\,\Phi_a(t_0), \qquad \varepsilon > 0,$$

$$\overset{\text{(Q5a)}}{=\!=} \exp(-iHt)\,\varepsilon \int_{-\infty}^{0} dt_0 \exp(\varepsilon t_0) \exp\,[i(H - E_a)t_0]\,\phi_a, \tag{Q9}$$

$$= \exp(-iHt)\,\frac{i\varepsilon}{E_a - H + i\varepsilon}\,\phi_a. \tag{Q9a}$$

Here the time $\varepsilon^{-1}$ is allowed to become very long (i.e., $\varepsilon \to 0$), but obviously must not be longer than the time for the colliding components of the system to cover the length of the normalization volume, i.e., $\varepsilon^{-1} < L/v$, $v$ being the relative velocity of the colliding components.# Thus the limiting process $\varepsilon \rightarrow 0$ is to be accompanied by that of $L \rightarrow \infty$ with the condition $1/\varepsilon L < 1/v$, and consequently $\lim\limits_{\substack{\varepsilon \to 0 \\ L \to \infty}} \dfrac{1}{\varepsilon L^3} = 0$. The "adiabatic switching" $\varepsilon \rightarrow 0$ and the normalization of $\phi_a$ in (Q9a) are to be understood in this sense.

From (Q9a), for $t = 0$, we obtain

$$\Psi_a^{(+)}(0) = \frac{-V + i\varepsilon + V}{E_a - H + i\varepsilon}\phi_a \overset{(Q6)}{=\!=} \phi_a + \frac{1}{E_a - H + i\varepsilon} V\phi_a, \qquad \text{(Q10)}$$

which can also be put in the form

$$\Psi_a^{(+)} = \phi_a + \frac{1}{E_a - H_0 + i\varepsilon} V\Psi_a^{(+)}. \qquad \text{(Q10a)}$$

In an entirely analogous manner, we can obtain

$$\Psi_a^{(-)} = \phi_a + \frac{1}{E_a - H_0 - i\varepsilon} V\Psi_a^{(-)}. \qquad \text{(Q11)}$$

$\Psi_a^{(+)}$, $\Psi_a^{(-)}$ are, respectively, the outgoing and ingoing wave eigenfunction of $H$ corresponding to the initial wave $\phi_a$. (Q10a), (Q11) are Equations (P33) and (P33a) obtained before.

The transition amplitude $f_{ba}(0)$ of (Q8) is, from (Q10a),

$$f_{ba}(0) = \langle\phi_b|\phi_a\rangle + \langle\phi_b| \frac{1}{E_a - H_0 + i\varepsilon} V\Psi_a^{(+)}\rangle$$

$$= \delta_{ba} + \frac{1}{E_a - E_b + i\varepsilon}\mathbf{T}_{ba},$$

where

$$\mathbf{T}_{ba} = \langle\phi_b, V\Psi_a^{(+)}(0)\rangle \qquad \text{(Q12)}$$

which is the same expression as (P36).

Gell-Mann and Goldberger give a derivation of the transition probability per unit time which is somewhat different from the procedure of Lippmann and Schwinger described in (P33)–(P36). We shall refer the reader to ref. 1 for further important details of the theory.

---

# If, instead of the eigenstates of $H_0$, we represent the spatially separated colliding components of the system in the initial state by wave packets, then $\varepsilon^{-1}$ is the length of the wave train divided by the group velocity $v$ (in the centre of mass system).

## 2. UNITARY OPERATOR, WAVE OPERATOR AND SCATTERING OPERATOR

Let $U(t, t_0)$ be the unitary (for finite times $t$, $t_0$) operator of (O39),

$$\Psi'(t) = U(t, t_0)\,\Psi'(t_0). \tag{Q13}$$

The properties (O38), (O40a, b, c) are

$$U(t, t_0)\,U^\dagger(t, t_0) = 1, \tag{Q14}$$

$$U(t, t_1)\,U(t_1, t_0) = U(t, t_0), \tag{Q15}$$

$$U(t, t_0)\,U(t_0, t) = 1, \tag{Q15a}$$

$$U(t, t) = 1, \tag{Q15b}$$

$$U^\dagger(t, t_0) = U(t_0, t). \tag{Q14a}$$

From (Q1) one has

$$\Psi(t) = \exp\left[-iH(t - t_0)\right]\Psi(t_0).$$

From this and (Q2), (Q12), one has

$$U(t, t_0) = \exp(iH_0t)\exp\left[-i(H_0 + V)(t - t_0)\right]\exp(-iH_0t_0). \tag{Q16}$$

Equations (O41), (O42) and (O45) are

$$i\,\frac{\partial U(t, t_0)}{\partial t} = \mathscr{H}(t)\,U(t, t_0) \tag{Q17}$$

$$U(t, t_0) = 1 - i\int_{t_0}^{t} dt'\,\mathscr{H}(t')\,U(t', t_0) \tag{Q18}$$

$$U(t, t_0) = 1 + i\int_{t}^{t_0} dt'\,U(t, t')\,\mathscr{H}(t'), \quad t_0 < t, \tag{Q19}$$

(Q18) can be solved by iteration as in (O51) and (O52).

The operators of great interest are $U(t, -\infty)$, $U(\infty, t)$, $U(\infty, -\infty)$, which are used in Lippmann and Schwinger's work in the preceding section. Some care is needed to pass from finite to infinite times.# If we define $U(t, -\infty)$ by the limit

$$U(t, -\infty) = \lim_{\varepsilon\to 0+}\int_{-\infty}^{\infty} dt'\,e^{\varepsilon t'}\,U(t, t'), \tag{Q20}$$

$$U(\infty, t) = \lim_{\varepsilon\to 0+}\int_{0}^{\infty} dt'\,e^{-\varepsilon t'}\,U(t', t), \tag{Q21}$$

---

# From the unitarity of $U(t, t_0)$ for finite times, it does not follow that $U(t, \infty)$, $U(t, -\infty)$ are unitary. See ref. 2, footnote 9a. A different treatment of $U$ in the limit of infinite times has been given by T. S. Yang (ref. 5).

then it can be shown that the following expressions are valid:

$$U(t, -\infty) = 1 - i\int_{-\infty}^{t} dt' \mathscr{H}(t')\, U(t', -\infty), \tag{Q22}$$

$$U(\infty, -\infty) = 1 - i\int_{-\infty}^{\infty} dt' \mathscr{H}(t')\, U(t', -\infty). \tag{Q23}$$

From (Q12), it is seen that $U(\infty, -\infty)$ is the $S$ matrix of (P22),

$$U(\infty, -\infty) = S. \tag{Q24}$$

We shall now relate the $U(t, -\infty)$, $U(0, -\infty)$ to the operators in (Q9) and (Q10). From the completeness relation for the eigenstates of $H_0$,

$$\int_j \phi_j\rangle\, dj \langle\phi_j = 1, \tag{Q25}$$

where the integration is taken over the complete set $\phi_j$, we have, on operating by $U(t, -\infty)$, by (Q16) and (Q20),

$$\begin{aligned} U(t, -\infty) &= \lim_{\varepsilon \to 0+} \int_{-\infty}^{\infty} \int dj\, dt_0 \exp(\varepsilon t_0) \\ &\quad \times \exp[iH_0 t - i(H_0 + V)(t - t_0) - iH_0 t_0]\, \phi_j\rangle\langle\phi_j \\ &\overset{(Q8)}{=\!=} \exp[iH_0 t - i(H_0 + V)t] \lim_{\varepsilon \leftarrow 0+} \int_j dj \frac{i\varepsilon}{E_j - H + i\varepsilon} \phi_j\rangle\langle\phi_j. \end{aligned} \tag{Q26}$$

From this and (Q10), we have

$$U(0, -\infty) = \int_j dj \Psi_j^{(+)}\rangle\langle\phi_j, \tag{Q27}$$

or

$$U(0, -\infty)|\phi_j\rangle = |\Psi_j^{(+)}\rangle. \tag{Q28}$$

The operator $U(0, -\infty)$ is one that, operating on $\phi_j$, gives the eigenstate of the total Hamiltonian $H = H_0 + V$ corresponding to outgoing wave condition and $\phi_j$ as the incident wave. It is identical with the wave matrix $\Omega_{(+)}$ of Møller (ref. 4)

$$U(0, -\infty) = \Omega_{(+)}. \tag{Q29}$$

Note that (Q27) follows directly from (O46) and (Q20).

In a parallel calculation, one obtains

$$U(0, +\infty)|\phi_j\rangle = |\Psi_j^{(-)}\rangle, \tag{Q30}$$

where $|\Psi_j^{(-)}\rangle$ is that given in (Q11). $U(0, +\infty)$ is identical with wave matrix

$$U(0, +\infty) = \Omega_{(-)}. \tag{Q31}$$

From (Q14) and (Q21) it is seen that

$$U(\infty, t) = U(\infty, 0)\, U(0, t). \tag{Q32}$$

From (Q20),

$$U(\infty, -\infty) = U(\infty, 0)\, U(0, -\infty). \tag{Q33}$$

From the unitarity of $U(t, t_0)$ for finite $t$, $t_0$, (Q13a),

$$U^{\dagger}(t, t_0) = U(t_0, t),$$

and by means of (Q20) or (Q21), it is readily seen that

$$U(\infty, 0) = U^{\dagger}(0, \infty), \tag{Q34}$$

$$U(-\infty, 0) = U^{\dagger}(0, -\infty), \tag{Q35}$$

so that the $S$ given by (Q24) and (Q33) can be expressed in terms of the $\Omega$ in (Q29) and (Q31),

$$S = \Omega^{\dagger}_{(-)}\Omega_{(+)}. \tag{Q36}$$

From (Q29) and (Q35), one has

$$U(-\infty, 0)\, U(0, -\infty) = \Omega^{\dagger}_{(+)}\Omega_{(+)}. \tag{Q37}$$

On using (Q25) and (Q28), one has

$$\begin{aligned} \Omega^{\dagger}_{(+)}\Omega_{(+)} &= \int \phi_i\rangle\, di \langle \phi_i \Omega^{\dagger}_{(+)} | \sum_j \Omega_{(+)} \phi_j\rangle\, dj \langle \phi_j \\ &= \int \phi_i\rangle\, di \langle \Psi_i^{(+)} | \Psi_j^{(+)}\rangle\, dj \langle \phi_j \\ &= \int \phi_i\rangle\, di \langle \phi_i = 1. \end{aligned} \tag{Q38}$$

Similarly, from (Q31), (Q34), we have

$$U(\infty, 0)\, U(0, \infty) = \Omega^{\dagger}_{(-)}\Omega_{(-)} = 1. \tag{Q39}$$

On the other hand, from (Q25), (Q28), (Q30), we have

$$\begin{aligned} \Omega_{(+)}\Omega^{\dagger}_{(+)} &= \int \Omega_{(+)}\, \phi_i\rangle\, di \langle \phi_i \Omega^{\dagger}_{(+)} \\ &= \int \Psi_i^{(+)}\rangle\, di \langle \Psi_i^{(+)}, \end{aligned} \tag{Q40}$$

$$\Omega_{(-)}\Omega^{\dagger}_{(-)} = \int \Psi_i^{(-)}\rangle\, di \langle \Psi_i^{(-)}.$$

If there are no bound states in $H$, the $\Psi_i^{(+)}$ ($\Psi_i^{(-)}$) then form a complete set and the sums in (Q40) are then unity. In this case, the $\Omega_{(+)}$, $\Omega_{(-)}$ are unitary. If there are bound states $\Psi_\alpha$, the $\Psi_i^{(+)}$ ($\Psi_i^{(-)}$) will not span a complete set. In this case we have, from (Q40),

$$\Omega_{(\pm)}\Omega^{\dagger}_{(\pm)} = 1 - \sum_\alpha \Psi_\alpha\rangle \langle \Psi_\alpha, \tag{Q41}$$

so that $\Omega\Omega^{+}$ is not unity and $\Omega_{(+)}$, $\Omega_{(-)}$ are not unitary.

The $S$ matrix can be shown, however, to be unitary. From (Q36), we have

$$S^\dagger = \Omega^\dagger_{(+)}\Omega_{(-)},$$

and, using (Q41),

$$\begin{aligned} S^\dagger S &= \Omega^\dagger_{(+)}\Omega_{(-)}\Omega^\dagger_{(-)}\Omega_{(+)} = \Omega^\dagger_{(+)}[1 - \sum \Psi_\alpha\rangle\langle\Psi_\alpha]\Omega_{(+)} \\ &= 1 - \Omega^\dagger_{(+)}[\sum_\alpha \Psi_\alpha\rangle\langle\Psi_\alpha]\Omega_{(+)}. \end{aligned} \tag{Q42}$$

On taking the $|\phi_j\rangle$ representative of this and using $\Omega_{(+)}|\phi_j\rangle = \Psi_j^{(+)}\rangle$ from (Q28), (Q29), we have $\langle\Psi_\alpha|\Psi_j^{(+)}\rangle = 0$ on account of the orthogonality of a bound state $\Psi_\alpha$ and a continuum state. Since the $\phi_j$ form a complete set, we have

$$S^\dagger S = 1. \tag{Q43}$$

By an entirely similar argument, one proves

$$SS^\dagger = 1. \tag{Q43a}$$

This result—the unitary of $S$—has been proved in (P23). It has been established in the time-independent theory by Møller before. We shall come back to this property of $S$ in Sect. W.

## 3. SCATTERING BY TWO POTENTIALS

### *i) Transition matrix element*

There are many physical situations in which one has to consider the action of two potentials. Examples are: (1) proton–nucleus scattering, with the nuclear and the Coulomb potential, (2) Bremsstrahlung with the Coulomb interactions and the interaction of the charged particles with the radiation field, (3) photo-electric effect, with the Coulomb interaction and the electron–radiation field interaction. As an example of the theory above, we shall treat the general problem where the Hamiltonian is

$$H = H_0 + U + V. \tag{Q44}$$

In the following, we shall use Equation (Q10a)

$$\Psi_a^{(+)} = \phi_a + \frac{1}{E - H_0 + i\varepsilon}(U + V)\Psi_a^{(+)}, \tag{Q45}$$

where $\phi_a$ is an eigenstate of $H_0$ of energy $E$, and $\Psi_a^{(+)}$ is an eigenstate of $H$ with outgoing wave condition and $\phi_a$ as incident wave.

The transition matrix element from a state $\phi_a$ to another state $\phi_b$ is given by (Q12)

$$\mathbf{T}_{ba} = (\phi_b, (U + V)\Psi_a^{(+)}). \tag{Q46}$$

Let $\chi_b^{(\pm)}$ be the eigenfunction of $H_0 + U$ for $\left\{\begin{matrix}\text{outgoing}\\ \text{ingoing}\end{matrix}\right\}$ wave condition and with $\phi_b$ (eigenfunction of $H_0$) as incident wave, i.e.,

$$\chi_b^{(\pm)} = \phi_b + \frac{1}{E - H_0 \pm i\varepsilon} U\chi_b^{(\pm)}. \tag{Q47}$$

Using (Q47) and (Q45) in (Q46), we find

$$\begin{aligned}
\mathbf{T}_{ba} &= \left(\chi_b^{(-)} - \frac{1}{E - H_0 - i\varepsilon} U\chi_b^{(-)}, (U + V)\Psi_a^{(+)}\right) \\
&= (\Psi_b^{(-)}, U\Psi_a^{(+)}) + (\Psi_b^{(-)}, V\Psi_a^{(+)}) \\
&\qquad - \left(\frac{1}{E - H_0 - i\varepsilon} U\chi_b^{(-)}, (E - H_0 + i\varepsilon)(\Psi_a^{(+)} - \phi_a)\right) \\
&= (\chi_b^{(-)}, U\Psi_a^{(+)}) + (\chi_b^{(-)}, V\Psi_a^{(+)}) - (U\chi_b^{(-)}, \Psi_a^{(+)}) + (U\chi_b^{(-)}, \phi_a) \\
&= (\chi_b^{(-)}, V\Psi_a^{(+)}) + (\chi_b^{(-)}, U\phi_a).
\end{aligned} \tag{Q48}$$

By (P41), this is equivalent to

$$\mathbf{T}_{ba} = (\chi_b^{(-)}, V\Psi_a^{(+)}) + (\phi_b, U\chi_a^{(+)}). \tag{Q48a}$$

Equation (Q45) can be rearranged into

$$(E - H_0 - U)\Psi_a^{(+)} = V\Psi_a^{(+)},$$

or

$$\Psi_a^{(+)} = \chi_a^{(+)} + \frac{1}{E - H_0 - U + i\varepsilon} V\Psi_a^{(+)}. \tag{Q49}$$

It can also be verified directly that

$$\Psi_a^{(+)} = \phi_a + \frac{1}{E - H_0 - U - V + i\varepsilon}(U + V)\phi_a, \tag{Q50}$$

and

$$\chi_a^{(+)} = \phi_a + \frac{1}{E - H_0 - U + i\varepsilon} U\phi_a, \tag{Q51}$$

and the solution of (Q49) is

$$\Psi_a^{(+)} = \chi_a^{(+)} + \frac{1}{E - H_0 - U - V + i\varepsilon} V\chi_a^{(+)}. \tag{Q52}$$

If $U$ is treated exactly (as implied by the use of the eigenfunction $\chi_a^{(+)}, \chi_b^{(-)}$ of $H - H_0 - U$) and if $V$ is small, we may approximate $\psi_a^{(+)}$ by $\chi_a^{(+)}$ in (Q48a) and obtain, to the first order in $V$,

$$\mathbf{T}_{ba} = (\chi_b^{(-)}, V\chi_a^{(+)}) + (\phi_b, U\chi_a^{(+)}). \tag{Q53}$$

The second term in (Q48) vanishes in many cases. For example, in the

Bremsstrahlung, $U$ is the Coulomb interaction and $V$ the charged particle-radiation field interaction. In the final state, there is a photon, but the state arising from the initial state $\phi_a$ through $U$ does not contain any photon. Hence $(\chi_b^{(-)}, U\phi_a) = 0$. It does not, however, vanish in the case of the combined Coulomb and nuclear scattering of a proton–nucleus collision.

### *ii) Ingoing wave as final state*

The first term in (Q48) is interesting in that the final state $\chi_b^{(-)}$ is an "ingoing wave". Since $\psi_b^{(-)}$ asymptotically is a superposition of a plane wave $\phi_b$ and an "ingoing" or convergent spherical wave, it seems rather odd at first thought that such a state $\psi_b^{(-)}$ should be used as the final state of the transition matrix element. That this can be made understandable is recently shown by Breit and Bethe (ref. 3). The key to the situation is the effect of the interference between the incident (plane) wave and the outgoing and ingoing spherical waves already discussed by Sommerfeld. The idea can be made clear by the simple case of scattering of a particle by a central potential. The eigenstate is always chosen to be one satisfying the outgoing wave condition which in the notations of § A, Chapter 1, is, asymptotically,

$$\rightarrow e^{i\mathbf{k}\cdot\mathbf{r}} + \frac{e^{ikr}}{r} f(\vartheta).$$

In directions for which $\mathbf{k}\cdot\mathbf{r} > 0$ and $\simeq kr$, the outgoing spherical wave and the incident plane wave have the same phase. Also, as seen in (A9) and (A10), $e^{i\mathbf{k}\cdot\mathbf{r}}$ behaves asymptotically as $e^{\pm ikr}/kr$. Hence the scattered and the incident plane wave are in a position to produce approximate constructive interference. On the other hand, an ingoing wave $(e^{-ikr}/r)f(\vartheta)$ will have an opposite phase relation with the incident wave $e^{i\mathbf{k}\cdot\mathbf{r}}$ and produce destructive interference.

Now consider the question of the *final state* of the matrix element $\mathbf{T}_{ba}$ in (Q48) for the transition amplitude. In the rough, first approximation one would use simply a "plane wave" state $\phi_b$ for the final state, which as seen from (Q47), is the "leading term" of $\chi_b^{(-)}$. To take into account the "distortion" of the plane wave $\phi_b$ by the potential $U$, one must use $\chi_b^{(+)}$ or $\chi_b^{(-)}$ for the final state in $\mathbf{T}_{ba}$.

If the outgoing wave solution $\psi_b^{(+)}$ is used for an inelastic scattering, both the plane wave part $e^{ik\cdot r}$ and the $e^{ikr}$ part "constructively" interfere with the unscattered wave packet, thereby resulting in the reduction of the intensity of the incident beam. A calculation of the transition matrix element $T_{ba}$ is possible, but the interpretation in terms of a differential cross section for the scattered particle having a definite momentum is lost.

If the ingoing $\psi_b^{(-)}$ is used (i.e., a plane wave distorted by an ingoing spherical part), the part $e^{-ikr}$ interferes "destructively" with the unscattered wave packet and effectively one is describing the inelastically scattered wave in terms of undistorted plane waves. These are suitable for the interpretation of particle flux in given directions.

### *iii) Pick-up processes*

As an example illustrating the scattering by two potentials, consider a class of problems that can be typified by the so-called "pick-up process" in which an incident neutron picks up a bound proton from the target and a deuteron emerges. In (Q44), let $K$ represent the kinetic energy operator of the proton and the neutron, $U$ the interaction of the proton with the rest of the target system, $V$ the proton–neutron interaction. Let $\psi_0$ be the eigenstate of $H_0 + U$

$$(H_0 + U)\,\psi_0 = E_i\psi_0 \tag{Q54}$$

representing the initial state of a bound proton and an incident neutron. Let $\psi_f$ be the eigenstate of $H_0 + V$,

$$(H_0 + V)\,\psi_f = E_f\psi_f \tag{Q55}$$

representing the final state of an emerging deuteron (i.e., the internal state of the deuteron and the motion of its center of mass). The eigenstate of the system $\psi^{(+)}$ is given by equations similar to (Q49) or (Q52), which in the present notation are

$$\psi^{(+)} = \psi_0 + \frac{1}{E_i - H_0 - U + i\varepsilon}\,V\psi^{(+)}, \tag{Q56}$$

$$\psi^{(+)} = \psi_0 + \frac{1}{E - H_0 - U - V + i\varepsilon}\,V\psi_0. \tag{Q57}$$

Note that $\psi_0$ here is different from $\chi_a^{(\pm)}$ in (Q47) since $\psi_0$ contains the *bound* state of the proton. On account of this difference, we cannot apply (Q48) directly but have to go back to (Q46)

$$\mathbf{T}_{ba} = \mathbf{T}_{f0} = (\psi_f,\, U\psi^{(+)}), \tag{Q58}$$

where $\psi_f$ is given by (Q55) and $\psi^{(+)}$ by (Q56).

It is possible to bring (Q58) into a form similar to (Q48) or (Q53). On using (Q57) this becomes

$$\mathbf{T}_{f0} = (\psi_f,\, U\psi_0) + \left(\frac{1}{E - H_0 - U - V - i\varepsilon}\,U\psi_f,\; V\psi_0\right).$$

Let $\psi^{(-)}$ be the ingoing wave eigenfunction of $H = K + U + V$,

$$\psi^{(-)} = \psi_f + \frac{1}{E - H_0 - V - i\varepsilon}\,U\psi^{(-)} \tag{Q59}$$

which is satisfied by [similar to (Q56) and (Q57)],

$$\psi^{(-)} = \psi_f + \frac{1}{E - H_0 - U - V - i\varepsilon} U\psi_f. \tag{Q59a}$$

Hence $\mathbf{T}_{f0}$ becomes

$$\mathbf{T}_{f0} = (\psi_f, U\psi_0) + (\psi^{(-)}, V\psi_0) - (\psi_f, V\psi_0). \tag{Q60}$$

On using (Q54), (Q55), we have (with $E_i = E_f = E$)

$$(\psi_f, U\psi_0) - (\psi_f, V\psi_0) = -(\psi_f, H_0\psi_0) + (H_0\psi_f, \psi_0)$$

which vanishes on applying Green's theorem and so on. Hence (Q58) and (Q60) can be written

$$\mathbf{T}_{f0} = (\psi_f, U\psi^{(+)}) = (\psi^{(-)}, V\psi_0). \tag{Q61}$$

That (Q61) differs in form from (Q48a) is due to the *differently* defined states in the two formulas.

It will be remembered that the expression (Q61) [also (Q48) or (Q48a)] is exact. In (Q61), $\psi^{(+)}$ involves the potential $V$ according to (Q56) or (Q57), and the $\psi^{(-)}$ involves $U$ according to (Q59) or (Q59a). The equivalence of the two apparently different matrix elements in (Q61) is a reminder of the complicated dependence of $\psi^{(+)}$, $\psi^{(-)}$, $\psi_f$, $\psi_0$ on $U$, $V$ and the asymptotic conditions. To obtain the Green's function in (Q57) or (Q59a) and to solve the equation are difficult for most problems. In any actual calculation of $\mathbf{T}_{f0}$, some approximate wave functions for $\psi^{(+)}$ (or $\psi^{(-)}$ may be tried. An iteration procedure of solving (Q56) or (Q59) will lead to $\mathbf{T}_{f0}$ to various orders in $U$ or $V$ [similar to (P43) for the simpler case of one interaction potential].

## 4. REARRANGEMENT COLLISIONS

In Sects. L and M, we have formulated the integral equation for collisions involving rearrangements [see (L26a) and (M14) for example]. We shall now give the transition matrix element $\mathbf{T}_{fi}$ of the rearrangement collision (refs. 7, 8)

$$A + BX \rightarrow B + AX, \tag{Q62}$$

where we shall assume for simplicity the target $X$ to be infinitely heavy (to avoid the corrections due to the relative motions, see refs. 7, 8). Let $K_A$, $K_B$ be the kinetic energy operators, and $V_A$, $V_B$ the potential energy of $A$, $B$ in the field of $X$. Let the total Hamiltonian be written in two alternative forms

$$\begin{aligned} H &= H_i + V_i = (K_A + K_B + V_B) + (V_A + V), \\ &= H_f + V_f = (K_B + K_A + V_A) + (V_B + V), \end{aligned} \tag{Q63}$$

with $H_i$, $H_f$ corresponding to the initial and the final state in (Q62), and $V$ standing for the interaction between $A$ and $B$. Let the free-particle systems be

$$(K_A - E_A)\,\varphi_A = 0, \qquad (K_B - E_B)\,\varphi_B = 0, \tag{Q64}$$

the $\varphi$ being plane waves. For the bound states in the field of $X$, let us write

$$(K_B + V_B - W_B)\,\xi_B = 0, \qquad (K_A + V_A - W_A)\,\xi_A = 0. \tag{Q65}$$

We have

$$E = E_A + W_B = E_B + W_A. \tag{Q66}$$

The scattering equations corresponding to $K_B + V_B$ and $K_A + V_A$ are [see, for example, (Q51)]

$$\begin{aligned} \chi_f^{(-)} &= \varphi_B + \frac{1}{E_B - K_B - V_B - i\varepsilon} V_B \varphi_B, \\ \chi_i^{(+)} &= \varphi_A + \frac{1}{E_A - K_A - V_A + i\varepsilon} V_A \varphi_A. \end{aligned} \tag{Q67}$$

The integral equations corresponding to (Q63) are [see Q10a) and (Q11)]

$$\begin{aligned} \Psi_i^{(+)} &= \phi_A \xi_B + \frac{1}{E - H_i + i\varepsilon} V_i \Psi_i^{(+)}, \\ \Psi_f^{(-)} &= \phi_B \xi_A + \frac{1}{E - H_f - i\varepsilon} V_f \Psi_f^{(-)}. \end{aligned} \tag{Q68}$$

From (Q65), (Q67) and (Q68), one readily obtains

$$\begin{aligned} \Psi_i^{(+)} &= \chi_i^{(+)} \xi_B + \frac{1}{E - (H - V) + i\varepsilon} V \Psi_i^{(+)}, \\ \Psi_f^{(-)} &= \chi_f^{(-)} \xi_A + \frac{1}{E - (H - V) - i\varepsilon} V \Psi_f^{(-)}. \end{aligned} \tag{Q69}$$

The transition matrix element $\mathbf{T}_{fi}$ from an initial state $\Phi_i$ to a final state $\Phi_f$ is given by (P36) or (P36a). When $\Phi_i = \phi_A \xi_B$ and $\Phi_f = \phi_B \xi_A$ are the eigenstates of two different Hamiltonians $H_i$ and $H_f$, then it can be shown (see ref. 9, for example) that

$$\begin{aligned} \mathbf{T}_{fi} &= (\Phi_f, V_f \Psi_i^{(+)}) = (\phi_B \xi_A, (V_B + V)\, \Psi_i^{(+)}) \\ &= (\Psi_f^{(-)}, V_i \Phi_i) = (\Psi_i^{(-)}, (V_A + V)\, \phi_A \xi_B). \end{aligned} \tag{Q70}$$

Let us take the first form in (Q70). By means of (Q68), (Q67), (Q65) and (Q66), it can be shown that

$$\mathbf{T}_{fi} = (\phi_B \xi_A, V_B \chi_i^{(+)} \xi_B) + (\chi_f^{(-)} \xi_A, V \Psi_i^{(+)}). \tag{Q71}$$

The first term vanishes since $\xi_A$, $\chi_i^{(+)}$ are the orthogonal eigenstates of $K_A + V_A$. Thus the exact theory gives

$$\mathbf{T}_{fi} = (\chi_f^{(-)} \xi_A, V \Psi_i^{(+)}). \tag{Q72}$$

Note that no ambiguity arises from the so-called prior-post interactions $V_i$ and $V_f$.

In (Q72), $\chi_f^{(-)}$ and $\Psi_i'^{(+)}$ are to be obtained from the solutions of (Q67) and (Q68). One may use the following approximations:

1) If for $\Psi_i'^{(+)}$ one uses only the first term in (Q69), then (Q71) becomes

$$\mathbf{T}_{fi} = (\phi_B\xi_A, V_B\chi_i^{(+)}\xi_B) + (\chi_f^{(-)}\xi_A, V\chi_i^{(+)}\xi_B). \tag{Q73}$$

The first term vanishes as before and the second term forms the so-called "distorted wave Born approximation" in the literature since $\chi_f^{(-)}$, $\chi_i^{(+)}$ being given by (Q67), take into account the effect of the field $V_B$, $V_A$ respectively.

2) If in (Q73) one further replaces $\chi_f^{(-)}$, $\chi_i^{(+)}$ by the first terms of (Q67), then

$$\mathbf{T}_{fi} = (\phi_B\xi_A, V\phi_A\xi_B), \tag{Q74}$$

which is the usual Born (plane wave) approximation but without the $V_B$ interaction which is usually included [see (L38, 39), for example]. The form (Q74) was first used by Brinkman and Kramers (ref. 87, Sect. M).

The question as to whether the interaction $V_B$ (or $V_A$) should be included in $\mathbf{T}_{fi}$ has been the subject of many discussions, especially for the charge-transfer process

$$H^+ + H \rightarrow H + H^+. \tag{Q75}$$

(See Sect. M.3, iii, and the brief summaries in refs. 87–96 at the end of Sect. M.) Since the matrix element containing $V_B$ in (Q71) vanishes in the exact theory, it would seem as if (Q74) should be regarded as the correct Born approximation. However, one may also look at the situation in the following manner. The replacement of $\chi_f^{(-)}$, $\chi_i^{(+)}$ in the second term of (Q71) by the plane waves $\phi_B$, $\phi_A$ respectively in (Q74) is an approximation which completely neglects the effect of the field of $X$ on the incident and the outgoing particle. But in a *consistent* approximation, one should also replace $\chi_i^{(+)}$ in the first term of (Q73) by $\phi_A$, and this makes the term different from zero. It is possible that this extra contribution, leading to the usual result

$$\mathbf{T}_{fi} = (\phi_B\xi_A, (V_B + V)\,\phi_A\xi_B), \tag{Q76}$$

may actually compensate for a part of the approximation made in the second term of (Q71). Thus even though the first term of (Q71) does vanish in the exact theory, there may still be some justification for including $V_B$ in the form (Q76). As a support of this view, we may note the good agreement between the experimental data on (Q75) and the calculations based on (Q76) (see refs. 91 and 94 at the end of Sect. M). Of course this "compensation" of approximations, if any, is different for different cases and is a function of the energy of collision.

**REFERENCES**


1. Gell-Mann, M. and Goldberger, M. L., Phys. Rev. **91**, 398 (1953), on which this section is based.
2. Ma, S. T., Phys. Rev. **87**, 652 (1952); **88**, 1211 (1952).
3. Breit, G. and Bethe, H. A., Phys. Rev. **93**, 888 (1954), discuss the ingoing-wave final state.
4. Møller, C., Kgl. Danske Videnskab. Selskab. Mat.-Fys. Medd. **23**, No. 1 (1945), "General properties of the characteristic matrix in the theory of elementary particles" treats the time-independent theory of scattering.
5. Yang, Tse-sen, Science Reports, Peking University, **4**, No. 2, 175 (1958). (In Chinese.) The paper treats the limits of $U(t, t_0)$ for $t_0 \to \pm\infty$ without making the adiabatic switching, by working with the Fourier transform of $e^{iHt}$ and the "$\varepsilon$-representation" of $U(t, t_0)$. The properties of the limiting process (Q22, 22a) of Gell-Mann and Goldberger are discussed. Also Heitler's equation of radiation damping is discussed.
6. Yang, Tze-sen, Science Reports, Peking University, **4**, No. 3, 293 (1958). It is shown that if the total Hamiltonian $H$ possesses a discrete spectrum (bound states), the solution in the interaction picture of the time-dependent Schrödinger equation with initial condition at $t \to \pm\infty$ is not unique.
7. Day, Rodberg, Snow and Sucher, Phys. Rev. **123**, 1051 (1961).
8. Bassel, R. H. and Gerjuoy, E., Phys. Rev. **117**, 749 (1960).
9. Gerjouy, E., Annals of Phys. **5**, 58 (1958).

section

# R Time-Dependent Theory: Method of Spectral Representation

In the preceding Chapters 1–3, the problem of scattering is treated by the time-independent, or stationary state, theory. In the preceding two sections, the problem is treated on the time-dependent Schrödinger equation. That the latter treatment leads to the same integral equation (P33), (P33a) as (B20) or (M17)–(M18) in the stationary state method indicates the equivalence in the results of the two methods. We have, however, given no "proof" of the complete equivalence, especially when in the time-dependent method such an artifice as the adiabatic switching-on and -off of an interaction has to be introduced. In the following, we shall give another time-dependent theory which employs the method of "spectral representation". This is done, at the expense of duplicating a lot of the results already presented, for the purpose of seeing the theory from a different point of view and of understanding a little more clearly the "irrelevance" of the "adiabatic switching" artifice.

## 1. SPECTRAL REPRESENTATION

The Hamiltonian of the system is $H = H_0 + V$ and the Schrödinger equation is

$$i\hbar \frac{\partial \Psi}{\partial t} = (H_0 + V)\Psi. \tag{R1}$$

The scattering problem is one in which, in the infinitely distant past $t \to -\infty$, the interaction $V$ vanishes and the system is described by

$$i\hbar \frac{\partial \Phi}{\partial t} = H_0 \Phi. \tag{R2}$$

i.e., as $t \to -\infty$, $\Psi$ will also give a solution of (R2). We shall assume in the present discussion that $H_0$ has no bound states but only a continuous spectrum

$$(H_0 - E)\, w_0(E, \alpha) = 0, \tag{R3}$$

where $\alpha$ stands for the totality of the quantum numbers of all other observables which together with $H_0$ form a complete commuting set. The $w_0(E, \alpha)$ are improper functions, and

$$\langle w_0(E, \alpha), w_0(E', \alpha')\rangle = \delta(E - E')\, \delta(\alpha, \alpha'), \tag{R4}$$

$\delta(\alpha, \alpha')$ being the Dirac $\delta$ functions for continuous $\alpha$, $\alpha'$, and the Kronecker $\delta$ when $\alpha$ takes on discrete values. $\langle \phi, \psi\rangle$ stands for the inner product of $\phi$ and $\psi$.

On expanding any state vector $\Psi$, we have

$$\Psi = \iint f(E, \alpha)\, w_0(E, \alpha)\, d\alpha\, dE, \tag{R5}$$

where

$$f(E, \alpha) = \langle w_0(E, \alpha), \Psi\rangle \tag{R6}$$

is the representative of $\Psi$ in the $H_0$-representation. If to every $\Psi$ it is possible to assign an $f(E, \alpha)$, the assignment will be called a spectral representation of $H_0$.

We shall now work, instead of the $\Psi$ and the $w_0$'s, with their representatives. The improper function $w_0(E', \alpha')$ has the representative

$$\langle w_0(E, \alpha), w_0(E', \alpha')\rangle = \delta(E - E')\, \delta(\alpha, \alpha'). \tag{R4}$$

This is regarded as the limiting form of a $f(E, \alpha)$ with a sharp peak at $E'$, $\alpha'$. We shall hence be dealing only with proper, quadratically integrable $f(E, \alpha)$'s.#

For the Hamiltonian $H = H_0 + V$, we shall assume that there may be bound states, but the $H$ has the same continuous spectrum as $H_0$. Let the equation analogous to (R3) be

$$(H - F)\, w(F, \beta) = 0. \tag{R7}$$

Then $F$ and $E$ lie in the same range

$$E_a \leqslant F \leqslant E_b, \qquad E_a \leqslant E \leqslant E_b, \tag{R8}$$

and the discrete (bound) states are $F_i < E_a$. On expanding $\Psi$ in terms of the $w(F, \beta)$, we have

$$\Psi = \iint g(F, \beta)\, w(F, \beta)\, d\beta\, dF, \tag{R9}$$

# This is equivalent to the introduction of a box for the system of size $L^3$, with $L \to \infty$ eventually (see Sect. Q.1).

where the integration over $F$ includes the summation over the discrete $F_i$, and

$$g(F, \beta) = \langle w(F, \beta), \Psi \rangle \tag{R10}$$

is the representative of $\Psi$ in the $H$-representation.

Let the representative of $w(F, \beta)$ in the $H_0$-representation be denoted by#

$$u(E\alpha | F\beta) = \langle w_0(E\alpha), w(F\beta) \rangle, \tag{R11}$$

so that

$$w(F, \beta) = \iint u(E\alpha | F\beta)\, w_0(E\alpha)\, d\alpha\, dE.$$

Then

$$\iint g(F\beta)\, w(F\beta)\, d\beta\, dF = \iiiint u(E\alpha | F\beta)\, w_0(E\alpha)\, g(F\beta)\, d\alpha\, dE\, d\beta\, dF.$$

This, together with (R5) and (R9), gives the relation

$$f(E, \alpha) = \iint u(E\alpha | F\beta)\, g(F\beta)\, d\beta\, dF, \tag{R12}$$

or

$$g(F, \beta) = \iint u(F\beta | E\alpha)\, f(E\alpha)\, d\alpha\, dE. \tag{R12a}$$

It will be convenient to define operators in the $H_0$ representation. Let $\psi$ be given by (R5). Furthermore let the vector $A\psi$ be written

$$A\psi = \iint \hat{f}(E, \alpha)\, w_0(E\, \alpha)\, dE\, d\alpha,$$

where $A$ is any operator. Then we define $A^{(E)}$ which is the operator $A$ in the $H_0$-representation by the relation

$$\hat{f}(E, \alpha) = A^{(E)} f(E, \alpha).$$

In particular, since

$$H_0\psi = \iint E f(E, \alpha)\, w_0(E\, \alpha)\, dE\, d\alpha,$$

$H^{(E)} = E$.

Let us find the integral equation to be satisfied by $u(E\alpha | F\beta)$. From (R7) and denoting by $H^{(E)}$, $V^{(E)}$ the operators $H$, $V$ as given in the $H_0$-representation, one has

$$(H^{(E)} - F)\, u(E\alpha | F\beta) = 0,$$

$$H^{(E)} = H^{(E)} + V^{(E)}, \qquad H_0^{(E)} u(E\alpha | F\beta) = E u(E\alpha | F\beta),$$

and

$$(F - E)\, u(E\alpha | F\beta) = V^{(E)}\, u(E\alpha | F\beta), \tag{R13}$$

---

# For simplicity, we shall in the following omit the comma between $E$ and $\alpha$, and that between $F$ and $\beta$, etc.

which must be satisfied for all $w(F\beta)$. Now $E$ is, by hypothesis (R8), in the continuum, and for discrete states $F_\iota$, the equation is

$$u(E\alpha|F_\iota\beta) = \frac{1}{F_\iota - E} V^{(E)} u(E\alpha|F_\iota\beta)$$

$$= \frac{1}{F_\iota - E} \int\int V^{(E)}(E\alpha|E'\alpha')\, u(E'\alpha'|F_\iota\beta)\, d\alpha'\, dE'. \qquad \text{(R14)}$$

$u(E\alpha|F_\iota\beta)$ is there a quadratically integrable function of the variables $E$, $\alpha$. The integral operator $V^{(E)}(E_\alpha|E'\alpha')$ is defined by

$$\langle w_0(E\alpha), Vw_0(E'\alpha')\rangle,$$

$$V^{(E)} f(E\alpha) = \int\int V^{(E)}(E\alpha|E'\alpha') f(E'\alpha')\, dE'\, d\alpha'.$$

For $F$ in the continuum, then $F - E$ may be zero. In this case, we have to introduce the $\delta$ function of Dirac through the formula

$$xk(x) = r(x),$$

$$k(x) = \text{const.}\ \delta(x) + \frac{P}{x} r(x),$$

where $P$ indicates "taking the principal value". Then

$$u(E\alpha|F\beta) = \lambda(E\alpha|F\beta)\,\delta(E - F) + \frac{P}{F - E} V^{(E)} u(E\alpha|F\beta)$$

$$= \lambda(E\alpha|F\beta)\,\delta(E - F) + \frac{P}{F - E} \cdot T(E\alpha|F\beta), \qquad \text{(R15)}$$

where

$$T(E\alpha|F\beta) = \int\int V^{(E)}(E\alpha|E'\alpha')\, u(E'\alpha'|F\beta)\, d\alpha'\, dE', \qquad \text{(R16)}$$

and $\lambda(E\alpha|F\beta)$ is a function as yet to be determined. [See (R30) below.]

## 2. SCATTERING

From (R3) to (R16), we have been dealing with time-independent quantities. Now we turn to the time-dependent theory, and introduce the following notation between $\Psi(t)$ and its representative in the $H_0$-representation by

$$\Psi(t) \underset{H_0}{\longleftrightarrow} f(E\alpha; t). \qquad \text{(R17)}$$

Similarly, for the representative of $\Psi(t)$ in the $H$-representation

$$\Psi(t) \underset{H}{\longleftrightarrow} g(F\beta; t) \qquad \text{(R18)}$$

(R12) then becomes

$$f(E\alpha;\,t) = \int\int u(E\alpha|F\beta)\, g(F\beta;\,t)\, d\beta\, dF. \tag{R19}$$

If we write $\Psi(t)$ in the form (putting $\hbar = 1$)

$$\Psi(t) = e^{-iHt}\,\Psi(0), \tag{R20}$$

its representative in $H$-representation is

$$g(F\beta;\,t) = e^{-iFt}\, g(F\beta;\,0), \tag{R21}$$

and (R19) becomes

$$e^{iEt} f(E\alpha;\,t) = \int\int u(E\alpha|F\beta) \exp\left[-i(F - E)t\right] g(F\beta;\,0)\, d\beta\, dF. \tag{R22}$$

If $u(E\alpha|F\beta)$ is known, it is possible to evaluate the limits

$$\lim_{t\to\pm\infty} e^{iEt} f(E\alpha;\,t) \equiv f_{\pm}(E\alpha) \tag{R23a}$$

and hence obtain the limits

$$\lim_{t\to\pm\infty} \exp\,(iH_0 t)\,\Psi(t) \equiv \Psi(\pm\infty) \tag{R23}$$

*if they exist*. (See below and Sect. S for the condition for their existence.) And if they exist, then we have obtained $\Psi(t)$ as $t \to +\infty$ in terms of $\Psi(t)$ for $t \to -\infty$, i.e., $\Psi(-\infty)$. Equation (R23) can be interpreted as saying that $\Psi(t)$ behaves like a solution of the unperturbed time-dependent Schrödinger equation for large $|t|$. The scattering process is then described by the scattering operator $S$

$$\Psi(+\infty) = S\Psi(-\infty) \tag{R24}$$

as in (R22) or (Q12) and (Q25).

The problem now is to investigate the conditions for the existence of the limits (R23). On separating the integration over $F$ in (R22) into a summation over the discrete states $F_i$ of $H$ and an integration over the continuum (R8), the former does not approach any limit but oscillates as $t \to \pm\,\infty$, whereas the latter integral does approach a limit (see R29–34 below). Thus for the limits (R23) to exist, it is necessary that

$$g(F_i\beta;\,0) = 0, \tag{R25}$$

which is, from (R10),

$$\langle w(F_i\beta), \Psi(0)\rangle = 0, \tag{R25a}$$

i.e., $\Psi(t)$ must be orthogonal to the discrete states of $H$.

We shall now proceed with the determination of the function $\lambda(E\alpha|F\beta)$ in (R15). In Chapter 1, Sect. 1, the eigenfunction for the scattering problem is determined by the asymptotic condition in coordinate space. Here we shall seek a solution $\Psi(t)$ of (R1) such that asymptotically for $t \to \pm\infty$, it is also a solution of (R2). In terms of its representatives, we have for finite times,

$$\Psi(t) \underset{H_0}{\overset{(R17)}{\longleftrightarrow}} f(E\alpha; t)$$

$$\Psi(t) \underset{H}{\overset{(R18)}{\longleftrightarrow}} g(F\beta; t) \overset{(R21)}{=\!=} e^{-iFt} g(F\beta; 0) \tag{R26a}$$

or, by (R23), (R25a), asymptotically,

$$\Psi(-\infty) \underset{H_0}{\longleftrightarrow} f_-(E\alpha), \qquad \Psi(0) \underset{H}{\longleftrightarrow} g(F\beta; 0). \tag{R26}$$

We now introduce a relation between $\Psi(-\infty)$ and $\Psi(0)$ [or alternatively, between $\Psi(+\infty)$ and $\Psi(0)$] by

$$f_-(E\alpha) = g(E\alpha; 0) \tag{R27}$$

[or $f_-(F\beta) = g(F\beta; 0)$]. This relation (or other choices) will determine the continuous eigenfunction of our problem as we shall show below. By (R27), we have from (R26a),

$$\Psi(t) \underset{H}{\longleftrightarrow} e^{-iFt} f_-(F\beta). \tag{R28}$$

On putting (R15) into (R22), we get

$$e^{iEt} f(E\alpha; t) = \int \lambda(E\alpha|E\beta)\, g(E\beta; t_0)\, d\beta$$
$$+ P\int\int \frac{T(E\alpha|F\beta)}{F-E} g(F\beta; 0) \exp[-i(F-E)t]\, d\beta\, dF. \tag{R29}$$

The $\lambda(E\alpha|F\beta)$ is to be so chosen that (R29) satisfies (R27), i.e.,

$$\lim_{t\to-\infty} e^{iEt} f(E\alpha; t) \equiv f_-(E\alpha) \overset{(R27)}{=\!=} g(E\alpha; 0). \tag{R30}$$

On taking the limit of (R29), the second integral can be written

$$\lim_{t\to-\infty} P\int \frac{T(E\alpha|F\beta)\, g(F\beta; 0) - T(E\alpha|E\beta)\, g(E\beta; 0)}{F-E} \exp[-i(F-E)t]\, d\beta\, dF$$
$$+ \lim_{t\to-\infty} \int T(E\alpha|E\beta)\, g(E\beta; 0)\, d\beta \cdot P\int_{E_a}^{E_b} \frac{\exp[-i(F-E)t]}{F-E}\, dF. \tag{R31}$$

The first term vanishes if $T(E\alpha|F\beta)\, g(F\beta; 0)$, as a function of $F$, has a derivative at $F = E$; for then the factor in front of $\exp[-i(F-E)t]$ is absolutely convergent, and the Fourier integral vanishes according to the Riemann–Lebesgue theorem. See footnote following (S33). The second integral in the second term in (R31) can be written

$$\lim_{t\to-\infty} \int_{(E_a-E)t}^{(E_b-E)t} \frac{e^{-i\xi}}{\xi}\, d\xi. \tag{R32}$$

For $E$ inside the spectrum of $H_0$, i.e., $E_a < E < E_b$# in (R8), the $\lim \int \cdots d\xi = i\pi$, and (R29), with (R30)–(R32), leads to the following equation for $\lambda(E\alpha|E\beta)$

$$\lambda(E\alpha|E\beta) = \delta(\alpha, \beta) - i\pi T(E\alpha|E\beta). \tag{R33}$$

With this, (R15) becomes

$$u_-(E\alpha|F\beta) = \delta(E - F)\,\delta(\alpha, \beta) + \gamma_-(F - E)\,T_-(E\alpha|F\beta), \tag{R34}$$

where $T_-$ is given in (R16) with $u_-$. The subscript $-$ reminds us that we have chosen the initial condition at $t \to -\infty$ in (R27). Here and in the following, we introduce the operator.

$$\gamma_\pm(x) = \lim_{\varepsilon \to 0^+} \frac{1}{x \pm i\varepsilon} = \pm i\pi\delta(x) + \frac{P}{x}, \tag{R36}$$

$P$ meaning "taking the principal value of".

To summarize: with the initial condition at $t \to -\infty$ prescribed by $\Psi(-\infty)$, or, by its representative $f_-(E\alpha)$,

$$\Psi(-\infty) \underset{H_0}{\longleftrightarrow} f_-(E\alpha), \tag{R26}$$

then we have

$$\Psi(t) \underset{H}{\longleftrightarrow} g(F\beta; t) \overset{(R28)}{=\!=} e^{-iFt} f_-(F\beta), \tag{R28}$$

$$\Psi(t) \underset{H_0}{\longleftrightarrow} f(E\alpha; t) \overset{(R22)}{\underset{(R27)}{=\!=}} \int\!\!\int u_-(E\alpha|F\beta)\, e^{-iFt} f_-(F\beta)\, d\beta\, dF, \tag{R37}$$

and

$$u(E\alpha|F_i\beta) = \frac{1}{F_i - E} \int\!\!\int V^{(E)}(E\alpha|E'\alpha')\, u(E'\alpha'|F_i\beta)\, d\alpha'\, dE', \tag{R14}$$

$$u_-(E\alpha|F\beta) = \delta(E - F)\,\delta(\alpha, \beta) + \gamma_-(F - E)\,T_-(E\alpha|F\beta). \tag{R34}$$

---

# For $E = E_a$, (R32) diverges; but the second term in (R31) will vanish if either

$$T(E_a\alpha|E_a\beta) = 0, \quad \text{meaning } V^{(E)}(E_a\alpha|E'\alpha') = 0, \tag{R35}$$

or

$$g(E_a\beta; 0) = 0, \tag{R35a}$$

which limits the $\Psi(-\infty)$ or $\Psi(0)$ that can be used. Similarly for $E = E_b$. The condition (R35) and $V^{(E)}(E_b\alpha|E'\alpha') = 0$ are closely connected with the requirement (R8).

### 3. SCATTERING OPERATOR $S$

From (R23a) and (R37) we have

$$\lim_{t\to\infty} e^{iEt} f(E\alpha; t) = f_+(E\alpha)$$
$$= \lim_{t\to\infty} \iint u_-(E\alpha|F\beta) \exp[-i(F - E)t] f_-(F\beta)\, d\beta\, dF. \tag{R38}$$

On substituting (R34) into this and calculating the integral containing $P/(F - E)$ as in (R31), one gets

$$f_+(E\alpha) = \iint S(E\alpha|E'\alpha') f_-(E', \alpha')\, d\alpha'\, dE', \tag{R39}$$

where

$$S(E\alpha|E'\alpha') = \delta(E - E')\,\delta(\alpha, \alpha') - 2\pi i\,\delta(E - E')\,T_-(E\alpha|E'\alpha'). \tag{R40}$$

On comparison with (P22), it is seen that $S(E\alpha|E'\alpha')$ are the matrix elements of the scattering operator $S$ in the $H_0$ representation. The first term $\delta(E - E')$ $\delta(\alpha, \alpha')$ represents the incoming wave while the second term represents the scattering wave. It is seen that the nonvanishing elements of $S$ connect states with the same energy. [Compare this with (P27)–(P40).]

Had we prescribed the initial condition at $t \to +\infty$ instead of $t \to -\infty$, then Equations (R26)–(R40) would have read

$$\Psi(+\infty) \underset{H_0}{\longleftrightarrow} f_+(E\alpha), \tag{R41}$$

$$\Psi(t) \underset{H}{\longleftrightarrow} g(F\beta; t) = e^{-iFt} f_+(F\beta), \tag{R42}$$

$$\Psi(t) \underset{H_0}{\longleftrightarrow} f(E\alpha; t) = \iint u_+(E\alpha|F\beta)\, e^{-iFt} f_+(F\beta)\, d\beta\, dF, \tag{R43}$$

$$u(E\alpha|F_i\beta) = \frac{1}{F_i - E} \iint V^{(E)}(E\alpha|E'\alpha')\, u(E'\alpha'|F_i\beta)\, d\alpha'\, dE', \tag{R44}$$

$$u_+(E\alpha|F\beta) = \delta(E - F)\,\delta(\alpha, \beta) + \gamma_+(F - E)\,T_+(E\alpha|F\beta), \tag{R45}$$

$$f_-(E\alpha) = \iint \hat{S}(E\alpha|E'\alpha') f_+(E\alpha')\, d\alpha'\, dE', \tag{R46}$$

$$\hat{S}(E\alpha|E'\alpha') = \delta(E - E')\,\delta(\alpha, \alpha') + 2\pi i\,\delta(E - E')\,T_+(E\alpha|E'\alpha'). \tag{R47}$$

#### i) Unitary of S

From the conservation of probability, we must have, on writing $\Psi_+ \equiv \Psi(\infty)$, $\Psi_- \equiv \Psi(-\infty)$,

$$\langle \Psi_+, \Psi_+ \rangle = \langle \Psi_-, \Psi_- \rangle. \tag{R48}$$

From (R39)

$$\Psi_+ = S\Psi_-, \tag{R49}$$

we have

$$\langle S\Psi_-, S\Psi_-\rangle = \langle \Psi_-, \Psi_-\rangle.$$

Hence

$$S^\dagger S = SS^\dagger = 1. \tag{R50}$$

Similarly for the $\hat{S}$ above. Thus $S$ is unitary, and also $\hat{S}$.

*ii) Reciprocity theorem*

From (R46), we have

$$\Psi_- = \hat{S}\Psi_+, \tag{R51}$$

and we wish to show that

$$\hat{S} = S^\dagger. \tag{R52}$$

This equation is

$$\hat{S}(E\alpha|E'\alpha') = S^\dagger(E\alpha|E'\alpha'),$$

or

$$\hat{S}(E\alpha|E'\alpha') = S^*(E'\alpha'|E\alpha).$$

From the definitional equations (R40) and (R47), this equation becomes

$$T_+(E\alpha|E\alpha') = T^*_-(E\alpha'|E\alpha). \tag{R53}$$

It can be shown, from the integral equations for $u_+$ and $u_-$, that this relation is true. Hence (R52) is true. But by (R50), $S^\dagger = S^{-1}$, hence

$$\hat{S} = S^{-1}. \tag{R54}$$

This is known as the reciprocity theorem. For further discussion of this theorem, see Chapter 6, Sect. W.

## 4. TRANSITION PROBABILITY

The probability of transition per unit time is defined by

$$w(E\alpha; t) = \frac{d}{dt}|f(E\alpha; t)|^2 \tag{R55}$$

which, by using (R37), is

$$w(E\alpha; t) = -i \iiiint u^*_-(E\alpha|F\beta)u_-(E\alpha|F'\beta')f^*_-(F\beta) \times f_-(F'\beta')(F' - F) \exp[-i(F' - F)t]\, d\beta'\, dF'\, d\beta\, dF$$

which can be written, on noting $F' - F = (F' - E) - (F - E)$ and using (R13) and the definition of $T(E\alpha|F\beta)$ in (R16),

$$w(E\alpha; t) = 2\,\mathrm{Im} \iiiint u^*_-(E\alpha|F\beta)\, T_-(E\alpha|F'\beta') \exp[-i(F' - F)t] \times f^*_-(F\beta) f_-(F'\beta')\, d\beta'\, dF'\, d\beta\, dF. \tag{R56}$$

If we take $\Psi(-\infty)$ as an eigenstate of $H_0$ derived from a proper state by a suitable limiting process [see lines following (R4a) and footnote], then

$$f_-(E\alpha) = \delta(E - E')\,\delta(\alpha, \alpha'), \tag{R57}$$

and $w(E\alpha; t)$ becomes

$$w(E\alpha; t) = 2\,\mathrm{Im}\, u_-^*(E\alpha|E'\alpha')\, T_-(E\alpha|E'\alpha').$$

On using (R34) for $u_-^*(E\alpha|E'\alpha')$, $\alpha' \neq \alpha$, we obtain

$$\begin{aligned} w(E\alpha; t) &= 2\,\mathrm{Im}\left[i\pi\delta(E - E') + \frac{P}{E' - E}\right]|T_-(E\alpha|E'\alpha')|^2 \\ &= 2\pi\delta(E - E')|T_-(E\alpha|E'\alpha')|^2, \end{aligned} \tag{R58}$$

which is the result (P35). (Note the difference in notations.)

This probability per unit time is constant independent of time so that it would seem that the probability would increase indefinitely with time. This result comes about because $\Psi(-\infty)$ has been chosen to be an eigenstate of $H_0$ which is an improper function. Let $\Psi(-\infty)$ now be taken to be a proper state (its representative $f(E\alpha)$ in $H_0$ having a sharp peak at $E'\alpha'$) (see ref. 3), and let us assume that in the vicinity of $F = E'$, $\beta = \alpha'$,

$$T_-(E\alpha|F\beta)\,f_-(F\beta) \cong T_-(E\alpha|E'\alpha')\,f_-(F\beta),$$

$$\frac{P}{F - E}\,T_-(E\alpha|F\beta)\,f_-(F\beta) \cong \frac{P}{E' - E}\,T_-(E\alpha|E'\alpha')\,f_-(F\beta).$$

Then, after some calculation, we obtain from (R56)

$$w(E\alpha; t) = 2\pi\,\mathrm{Re}\,|T_-(E\alpha|E'\alpha')|^2\, h^*(E) \times \int_{E_a}^{E_b} \exp\left[-i(F' - E)t\right] h(F')\, dF', \tag{R59}$$

where

$$h(E) = \int f_-(E\alpha)\, d\alpha. \tag{R60}$$

[(R59) will reduce to (R58) if (R57) is used.] If $f_-(E\alpha)$, and hence $h(E)$, are bounded and absolutely convergent in the range $E_a \leqslant E \leqslant E_b$, then, by the Riemann–Lebesgue theorem for Fourier transforms, one obtains

$$\lim_{t\to\pm\infty} w(E\alpha; t) = 0. \tag{R61}$$

This ensures that the probability does not grow with time indefinitely.

In (R59), the $\delta(E - E')$ of (R58) is absent and the "conservation of energy" is only approximately ensured by the sharply peaked $f(E, \alpha)$, and hence $h(E)$, at $E = E'$, $\alpha = \alpha'$. Let us assume that $h(E)$ has a width $\delta E$, i.e.,

$$\begin{aligned} h(E) &= K \quad \text{for} \quad E' < E < E' + \delta E \\ &= 0 \quad \text{otherwise.} \end{aligned} \tag{R62}$$

Then (R59) becomes

$$w(E\alpha;\, t) = -2\pi \ \mathrm{Im} \frac{K}{t} |T_{-}(E\alpha|E'\alpha')|^2 \times h^*(E) \exp[-i(E - E')t] [\exp(-it\delta E) - 1]. \quad \text{(R63)}$$

For $w(E\alpha;\, t)$ to be approximately equal to its maximum (constant) value, $t$ must be small such that

$$|t| < \tau, \qquad \tau\delta E < 1. \quad \text{(R64)}$$

[From (R62), $(E - E')\tau < 1$ is also implied], and for this $|t| < \tau$,

$$w(E\alpha;\, t) = 2\pi|h(E)|^2 \, |T_{-}(E\alpha|E'\alpha')|^2 \, \delta E. \quad \text{(R65)}$$

This result reflects the uncertainty relation between the energy $E$ of the initial state $\Psi(-\infty)$ and the "lifetime of the state".

## 5. ADIABATIC SWITCHING

To investigate the equivalence between the adiabatic switching artifice in the usual time-dependent treatment [see (P25)] and the spectral representation method, let the Hamiltonian $H$ in (R1) be written

$$H_\varepsilon = H_0 + e^{-\varepsilon|t|}\, V, \quad \varepsilon \geqslant 0, \quad \text{(R66)}$$

and the Schrödinger equation (R1)

$$i\hbar \frac{\partial \Psi_\varepsilon}{\partial t} = H_\varepsilon \Psi_\varepsilon(t). \quad \text{(R67)}$$

The representative of $\Psi(t)$ in the $H_0$-representation then satisfies

$$i\hbar \frac{\partial f_\varepsilon(E\alpha;\, t)}{\partial t} = Ef_\varepsilon(E\alpha;\, t) + e^{-\varepsilon|t|} \int\!\!\int V^{(E)}(E\alpha|E'\alpha') f_\varepsilon(E'\alpha';\, t)\, d\alpha'\, dE'. \quad \text{(R68)}$$

The representative of the solution of $H_0$ is

$$f(E\alpha;\, t) = e^{-iEt} f_{-}(E\alpha), \quad \text{(R69)}$$

and we shall take $f_{-}(E\alpha)$ as the initial condition

$$\lim_{t\to-\infty} e^{iEt} f_\varepsilon(E\alpha;\, t) = f_{-}(E\alpha). \quad \text{(R70)}$$

Let us introduce an integral operator with the kernel $v_{\varepsilon,t}(E\alpha|E'\alpha')$ such that#

$$e^{iEt} f_\varepsilon(E\alpha;\, t) = \int\!\!\int v_{\varepsilon,t}(E\alpha|E'\alpha') f_{-}(E'\alpha')\, d\alpha'\, dE'. \quad \text{(R71)}$$

# This kernel $v_{\varepsilon,t}$ is equivalent to the operator $U_+$ in (P5).

The initial condition (R70) can then be expressed as

$$\lim_{t\to-\infty} v_{\varepsilon,t}(E\alpha|E'\alpha') = \delta(E - E')\,\delta(\alpha, \alpha'). \tag{R72}$$

Substituting (R71) into (R68) gives the equation for $v_{\varepsilon,t}$ ($\hbar = 1$)

$$\frac{\partial}{\partial t} v_{\varepsilon t}(E\alpha|E'\alpha') = -i \iint V^{(E)}(E\alpha|E''\alpha'') \times \exp\,[i(E - E'')t - \varepsilon|t|]\, v_{\varepsilon,t}(E''\alpha''|E'\alpha')\, d\alpha''\, dE''.$$

On integrating with the initial condition (R72)

$$v_{\varepsilon t}(E\alpha|E'\alpha') = \delta(E - E')\delta(\alpha, \alpha') - i \int_{-\infty}^{t} dt' \iint V^{(E)}(E\alpha|E''\alpha'') \times \exp\,[i(E - E'')t' - \varepsilon|t'|]\, v_{\varepsilon,t'}(E''\alpha''|E'\alpha')\, d\alpha''\, dE''. \tag{R73}$$

On introducing another kernel $W_{\varepsilon,t}$ by

$$v_{\varepsilon t}(E\alpha|E'\alpha') = W_{\varepsilon t}(E\alpha|E'\alpha') \exp\,[i(E - E')t], \tag{R74}$$

substituting this into (R73), integrating over $t'$ by parts and finally taking the limit as $\varepsilon \to 0$, we obtain

$$\begin{aligned} W_{0,t}&(E\alpha|E'\alpha') \\ &= \delta(E - E')\,\delta(\alpha, \alpha') + \left[-i\pi\delta(E' - E) + \frac{P}{E' - E}\right] \\ &\qquad \times \iint V^{(E)}\,(E\alpha|E''\alpha'')\, W_{0,t}(E''\alpha''|E'\alpha')\, d\alpha''\, dE'' \\ &\quad + i \exp\,[-i(E - E')t] \lim_{\varepsilon\to 0} \frac{1}{i(E - E') + \varepsilon} \\ &\qquad \times \int_{-\infty}^{t} dt' \iint V^{(E)}(E\alpha|E''\alpha'') \exp\,[i(E - E')t' - \varepsilon|t'|] \\ &\qquad \times \frac{\partial}{\partial t'} W_{\varepsilon,t'}(E''\alpha''|E'\alpha')\, d\alpha''\, dE'' \end{aligned} \tag{R75}$$

On comparison with (R34), it is seen that $W_{0,t}(E\alpha|E'\alpha')$ satisfies the same integral equation as $u_{-}(E\alpha|E'\alpha')$, if

$$\lim_{\varepsilon\to 0} \frac{1}{i(E - E') + \varepsilon} \int_{-\infty}^{t} dt \iint d\alpha''\, dE''\, V^{(E)}(E\alpha|E''\alpha'') \times \exp\,[i(E - E'')t' - \varepsilon|t'|] \frac{\partial}{\partial t'} W_{\varepsilon,t'}(E''\alpha''|E'\alpha') = 0. \tag{R76}$$

In terms of $W_{\varepsilon,t}$, (R71) is

$$f_{\varepsilon}(E\alpha; t) = \iint W_{\varepsilon,t}(E\alpha|E'\alpha')\, e^{-iE't} f_{-}(E'\alpha')\, d\alpha'\, dF'. \tag{R77}$$

If (R76) is satisfied, then, on taking the limit $\varepsilon \to 0$, (R77) will become

$$f_0(E\alpha; t) = \iint u_-(E\alpha|E'\alpha')\, e^{-iE't} f_-(E'\alpha')\, d\alpha'\, dF'. \qquad \text{(R78)}$$

On comparison with (R34), one will then obtain

$$f_0(E\alpha; t) = f(E\alpha; t) \underset{H_0}{\longleftrightarrow} \Psi(t), \qquad \text{(R79)}$$

i.e., $f_0(E\alpha; t) = \lim_{\varepsilon \to 0} f_\varepsilon(E\alpha; t)$ in the adiabatic switching treatment is equal to the $f(E\alpha; t)$ in the spectral representation treatment. Hence the equivalence of the two treatments will be established if (R76) is satisfied. (R76) is a necessary and sufficient condition.

A sufficient condition is

$$\lim_{\varepsilon \to 0} \frac{\partial}{\partial t} W_{\varepsilon,t}(E\alpha|E'\alpha') = 0. \qquad \text{(R80)}$$

We shall not give the details here, but it can be shown that the condition (R80) is satisfied if the series expansion of $u_-$ in powers of the potential $V$ is convergent.# Since the condition is only a sufficient one, it is possible that the adiabatic switching treatment is valid for a much larger class of Hamiltonian $H$.

## 6. EQUIVALENCE OF STATIONARY-STATE AND TIME-DEPENDENT METHODS

We shall carry out an explicit calculation in the simple case of the scattering of a particle by a potential $U(x)$. In the stationary state treatment, the Schrödinger equation (A2)

$$(\nabla^2 + k^2)\, \psi(\mathbf{x}|\mathbf{k}) = U(\mathbf{x})\, \psi(\mathbf{x}|\mathbf{k}) \qquad \text{(R81)}$$

can be written [see (B14)]

$$\psi(\mathbf{x}|\mathbf{k}) = \phi(\mathbf{x}|\mathbf{k}) - \frac{1}{4\pi} \int \frac{\exp(ik|\mathbf{x} - \mathbf{x}'|)}{|\mathbf{x} - \mathbf{x}'|}\, U(\mathbf{x}')\, \psi(\mathbf{x}'|\mathbf{k})\, d\mathbf{x}' \qquad \text{(R82)}$$

$$\phi(\mathbf{x}|\mathbf{k}) = (2\pi)^{-3/2}\, e^{i\mathbf{k}\cdot\mathbf{x}}.$$

In the spectral representation treatment, we have from (R34)

$$u_-(E, \vartheta\varphi|F, \lambda\nu) = \delta(E - F)\, \delta(\vartheta - \lambda)\, \delta(\varphi - \nu) + \gamma_-(F - E)$$
$$\times \iint U(E\vartheta\varphi|E'\vartheta'\varphi')\, u_-(E'\vartheta'\varphi'|F\lambda\nu)\, d\vartheta'\, d\varphi'\, dE', \qquad \text{(R83)}$$

---

# One could start from (R73) [transformed into an equation for $W_{\varepsilon t}$ by means of (R74)], solving it by iteration and passing to the limit $\varepsilon \to 0$, and obtain the sufficient condition in terms of the convergence of the series development of $W_{0,t}$ which is the same as $u_-$. See ref. 4.

where $\vartheta$, $\varphi$ stand for the directions of the momentum $\mathbf{k}$, $E = k^2$, and $\lambda$, $\nu$ the directions of the momentum $\mathbf{F}$, $|F| = k^2$.

We shall show that $u_-$ is the representative of $\psi(\mathbf{x}|\mathbf{k})$ in the $H_0$ representation, i.e.,

$$u_-(E\vartheta\varphi|F\lambda\nu) = \int \phi^*(\mathbf{x}|E\vartheta\varphi)\,\psi(\mathbf{x}|F\lambda\nu)\,d\mathbf{x}, \tag{R84}$$

or, on using the orthogonality of the $\phi(\mathbf{x}|E\vartheta\varphi)$,

$$\psi(\mathbf{x}|F\lambda\nu) = \int_0^\infty dE \int_0^\pi d\vartheta \int_0^{2\pi} d\varphi\,\phi(\mathbf{x}|E\vartheta\varphi)\,u_-(E\vartheta\varphi|F\lambda\nu). \tag{R85}$$

Using (R83) in (R85), one obtains

$$\psi(\mathbf{x}|F\lambda\nu) = \phi(\mathbf{x}|F\lambda\nu) + \int_0^\infty dE \int_0^\pi d\vartheta \int_0^{2\pi} d\varphi\,\phi(\mathbf{x}|E\vartheta\varphi)\,\gamma_-(F - E)\,T_-(E\vartheta\varphi|F\lambda\nu), \tag{R86}$$

where

$$T_-(E\vartheta\varphi|F\lambda\nu) = \int_0^\infty dE' \int_0^\pi d\vartheta' \int_0^{2\pi} d\varphi'\,U(E\vartheta\varphi|E'\vartheta'\varphi')\,u_-(E'\vartheta'\varphi'|F\lambda\nu). \tag{R87}$$

Now

$$U(E\vartheta\varphi|E'\vartheta'\varphi') = \int \phi^*(\mathbf{x}'|E\vartheta\varphi)\,U(x')\,\phi(\mathbf{x}'|E'\vartheta'\varphi')\,d\mathbf{x}', \tag{R88}$$

and

$$\iiint \phi^*(\mathbf{x}|E'\vartheta'\varphi')\,\phi(\mathbf{x}'|E'\vartheta'\varphi')\,dE'\,d\vartheta'\,d\varphi = \delta(\mathbf{x} - \mathbf{x}'). \tag{R89}$$

With these relations, using (R84) in (R87), one obtains

$$T_-(E\vartheta\varphi|F\lambda\nu) = \int \phi^*(\mathbf{x}'|E\vartheta\varphi)\,U(x')\,\psi(\mathbf{x}'|F\lambda\nu)\,d\mathbf{x}'. \tag{R90}$$

[This is the expression (P36).]

On putting (R90) into (R86), the second term is seen to contain the integral

$$\int_0^\infty dE \int_0^\pi d\vartheta \int_0^{2\pi} d\varphi\,\phi^*(\mathbf{x}'|E\vartheta\varphi)\,\gamma_-(F - E)\,\phi(\mathbf{x}|E\vartheta\varphi). \tag{R91}$$

With $\gamma_-$ given by (R35), namely

$$\gamma_-(F - E) = \lim_{\varepsilon \to 0+} \frac{1}{F - E + i\varepsilon},$$

and

$$\phi(\mathbf{x}|E\nu\varphi) = \frac{1}{(2\pi)^{3/2}}\,e^{i\mathbf{k}\cdot\mathbf{x}},$$

it can be shown [see the relations (B12)–(B13) of Chapter 1] that the integral (R91) is

$$\int \phi^*(\mathbf{x}'|\mathbf{k})\, \gamma_-(F - E)\, \phi(\mathbf{x}|\mathbf{k})\, d\mathbf{k} = -\frac{1}{4\pi} \frac{\exp(ik|\mathbf{x} - \mathbf{x}'|)}{|\mathbf{x} - \mathbf{x}'|}. \qquad \text{(R92)}$$

Hence (R86) becomes

$$\psi(\mathbf{x}|F\lambda\nu) = \phi(\mathbf{x}|F\lambda\nu) - \frac{1}{4\pi} \int \frac{\exp[ik|\mathbf{x} - \mathbf{x}'|]}{|\mathbf{x} - \mathbf{x}'|} U(x')\, \psi(\mathbf{x}'|F\lambda\nu)\, d\mathbf{x}'$$

which is the Schrödinger equation (R82) obtained by the stationary state method.

To complete the "proof" that the stationary state and the time-dependent method give identical descriptions of the scattering problem, we may just remark that the $S$ operator defined in the time-dependent theory is identical with that defined in the time-independent theory [see Chapter 6, also (A34)]. An explicit, but somewhat lengthy, demonstration of this identity is possible, but will not be given here.

## REFERENCES

The treatment of the time-dependent theory in this section is in essence that of Friedrichs. It is based on an article by

1. Moses, H. E., Nuovo Cimento **1**, 103 (1955).
2. Friedrichs, K. O., Comm, on Applied Math. **1**, 361 (1948), treats the time-dependent theory by the method of spectral representation.
3. Friedrichs, K. O., Nach. Akad. Wiss. Göttingen, Math.-Phys. Klasse IIa, 7, 43 (1952).
4. Synder, H. S., Phys. Rev. **83**, 1154 (1951), treats the problem of adiabatic switching for the first time for systems possessing continuous spectrum. The condition found for the adiabatic switching to be valid is expressed in terms of the convergence of an iteration solution of the time-dependent Schrödinger equation. This, as seen in (R80), seems only a sufficient condition.
5. For another treatment of the question of adiabatic switching, see ref. 5 of Section Q.

section

# S Mathematical Theory of Scattering Operator

The time-dependent theories of collisions in the preceding several sections have been based on physically "acceptable assumptions" and on "plausible" arguments. We shall now start from a mathematically more rigorous basis. We do not want to use a "plane wave" state because it does not belong to the usual Hilbert space, and because it is actually not necessary to use such an unphysical state. The actual state is a wave packet and normalizable.

The procedure of developing the theory is as follows. Firstly, two assumptions are made, and it will be investigated how the usual conclusions are derived from these assumptions only. Secondly, we like to see for what Hamiltonians are the assumptions valid. We shall follow Kato's unpublished lecture in the first two subsections.# Part 1 contains the theory developed by Jauch, ref. 1.

## 1. AXIOMATIC TREATMENT

Let the Hamiltonian of the system be $H$, and the unperturbed Hamiltonian be $H_0$. $H$ and $H_0$ are assumed to be self-adjoint. ($H^\dagger = H$, $H_0^\dagger = H_0$). Consider the state $f$ which belongs to the Hilbert space $\mathfrak{H}$. The evolution of $f$ in time is given by ($\hbar = 1$)

$$e^{-itH}f \tag{S1}$$

in the Schrödinger picture. The state (S1) will behave as if it were free from the interaction, if sufficient time has passed. That is,

$$\exp(-itH)f \sim \exp(-itH_0)\,g, \quad t \to +\infty. \tag{S2}$$

#We would like to thank Professor T. Kato for permission to quote from his lectures.

In the same way $f$ will behave as free in the infinite past,

$$\exp(-itH)f \sim \exp(-itH_0)\,h, \quad t \to -\infty. \tag{S3}$$

Now the scattering operator is defined by

$$Sh = g. \tag{S4}$$

Let us formulate the above statement in a mathematically rigorous way. The conditions (S2) and (S3) are equivalent to

$$\|\exp(-itH)f - \exp(-itH_0)\,g\| \to 0, \quad t \to +\infty, \tag{S5a}$$

$$\|\exp(-itH)f - \exp(-itH_0)\,h\| \to 0, \quad t \to -\infty, \tag{S5b}$$

where $\| \quad \|$ is the norm. If we require that this condition be valid for all $f$ which belong to $\mathfrak{H}$, we shall meet some contradictions. Therefore it is assumed that the condition (S5) is valid only for $f$ which belong to a subspace **R** of $\mathfrak{H}$. But it is also assumed that, if $g$ and $h$ are varied in the whole space of $\mathfrak{H}$, $f$ is varied in the whole space of **R**. (S5) can be rewritten in a more convenient form; for arbitrary $g$ and $h$ which belong to $\mathfrak{H}$, there exists $f$ which satisfies the following conditions:

$$\begin{aligned} &\|\exp(itH)\exp(-itH_0)\,g - f\| \to 0, \quad t \to +\infty, \\ &\|\exp(itH)\exp(-itH_0)\,h - f\| \to 0, \quad t \to -\infty. \end{aligned} \tag{S6}$$

These conditions can be written in the following operator form#

$$\begin{aligned} &\exp(itH)\exp(-itH_0) \to \Omega_-, \quad t \to +\infty, \\ &\exp(itH)\exp(-itH_0) \to \Omega_+, \quad t \to -\infty. \end{aligned} \tag{S7}$$

(S7) is our "first assumption" (I). $\Omega_\pm$ is the wave operator.

Let the whole of $f$ corresponding to the whole of $g$ which belongs to $\mathfrak{H}$ be called the "range" of $\Omega_-$. The range of $\Omega_+$ is also analogously defined. That $f$ is common in (S5) is expressed by the following "second assumption" (II.) The ranges of $\Omega_+$ and $\Omega_-$ are the same. This is called **R**.

The two assumptions made above are natural from the physical point of view, but far from clear mathematically. We shall examine later for what Hamiltonian each of these assumptions is valid. The following conclusions come from the assumptions.

First of all, $\Omega_+$ and $\Omega_-$ are isometric, that is,

$$(g, \Omega_-^\dagger \Omega_- g)^{1/2} = \|\Omega_- g\| = \|g\|, \qquad \|\Omega_+ h\| = \|h\|, \quad \text{for all} \quad g, h \subset \mathfrak{H}. \tag{S8}$$

(S8) is clear by observing that $\Omega_-$ (or $\Omega_+$) is the limit of the unitary operator

# The $\Omega_+$, $\Omega_-$ here are the $\Omega_{(+)}$, $\Omega_{(-)}$ in (Q29), (Q31) respectively.

$\exp(itH)\exp(-itH_0)$. But $\Omega_-$ (or $\Omega_+$) is not necessarily unitary. (S8) only means that

$$\Omega_+^\dagger \Omega_+ = 1, \qquad \Omega_-^\dagger \Omega_- = 1 \tag{S9}$$

but not $\Omega_\pm \Omega_\pm^\dagger = 1$. † indicates the adjoint operator. We put

$$\Omega_\pm \Omega_\pm^\dagger = P. \tag{S10}$$

$P$ is the projection operator onto **R**, because $\Omega_\pm$ always map $g \subset \mathfrak{H}$ onto **R**.

According to (S6), $h$ is mapped by $\Omega_+$ onto **R**, and further becomes $g$ by $\Omega_-^{-1}$. Since $\Omega_-^{-1}$ is defined only in **R**, we will use $\Omega_-^\dagger$ instead of $\Omega_-^{-1}$ for convenience of later discussion.

$$g = \Omega_-^\dagger \Omega_+ h. \tag{S11}$$

By definition (S4), we have

$$S = \Omega_-^\dagger \Omega_+. \tag{S12}$$

Since (S12) represents the isometric correspondence from $\mathfrak{H}$ to $\mathfrak{H}$, $S$ is unitary.

$\Omega_\pm$ has the following property,

$$\exp(-itH)\,\Omega_\mp = \Omega_\mp \exp(-itH_0), \qquad -\infty < t < +\infty. \tag{S13}$$

(S13) can be verified as follows. Multiplying (S7) by $\exp(isH)$ and $\exp(-isH_0)$, $(-\infty < s < +\infty)$ from the left and right respectively, we have

$$\exp[i(s+t)H]\exp[-i(s+t)H_0] \to \exp(isH)\,\Omega_- \exp(-isH_0), \quad t \to +\infty. \tag{S14}$$

The left-hand side of (S14) approaches $\Omega_-$ irrespective of the value of $s$, hence we get (S13) from (S14). Taking the adjoint of Equation (S13), we have

$$\Omega_\mp^\dagger \exp(itH) = \exp(itH_0)\,\Omega_\mp^\dagger, \qquad -\infty < t < +\infty. \tag{S15}$$

Therefore, by using (S10),

$$P\exp(itH) = \Omega_-\Omega_-^\dagger \exp(itH) = \Omega_- \exp(itH_0)\,\Omega_-^\dagger = \exp(itH)\,P. \tag{S16}$$

$P$ thus commutes with $e^{itH}$. In other words, **R** is an invariant subspace with respect to $H$. Similarly we get

$$S\exp(itH_0) = \exp(itH_0)\,S. \tag{S17}$$

This means that $S$ commutes with $H_0$. If there is no degeneracy, $S$ can be expressed as a function of $H_0$. (S8)–(S17) are the main results derived from the two assumptions, and some of them have already been explained in Sects. P–R.

It should be noted here that the difference of $\mathfrak{H}$ and **R** is due to the "bound states". See (Q41).

## 2. VALIDITY OF THE TWO ASSUMPTIONS

Does the system which satisfies the two assumptions exist actually? We do not know the general conditions about it, but some results have been obtained for several simple cases. Kato reviews this in the following way.

**a)** Friedrichs (ref. 6) considers the following Hamiltonian

$$H = H_0 + \varepsilon V, \qquad H_0 = \lambda x \quad \text{for} \quad a \leqslant \lambda \leqslant b. \tag{S18}$$

That is, $H_0 f(\lambda) = \lambda f(\lambda)$. $V$ is the hermitian integral operator which satisfies the condition

$$\frac{|V(\lambda, \mu) - V(\lambda', \mu')|}{|\lambda - \lambda'|^\alpha + |\mu - \mu'|^\alpha} \leqq C, \quad 0 < \alpha < 1.$$

This condition on $V(\lambda, \mu)$ is a little stronger than that $V(\lambda, \mu)$ be merely "continuous". It is also assumed that $V(\lambda, \mu)$ is zero at the boundary of the square region [$(a, a)$, $(a, b)$, $(b, b)$ and $(b, a)$].

It can be proved that the assumptions I and II are both correct for sufficiently small $\varepsilon$. He shows that $\mathbf{R} = \mathfrak{H}$, hence $\Omega_\pm$ is unitary.

**b)** The condition for $H_0$ in **a)** is too strong. Aronszajn (ref. 7) imposes a less stringent restriction on $H_0$, namely, $H_0$ is assumed to have an "absolute continuous spectrum". "Absolute continuous" means that the spectrum has a "density". The continuous spectrum which is not absolute continuous is called "singular continuous".# On the other hand, a stronger condition on $V$

# Example of $H_0$, which has a singular continuous spectrum. Consider the problem of finding a non-decreasing continuous function $f(x)$ which has the largest length of curved line which connects the origin and $f(x = 1) = 1$. If we connect (0, 0) and (1, 0), and (1, 0) and (1, 1), the total length is 2. But this is not a continuous function. Some continuous function $f(x)$ with the length 2 can be constructed as follows. Divide $x = 0$ to $x = 1$ into three equal parts, and let the value of $f(x)$ be $\frac{1}{2}$ for the middle part. Divide each of the remaining parts into three equal parts, and let the value of $f(x)$ in the middle part of each be the average value, namely $\frac{1}{4}$, of the values 0 and $\frac{1}{2}$ already fixed on each side. Continue the same procedure at infinitum. Thus constructed, $f(x)$ is continuous according to the definition in the usual differential (and integral) calculus, and has the length 2. We define the norm by $\|\varphi\|^2 = \int_0^1 \varphi(x)|^2\, df(x)$, and set $H_0\varphi(x) = x\varphi(x)$. This $H_0$ has a singular continuous spectrum, because the derivative of $f(x)$ is zero for almost all points $x$.

Usually we assume that there is no singular continuous spectrum. But this is not always true. Aronszajn (ref. 7) has given an example of singular continuous spectrum for the Hamiltonian, $H = -(d^2/dx^2) + V(x)$, $0 \leqq x \leqq \infty$. For the Hamiltonian taken in **e)**, we do not know whether there is no singular continuous spectrum. If $V(x, y, z)$ satisfies the condition described in **g)** below, Ikebe (ref. 3) proves that the singular continuous spectrum does not appear. (Ikebe proves the expansion theorem by eigenfunctions of $H$ taken in **e)** below and discusses $S$ matrix by using it.)

is imposed, that is, $V$ is a degenerate integral operator. This means that the "range" of $V$ is of finite dimension,

$$V(\lambda, \mu) = \sum_{n=1}^{N} \lambda_n \phi_n(\lambda)\, \phi_n^*(\mu). \tag{S19}$$

For these $H_0$ and $V$, the assumptions are valid.

**c)** The restrictions on $V$ in **b)** can be reduced a little bit. $H_0$ and $H$ are assumed to have only absolute continuous spectra. The condition on $V$ imposed by Rosenblum (ref. 8) is

$$V(\lambda, \mu) = \sum_{n=1}^{\infty} \lambda_n \phi_n(\lambda)\, \phi_n^*(\mu), \tag{S20}$$

where $\sum_n |\lambda_n| < \infty$, $\|\phi_n\| = 1$. Under these conditions I and II are satisfied.

**d)** Let the subspace for which $H_0$ (and $H$) has an absolute continuous spectrum be $\mathbf{R}_0$ and $\mathbf{R}$). $V$ is assumed to have the form (S20). It is seen from (S5) or (S6) that, if $g$ belongs to $\mathbf{R}_0$, $f$ belongs to $\mathbf{R}$, and *vice versa*. If $\mathbf{R}_0 = \mathfrak{H}$, I and II are valid, according to T. Kato (ref. 4).

**e)** For the very special case: $H_0 = -\Delta$, $V = V(x, y, z)$

$$\int |V(x, y, z)|^2\, d^3x < \infty, \tag{S21}$$

Cook (ref. 5) was able to prove the first assumption, that is, $\Omega_\pm$ does exist. This is interesting physically. Jauch and Zinnes (ref. 5) and Hack (ref. 5) obtain the same conclusion for $V = \text{const} \times r^{-\beta}$ $(1 < \beta < 3/2)$, and for $V \rightarrow O(r^{-\alpha})$, $r \rightarrow \infty$, $\alpha > 0$, respectively. But since the second assumption has not been proved yet, it is not certain whether $S$ is unitary.#

**f)** If $V$ satisfies the following condition besides (S21)

$$\int |V(x, y, z)|\, d^3\mathbf{x} < \infty, \tag{S22}$$

I and II are both valid. If $V$ decreases faster than $1/r^3$ at large distances, and has a weaker singularity than $1/r^{3/2}$ at the origin, the assumptions (S21) and (S22) may be satisfied. This statement is a consequence of the more general result obtained by Kuroda (ref. 2) by reducing the restriction on $V$ in Kato's treatment which is described in **d)**.

**g)** If, besides the condition (S21), $V(x, y, z)$ is $0(r^{-2-\varepsilon})$ at large distances and is Hoelder continuous except for a finite number of singularities, Ikebe (ref. 3) shows that I and II are both valid. This restriction on $V$ is, apart from

---

# Kato and Kuroda (ref. 9) have given an example in which the wave operator $\Omega$ exists but $S$ is not unitary.

the continuity assumption of $V$, weaker than that of Kuroda ($\varepsilon > 0$). $V$ is not necessarily spherically symmetric.

**h)** Green and Lanford (ref. 11) have recently improved the result above in the case of central potentials; that is, the two assumptions are valid if $V$ is $0(r^{-1-\varepsilon})$ at large distances and $0(r^{-2+\varepsilon})$ at the origin.# Since their approach may be more familiar to the physicist, a brief outline of their method will be given in the following.

## 3. CENTRAL POTENTIAL

The eigenfunction expansion for the Schrödinger equation is first summarized, and then it will be shown how the Equations (S6) etc., are derived.

### *i) Eigenfunction expansion*

The expansion theorem in terms of the eigenfunctions of physical observables is well-known in quantum mechanics. Consider the Hamiltonian

$$H = H_0 + V(r), \quad H_0 = -\Delta/2m, \quad r = \sqrt{x^2 + y^2 + z^2}.$$

Any square integrable function $\varphi(x, y, z)$ can be expanded in terms of the eigenfunctions of the Hamiltonian as follows:

$$\varphi(\mathbf{r}) = \sum_{l,m} r^{-1} Y_l^m(\vartheta, \phi)\left[\sum_n C_{lmn}\psi_{ln}(r) + \int_0^\infty C_{lm}(k)\,\psi_l(r, k)\,dk\right], \tag{S23}$$

where $\psi_{ln}(r)$ is the normalized solutions of the equation

$$\left[-\frac{d^2}{dr^2} + 2mV(r) + \frac{l(l+1)}{r^2}\right]\psi_{ln}(r) = k_{ln}^2\psi_{ln}(r) \tag{S24}$$

for $k_{ln}^2 \leqq 0$. $\psi_l(r, k)$ is the solution for $k > 0$, and has the normalization

$$\psi_l(r, k) \to (2/\pi)^{1/2} \sin\,[kr - l\pi/2 + \delta_l(k)], \quad r \to \infty. \tag{S25}$$

The "coefficients" $C_{lmn}$ and $C_{lm}(k)$ are given by

$$\begin{aligned} C_{lmn} &= \int_0^\infty u_{lm}(r)\,\psi_{ln}(r)\,dr, \\ C_{lm}(k) &= \int_0^\infty u_{lm}(r)\,\psi_l(r, k)\,dr, \end{aligned} \tag{S26}$$

---

# The conditions are more precisely specified by

$$\int_0^R rV(r)\,dr < \infty, \qquad \int_R^\infty V(r)\,dr < \infty$$

and either

$$\int_R^\infty \left|\int_r^\infty V(s)\,ds\right|^2 dr < \infty,$$

or

$$V(r) = O(r^{-1-\varepsilon}) \quad \text{as} \quad r \to \infty.$$

where $u_{lm}(r)$ is the radial wave function defined by

$$\varphi(\mathbf{r}) = \sum r^{-1} u_{lm}(r)\, Y_l^m(\vartheta, \phi).$$

(S23) and (S26) establish a one–one correspondence between $\varphi(\mathbf{r})$ and the set $\{C_{lmn}, C_{lm}(k)\}$. This will be expressed by

$$F(\varphi) = \{C_{lmn}, C_{lm}(k)\}.$$

The transform $F$ with $V(r) = 0$ is denoted by $F_0$. If $V = 0$ there are no bound states so that $C_{lmn} = 0$.

The radial expansion theorem was proved by Kodaira (ref. 12) under the condition that $V(r)$ is continuous on $(0, \infty)$, and that $V(r) = O(r^{-2+\varepsilon})$ as $r \to 0$, and $V(r) = 0(r^{-1-\varepsilon})$ as $r \to \infty$ $(\varepsilon > 0)$. These are the conditions under which the following discussions are also valid.

Since $F$ is defined in terms of the eigenfunction of $H$, the operation of $H$ on $\varphi$ should correspond to the multiplication by $(k^2/2m)$ of the constants $\{C_{lmn}, C_{lm}(k)\}$.

$$F(H\varphi) = \{(k_{ln}^2/2m)C_{lmn}, (k^2/2m)C_{lm}'(k)\}.$$

Similarly we have

$$F_0(H_0\varphi) = \{0, (k^2/2m)C_{lm}'(k)\}, \text{ where } F_0(\varphi) = \{0, C_{lm}'(k)\}.$$

It can also be shown that

$$F(e^{-iHt}\varphi) = \{\exp(-ik_{ln}^2 t/2m)C_{lmn}, \exp(-ik^2 t/2m)C_{lm}(k)\},$$

and the corresponding formula for $H_0$.

### *ii) Asymptotic limits*

We shall now prove the two assumptions: I. (S6) or (S7) is valid and II. the ranges of $\Omega_+$ and $\Omega_-$ are the same. Let (any wave packet) $\Psi$ be defined by [see (R25)]

$$F(\Psi) = \{0, C_{lm}(k)\}, \tag{S27}$$

namely, $\Psi$ is orthogonal to the subspace spanned by the bound states. We shall consider the functions $\Psi_t$ and $\Phi_t$ defined by

$$\Psi_t = e^{-iHt}\Psi = \sum r^{-1}\, Y_l^m(\vartheta, \phi)\, u_{lm}(r, t) \tag{S28}$$

$$\Phi_t = \sum r^{-1}\, Y_l^m(\vartheta, \phi)\, v_{lm}(r, t), \tag{S29}$$

where

$$u_{lm}(r, t) = \int_0^\infty \exp(-ik^2 t/2m) C_{lm}(k)\, \psi_l(r, k)\, dk, \tag{S30}$$

$$v_{lm}(r, t) = \int_0^\infty \exp(-ik^2 t/2m) C_{lm}(k)\, \phi_l(r, k)\, dk, \tag{S31a}$$

$$\phi_l(r, k) = (2/\pi)^{1/2} \sin[kr - l\pi/2 + \delta_l(k)]. \tag{S31b}$$

(S31b) should be compared with (S25).

It is now possible to show that

$$\lim_{|t|\to\infty} \|\Psi_t - \Phi_t\| = 0. \tag{S32}$$

The norm is calculated according to

$$\|\Psi_t = \Phi_t\|^2 = \sum_{l,m} \int_0^\infty |u_{lm}(r, t) - v_{lm}(r, t)|^2\, dr.$$

From (S30) and (S31a) we have

$$u_{lm}(r, t) - v_{lm}(r, t) = \int_0^\infty \exp(-ik^2t/2m) C_{lm}(k)[\psi_l(r, k) - \phi_l(r, k)]\, dk. \tag{S33}$$

The essential point of proving (S32) is to show that

$$\lim_{|t|\to\infty} [u_{lm}(r, t) - v_{lm}(r, t)] = 0$$

for all $0 \leqslant r < \infty$. This is done with the aid of the Riemann–Lebesque theorem,# namely, for very large $t$ the factor $\exp(-ik^2t/2m)$ oscillates very rapidly so that the contributions from the integrand with different $k$ are cancelled out and the integral vanishes.

We shall show that $\Phi_t$ approaches an outgoing and ingoing part as $t \to +\infty$ and $-\infty$, respectively. Defining $\Phi_t^+$ and $\Omega_t^-$ by

$$\Phi_t^\pm = \sum r^{-1}\, Y_l^m(\vartheta, \phi)\, \phi_{lm}^\pm(r, t),$$

$$\phi_{lm}^\pm(r, t) = \int_0^\infty \exp(-ik^2t/2m) C_{lm}(k) \tag{S34a}$$

$$\times (2\pi)^{-1/2} \exp\{\pm i[kr - (l + 1)\pi/2 + \delta_l(k)]\}\, dk. \tag{S34b}$$

From (S29), (S31) and (S34) we have

$$\Phi_t = \Phi_t^+ + \Phi_t^-. \tag{S35}$$

Consider the norm of $\Phi_t^-$ in the limit: $t \to +\infty$

$$\|\Phi_t^-\| = \sum_{l,m} \int_0^\infty |\phi_{lm}^-(r, t)|^2\, dr. \tag{S36}$$

The integrand on the right of (S36) has the factor:

$$\int_0^\infty \exp(-ik^2t/2m - ikr) \cdot C_{lm}(k) \exp[-i\delta_l(k)]\, dk. \tag{S37}$$

---

# If $f(x)$ is integrable over $(a, b)$, then as $t \to \pm\infty$,

$$\int_a^b f(x)\, e^{ixt}\, dx \to 0.$$

The integral (S37) tends toward zero by the Riemann–Lebesque theorem as $t \to +\infty$. (The theorem is not applicable for $t \to -\infty$. See ref. 11.) Therefore

$$\lim_{t \to +\infty} \|\Phi_t^-\| = 0. \tag{S38}$$

Similarly we have

$$\lim_{t \to -\infty} \|\Phi_t^+\| = 0. \tag{S39}$$

From (S32), (S35), (S38) and (S39) we see that $\Phi_t^+$ and $\Phi_t^-$ represent the outgoing and ingoing part of $\Psi_t$.

By using (S28), (S32), (S35), (S38) and (S39) we get

$$\lim_{t \to \pm\infty} \|[\exp(-iHt)]\Psi - \Phi_t^\pm\| = 0. \tag{S40}$$

$\Phi_t^\pm$ can be written as $[\exp(-iH_0t)]\Phi^\pm$, where

$$\Phi^\pm = \sum r^{-1}\, Y_l^m(\vartheta, \phi)\, \phi_{lm}^\pm(r), \tag{S41}$$

$$\phi_{lm}^\pm(r) = \int_0^\infty (2\pi)^{-1/2}\, C_{lm}(k) \exp\{\pm i[kr - (l+1)\pi/2 + \delta_l(k)]\}\, dk.$$

(S40) becomes

$$\lim_{t \to \pm\infty} \|[\exp(-iHt)]\Psi - [\exp(-iH_0t)]\Phi^\pm\| = 0.$$

Thus (S5a) and (S5b) are proved. The wave operators, $\Omega_+$ and $\Omega_-$, are defined by [see (S7)]

$$\Omega_-\Phi_+ = \Psi, \qquad \Omega_+\Phi^- = \Psi.$$

In the course of the derivation $\Psi$ and $\Phi^\pm$ are explicitly given by (S27) and (S41). Therefore it is clear that the ranges of $\Omega_+$ and $\Omega_-$ are the same.

We have sketched a rough outline of Green and Lanford's method. For the mathematically rigorous conditions for the various proofs and on the definition of self-adjoint operators $H$ and $H_0$, the reader is referred to their original paper.

## REFERENCES


1. Jauch, J. M., Helv. Phys. Acta **31**, 127 (1958). Mathematical formulation of the theory of scattering operator with physical interpretation.

2. Kuroda, S. T., Nuovo Cimento **12**, 431 (1959); J. Math. Soc. Japan, **11**, 247 (1959). Existence and the unitary property of scattering operators. (Subsection f.)

3. Ikebe, T., Archive for Rational Mechanics and Analysis, **5**, 1 (1960). (Subsections f and g.)

4. Kato, T., Journ. Math. Soc. Japan **9**, 239 (1957); Proc. Japan Acad. **33**, 260 (1957). (Subsection d.)

5. Cook, J. M., J. Math. Phys. Mass. Inst. Tech. **36,** 82 (1957); Jauch, J. M. and Zines, I. I., Nuovo Cimento **11,** 553 (1959); Hack, M. N., Nuovo Cimento **9,** 731 (1958). (Subsection e.)

6. Friedrichs, K. O., Comm. Pure Appl. Math. **1,** 361 (1948). (Subsection a.)

7. Aronszajn, N., Amer. J. Math. **79,** 597 (1957). (Subsection b.)

8. Rosenblum, M., Pacific Journ. Math. **7,** 997 (1957). (Subsection c.)

9. Kato, T. and Kuroda, T., Nuovo Cimento **14,** 1102 (1959). (Subsection e.)

10. Hack, M. N., Nuovo Cimento **13,** 231 (1959); Jauch, J. M., Helv. Phys. Acta **31,** 661 (1958). Mathematical treatment of scattering when composite particles exist.

11. Green, T. A. and Lanford, O. E., J. Math. Phys. **1,** 139 (1960). (Subsection 3.)

12. Kodaira, K., Am. J. Math. **71,** 921 (1949).

chapter

5

# Nuclear Reactions

The study of nuclear scatterings (including reactions) is one of the most important subjects in nuclear physics. The research in this field may be roughly divided into two categories, namely, the investigation of the law of nuclear forces between two nucleons (discussed in Chapter 2), and the study of the reaction (which will include elastic, inelastic, rearrangement, and radiative capture) mechanisms and the nuclear structure.

In the studies of nuclear reactions, it has been found that they can be grouped in a few general categories each of which is characterized by certain features and is conveniently describable in terms of a specific theoretical picture or model. There is a large class of observed reactions which are characterized, among other features, by a resonance phenomenon (namely, sharp maxima in the cross sections at certain energies of the collision). These reactions are best understood on the basis of the "compound nucleus" theory. The first theoretical treatment of the resonance phenomenon is that of Breit and Wigner (the one-level formula). More general theories, the many level formula of Kapur and Peierls and the $R$ matrix of Wigner and Eisenbud, have been developed. There is another large class of observed reactions which are called the "direct reactions" and which include the "stripping" and the "pick-up" processes. They are characterized by certain angular distributions of the scattered or outgoing particles that can best be understood by regarding the reactions as having involved only the interaction between the incident particle and the "outer" nucleons of the target nucleus. There is another class of reactions, namely, the excitation of a nucleus by its Coulomb interaction with the incident particle.

The literature in the experimental and the theoretical aspects of each one of these classes of nuclear reactions is very big. It is not the aim of the present work to treat the many topics of nuclear reactions. We shall attempt only to give a brief account of the main features of the "compound nucleus" processes and the "direct processes" and the theoretical methods for treating them. Thus in Sect. T we shall take up the "compound nucleus" processes. This section may be regarded as a summary of the basic facts and ideas in preparation for the more general theoretical treatments by means of the $S$ and the $R$ matrices in Sects. W and X. These two sections should perhaps be read together with Sect. T, or might even precede Sect. T, except that, without the physical problems presented by the reactions mentioned in Sect. T, the theory of the $R$ matrix will be too abstract.

In Sect. U, we give a brief account of the phenomenological method of describing certain nuclear collisions by means of an empirical complex potential (which has been treated in Sect. N in another connection).

In Sect. V, we describe the simple form of the theory of direct processes, such as the deuteron stripping and the pick-up processes.

As emphasized above, we include the cursory treatments in these sections mainly to indicate the nature of the methods for treating different nuclear collision processes. For further detailed accounts of the theoretical developments and the experimental data, the reader must be referred to the various recent review articles quoted at the end of each section.

section

# T Resonance Reactions

This section will deal with the phenomena of sharp, isolated resonance reaction. The concept of the compound nucleus plays an important role in this case.

## 1. EMPIRICAL FACTS

It has long been known that some reaction cross sections are abnormally large for particular energies, $E_{res}$, of the incident particle. This phenomenon was first noticed by several workers in 1935 who observed sharp isolated resonance peaks for slow neutrons. The reaction cross section near the peak can be represented by

$$\sigma = \frac{\text{const.}}{(E_{res} - E)^2 + \frac{1}{4}\Gamma^2}. \tag{T1}$$

$\Gamma$ is the width of the resonance level at $E_{res}$, which will be shown later to be the sum of the partial widths $\Gamma_\alpha$ of all possible reactions.

Such a law as (T1) applies to most of the low energy reaction phenomena. The general nature of the many reactions can be summarized by the following empirical results. For heavy nuclei (mass number $A \gtrsim 100$), the resonance radiative capture of neutrons (below 1 KeV), and the elastic scattering of neutrons (below 1 MeV) are the main processes. The spacing of resonance peaks (corresponding to the same $J$ and parity) is of the order of $\gtrsim 10^2 \sim 10^3$ eV. The width $\Gamma_c$ ($c$ for radiative capture) is of the order of 0.1 eV and does not depend on the incident energy, while $\Gamma_n$ ($n$ for elastic scattering of neutron) is of the order of 0.1 eV at a KeV and 1 eV at 100 KeV.

$\Gamma_n$ is roughly proportional to the square root of the neutron energy. Inelastic scatterings of neutrons, and neutron emitting reactions caused by charged particles become detectable above, say, 0.5 MeV. For intermediate nuclei (for example, $A \sim 30$), the predominant reaction below 1 MeV is the elastic scattering of neutrons. See Fig. 1 on p. 376 (ref. 12). The level spacing is of the order of $10^2 \sim 10^3$ KeV, and the width $\Gamma_n$ is of the order of $1 \sim 10$ KeV at 100 KeV, and $1 \sim 10^2$ KeV at 1 MeV. Neutron emitting reactions caused by charged particles can sometimes be detected below 1 MeV. Radiative capture is less important.

For energies above 1 MeV, many other types of reactions are possible. The "stripping" $(d, p)$ reaction (i.e., deuteron incident and proton emergent) is sometimes important even for very low energies because of the loosely bound nature of the deuteron. This reaction will be discussed separately in Section V.

## 2. COMPOUND NUCLEUS MODEL

Historically, the resonance peak of nuclear reaction and the large radiative capture cross section were interpreted by the compound-nucleus model proposed by Bohr. According to this theory the sharp peak of the reaction cross section is due to the existence of a quasi-stable nuclear state for the particular value of $E_{res}$. When the particle $A$ (for example, neutron) with this energy approaches the target nucleus $B$, to within the range of nuclear forces, a compound nucleus $N$ is formed. This compound nucleus will then disintegrate into two (or more than two) products $C$ and $D$. Thus the nuclear reaction is regarded as consisting of two successive steps,

$$A + B \rightarrow N, \qquad N \rightarrow C + D.$$

Bohr assumes that the decay of the compound nucleus $N$ (the second step above) is independent of the process by which the compound nucleus is formed (i.e., the first step). The argument for this assumption is that on account of the strong, short-ranged nuclear forces, the incident particle $A$ rapidly shares its energy with all the particles of the system $A + B$, forming the compound nucleus $N$ which can be considered as one of the excited states of the nucleus composed of $A$ and $B$. (This is usually a highly excited state.)

The metastable nature of such a decaying state as $N$ may be represented by a complex energy eigenvalue $(E_d \equiv E_0 - \frac{1}{2}i\Gamma_d)$, as will be discussed in Sect. W. The mean lifetime $\tau$ equals $\hbar/\Gamma_d$.

$$\Gamma_d = \hbar/\tau. \tag{T2}$$

It will be shown in the following that the decay width $\Gamma_d$ is actually the same as the $\Gamma$ appearing in the cross section formula (T1) for resonance reaction.

The width of 0.1 eV corresponds to $\tau \sim 10^{-4}$ sec. This lifetime is long enough to define a compound state, compared with the $\sim 2 \times 10^{-20}$ sec. taken by an incident neutron with energy 1 KeV to traverse a nucleus, namely, $10^{-12}$ cm for a heavy nucleus.

### 3. ELASTIC SCATTERINGS OF *S*-WAVE NEUTRONS

For simplicity, consider the case where only the elastic scattering of $s$ neutrons is possible. Let us assume that the incoming neutron does not interact with the target nucleus for $r > R$, $R$ being the radius of a "nuclear sphere" with a sharp boundary. The radial wave function $u(r)$ of the relative motion, $\Psi = \{u(r)/r\}\, Y_{0,0}$; $r > R$, satisfying the equation

$$(d^2/dr^2 + k^2)\, u = 0, \quad r > R,$$

can be written in terms of the scattering matrix $S$ (A34)

$$u = A[e^{-ikr} - Se^{ikr}]; \quad r > R. \tag{T3}$$

$S$ can be determined in principle by solving the Schrödinger equation inside the nucleus, but $S$ can also be expressed by the logarithmic derivative, $f$, of $u$ at the nuclear surface $r = R$,

$$f = \frac{R}{u}\frac{du}{dr} \quad \text{at} \quad r = R. \tag{T4}$$

From (T3) and (T4), we get

$$S = \frac{f + ikR}{f - ikR} \exp(-2ikR). \tag{T5}$$

It is clear that, if $f$ is real, $S$ is a number of modulus 1, and there is no reaction (except elastic scattering), (see Sect. A). On using (T5) the scattering cross section is obtained from (A37),

$$\sigma_s = \frac{\pi}{k^2} |A_{res} + A_{pot}|^2 \tag{T6}$$

with

$$A_{res} = -2ikR/(f - ikR), \qquad A_{pot} = \exp(2ikR) - 1. \tag{T7}$$

$A_{pot}$ is called the "potential scattering" term, and represents the elastic scattering from a hard core (i.e., impenetrable sphere) of radius $R$. This can be seen by considering that, for the hard core scattering, $f$ in (T4) would be infinitely large (because of $u(R) = 0$) and $A_{res}$ in (T6) vanishes. $A_{res}$ will be shown below to give a resonance phenomenon.

The sharp separation of the scattered amplitude into $A_{res}$ and $A_{pot}$ is possible only because we have assumed the model of a sharply defined radius $R$ for the interaction. While such a separation is not realistic, this model is very useful for bringing out certain features of nuclear reactions.

### 4. CONDITION OF RESONANCE

We shall now consider a qualitative feature of the inside wave function. The nucleus may be replaced by an optical potential (see following Sect. U) in the first approximation. Then the neutron wave function inside the nucleus may be represented by

$$u_{\text{in}}(r) = C \cos [Kr + \alpha(E)].$$

$K$ is the wave number of the neutron inside the nucleus,

$$K = \sqrt{2m(E + V_0)}/\hbar,$$

where $V_0$ is estimated as 30 ~ 40 MeV, and $\hbar^2k^2/2m = E$ = energy of the incident neutron. The outside wave function (T3) can be rewritten as

$$u_{\text{out}}(r) = D \cos [kr + \beta(E)].$$

Equating both functions and their derivatives at $r = R$, we have

$$\frac{C}{D} = \sqrt{\frac{(KR)^2 + f^2}{(kR)^2 + f^2}} \times \frac{k}{K}, \quad f = -KR \tan [KR + \alpha(E)]. \tag{T8}$$

When $f$ is of order of magnitude of $KR$, we have $C/D \sim k/K$, i.e., the incident wave does not enter much into the nucleus if the incident energy is much smaller than the kinetic energy of the neutron inside the nucleus. $C/D$ is largest and equals 1 for $f = 0$. This particular state corresponds to "resonance" for which the condition is

$$f(E_{res}) = 0. \tag{T9}$$

In the neighborhood of $E_{res}$, $f(E)$ can be expanded as,

$$f(E) = (E - E_{res}) \left.\frac{df}{dE}\right|_{E=E_{res}} + \cdots. \tag{T10}$$

If $\Gamma$ is defined by

$$\Gamma \equiv -2kR \Big/ \left(\frac{df}{dE}\right)_{E=E_{res}}, \tag{T11}$$

the resonance amplitude $A_{res}$ in (T7) becomes

$$A_{res} = \frac{i\Gamma}{(E - E_{res}) + i\Gamma/2}.$$

Therefore the total elastic cross section is given by

$$\sigma = \frac{\pi}{k_\alpha^2} \left| A_{pot} + \frac{i\Gamma}{(E - E_{res}) + i\Gamma/2} \right|^2. \tag{T12}$$

In the vicinity of $E = E_{res}$, $A_{pot}$ is relatively small compared with the resonance part. If $A_{pot}$ is omitted (T12) becomes

$$\sigma = \frac{\pi}{k_\alpha^2} \frac{\Gamma^2}{(E - E_{res})^2 + \Gamma^2/4}. \tag{T13}$$

## 5. CONNECTION WITH DECAY WIDTH $\Gamma_d$

The condition that $C/D$ be maximum at resonance has led to the correct resonance formula (T1). It may be interesting to note that, in the neighborhood of $E_{res}$, if $(f/KR)^2$ is negligible compared with unity in (T8), $|C/D|^2$ becomes

$$|C/D|^2 = \frac{(\frac{1}{2}\Gamma)^2}{(E - E_{res})^2 + (\frac{1}{2}\Gamma)^2}.$$

Thus, in the scattering process, the incident neutron penetrates into the nucleus appreciably only for $E_{res} - \Gamma < E < E_{res} + \Gamma$. If the energy is outside of this range the incident wave is mainly reflected by the nuclear surface, and the nuclear scattering resembles the hard core scattering with radius $R$.

We have mentioned in 2 that according to the Bohr theory the resonance scattering is due to the existence of the quasi-stable compound-nucleus state. Consider a decaying state. The radial wave function has now outgoing waves only. From the wave function $\psi = c \exp i[kr - (E_d/\hbar)t]$ for a decaying state, and the logarithmic derivative at $R$, we get, for a decaying state,

$$f(E_d) = ikR. \tag{T14}$$

Under the assumption that the imaginary part of $E_d$ is small compared with the real part of $E_d$ in magnitude, $k$ in (T14) is nearly the same as $k$ in (T11). Again expanding $f(E_d)$ around $E = E_{res}$, and by using (T9) and (T10), we have

$$f(E_d) = (E_d - E_{res}) \left.\frac{df}{dE}\right|_{E=E_{res}} + \cdots. \tag{T14a}$$

On substituting (T11) into (T14a) and using (T14), we obtain for the imaginary part of $E_d$

$$\text{Im}\,(E_d) = -\Gamma/2.$$

The real part of $E_d$ coincides with $E_{res}$. Thus the eigenvalue of the decaying state, $E_d \equiv E_0 - i(\Gamma_d/2)$, is

$$E_d = E_{res} - i(\Gamma/2).$$

Thus it is proved that the resonance width $\Gamma$ is identical with the decay width of the compound nucleus. In a closely parallel way we can show that the total width $\Gamma$ appearing in (T1) is equal to the decay width of the corresponding compound nucleus.

## 6. REACTION WIDTH

From the time dependence $\exp(-iEt/\hbar)$ of the wave function, it follows that $\Gamma/\hbar$ gives the decay rate. Let $\Gamma_\beta$ be the width for the decay in a particular mode, or channel, $\beta$. (See Chapter 6, Sect. W.) $\beta$ specifies the nature of decay particles (including $\gamma$-ray), decay energy, the various angular momentum states and so on. The total decay rate is the sum of $\Gamma_\beta$ for all possible decays,

$$\Gamma = \sum_\beta \Gamma_\beta.$$

The "partial width" $\Gamma_\beta$ is proportional to the following three factors $P$, $Q$, $S$.

**i)** *Penetration factor P* through a barrier. The ejected particle $C$ must penetrate the centrifugal barrier (and the Coulomb barrier if $C$ is a charged particle [see (A51)].

**ii)** *Transmission factor Q* through the nuclear surface. Suppose that there is no centrifugal or Coulomb barrier and that the particle $C$ has sufficient energy to leave the nucleus. The particle $C$ can still be reflected at the nuclear surface because of the sudden change of the "index of refraction" at the surface. If $T$ is the transmission probability, then $T = PQ$.

**iii)** *Separable probability S*. The compound nucleus is represented mainly by a wave function corresponding to a uniform distribution of the total energy over all the particles. However, there is a chance ($S$) that a sufficient energy is concentrated in only one particle (or group of particles) $C$ so that $C$ can go out from the compound system. Let us estimate these factors.

We assume some simplified wave function for a neutron which is just inside the nuclear surface and ready to go out of the surface. Consider the radial part of an $l$-wave in an attractive square well potential with depth $V_0$ and range $R$ and a centrifugal barrier outside the nucleus. The outgoing wave with amplitude $a$ is reflected by the boundary $r = R$ and by the centrifugal potential. Let the amplitude of the reflected wave be $b$. The inside wave function then becomes

$$u_l(r) = ae^{iKr} - be^{-iKr}, \qquad K = \sqrt{k^2 + V_0}; \quad r < R. \tag{T15}$$

For $r > R$, let the wave function be written

$$u_l(r) = Cu_l^{(+)}(r); \quad r > R, \tag{T16}$$

where $u_l^{(+)}(r)$ is a solution of

$$\frac{d^2u_l}{dr^2} + \left(k^2 - \frac{l(l+1)}{r^2}\right)u_l(r) = 0, \tag{T16'}$$

namely,

$$u_l^{(+)}(r) = ikr[j_l(kr) + in_l(kr)] \equiv i[F_l(kr) + iG_l(kr)]$$
$$= ikrh_l^{(1)}(kr) \to \exp\left[i\left(kr - \frac{l}{2}\pi\right)\right], \quad r \to \infty, \qquad \text{(T16a)}$$

where $j_l$, $n_l$, $h_l^{(1)}$ are Bessel functions as defined in (A11). On denoting the real and the imaginary part of the logarithmic derivative by

$$\left[\frac{R}{u_l^{(+)}(r)} \frac{du_l^{(+)}(r)}{dr}\right]_{r=R} \equiv \rho + i\kappa \qquad \text{(T17)}$$

and equating this to the logarithmic derivative of $u_l(r)$ in (T15), we get

$$\frac{b}{a} = \frac{\rho_l + i(\kappa_l - KR)}{\rho_l + i(\kappa_l + KR)} \exp(2iKR).$$

The transmission probability $T$ of the neutron is then#

$$T = \frac{|a|^2 - |b|^2}{|b|^2} = \frac{4kKv_l}{K^2 + (2kK + k^2w_l)\, v_l}, \qquad \text{(T18)}$$

where

$$v_l \equiv [F_l^2(kR) + G_l^2(kR)]^{-1}, \qquad k^2w_l \equiv \left[\left(\frac{dF_l}{dr}\right)^2 + \left(\frac{dG_l}{dr}\right)^2\right]_{r=R}. \qquad \text{(T19)}$$

In (T18), use has been made of the property that $F_l$, $G_l$ being two independent solutions of the equation for $u_l$, $F_lG_l' - F_l'G_l = \text{constant} = k$.

For $k^2/K^2 \ll 1$,## (T18) reduces to

$$T \simeq \frac{4kv_l}{K}. \qquad \text{(T20)}$$

The functions $v_l$, $w_l$ for $l = 0, 1$ are ###

$$v_0 = 1, \qquad v_1 = \frac{(kR)^2}{1 + (kR)^2},$$

$$w_0 = 1, \qquad w_1 = \frac{1}{(kR)^2} + \left[1 - \frac{1}{(kr)^2}\right]^2.$$

The penetration factor $P$ due to the centrifugal barrier should be 1 for $l = 0$.

---

# For incoming charged particles, $u_l^{(+)}(r)$ is a linear combination of the regular and irregular solutions of the Coulomb field (see ref. 12 of Sect. A and Appendix to Sect. E and ref. 4 there), and the expression for $T$ is more complicated. Also ref. 1, p. 362.

## On the optical model, $V_0 \simeq 30$ MeV, and $KR \sim 5$ for atomic weight $A \gtrsim 100$.

### For charged particles,

$$v_0 = \frac{1}{v} \exp[-\,2\pi z_i z_t e^2/\hbar v],$$
$$z_i z_t > 0, \qquad v \to 0.$$

From $T = PQ$, the transmission factor $Q$ can now be written for $l = 0$ as

$$Q = \frac{4kK}{(K + k)^2} \sim \frac{4k}{K}. \tag{T21}$$

On combining (T18), (T20) and (T21), the penetration factor $P$ is seen to be roughly proportional to $v_l$

$$P \sim v_l. \tag{T22}$$

We have no actual methods for estimating the "separable probability" $S$; but it obviously depends on the nature of the compound-nucleus state and the nature of the channel $\beta$ of the decay. We shall introduce the "reduced width", $\gamma^2$, in this connection,

$$\Gamma_\beta \equiv k v_\beta \gamma_\beta^2. \tag{T22a}$$

Since the transmission factor $Q$ is proportional to $k$ [see (T21)] and $v_\beta$ is the penetration factor, it is seen that the reduced width $\gamma_\beta^2$ is proportional to the separability factor $S$.

Let us estimate the order of magnitude of $\gamma_\beta^2$. First consider that the resonance levels are evenly spaced with spacing $D$. For the energy range in which $K$ can be considered as a constant, the phase $\alpha(E)$ in (T8) will be a linear function of the incident energy $E$. Considerable deviation from linearity is quite possible depending on the particular nature of the compound level, but we can assume that the linear dependence will give an average magnitude of $\gamma_\beta^2$. Thus, $f$ in (T8) may be written

$$f = -KR \tan (a + bE),$$

where $a$ and $b$ are constants. Since the position of the level is given by the condition (T9), $b$ is connected with the level spacing $D$ as

$$b = \pi/D.$$

On using (T11), $\Gamma$ is given by

$$\Gamma \sim -2kR/(-KRb) = (4k/K)(D/2\pi). \tag{T23}$$

The formula (T8) is only applicable to the $s$ scattering of neutron, and (T23) must be multiplied by the penetration of factor $v$,

$$\Gamma_\beta \sim (4k/K)\, v_\beta(D/2\pi). \tag{T24}$$

The reduced width $\gamma_\beta^2$ is therefore expected to be of the order of magnitude:

$$\gamma_\beta^2 \sim \frac{4}{K}\cdot\frac{D}{2\pi}. \tag{T25}$$

$\gamma_\beta^2$ is, as seen from the definition (T22a), independent of the penetrability and the reflection factor at the nuclear surface, and may be considered to characterize the nature of the compound nucleus state itself. We shall introduce the "strength function $\xi$"

$$\xi_\beta \equiv \gamma_\beta^2/D \sim 2/(\pi K). \tag{T26}$$

The numerical value of the statistical average, $2/(\pi K)$, is about $6 \times 10^{-14}$ cm for an actual nucleus. (For further detail on $\xi$ see the following section on the optical model.)

It will be shown that the reduced width $\gamma_\beta^2$ can be expressed in terms of the wave amplitude at the nuclear surface. This fact may be inferred from the fact that $\gamma_\beta^2$ is proportional to the separability of the decay particle from the compound state. The decay probability per unit time into the particular mode (i.e., channel) $\beta$ is $\Gamma_\beta/\hbar$, which is equal to the number of decay particles through the sphere of radius $R$.

$$\Gamma_\beta/\hbar = \frac{\hbar}{2im_\beta} \int \left( u_\beta^{(+)*} \frac{du_\beta^{(+)}}{dr} - u_\beta^{(+)} \frac{du_\beta^{(+)*}}{dr} \right)_{r=R} |Y_{lm}|^2 \, d\Omega. \tag{T27}$$

$u_\beta^{(+)}$ is the radial wave function of the outgoing particle as (T16). The evaluation of the integral (T27) gives

$$\Gamma_\beta = (\hbar^2 k v_\beta/m_\beta)|u_\beta^{(+)}(R)|^2,$$

where $v_\beta$ is the penetration factor defined in (T19); $m_\beta$, reduced mass. The reduced width, $\gamma_\beta^2$, is thus given by $u_\beta^{(+)}(R)$,

$$\gamma_\beta^2 = \frac{\hbar^2}{m_\beta} |u_\beta^{(+)}(R)|^2 \sim \frac{4}{K} \cdot \frac{D}{2\pi}. \tag{T28}$$

The compound nucleus wave function $\psi$ can be expanded in terms of the complete set of wave functions $\phi_\beta(\boldsymbol{\xi})$ of the residual nucleus and the emitted particle,

$$\psi(\boldsymbol{\xi}, \mathbf{r}) = \sum_\beta \psi_\beta(\mathbf{r}) \, \phi_\beta(\boldsymbol{\xi}). \tag{T29}$$

$\psi_\beta(\mathbf{r})$ is the wave function of their relative motion of which the radial part is $u_\beta(\mathbf{r})$. $\psi_\beta(\mathbf{r})$ may have the form $u_\beta^{(+)} Y_{lm}/r$, if $\psi_\beta(\mathbf{r})$ corresponds to a decay channel. If $\psi(\boldsymbol{\xi}, \mathbf{r})$ can be written, as a good approximation, as the product of $u_{\beta_0}^{(+)}$ and $\phi_0(\boldsymbol{\xi})$ (single particle model, subscript 0 indicates the ground state of residual nucleus),

$$\psi(\boldsymbol{\xi}, \mathbf{r}) = u_{\beta_0}^{(+)}(r) \, Y_{lm}(\vartheta, \phi) \, \phi_0(\boldsymbol{\xi})/r,$$

the wave function $u_{\beta_0}^{(+)}$ will have the order of magnitude $R^{-1/2}$,

$$\int_{r<R} |\psi|^2 \, d\mathbf{r} \, d\boldsymbol{\xi} = \int_0^R |u_\beta^{(+)}(r)|^2 \, dr = 1.$$

Therefore the maximum value of $\gamma_\beta^2$ (i.e., "single particle width") has the order of magnitude:

$$\gamma^2_{\beta(\max)} = \frac{\hbar^2}{m_\beta R}. \tag{T30}$$

The ratio of $\gamma_\beta^2$ to $\gamma^2_{\beta(\max)}$ gives us a rough idea about the amount the single particle wave function for the particular decay process is included in the compound nucleus wave function $\psi(\boldsymbol{\xi}, \mathbf{r})$.

## 7. THE BREIT-WIGNER DISPERSION FORMULA

The elastic cross section of $s$-wave neutron, given by (T12) and (T13), will be extended to cover the nuclear reaction in which several final channels are possible.

Consider the neutron induced reaction first.# The cross section formula will be derived by means of the logarithmic derivative of the wave function at the nuclear surface as in 3. The incident beam is described by the plane wave [see (A2)],

$$\psi \to \sum_l \frac{(2l+1)i^{l+1}}{2kr}\left\{\exp\left[-i\left(kr - \frac{l\pi}{2}\right)\right] - \exp\left[i\left(kr - \frac{l\pi}{2}\right)\right]\right\} \times P_l(\cos\vartheta), \quad (r \to \infty). \tag{T31}$$

The plane wave is modified by the nuclear interaction, and the outgoing amplitude and the phase are changed accordingly,

$$\psi \to \sum_l \frac{(2l+1)\, i^{l+1}}{2kr}\left\{\exp\left[-i\left(kr - \frac{l\pi}{2}\right)\right] - |\eta_l| \exp(2i\delta_l) \times \exp\left[i\left(kr - \frac{l\pi}{2}\right)\right]\right\} P_l(\cos\vartheta), \quad (r \to \infty). \tag{T32}$$

If $|\eta_l| = 1$, only elastic scattering is possible. If $|\eta_l| < 1$, the outgoing flux decreases compared with the outgoing term of the plane wave, that is, some of the incident particle is removed to different channels, and this causes (non-elastic) reactions.## $|\eta_l| \exp(2i\delta_l)$ will be written simply as $\eta_l$ hereafter, which will be called the "amplitude factor".

The elastically scattered wave $\psi_{sc}$ is the difference between (T31) and (T32),

$$\psi_{sc} = \sum \frac{(2l+1)\, i^{l+1}}{2kr}(1 - \eta_l) \exp\left[i\left(kr - \frac{\pi l}{2}\right)\right] P_l(\cos\vartheta).$$

---

# For the charged particle induced reaction, the modification of the theory is easy, but it is convenient to make use of the numerical tables given by Bloch *et al.*, Rev. Mod. Phys. **23,** 147 (1951). Also see ref. 6 in Appendix to Sect. E.

## But see the discussion preceding (W9).

Hence the elastic total cross section is given by

$$\sigma_{sc} = \sum_l \frac{(2l + 1)\pi}{k^2} |1 - \eta_l|^2. \tag{T33}$$

We shall call the non-elastic cross section (including inelastic, capture and rearrangement collision) collectively the "reaction cross section $\sigma_r$". $\sigma_r$ is obtained by counting the difference between the incoming flux (not incident flux) and the outgoing flux in the incident channel, $\alpha$, and dividing the difference by the incident flux $v$ (velocity),

$$\sigma_r = \sum_l \frac{(2l + 1)\pi}{k^2} (1 - |\eta_l|^2). \tag{T34}$$

We shall express $\eta_l$ in terms of the logarithmic derivative $f_l$,

$$f_l = \frac{R}{u_l}\frac{du_l}{dr}\bigg|_{r=R}. \tag{T35}$$

$u_l$ is a linear combination of the outgoing and incoming radial functions, $u_l^{(+)}$ and $u_l^{(-)}$,

$$u_l(r) = au_l^{(-)}(r) + bu_l^{(+)}(r), \quad \eta_l = -a/b, \tag{T36}$$

where $u^{(+)}$ is expressed by $F_l(kr)$ and $G_l(kr)$ defined in (T16a).

$$\begin{aligned} u_l^{(\pm)}(r) &= -G_l(kr) \pm iF_l(kr) \\ &= kr[-n_l(kr) \pm ij_l(kr)]. \end{aligned} \tag{T37}$$

From (T35) and (T36), we get

$$\eta_l = \frac{f_l - \rho_l + i\kappa_l}{f_l - \rho_l - i\kappa_l} \exp(2i\xi_l), \tag{T38}$$

where $e^{2i\xi} \equiv u_l^{(-)}(R)/u_l^{(+)}(R)$, and $\rho_l$, $\kappa_l$ are defined in (T17). $\kappa_l$ is just $kRv_l$. (T38) reduces to $S_l$ in (T5) because of $\rho_0 = 0$. The elastic cross section can be derived from (T38),

$$\sigma_{sc} = \sum_l \frac{(2l + 1)\pi}{k^2} |A_{pot}^l + A_{res}^l|^2, \tag{T39}$$

where

$$A_{pot}^l = \exp(-2i\xi_l) - 1, \tag{T40}$$

$$A_{res}^l = \frac{-2i\kappa_l}{[\mathrm{Re}(f_l) - \rho_l] + i[\mathrm{Im}(f_l) - \kappa_l]}. \tag{T41}$$

The equations (T39), (T40) and (T41) are direct extensions of Equations (T6) and (T7). $A_{pot}^l$ is again the scattering amplitude from the impenetrable sphere of radius $R$. The magnitude of $A_{pot}^l$ is very small in low energy scattering for $l \geqslant 1$ because of the centrifugal barrier. The reaction cross section $\sigma_r$ is also derivable by using (T38),

$$\sigma_r = \sum_l \frac{(2l+1)\,\pi}{k^2} \frac{-4\kappa_l \operatorname{Im}(f_l)}{[\operatorname{Re}(f_l) - \rho_l]^2 + [\operatorname{Im}(f_l) - \kappa_l]^2}. \tag{T42}$$

The logarithmic derivative $f_l$ is no longer a real function of energy, hence the condition of resonance (T9) can not be directly applied to the present case. The formal resonance energy $E_s$ is defined by

$$\operatorname{Re}[f_l(E_s)] = 0. \tag{T43}$$

On expanding $f_l(E)$ about $E = E_s$,

$$f_l(E) = (E - E_s)\left(\frac{\partial f_l}{\partial E}\right)_{E=E_s} - ia + \cdots, \tag{T44}$$

where $-ia$ is the imaginary part of $f_l(E_s)$, and defining $\Gamma_\alpha$ and $\Gamma_r$ by

$$\Gamma_\alpha \equiv -2\kappa \Big/ \left(\frac{\partial f}{\partial E}\right)_s, \qquad \Gamma_r \equiv -2a \Big/ \left(\frac{\partial f}{\partial E}\right)_s, \tag{T44$'$}$$

the resonance amplitude $A^l_{res}$ can be rewritten as

$$A^l_{res} = \frac{i\Gamma_\alpha}{E - E_{res} + i\Gamma/2}, \tag{T45}$$

where $\Gamma \equiv \Gamma_\alpha + \Gamma_r$, and $E_{res} = E_s + \Delta s$, $\Delta s \equiv \rho_l/(\partial f/\partial E)_s$. $\Gamma$ is called the total width. $\Delta s$ is a small quantity and represents the difference of the actual resonance energy $E_{res}$ and the formal definition of resonance energy $E_s$, (T43). $\Delta s$ vanishes for $l = 0$. The elastic cross section is reduced by (T45) to

$$\sigma^{(0)}_{sc} = \frac{\pi}{k^2}\left|[\exp(2ikR - 1] + \frac{i\Gamma_\alpha}{E - E_{res} + i\Gamma/2}\right|^2; \quad \text{for} \quad l = 0, \tag{T46}^{\#}$$

$$\sigma^{(l)}_{sc} = \frac{(2l+1)\,\pi}{k^2} \frac{\Gamma_\alpha^2}{(E - E_{res})^2 + (\Gamma/2)^2}; \qquad \text{for} \quad l \geqslant 1. \tag{T47}$$

$A_{pot}$ has been omitted for $l \geqslant 1$. By using (T44), (T44′) and (T42), we obtain the reaction cross section,

$$\sigma^{(l)}_r = \frac{(2l+1)\,\pi}{k^2} \frac{\Gamma_\alpha \Gamma_r}{(E - E_{res})^2 + (\Gamma/2)^2}, \quad \text{for all } l. \tag{T48}$$

Sharp resonance occurs for a particular $l$ at the particular energy $E^{(l)}_{res}$.

---

# (T46) can be equivalently rewritten

$$\sigma^{(0)}_{sc} = \frac{4\pi}{k^2}\left|e^{ikR}\sin kR + \frac{\frac{1}{2}\Gamma_\alpha}{E - E_{res} + i\Gamma/2}\right|^2. \tag{T46a}$$

Equations (T46), (T47) and (T48) are the one-level Breit–Wigner formulas. The reaction width $\Gamma_r$ is the sum of the partial width $\Gamma_\beta$ ($\beta \neq \alpha$),

$$\Gamma_r = \sum_{\beta \neq \alpha} \Gamma_\beta, \qquad \Gamma = \Gamma_\alpha + \Gamma_r. \tag{T49}$$

The reaction cross section from the channel $\alpha$ into the channel $\beta$ is, analogously to (T48),

$$\sigma_{\alpha\beta}^{(l)} = \frac{(2l+1)\,\pi}{k^2}\,\frac{\Gamma_\alpha\Gamma_\beta}{(E-E_{res})^2 + (\Gamma/2)^2}. \tag{T50}$$

The formulas (T47), (T48) and (T50) can be interpreted as follows. The cross section of the compound nucleus formation is

$$\sigma_{\alpha c}^{(l)} = \frac{(2l+1)\,\pi}{k^2}\,\frac{\Gamma_\alpha\Gamma}{(E-E_{res})^2 + (\Gamma/2)^2}. \tag{T51}$$

The $\gamma$-ray decay is also included in the decay channel $\beta$. The predominant radiative capture of neutrons for heavy nucleus is expressed by the radiative decay width $\Gamma_\gamma$ being much larger than the particle width $\Gamma_\alpha$ and $\Gamma_\beta$ ($\beta \neq \gamma$, $\neq \alpha$). The $\Gamma_\alpha$ is small because $k/K$, $v_\alpha$ and $D$ [see (T24)] are small, and $\Gamma_\beta$ is smaller than $\Gamma_\alpha$ because of the Coulomb penetration factor for charged particles. In the actual application of the theory, the $E_{res}$ and the widths are considered as adjustable empirical parameters.

In the elastic scattering (especially $l = 0$), the cross section consists of three parts; the resonance scattering, the potential scattering and the interference term between the resonance and potential scatterings. On account of this interference there is a minimum of the scattering cross section on the low energy side of a resonance, and a slower decrease on the high energy side. The total cross section shown in Fig. 1 is due mainly to the elastic scattering, and the resonances at $E = 0.11$, 0.38 and 0.70 MeV show the interference effect. Other resonances correspond to $l$ higher than zero.

In the above derivation of the formulas, we have not used the compound nucleus theory. Nevertheless the formulas have always been interpreted on the compound nucleus picture. However, the more rigorous theory developed by Wigner and others shows that the sharp resonance reaction is the consequence of the formation of the compound nucleus state which is independent of the process by which the state is formed, provided that the one level approximation in the $R$ matrix formation is valid. The fact that the compound nucleus state formed has no memory of the history of the formation process is the consequence of the strong nuclear interaction. For details, see Sect. X on $R$ matrix.

It should be noticed that the target nucleus is not always "black". The nucleus is black if an incident particle with impact parameter less than the nuclear radius is completely absorbed, and the formation of a compound nucleus by absorbing the incident particle is possible only for incident energies which are very near the resonance energy $E_{res}$ and for a particular value of the angular momentum. For other cases the incident particle is just reflected at the nuclear surface by the sudden change of the nuclear optical potential.

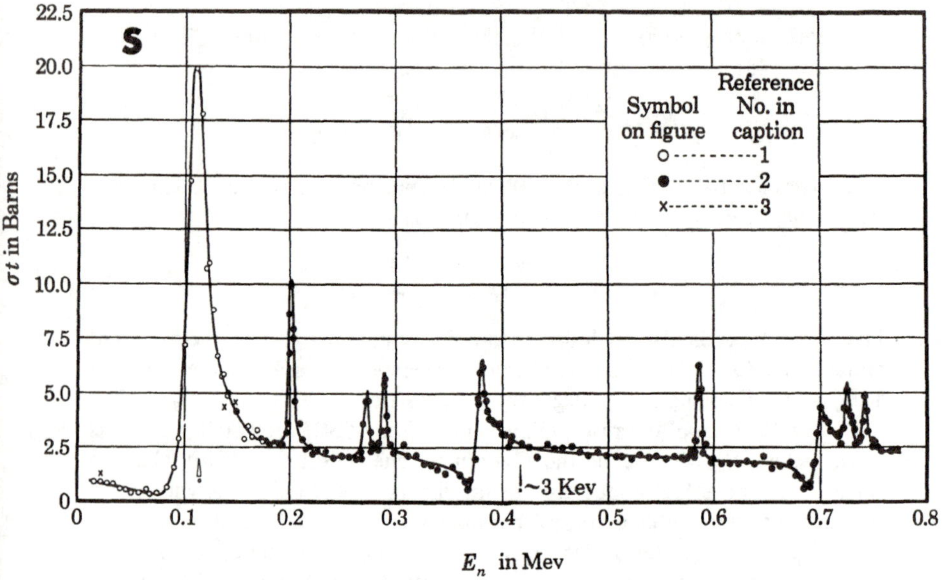


**Figure 1.** Neutron total cross section by sulfur (ref. 12).
(1) Adair, Bockelman, and Peterson, *Phys. Rev.* **76**, 308 (1949);
(2) Peterson, Barschall, and Bockelman, *Phys. Rev.* **79**, 593 (1950);
(3) Fields, Russell, Sachs, and Wattenberg, *Phys. Rev.* **71**, 508 (1947).

It seems that the compound nucleus model and the shell model have each its own domain of applicability. The optical model of nuclear reaction and the theory of direct reaction will give us further understanding about nuclear processes.

For the angular distribution formula, see Sects. W and X on $S$ matrix and $R$ matrix.

## 8. GENERAL BEHAVIOR OF THE CROSS SECTION

In this subsection the general behavior of the cross section, especially at the threshold, will be considered.

The scattering and the reaction cross section are given by (T33) and (T34),

$$\sigma_{sc}^{(l)} = \frac{(2l+1)\pi}{k^2}|1-\eta_l|^2, \qquad \sigma_r^{(l)} = \frac{(2l+1)\pi}{k^2}(1-|\eta_l|^2).$$

By the conservation of flux $\eta_l$ is restricted to $|\eta_l| \leqslant 1$. The maximum scattering cross section will be attained if $\eta_l = -1$. The maximum value is $\sigma_{sc(\max)}^{(l)} = 4(2l + 1)\,\pi/k^2$). The reaction cross section then vanishes. For $\eta_l = 1$, both $\sigma_{sc}^{(l)}$ and $\sigma_r^{(l)}$ vanish. On the other hand the reaction cross section has the maximum value $\sigma_{r(\max)}^{(l)} = (2l + 1)\,\pi/k^2$ for $\eta_l = 0$. In this case the scattering cross section is $\sigma_{sc}^{(l)} = (2l + 1)\,\pi/k^2$. The reaction cross section is not determined by the scattering cross section in general and *vice versa*, except in the special cases considered as above. However the maximum and the minimum values of $\sigma_{sc}^{(l)}$ are not independent# of the value of $\sigma_r^{(l)}$. Similarly the maximum value of $\sigma_r^{(l)}$ is restricted by the value of $\sigma_{sc}^{(l)}$.

The $\sigma_{r(\max)}^{l}$ corresponds to $\eta_l = 0$, and this means that the incident $l$-wave is absorbed completely (i.e., no outgoing wave in the incident channel). The elastic scattering caused by a "black" body is called the "shadow scattering".

Let us assume that the black body (radius $R$) is much larger than the wave length (divided by $2\pi$), $\lambdabar = 1/k$, and that all the partial waves are absorbed when $l < R/\lambdabar$ (see first footnote of Sect. W),

$$\eta_l = 0, \qquad l\lambdabar < R,$$

$$\eta_l = 1, \qquad l\lambdabar > R.$$

Therefore we have the cross sections for the elastic and the absorption (i.e., reaction) processes,

$$\sigma_r = \sigma_{sc} = \sum_{l=0}^{kR} \frac{(2l + 1)\,\pi}{k^2} \cong \pi R^2.$$

The absorption cross section is identical with the (classical) geometrical cross section. The diffracted waves are limited within an angle of $\sim \lambdabar/R = 1/kR$, and they smear out the shadow at a distance $\sim R^2/\lambdabar$ behind the body. For incident nucleon of energy $x$-MeV, we have $k \sim 2.2\sqrt{x} \times 10^{12}\ \text{cm}^{-1}$. Taking the nuclear diameter as $R \sim 8 \times 10^{-13}$ cm, we have, $kR \sim 2\sqrt{x}$. This forward diffraction (shadow scattering) wave can be detected in properly arranged experiments.

We will finally remark on the qualitative features of the cross sections for various reactions.

The elastic scattering cross section of a neutral particle will be constant in the region of very low energies, provided that no resonance level lies at zero incident energy. This is the consequence of the formula (T46), where both the

---

# The maximum and minimum values of $\sigma_{sc}$ are given by

$$\sigma_{sc} = [2(2l + 1)\,\pi/k^2] - \sigma_r \pm [2(2l + 1)\,\pi/k^2]\,\sqrt{1 - [k^2\sigma_r/(2l + 1)\pi]}.$$

potential scattering amplitude and $\Gamma_\alpha$ are proportional to $k$. If the neutral particle has an angular momentum $l$ ($l \neq 0$), $\Gamma_\alpha$ is proportional to $k^{2l+1}$ [see (T22a), (T20) or Table II of Sect. X]. Therefore we see from (T47) that the cross section is proportional to $k^{4l}$. The elastic scattering of charged particle (with the *repulsive* Coulomb potential) is practically identical with the Coulomb scattering cross section for very small velocities. The angular distribution is more sensitive to the nuclear effect than the total cross section at small energies. If the charges of the incident particle and of the target are opposite, the exponential term in the penetration factor drops out. [See (A51).] The cross section is thus proportional to $k^{-2}$. If the outgoing particle has a different $l'$ than the incident $l$, the cross section of a neutral particle is proportional to $k^{2l+2l'}$.

Exothermic reactions: $\Gamma_\beta$ in (T50) is constant for small changes of the incident energy in exoenergic reaction. $\Gamma_\alpha$ is proportional to $k^{2l+1}$ for neutral particles with angular momentum $l$, hence $\sigma_{\alpha\beta}^{(l)}$ is proportional to $k^{2l-1}$. One example is found in the $1/v$ law of the photo-absorption of slow neutrons where the incident neutron is absorbed from $s$ wave. The Coulomb penetration factor $v_\beta$ must be used for charged incident particles. The factor for a repulsive Coulomb potential is roughly characterized by $v^{-1} \exp\{-2\pi z_i z_t e^2/(\hbar v)\}$ ($v$ = velocity, $z_i$ and $z_t$ the charge number of the incident and the target particles) for low energies. From (T22a) and (T50) we see that $\sigma$ (repulsive) $\propto k^{-2} \exp\{-2\pi z_i z_t e^2/(\hbar v)\}$ and $\sigma$(attractive) $\propto k^{-2}$.

Endothermic reactions: Consider the case when the incident energy is high enough to cause the endoenergic reaction, but the outgoing particle energy is small. Then $\Gamma_\alpha$ and $k^2$ in (T50) are constant. $\Gamma_\beta$ is proportional to $k_\beta^{2l+1}$ for neutral particle. The reaction cross section ending in neutral particle emission is thus proportional to (velocity)$^{2l+1}$ of the emitted particle. For an emitted charged particle, the penetration factor must be used.

The behavior of the cross section of these endothermic reactions can be derived by the detailed balance theorem from the inverse (exothermic) reactions. [See (W42).] Since the velocity of the incident particle is considered to be constant near the threshold, the energy dependence of the cross section is obtained by multiplying the square of the velocity of the emitted particles. Thus $\sigma$(neutral) $\propto k^{2l+1}$, $\sigma$(repulsive) $\propto \exp\{-2\pi z_i z_t e^2/(\hbar v)\}$ and $\sigma$ (attractive) $\propto$ constant.

The qualitative behavior of the cross section is usually obtained by considering $l = 0$.

We shall remark briefly on an anomalous nature of the elastic scattering cross section at the threshold where a new reaction becomes possible: $a + A \rightarrow b + B$. The scattering matrix as defined in (H41) will be used conveniently. [Also see (W7b), (W9) and (W10).] The subscripts 1 and 2 refer to the original and the newly created channels (=type of reactions) respectively. Suppose that one (or both) of $b$ and $B$ is neutral and has no relative angular momentum. Since we know that the new reaction cross section is proportional to $k_2$ (=the wave number of emitted particle), we have

$$S_{12} \propto k_2^{1/2}.$$

By virtue of the conservation of flux the following relation holds [see (W12)]

$$S_{11}^* S_{11} + S_{12}^* S_{12} = 1.$$

Only the linear term in $k_2$ will be retained in the following. We get from $S_{11} = A(1 - ak)$ where $|A| = 1$,

$$\begin{aligned} \sigma_{11} &= \frac{\pi}{k_1^2} |S_{11} - 1|^2 \\ &= \sigma_{11}^{(0)} + ak \cos \vartheta, \qquad k_2^2 > 0 \\ &= \sigma_{11}^{(0)} - ia|k| \sin \vartheta, \quad k_2^2 < 0, \end{aligned}$$

where $\sigma_{11}^{(0)}$ is the elastic cross section at $k_2 = 0$. The constant $a$ is assumed to be positive without loss of generality. Therefore, if $\vartheta$ is in the first, second, third

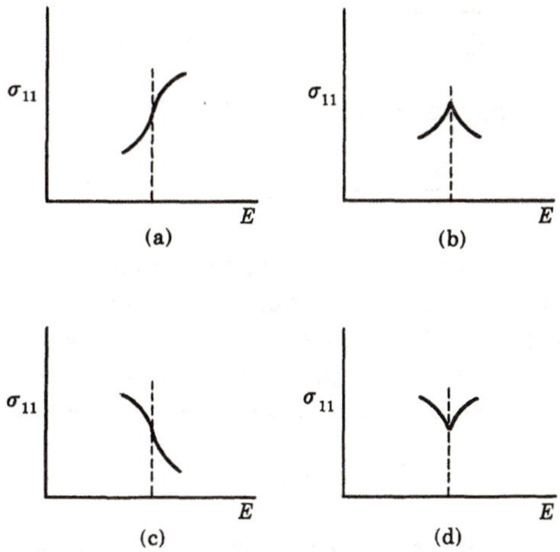


**Figure 2.** Cross section at threshold (see text).

and fourth quadrant, the elastic cross section behaves as shown in Figs. 2a, b, c and d. respectively. The "cusp" behavior (as in Fig. 2b) was noticed by Wigner (ref. 9). For a neutral particle with $l \geqslant 1$ and for equal-charge particles, the discontinuity of $d\sigma_{11}/dE$ at the threshold does not occur.

## 9. MANY-LEVEL FORMULA OF KAPUR AND PEIERLS

The one-level resonance formula has been derived in subsect. 7 above. If the level spacing is not very large compared with the total width, the one-level formula is not accurate enough to describe the reaction. We shall now give an elementary account of the many-level theory of Kapur and Peierls.

To single out the main features of the theory, we shall first consider the simplest case; that is, $s$ wave scattering of neutrons by a static potential $V(r)$. The Schrödinger equation for the radial part $u(r)$ of the wave function is

$$\frac{\hbar^2}{2M}\frac{d^2u}{dr^2} + [E - V(r)]u = 0. \tag{T52}$$

Suppose that $V(r)$ is zero for $r$ larger than the nuclear radius $R$. The solution of (T52) outside the nuclear surface has the form

$$u(r) = \frac{\sin kr}{k} + fe^{ikr}; \quad r > R, \tag{T53}$$

where $f$ is the $s$-wave part of the scattering amplitude. Now we define the "compound state $u_m$", which is the solution of the Schrödinger equation

$$\frac{\hbar^2}{2M}\frac{d^2u_m}{dr^2} + (E_m^{(s)} - V)\, u_m = 0, \tag{T54}$$

subject to the special boundary condition

$$\frac{du_m}{dr} = iku_m, \quad \text{at} \quad r = R, \tag{T55}$$

where $k^2 = 2ME/\hbar^2$, $E$ being the energy of the incident neutron. This definition of the compound nucleus is of an approximate nature, because the wave function of the compound nucleus in the proper meaning should not depend on the incident energy. The eigenvalues $E_m^{(s)}$ are complex. The superscript ($s$) indicates the potential model. From (T54) and (T55), we have

$$\int_0^R u_n(r)\, u_m(r)\, dr = \delta_{mn}. \tag{T56}$$

It must be noted that (T56) is different from the usual orthogonality relation in that $u_n(r)$ is used instead of $u_n^*(r)$. On expanding $u(r)$ in terms of $u_m(r)$,

$$u(r) = \sum a_m u_m(r), \quad r < R \tag{T57}$$

and using (T52) and (T54), we have

$$(E - E_m)\, a_m = -\frac{\hbar^2}{2M}\left(u_m\frac{du}{dr} - u\frac{du_m}{dr}\right)\bigg|_{r=R}. \tag{T58}$$

By using (T53) and (T55), this becomes,

$$(E - E_m)\, a_m = -\frac{\hbar^2}{2M}\left(\frac{du}{dr} - iku\right) u_m\bigg|_{r=R} = -\frac{\hbar^2}{2M} e^{-ikR} u_m(R). \tag{T59}$$

From (T53) and (T57), we have

$$\sum a_m u_m(R) = \frac{\sin kR}{k} + fe^{ikR}.$$

On substituting $a_m$ from (T59) into this, we have

$$f = \frac{1}{k}e^{-ikR}\left[\sum_m \frac{\hbar^2 k}{2M} e^{-ikR} \frac{u_m^2(R)}{E_m - E} - \sin kR\right]. \tag{T60}$$

If we define the (single particle) width $B_m^{(s)}$ by

$$B_m^{(s)} \equiv \frac{\hbar^2 k}{M} u_m^2(R). \tag{T61}$$

Equation (T60) is reduced to

$$f = -\frac{\exp(-2ikR)}{k}\left[e^{ikR}\sin kR + \sum_m \frac{\frac{1}{2}B_m^{(s)}}{E - E_m^{(s)}}\right]. \tag{T62}$$

The cross section $\sigma$ is equal to $4\pi|f|^2$. $B_m^{(s)}$ in (T61) is not identical with the width $\Gamma$ introduced in subsect. 7, because $B_m^{(s)}$ is complex in general. But for small $k$, the $u_m$ is seen from (T55) to be nearly real, and hence the imaginary part of $B_m^{(s)}$ is small as seen from (T61). Let us consider the imaginary part of $E_m^{(s)}$. From the Equations (T54) and (T55), we have

$$(E_m^{(s)*} - E_m^{(s)}) \int_0^R u_m u_m^* \, dr = \frac{i\hbar^2 k}{m} u_m^*(R)\, u_m(R).$$

On writing

$$E_m^{(s)} \equiv \varepsilon_m^{(s)} - \frac{i}{2}\Gamma_m^{(s)},$$

we have

$$\Gamma_m^{(s)} = \frac{\hbar^2 k}{m} \frac{|u_m(R)|^2}{\int_0^R |u_m(r)|^2\, dr}.$$

On comparing this with (T61) it is seen that $B_m^{(s)}$ is very similar to $\Gamma_m^{(s)}$. Since $u_m$ is almost real for very small $k$, $B_m^{(s)}$ and $\Gamma_m^{(s)}$ are nearly equal in this case,

$$B_m^{(s)} \cong \Gamma_m^{(s)} \quad \text{for low energies.}$$

The cross section is, from (T62),

$$\sigma = \frac{4\pi}{k^2}\left|e^{ikR}\sin kR + \sum_m \frac{\frac{1}{2}B_m^{(s)}}{E - \varepsilon_m^{(s)} + \frac{1}{2}i\Gamma_m^{(s)}}\right|^2. \tag{T63}$$

If a single term from the summation in (T63) is retained, and if $B^{(s)}$ is replaced by $\Gamma^{(s)}$, we then obtain the formulas (T46) and (T46a).

We shall not enter any further into the discussion for the general reaction, but refer the reader to the original paper. Only the final formula of the Kapur and Peierls theory for the more complicated reaction phenomena will be given. Let $f_{\alpha\alpha}$ and $f_{\alpha\beta}$ be the amplitude for elastic scattering in channel $\alpha$ and for reaction from channel $\alpha$ into channel $\beta$ respectively. The elastic and reaction cross sections (for $\alpha \to \beta$) are $4\pi|f_{\alpha\alpha}|^2$ and $(4\pi k_\beta/k_\alpha)|f_{\alpha\beta}|^2$ respectively. $f_{\alpha\alpha}$ and $f_{\alpha\beta}$ are given by

$$f_{\alpha\alpha} = -\frac{(2l_\alpha + 1)^{1/2}\, i^{l\alpha}}{k}\, g_\alpha(R)\left[\frac{i}{2}\{1 - g_\alpha^{-1}(R)\} + \sum_m \frac{\frac{1}{2}B_{\alpha m}}{E - \varepsilon_m + \frac{1}{2}i\Gamma_m}\right], \quad \text{(T64)}$$

$$f_{\alpha\beta} = (2l_\alpha + 1)\, e^{ic} \sum_m \frac{P_{\alpha m}P_{\beta m}}{E - \varepsilon_m + \frac{1}{2}i\Gamma_m}, \quad \text{(T65)}$$

where $g_\alpha(R) = u_\alpha^-(R)/u_\alpha^+(R)$, $u_\alpha^\pm(r)$ being the outgoing and incoming solutions outside the nuclear surface, and have the asymptotic form $\exp[\pm i(kr - l_\alpha\pi/2)]$, and $c$ is a real constant. $B_{\alpha m}$, $P_{\alpha m}$, $P_{\beta m}$ are complex in general, but nearly real for low energies.

$$P_{\alpha m}^2 \cong \Gamma_{\alpha m}, \qquad P_{\beta m}^2 \cong \Gamma_{\beta m}.$$

The one-level formulas, (T46), (T47) and (T48), are derivable from (T64) and (T65) as special cases. Although $\Gamma_m$ is energy dependent and $B_m$, $P_m$ are complex in general, the fact that the amplitude $f$ itself can be expressed by a simple sum of many-level terms is an advantageous point in the Kapur–Peierls theory. See Sect. X on the $R$ matrix formalism.

## REFERENCES

Some of the standard books on theoretical nuclear physics are:

1. Blatt, J. M. and Weisskopf, V. F., *Theoretical Nuclear Physics*, John Wiley and Sons, New York (1952), see especially Chapters 8 and 10.

2. Bethe, H. A., Rev. Mod. Phys. **9**, 69 (1937); Bethe, H. A. and Bacher, R. F., Rev. Mod. Phys. **8**, 82 (1936).

3. Breit, G., article in Handbuch der Physik, Vol. **41/1**, Springer (1959). This monograph gives a comprehensive treatment of many topics and a full bibliography.

4. Feshbach, H., article on the compound nucleus, in *Nuclear Spectroscopy*, ed. by F. Ajzenberg-Selove, Acad. Press, New York (1960).

5. Moore, R. G. Jr., Review Mod. Phys. **32**, 101 (1960). Cross section in compound nucleus processes.

Original papers on compound nucleus model are:

6. Bohr, N., Nature **137**, 344 (1936); Wigner, E. P. and Breit, G., Bull. Amer. Phys. Soc. Feb. 1936; Briet, G. and Wigner, E. P., Phys. Rev. **49**, 519, 642 (1936); Bohr, N. and Kalckar, F., Kgl. Danske Videnskab. Selskab. Mat-fys. Medd. **14**, 10 (1937). Breit and Wigner's paper derives the "resonance formula".

7. Weisskopf, V. F. and Ewing, D. H., Phys. Rev. **57**, 472, 935 (1940); Hauser, W. and Feshbach, H., Phys. Rev. **87**, 366 (1952). Compound nucleus theories with statistical assumption.

8. Feshbach, H., Peaslee, D. C. and Weisskopf, V. F., Phys. Rev. **71**, 145 (1947). The contents in subsects. 3 to 7 are based on this paper.

9. Wigner, E. P., Phys. Rev. **73**, 1002 (1948), qualitative behavior of cross sections at the threshold. See also ref. 10 below.

10. Breit, G., Phys. Rev. **107**, 1612 (1957); Baz, A. I., Soviet Physics JETP **6**, 709 (1958); Newton, R. G., Annals of Phys. **4**, 29 (1958); Yamaguchi, Y., Supplm. Prog. Theor. Phys. (Kyoto), No. 7 (1959).

11. Kapur, P. L. and Peierls, R., Proc. Roy. Soc. **A166**, 277 (1938). The original paper on the many-level formula of Kapur and Peierls. See also Siegert, A. J., Phys. Rev. **56**, 750 (1939); Peierls, R. E., Proc. Cambridge Phil. Soc. **44**, 242 (1948).

12. Adair, R. K., Rev. Mod. Phys. **22**, 249 (1950). See Fig. 1.

The reader is also referred to references of Sect. X.

section

# U Optical Model

For "strong" interactions between the incident nucleon and the target nucleus, the reaction (excluding the forward elastic) cross section can be expected to be $\sim\pi(R + \lambda\!\!\!^{-})^2$, where the $\lambda\!\!\!^{-}$ part is the wave mechanical correction to the classical geometric cross section $\pi R^2$. Experiments at high energies ($\gtrsim 15$ MeV) confirm this consideration. In the low energy region, the most characteristic nuclear collisions are resonance scattering and reactions. The total cross section averaged over many resonance levels can be expected to be a smooth monotonically decreasing function of energy $E$ of the incident particle for strong interactions. Measurements, however, of the neutron total cross section $\sigma$ by Barshall and his collaborators and others at low energies ($\leqslant$ several MeV) show that, while $\sigma$ does vary slowly and regularly with the energy and $A$, it deviates definitely from $\pi(R + \lambda\!\!\!^{-})^2$. The $\sigma - E$ and $\sigma - A$ curves do not seem to depend on such properties of the nucleus as "shell structure", but only on the nuclear radius $R$. To account for these features of the cross section, Feshbach, Porter and Weisskopf# suggest the theory of optical potential scattering.

The reaction amplitude (including elastic scattering) may be regarded as consisting of the following parts: (i) The reflected wave by the nuclear surface. This is the potential scattering $A_{pot}$ in (T40). (ii) The incident wave enters and goes through the nucleus. (i) and (ii) are collectively called the "shape elastic scattering". (iii) The incident particle collides with a nucleon inside the nucleus (or a few nucleons) and causes elastic scattering and reactions in general. This is called the "direct process". (iv) The incident particle enters

# This work (ref. 4) will be referred to as *FPW* in the following.

the nucleus, distributing its energy to many nucleons, forming a compound nucleus. For (nearly) monochromatic incident particles, the processes (ii), (iii), (iv) correspond to the resonance part $A_{res}$ in (T45). [(T45) obtained by the one-level approximation represents, however, (iv) only. See below.]

When the incident beam has an energy spread greater than the level spacing the resonance peaks are smeared out. Consider the nuclear reaction process initiated by a wave packet with an energy spread $\Delta E \sim \hbar/\Delta t$. The wave scattered by the nuclear potential will first come out from the target, and then, the waves due to direct nuclear reactions and finally the decay waves from the compound nucleus. Since $\Delta E \approx \hbar/\Delta t$ is much larger than the level width $\Gamma \approx \hbar/\tau$ ($\tau \equiv$ lifetime of the compound nucleus), the decay waves do not interfere with the waves scattered by the nuclear potential. It is assumed that the scattered amplitude by the nuclear potential does not interfere with that due to the direct nuclear reaction too, although this may not be justified in general.

The shape elastic scatterings due to (i) and (ii) may be described by a (real) nuclear potential. But, an imaginary part of the potential is necessary in order to take into account the processes of (iii) and (iv). The wave function is obtained by solving the Schrödinger equation (A2) with this complex nuclear potential (i.e., optical potential). The scattering amplitude by the optical potential will give the "shape elastic" scattering cross section. Because of the (negative) imaginary part of the optical potential, the outgoing scattering flux is less than the incoming flux. The decrease of flux gives us the absorption cross section, ($\sigma_c$). We can not give a description on the optical model alone how the absorbed particles are re-emitted and cause the various nuclear reactions (see later discussion). The optical model also can not describe the closely spaced sharp resonance phenomena, but rather give us the cross sections averaged over many sharp resonances.

The imaginary part of the optical potential, introduced to account for (iii) and (iv), is expected to be small for low incident energies. The following consideration was probably first given by Fermi (1949). The nucleons in a nucleus may be regarded, in the first approximation, as a degenerate Fermi gas. The states within the Fermi surface are occupied. Suppose an incident particle of momentum $p_i$ collides with a nucleon of momentum $p$, and after the scattering these particles have the momenta $p_f$, $p'$. The Pauli principle forbids the transition unless $p_f$, $p'$ are larger than the Fermi maximum momentum. The result is as if the interaction of the incident particle with the nucleons in the nucleus were greatly reduced. This is also the basic reason for the justification of the independent-particle model of the nucleus.

## I. SQUARE WELL MODEL

The optical potential used in *FPW* is a square well with a constant imaginary part,

$$V(r) = \begin{cases} -(V + iW), & r < R, \\ 0, & r > R. \end{cases} \tag{U1}$$

The values of $V$ and $W$ are adjusted so as to fit the experimental (total and differential) cross sections (see Fig. 1). The nuclear radius $R$ is taken to be

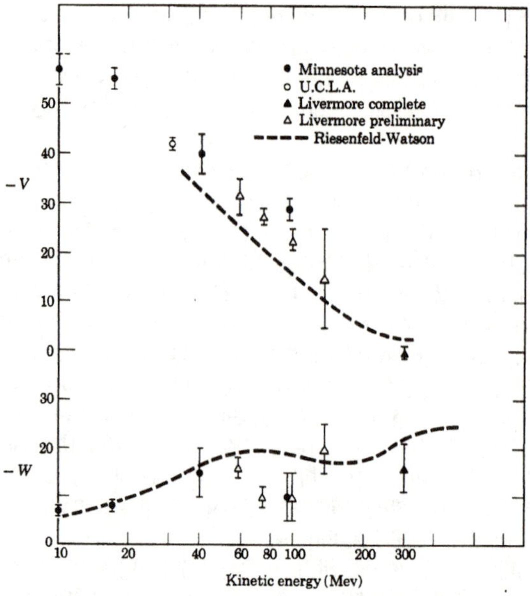


**Figure 1.** Real and imaginary parts of the central part of the optical potential as a function of proton energy. The adopted constants in (U17a) are $R = 1.25\ A^{1/3} \times 10^{-13}$ cm, $d = 0.65 \times 10^{-13}$ cm. The Coulomb potential is added to (U17a). The energy dependence is qualitatively reproduced by the theory of Riesenfeld and Watson (*Phys. Rev.* **102**, 1157, 1956). (See also Sec. N.) [From Ref. 9, Glassgold, A. E., *Review Mod. Phys.* **30**, 419 (1958).]

about $1.3 \sim 1.4A^{1/3} \times 10^{-13}$ cm ($A$ = mass number). The mean free path of the incident nucleon inside the nucleus is about $2 \times 10^{-12}$ cm for 5 MeV. We shall call $\zeta_l \equiv \exp[2i\delta_l(op)]$ the scattering amplitude factor, where $\delta_l(op)$ is the (complex) phase shift due to the optical potential, which is obtained from the asymptotic form of the solution of the Schrödinger equation (A2) with the optical potential (U1). The differential cross section of the (shape) elastically scattered neutron is given by (with normalized $\Psi_{l,0}$)

$$\frac{d\sigma_{se}}{d\Omega} = \frac{\pi}{k^2}\left|\sum_l (2l+1)^{1/2}(1-\zeta_l)\,Y_{l,0}(\vartheta)\right|^2. \tag{U2}$$

For an incident proton, we must include the Coulomb potential outside the nuclear surface. The differential cross section for proton is given by

$$\frac{d\sigma_{se}}{d\Omega} = \left|\frac{Ze^2}{2mv^2}\exp\left[-2i\gamma\ln\sin\left(\frac{\vartheta}{2}\right)\right]\Big/\sin^2\left(\frac{\vartheta}{2}\right)\right.$$

$$\left.-\frac{i\sqrt{\pi}}{k}\sum_l\sqrt{2l+1}\exp\left[2i(\sigma_l-\sigma_0)\right](1-\zeta_l)\,Y_{l,0}(\vartheta)\right|^2 \tag{U3}$$

where

$$\exp\left[2i(\sigma_l-\sigma_0)\right] = \frac{(l+i\gamma)(l-1+i\gamma)\ldots(1+i\gamma)}{(l-i\gamma)(l-1-i\gamma)\ldots(1-i\gamma)}, \qquad \gamma = \frac{Ze^2}{\hbar v}.$$

The $\zeta_l$ in (U2) and (U3) are approximately equal.

Figs. 2 and 3 show the calculated and the observed neutron total cross section as a function of energy and mass number $A$. The cross section has slowly varying maxima and minima. This gross structure can be understood by considering the ("giant", namely, not closely spaced, sharp) resonance condition. For $l = 0$, giant resonances occur when [assuming that $|W| \ll |V|$ in (U1)]

$$\sqrt{K^2+k^2}\,R \sim (n+\tfrac{1}{2})\,\pi, \quad K^2 = 2mV/\hbar^2. \tag{U4}$$

The condition (U4) is fulfilled for $A \sim 13$ ($n = 1$), $\sim 60$ ($n = 2$), $\sim 160$ ($n = 3$), ... in the low energy region. If $W = 0$, the cross section would become very large at $k^2 \to 0$ at these mass numbers. The $p$ resonances occur at mass numbers $A_1$ which lie between the $A$'s above. At zero energy, the $p$-wave cross section vanishes, but rises to a pronounced peak at very low energies. The uneven behavior of the cross section is thus understood by the partial wave resonance effect, although such resonances are diminished a little by the existence of the imaginary part $W$.

Fig. 4 shows the differential cross section of neutron elastically scattered by several nuclei. One sees a diffraction pattern clearly. One expects the cross section maxima to appear at angles $\vartheta_n$ satisfying the relation

$$2R\sin\vartheta_n \cong n\lambda.$$

Thus the elastic scattering is a diffraction scattering characterized by the geometry of the nucleus, and is called the "shape elastic scattering" by Feshbach, Porter and Weisskopf.

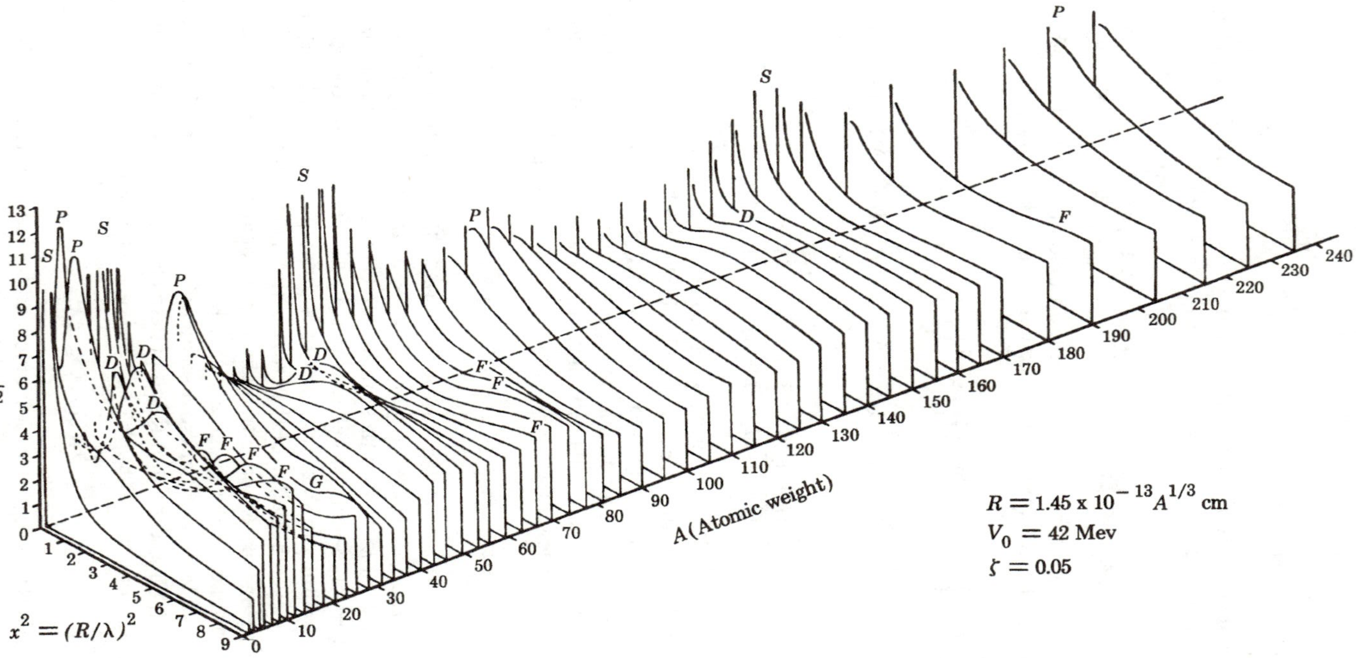


**Figure 2.** Calculated neutron total cross sections as a function of energy and mass number for a well depth $V_0 = 42$ MeV, radius $R = 1.45 \times 10^{-13} A^{1/3}$, $\zeta = 0.05$. The energy is expressed in terms of

$$x^2 = \left[A^{2/3} \frac{A}{10(A-1)}\right]\varepsilon.$$

$\varepsilon$ = energy in lab. system in MeV (ref. 4).

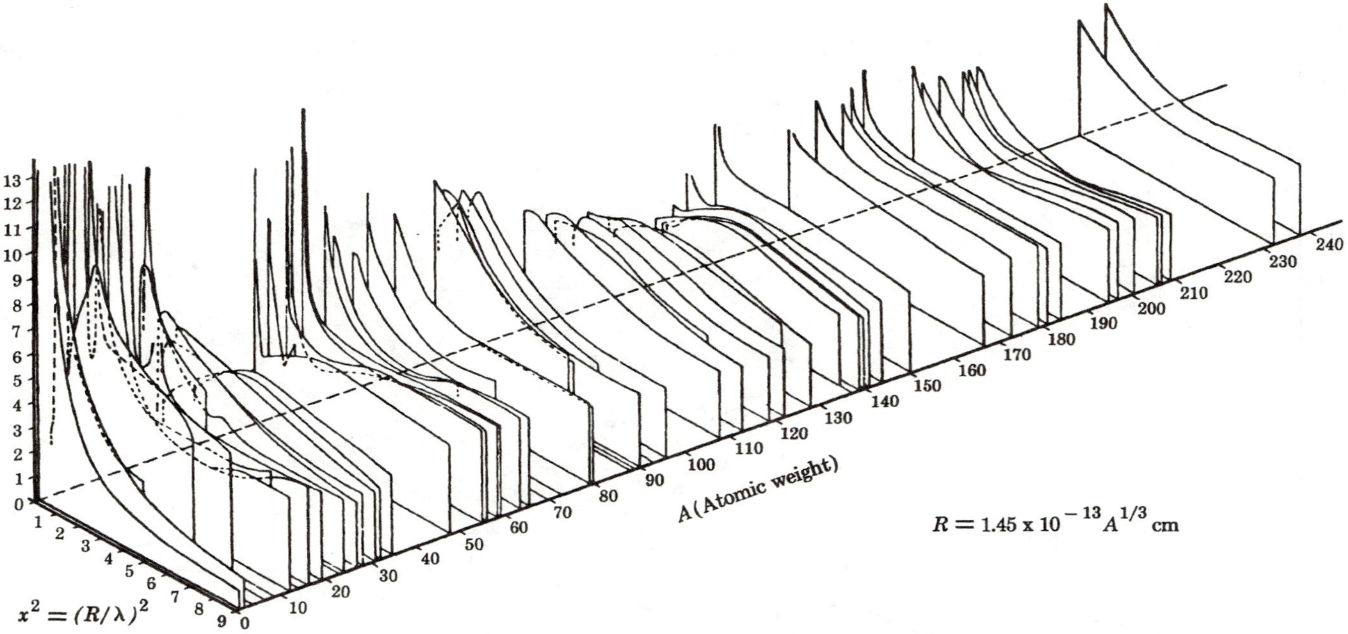


**Figure 3.** Observed neutron total cross sections. See caption for Figure 2 above (ref. 4).

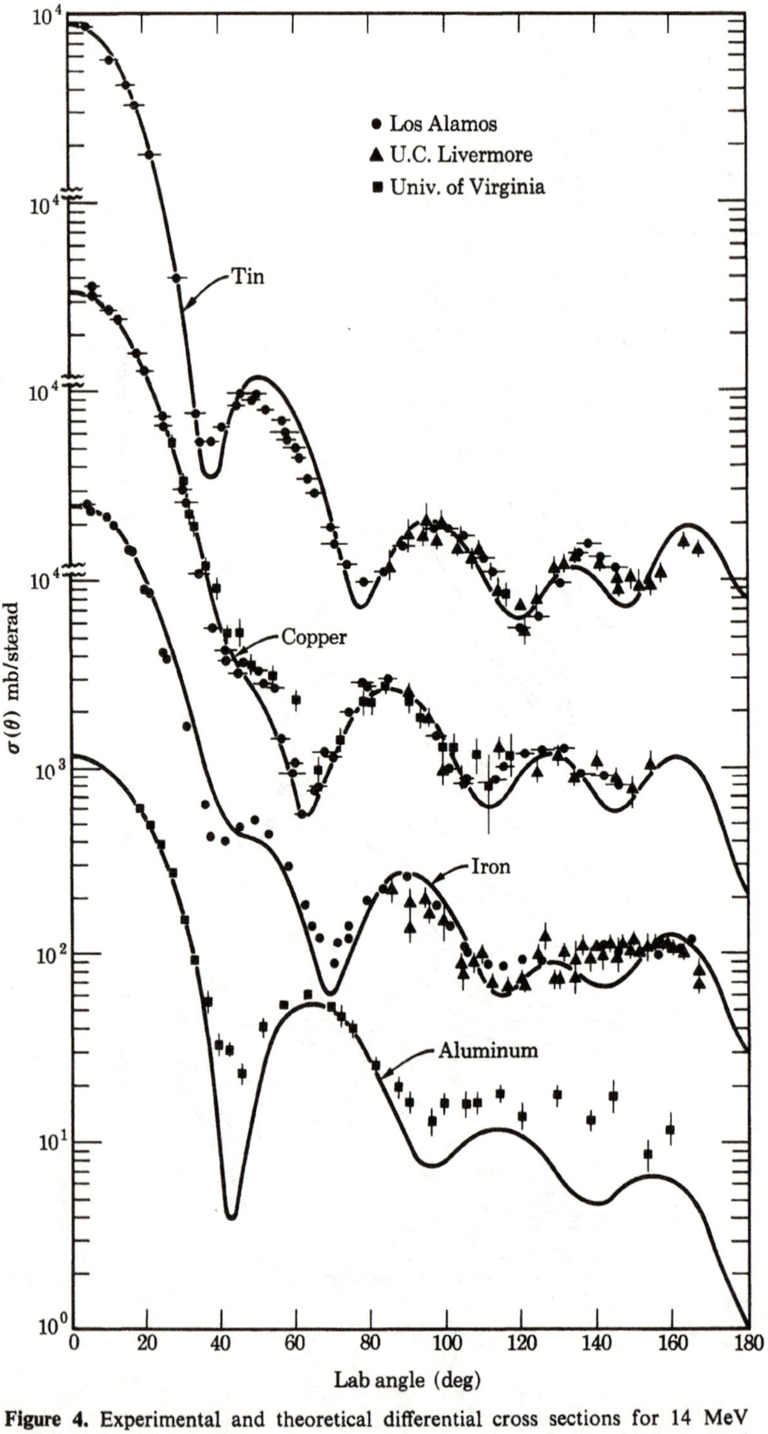


**Figure 4.** Experimental and theoretical differential cross sections for 14 MeV neutrons scattered from Sn, Cu, Fe, Al (observed data not completely corrected for multiple scattering) (ref. 10).

To get, on the optical model, the elastic cross section which can be compared with the experimental data, we must add to the "shape elastic scattering" $\sigma_{se}$ in (U2), or (U3), the contributions $\sigma_{ce}$ that come from the (elastic) outgoing waves arising from the processes (iii) and (iv). If $\sigma_c^l$ is the absorption cross section of (iii) and (iv), and $\omega_l$ is the probability that the absorbed particle is reemitted without changing the state of the nucleus, then

$$d\sigma_{ce} = \sum_l \frac{2l+1}{2\pi} [P_l(\cos\vartheta)]^2 \omega_l \sigma_c^l \, d\Omega.$$

This $\sigma_{ce}$ will be called the "compound elastic scattering"; it is usually a small correction to $\sigma_{se}$. We shall assume that the compound elastic scattering does not interfere with the shape elastic scattering.

$\sigma_c$ can be calculated from the imaginary part of the optical potential.

## 2. CONNECTION BETWEEN THE ACTUAL SCATTERING AMPLITUDE AND THE AVERAGE AMPLITUDE

When only elastic scattering is energetically possible, the phase shift $\delta_l$ increases by about $\pi$ whenever the energy $E$ of the incident particle increases by an amount equal to the spacing of the resonance levels. The amplitude factor $\eta_l = \exp[2i\delta_l]$ is thus a rapidly varying function of $E$. On the other hand, the factor $\zeta_l = \exp[2i\delta_l(op)]$ calculated from the optical potential is a slowly varying function of $E$. It is shown by Feshbach, Porter and Weisskopf that $\zeta_l$ can be considered as the averaged $\eta_l$ over the energy spread $\Delta E$ of the incident particles. We shall assume that $\Delta E$ contains many resonance levels, and that $\Delta E/E \ll 1$ so that slowly varying functions of $E$ (including $k^2$) are practically constant over $\Delta E$. The average over $\Delta E$ of $\eta_l(E)$ is then

$$\langle\eta_l(E)\rangle \equiv \frac{1}{\Delta E}\int_{E-\frac{1}{2}\Delta E}^{E+\frac{1}{2}\Delta E} \eta_l(\varepsilon')\, d\varepsilon'. \tag{U5}$$

The average elastic-scattering and reaction cross sections are [see (T33, 34)]

$$\langle\sigma_{el}^{(l)}\rangle = \frac{\pi(2l+1)}{k^2}\langle|1-\eta_l|^2\rangle \tag{U6}$$

$$= \frac{\pi(2l+1)}{k^2}\{|1-\langle\eta_l\rangle|^2 - |\langle\eta_l\rangle|^2 + \langle|\eta_l|^2\rangle\}$$

$$\langle\sigma_r^{(l)}\rangle = \frac{\pi(2l+1)}{k^2}(1-\langle|\eta_l|^2\rangle), \tag{U7}$$

and the total cross section $\sigma_t$ is given

$$\langle\sigma_t^{(l)}\rangle = \langle\sigma_{el}^{(l)}\rangle + \langle\sigma_r^{(l)}\rangle \equiv \frac{\pi(2l+1)}{k^2}\{|1-\langle\eta_l\rangle|^2 + 1 - |\langle\eta_l\rangle|^2\}. \tag{U8}$$

$\langle\sigma_{el}^{(l)}\rangle$ can be divided into two parts

$$\langle\sigma_{el}^{(l)}\rangle = \sigma_{se}^{(l)} + \sigma_{ce}^{(l)}, \tag{U9}$$

$$\sigma_{se}^{(l)} \equiv \frac{\pi(2l+1)}{k^2}|1-\langle\eta_l\rangle|^2, \tag{U10}$$

$$\sigma_{ce}^{(l)} \equiv \frac{\pi(2l+1)}{k^2}\{\langle|\eta_l|^2\rangle - |\langle\eta_l\rangle|^2\}. \tag{U11}$$

From (U7) and (U11), we obtain the sum

$$\sigma_c^{(l)} \equiv \sigma_{ce}^{(l)} + \langle\sigma_r^{(l)}\rangle = \frac{\pi(2l+1)}{k^2}\{1 - |\langle\eta_l\rangle|^2\}, \tag{U12}$$

so that

$$\langle\sigma_t^{(l)}\rangle = \sigma_{se}^{(l)} + \sigma_c^{(l)}.$$

To interpret the $\sigma_{ce}^{(l)}$ defined in (U11), let us express $|\langle\eta_l\rangle|^2$, $\langle|\eta_l|^2\rangle$ in terms of the resonance scattering parameters. For simplicity, consider $l = 0$. $\eta_0$ is given by

$$\eta_0 = \exp(-2ikR')\left(1 - \frac{i\Gamma_\alpha^s}{E - E_s + \frac{i\Gamma^s}{2}}\right), \tag{U13}$$

where $\Gamma^s$, $\Gamma_\alpha^s$ are the total width and the partial width for elastic scattering in the region $D_s$ containing one resonance $E_s$, and $R'$ is the sum of the nuclear radius $R$ and some corrections coming from far-away levels. [For details about (U13), see FPW.] Then

$$\langle\eta_0\rangle = \exp(-2ikR')\left(1 + \pi\langle\frac{\Gamma_\alpha}{D}\rangle\right) \tag{U14}$$

and

$$\sigma_{ce}^{(0)} = \frac{2\pi^2}{k^2}\langle\frac{\Gamma_\alpha^2}{D\Gamma}\rangle. \tag{U15}$$

The elastic scattering cross section $\sigma_{el}^{(0)}$ in the region $D_s$ is given by (U6) and (U13),

$$\sigma_{el}^{(0)} = \frac{\pi}{k^2}\left|\exp(2ikR') - 1 + \frac{i\Gamma_\alpha^s}{E - E_s + \frac{i\Gamma^s}{2}}\right|^2. \tag{U16}$$

The cross section corresponding to the resonance amplitude alone in (U16) is

$$\sigma^{(0)} = \frac{\pi}{k^2}\cdot\frac{(\Gamma_\alpha^s)^2}{(E - E_s)^2 + \left(\frac{\Gamma^s}{2}\right)^2}. \tag{U17}$$

It is easy to verify that $\sigma_{ce}^{(0)}$ in (U15) is the average of (U17). Therefore $\sigma_{ce}^{(l)}$ defined in (U11) represents the cross section that the incident particle is absorbed by the nucleus and reemitted without changing the state of the (target) nucleus, that is, the "compound elastic scattering" mentioned in the last paragraph of subsect. 1 above.

Accordingly, $\sigma_{ce}^{(l)}$ defined in (U10) represents the cross section that the incident particle is just scattered by the nucleus without being absorbed, which is the "shape elastic scattering" cross section. On the basis of the optical potential model it is clear from (U10) and (U2) that $\langle\eta_l\rangle$ should be identical with $\zeta_l$ given by the optical potential in (U2). We now know that $\sigma_c^{(l)}$

is the absorption cross section in the optical model. As the energy increases, the width $\Gamma$ becomes greater than the level spacing $D$, and the amplitude factor $\eta$ does not vary rapidly with energy. $\langle\eta_l\rangle$ will then become the actual $\eta_l$, and $\sigma_{ce}^{(l)}$ becomes very small. It is interesting that the average amplitude factor itself can be physically interpreted in such a way.

### 3. OPTICAL POTENTIAL WITH DIFFUSE BOUNDARY

To achieve a better representation of the experimental data, especially the angular distribution, the sharp edge of the square-well potential (U1) can be rounded off to reduce the reflection at the boundary of the potential. One form of "rounded-edge" or "diffuse boundary" potential usually employed is

$$V(r) = -(V + iW)\left[1 + \exp\left(\frac{r - R}{d}\right)\right]^{-1}. \tag{U17a}$$

The imaginary part $W$ of the optical potential (U1) is not necessarily proportional to the real part. Since the effective interaction between the incident particle with the nucleus is "reduced" by the Pauli principle [see paragraph preceding (U1)], one may picture the imaginary part $W$ as being stronger near the edge than in the interior of the nucleus. This change in the shape of $W(r)$ does not change the elastic scattering appreciably, but affects the reaction cross section, especially that of the direct processes (see next section).

To explain the experimental data of polarization, we further add an optical potential containing the spin-orbit coupling (see the subsect. 5 of Sect. J),

$$\left(\frac{\hbar}{mc}\right)^2 (V' + iW')\frac{1}{r}\frac{d}{dr}\left[1 + \exp\left(\frac{r - R}{d}\right)\right]^{-1} \boldsymbol{\sigma}\cdot\mathbf{L}. \tag{U17b}$$

The depth $V'$ is about 10 MeV for low energies, and decreases to 1 ~ 2 MeV at 300 MeV, while $W'$ is not appreciably different from zero at low energies, but is about $-1$ MeV or so for energies higher than 100 MeV. The experimental data on elastic, reaction and polarization experiments have not yet determined the shape of the surface, the shape of the central region of optical potentials and the radial shape of the spin-orbit potential. The experimental data are insensitive to these details. (U17a) and (17b) are, however, conventional forms for optical potentials, and reproduce the experimental data well except for some of the neutron polarization experiments at low energies. If the imaginary part of the central optical potential is assumed to be stronger at the surface, the following form is usually used instead of $-iW\{1 + \exp[(r - R)/\alpha]\}^{-1}$:

$$iW \exp[-(r - R)^2/b^2],$$

where $b$ is $1.0 \sim 1.2 \times 10^{-13}$ cm.

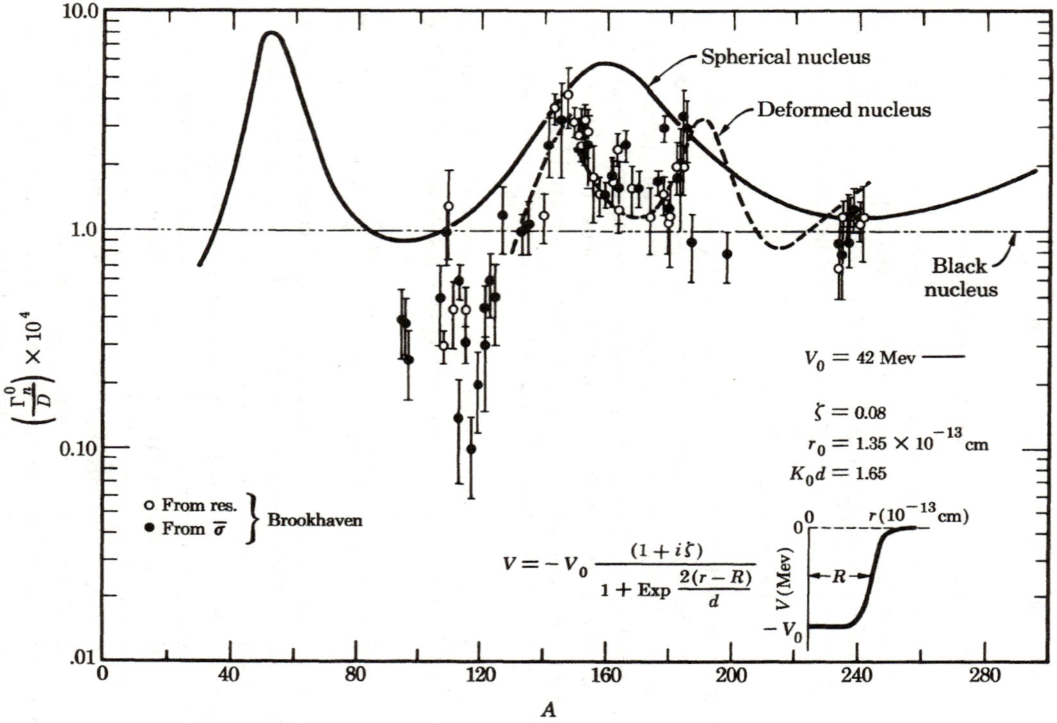

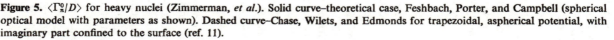

**Figure 5.** $\langle \Gamma_n^0 / D \rangle$ for heavy nuclei (Zimmerman, *et al.*). Solid curve–theoretical case, Feshbach, Porter, and Campbell (spherical optical model with parameters as shown). Dashed curve–Chase, Wilets, and Edmonds for trapezoidal, aspherical potential, with imaginary part confined to the surface (ref. 11).

Some nuclei are not spherically symmetric, and the use of a "deformed optical potential" for such nuclei gives a better fit to the experimental data for the average width $\langle\gamma_n^2/D\rangle$ and $R'$.

Since the absorption cross section $\sigma_c^{(0)}$ (U12) is obtained from (U15) by multiplying by $\Gamma/\Gamma_\alpha$, we have

$$\sigma_c^{(0)} = \frac{2\pi^2}{k^2}\left\langle\frac{\Gamma_\alpha}{D}\right\rangle = \frac{2\pi^2 v_l}{k}\left\langle\frac{\gamma_\alpha^2}{D}\right\rangle.$$

From (U12) for $\sigma_c^{(0)}$, with $\langle\eta_l\rangle = \zeta_l$ given by the optical potential, we can calculate $\langle\gamma_\alpha^2/D\rangle$. The result is shown in Fig. 5, in which the values derived from the experimental data are indicated by vertical bars. It is seen that the curve calculated with a deformed optical potential gives a better fit.

By comparing Fig. 5 and Figs. 3 and 4, it is seen that $\langle\gamma_\alpha^2/D\rangle_{l=0}$ shows a similar giant resonance (with a maximum approximately at $A \sim 52$ and $\sim 162$) as the shape elastic $\sigma_{se}^{(0)}$. To understand this, let us evaluate $\langle\gamma_\alpha^2/D\rangle$ for a typical model.

The wave function $\phi$ corresponding to one (closely spaced) resonance energy can be expanded in terms of the wave functions $\psi_n$ which are the products of the target nucleus wave function and the wave function of the internal degrees of freedom of the corresponding emitted particle,

$$\phi = \phi_0(\mathbf{r})\,\psi_0 + \sum_{n\neq 0}\phi_n(\mathbf{r})\,\psi_n. \tag{U18}$$

The coefficients $\phi_n(\mathbf{r})$ are functions of the relative coordinates of the emitted particle and the nucleus. According to the compound–nucleus model, $\gamma_\alpha^2/D$ will be of the order of magnitude of $1/K$ (see T28) with random fluctuation. Since $\gamma_\alpha^2/D$ is proportional to the existence probability of a single particle wave function at the nuclear surface, the experimental fact that $\langle\gamma_\alpha^2/D\rangle$ shows giant resonance suggests that the degree of mixing of the first term in (U18) exhibits the giant-resonance-like behavior statistically.

If we add the constant imaginary potential for $r < R$ to the real square well in (T52), the eigenvalue of $E_m = \varepsilon_m - \frac{1}{2}i\Gamma_m^{(s)}$ is only changed to $\bar{E}_m = E_m - iW = \varepsilon_m - i(\frac{1}{2}\Gamma_m^{(s)} + W)$, $(s)$ denoting the single particle quantity. That is, the imaginary part of the optical potential broadens the (single particle) resonance level. It can be shown (ref. 1a, p. 904) that, if $E$ is in the neighborhood of $\varepsilon_n$,

$$\pi\left\langle\frac{\Gamma}{D}\right\rangle = \frac{\Gamma_n^{(s)}(\frac{1}{2}\Gamma_n^{(s)} + W)}{(E - \varepsilon_n)^2 + (\frac{1}{2}\Gamma_n^{(s)} + W)^2}. \tag{U19}$$

Thus, the width of the giant resonance is the sum of the single partial width and twice the imaginary potential.

### REFERENCES

The following are recent review articles on the optical potential.

**1a.** Brown, G. E., Rev. Mod. Phys. **31**, 893 (1959). Emphasis is placed on theoretical foundations of the model.

**1b.** Breit, G., article in Handbuch der Physik, Vol. **41/1**, Springer (1959).

**1c.** Feshbach, H., article in *Nuclear Spectroscopy*, Ed. by F. Ajzenberg-Selove, Acad. Press, New York (1960).

Complex potentials for representing nuclear scattering with absorption have been used as early as 1935 by

**2.** Ostrofsky, M., Breit, G. and Johnson, D. P., Phys. Rev. **49**, 22 (1935).

**3.** Fernbach, S., Serber, R. and Taylor, T. B., Phys. Rev. **75**, 1352 (1949). Complex potential and complex phase shifts are used for high energy nuclear scattering.

**4.** Feshbach, H., Porter, C. and Weisskopf, V. F., Phys. Rev. **96**, 448 (1954). The optical complex potential model for low energy scattering. See also Adair, R. K., Phys. Rev. **94**, 737 (1954); Woods, R. D. and Saxon, D. S., Phys. Rev. **95**, 577 (1954). The last paper introduces the potential of (U17a), the so-called Saxon potential.

**5.** Lane, A. M., Thomas, R. G. and Wigner, E. P., Phys. Rev. **98**, 693 (1955); Scott, J. M., Phil. Mag. **45**, 1322 (1954); Wigner, E. P., Science **120**, 790 (1954); Vogt, E., Phys. Rev. **101**, 1792 (1956); Bloch, C., Nuclear Physics **4**, 503 (1957). Discussions connected with the last part of this section.

**6.** Brueckner, K. A., Phys. Rev. **97**, 1353 (1955); Riesenfeld, W. B. and Watson, K. M., Phys. Rev. **102**, 1157 (1956). The potential depths are derived as functions of energy from two-body nuclear forces.

**7.** Sasakawa, T., Progr. Theor. Phys. (Kyoto) Supplement 8 (1959). Time dependent theory of nuclear reactions.

**8.** Proc. International Conference on Nuclear Optical Model, Florida State University, Tallahassee (1959).

**9.** Glassgold, A. E., Rev. Mod. Phys. **30**, 419 (1958). See Fig. 1.

**10.** Fernbach, S., Rev. Mod. Phys. **30**, 414 (1958). See Fig. 4.

**11.** Seth, K. K., Rev. Mod. Phys. **30**, 442 (1958). See Fig. 5.

section

# V

# Deuteron Stripping Reaction and Other Direct Processes

In the preceding section it has been remarked that nuclear reactions are caused by both the compound process and the direct process. We can consider the shape elastic scattering as a direct process in the sense that the formation of the compound nucleus is not involved in that scattering. The direct reaction (in the usual sense) is, however, caused by the absorbed particle, which is subsequently emitted from the surface region without having gone deeply into the nucleus. The most striking example of the direct process may be the deuteron stripping (and also pick-up) reactions. Consider the following rearrangement collision ($d$ = deuteron, $p$ = proton, $n$ = neutron),

$$d + A \rightarrow p + B, \quad \text{and} \quad d + A \rightarrow n + C.$$

These are usually indicated by $(d, p)$ and $(d, n)$ briefly. The binding energy of the deuteron is small ($\sim$2.2 MeV) compared with the binding energy per nucleon in nucleus ($\sim$roughly 8 MeV). The size of the deuteron is also "large". Therefore, it is quite possible that one of the two nucleons is captured at the surface of the nucleus and the other just goes away. This is the stripping process of the deuteron. (Serber, ref. 5.) Historically, the deuteron reactions at very low energies are first discussed by Oppenheimer and Philips (ref. 4). For a very slow deuteron the Coulomb barrier of the nucleus prevents the proton from approaching the target, and only the neutron is likely to be captured. The computed cross section of $(d, p)$ reaction on this idea is much larger (as indicated by experiments) than that computed by the compound nucleus model according to which the whole deuteron must go into the target and subsequently the proton must go out through the Coulomb barrier. For higher energies S. T. Butler gives the differential cross section formula based on the "stripping assumption" (see below) and points out that the angular distribution determines the orbital angular momentum of the captured neutron (or proton). In the following, we shall give an account of Butler's theory (ref. 3). The pick-up process $(p, d)$ is the inverse process of the stripping $(d, p)$ reaction. Since the approximations involved in the derivation of the cross section formula are more clearly brought out in $(p, d)$ reaction than in $(d, p)$ case, we shall first consider the $(p, d)$ process.

## 1. (p, d) REACTION

The energies of the incident proton are of the order of 10 ~ 30 MeV, and after the collision the residual nucleus may be excited. The cross section for such a process is reasonably accounted for on the assumption of the formation of a compound nucleus. However, for the case of the residual nucleus in a low energy state, the observed cross section is larger than the calculated one. Probably in such cases direct nuclear processes play an important role since the incident proton will interact with a neutron at the surface of the initial nucleus and will not excite the nucleus greatly.

We shall calculate the cross section that the proton picks up a neutron through the interaction $V_{pn}$ between the proton and the neutron, and leaves the final nucleus in the state $f$. Let $\mathbf{k}_p$ and $\mathbf{k}_d$ be the wave vectors of the incident proton and outgoing deuteron, and let $\mathbf{r}_n$ be the coordinate of the neutron which is to be picked up and $\xi$ the coordinates of the final nucleus. For simplicity we first assume spinless particles and infinitely heavy nucleus. The refinement on this point and also the approximations involved will be discussed in 2. The matrix element of the process can be written as,

$$M(\mathbf{k}_p, \mathbf{k}_d) = \int d\tau\, \psi_i(\boldsymbol{\xi}, \mathbf{r}_n, \mathbf{r}_p, \mathbf{k}_p)\, V_{np}\, \psi_f^*(\boldsymbol{\xi}, \mathbf{r}_n, \mathbf{r}_p; \mathbf{k}_d), \tag{V1}$$

where $d\tau = d\mathbf{r}_p\, d\mathbf{r}_n\, d\boldsymbol{\xi}$. In the first approximation, the initial and final wave functions, $\psi_i$ and $\psi_f$, are assumed to be

$$\psi_i = v_i(\boldsymbol{\xi}, \mathbf{r}_n) \exp(i\mathbf{k}_p\cdot\mathbf{r}_p), \qquad \psi_f = u_f(\boldsymbol{\xi})\, \phi(|\mathbf{r}_p - \mathbf{r}_n|) \exp[i\mathbf{k}_d\cdot(\mathbf{r}_p + \mathbf{r}_n)/2],$$

where $v_i(\boldsymbol{\xi}, \mathbf{r}_n)$ and $u_f(\boldsymbol{\xi})$ are the normalized wave functions of the initial and final nuclei, and $\phi$ is the internal wave function of the deuteron. On expanding $v_i$,

$$v_i(\boldsymbol{\xi}, \mathbf{r}_n) = \sum_j u_j(\boldsymbol{\xi})\, F_j(\mathbf{r}_n), \tag{V2}$$

we get

$$M = \int d\mathbf{r}_n\, d\mathbf{r}_p\, F_f(\mathbf{r}_n)\, \phi(\mathbf{r}_{pn})\, V_{pn} \exp(i\mathbf{k}_p\cdot\mathbf{r}_p)\cdot\exp[-i\mathbf{k}_d\cdot(\mathbf{r}_p + \mathbf{r}_n)/2]. \tag{V3}$$

For the interaction $V_{pn}$, which is short-ranged, we assume the following form

$$V_{pn} = V_0(\mathbf{r}_n)\, \delta(\mathbf{r}_p - \mathbf{r}_n). \tag{V4}$$

$F_f(\mathbf{r}_n)$ may be expanded,

$$F_f(\mathbf{r}_n) = \sum_{l,m} f_f(r_n, l, m)\, Y_{lm}(\vartheta_2, \varphi_o), \tag{V5}$$

where $\vartheta_n$ and $\varphi_n$ are spherical polar coordinates of $\mathbf{r}_n$ with respect to $\mathbf{k}_d - \mathbf{k}_p \equiv \mathbf{Q}$ (see below). If only one term in (V5) contributes mainly to the integral, $M$ becomes

$$M = \int d\mathbf{r}_n f_f(r_n, l, m)\, Y_{lm}(\vartheta_n, \varphi_n)\, \phi(0)\, V_0(r_n) \exp(-i\mathbf{Q}\cdot\mathbf{r}_n). \tag{V6}$$

As discussed in Sect. U, we shall take the imaginary part of the optical potential to be large near the surface region. Hence, the direct reaction will take place mainly near the surface. Therefore it will be a good approximation to assume that the effective interaction $V_0(r_n)$ has the form,

$$V_0(r_n) = \begin{cases} V_0, & r_n \geqslant r_0 \\ 0, & r_n < r_0 \end{cases}, \tag{V7}$$

where $r_0$ is of the order of nuclear radius. For $r_n > r_0$, $f_f$ can be replaced by

$$f_f(r_n, l, m) = A(l, m)\, h_l^{(1)}(i\kappa r_n), \quad r_n \geqslant r_0, \tag{V8}$$

where $h_l^{(1)}$ is the spherical Hankel function of the first kind. [See footnote to (A8).] The form (V8) represents an exponentially decreasing function with the binding energy $\hbar^2\kappa^2/(2m)$, $m$ being the nucleon mass. On substituting (V7) and (V8) into (V6), we get

$$M = V_0 A(l, m)\, \phi(0) \int_{r_n \geqslant r_0} d\mathbf{r}_n \exp(-i\mathbf{Q}\cdot\mathbf{r}_n)\, Y_{lm}(\vartheta_n, \varphi_n)\, h_l^{(1)}(i\kappa r_n). \tag{V9}$$

By using (A8), (V9) is readily reduced to

$$M = c_l \delta_{m,0} \int_{r_0}^{\infty} dr_n r_n^2\, h_l^{(1)}(i\kappa r_n)\, j_l(Q r_n), \tag{V10}$$

where $j_l$ is the spherical Bessel function [see footnote to (A8)], and $c_l = \{4\pi(2l+1)\}^{1/2} i^{-l} V_0 A(l, 0)\phi(0)$. If one notices that $rh_l^{(1)}(i\kappa r)$ and $rj_l(Qr)$ are the solutions of equation (T16′) with $k^2 = -\kappa^2$ and $= Q^2$ respectively, one can get the relation,

$$\int_{r_0}^{\infty} r^2 h_l^{(1)} j_l\, dr = r_0^2 \left( j_l \frac{\partial h_l^{(1)}}{\partial r} - h_l^{(1)} \frac{\partial j_l}{\partial r} \right)\Bigg|_{r_0} \Bigg/ (\kappa^2 + Q^2). \tag{V11}$$

On substituting (V11) into (V10) we finally have the formula for the cross section,

$$\begin{aligned} \sigma(\mathbf{k}_p \to \mathbf{k}_d) &= \frac{2m^2}{(2\pi\hbar^2)^2} \frac{k_d}{k_p} |M|^2 \\ &= \frac{m^2 r_0^4 |c_l|^2 k_d}{2\pi^2\hbar^4 k_p (Q^2 + \kappa^2)^2} \left[ j_l(Qr_0) \frac{\partial h_l^{(1)}(i\kappa r_0)}{\partial r_0} - \frac{\partial j_l(Qr_0)}{\partial r_0} h_l^{(1)}(i\kappa r_0) \right]^2. \end{aligned} \tag{V12}$$

The first term containing $j_l(Qr_0)$ is usually much larger than the second term. Therefore the angular distribution is characterized by the factor $j_l^2(Qr_0)$ which shows a striking diffraction pattern determined by the angular momentum $l$ of the neutron (which is picked up by the incident proton) in the initial nucleus (see Figs. 1 and 3 below). It is noted that if the interaction takes place exclusively at $r = r_0$, that is,

$$V_0(r_n) = \text{const}\ \delta(r_n - r_0) \tag{V13}$$

the differential cross section would be directly proportional to $j_l^2(Qr_0)$, apart from the slowly varying function of angle, $(Q^2 + \kappa^2)^{-2}$.

## 2. (*d*, *p*) REACTION

(*p*, *d*) reactions are usually endothermic (that is, the kinetic energy of the emerging deuteron is less than the energy of incident proton), and for moderately high energy proton (say 10 ~ 30 MeV) the main reactions are probably other than (*p*, *d*) process. On the other hand (*d*, *p*) is exothermic, and this is one of the most frequent reactions for 0 ~ 20 MeV. Therefore (*d*, *p*) is more conveniently employed to investigate nuclear phenomena. The cross section $\sigma(\mathbf{k}_d, \mathbf{k}_p)$ of (*d*, *p*) is obtained from the corresponding $\sigma(\mathbf{k}_p, \mathbf{k}_d)$ by applying the detailed balance relation [see (W42)]. However, we shall start from the exact treatment and note the nature of the approximations made in the preceding section. The assumption (V4) will not be used throughout.

The total wave function $\psi$ satisfies the Schrödinger equation

$$(H_\xi + K_p + K_n + V_{\xi p} + V_{\xi n} + V_{pn})\,\psi(\boldsymbol{\xi}, \mathbf{r}_p, \mathbf{r}_n) = E\psi. \tag{V14}$$

$H_\xi$ is the Hamiltonian of the initial nucleus, and $K$ and $V$ are the kinetic energy and the interaction operators. $E$ is the sum of the internal energy of the initial nucleus $E_\xi$, the kinetic energy of the deuteron $\hbar^2k_d^2/(2m_d)$, and the internal (binding) energy of the deuteron $-\hbar^2\gamma^2/m$. The (normalized) wave function of the final nucleus $v_f(\boldsymbol{\xi}, \mathbf{r}_n)$ is the solution of the equation:

$$(H_\xi + K_n + V_{\xi n})\,v_f(\boldsymbol{\xi}, \mathbf{r}_n) = \left(E_\xi - \frac{\hbar^2\kappa^2}{2m}\right)v_f, \tag{V15}$$

where $\hbar^2\kappa^2/2m$ is the binding energy of the neutron to the initial nucleus. $\psi(\boldsymbol{\xi}, \mathbf{r}_p, \mathbf{r}_n)$ can be expanded as

$$\psi(\boldsymbol{\xi}, \mathbf{r}_p, \mathbf{r}_n) = \sum_s \phi_s(\mathbf{r}_p)\,v_s(\boldsymbol{\xi}, \mathbf{r}_n). \tag{V16}$$

From (V16), we have

$$\phi_f(\mathbf{r}_p) = \int d\mathbf{r}_n\, d\xi\, v_f(\boldsymbol{\xi}, \mathbf{r}_n)\,\psi(\boldsymbol{\xi}, \mathbf{r}_p, \mathbf{r}_n). \tag{V17}$$

Using (V14)–(V17) we obtain

$$\left(K_p - \frac{\hbar^2 k_p^2}{2m}\right) \phi_f(\mathbf{r}_p) = -\int d\boldsymbol{\xi}\, d\mathbf{r}_n v_f^*(\boldsymbol{\xi}, \mathbf{r}_n)(V_{np} + V_{\xi p})\, \psi(\boldsymbol{\xi}, \mathbf{r}_p, \mathbf{r}_n)$$
$$= -\sum_s \left[\int d\boldsymbol{\xi}\, d\mathbf{r}_n\, v_f^*(\boldsymbol{\xi}, \mathbf{r}_n)(V_{np} + V_{\xi p})\, v_s(\boldsymbol{\xi}, \mathbf{r}_n)\right] \phi_s(\mathbf{r}_n). \tag{V18}$$

On using the outgoing wave Green function, (V18) can be rewritten as

$$\phi_f(\mathbf{r}) = -\frac{m}{2\pi\hbar^2} \int d\mathbf{r}_p\, d\mathbf{r}_n\, d\boldsymbol{\xi}\, \frac{\exp(ik_p|\mathbf{r}_p - \mathbf{r}|)}{|\mathbf{r}_p - \mathbf{r}|}\, v_f^*(V_{np} + V_{\xi p})\, \psi. \tag{V19}$$

The cross section is the ratio of the outgoing current and the deuteron current density,

$$\sigma(\mathbf{k}_d, \mathbf{k}_p) = (\hbar R^2/2mi)\left(\phi_f^* \frac{\partial \phi_f}{\partial r_p} - \phi_f \frac{\partial \phi_f^*}{\partial r_p}\right)\Bigg|_{r_p = R} \Bigg/ \left(\frac{\hbar k_d}{m_d}\right)$$
$$= \frac{2m^2 k_p}{(2\pi)^2 \hbar^4 k_d} \left| \int d\mathbf{r}_p\, d\mathbf{r}_n\, d\boldsymbol{\xi} \exp(-i\mathbf{k}_p\cdot\mathbf{r}_p)\, v_f^*(V_{np} + V_{\xi p})\psi \right|^2. \tag{V20}$$

According to the detailed balance relation (W42), we have

$$\sigma(\mathbf{k}_d, \mathbf{k}_p) = (k_p/k_d)^2\, \sigma(\mathbf{k}_p, \mathbf{k}_d). \tag{V21}$$

By comparing (V12) and (V20) we verify that the integral included in (V20) is the exact form of $M^*$. Therefore equation (V3) contains the following approximations: 1) Neglect of $V_{\xi p}$ interaction, 2) Replacement of $\psi$ by $\phi(r_{pn}) \exp[i\mathbf{k}_d\cdot(\mathbf{r}_p + \mathbf{r}_n)/2]\, u_i(\xi)$. The final cross section is the consequence of further approximations: 3) The interaction $V_{np}$ is extremely short-ranged [see (V4)]; 4) Only one term in (V5) contributes to the cross section, 5) The neutron inside the radius $r_0$ does not interact with the incident proton [for $(p, d)$ reaction] or with the outgoing proton [for $(d, p)$ reaction], see (V7). The approximation 1) and 2) are valid only when the nucleus $\xi$ does not distort the incident and the outgoing waves appreciably. Therefore these two approximations are not valid when resonance reaction has an influence on the process. When the average cross section which covers many closely spaced resonance levels is measured, the approximation 1) and 2) are not valid in the region of resonance in the gross structure problem (see Fig. 3 in Sect. U). If the incident beam is monochromatic, similar argument applies to the individual resonance. When the Coulomb barrier is higher than the deuteron kinetic energy, the plane wave is not a good approximation. The approximations 1), 2) and 5) are essential for obtaining a closed form for the stripping (or pick-up) cross section. On the other hand the assumptions 3) and 4) are not essential, and will not be used to derive the final formula (V32).

The integral appearing in (V20) is reduced by the assumptions 1), 2) and 5) to

$$M^* = \int d\mathbf{r}\, V_{np}(\mathbf{r})\, \phi(\mathbf{r}) \exp(i\mathbf{k}\cdot\mathbf{r}) \int d\xi \int_{r_n \geqslant r_0} d\mathbf{r}_n\, u_i(\xi)\, v_f^*(\xi, \mathbf{r}_n) \exp(i\mathbf{Q}\cdot\mathbf{r}_n), \tag{V22}$$

where $\mathbf{r} = \mathbf{r}_p - \mathbf{r}_n$, $\mathbf{k} = (\mathbf{k}_d/2) - \mathbf{k}_p$, $\mathbf{Q} = \mathbf{k}_d - \mathbf{k}_p$, and $u_i(\xi)$ is the wave function of the initial nucleus. $\phi(\mathbf{r})$ is the solution of the equation:

$$\left(-\frac{\hbar^2}{m}\nabla_r^2 + V_{pn}(r)\right)\phi(\mathbf{r}) = -\frac{\hbar^2\gamma^2}{m}\phi(\mathbf{r}). \tag{V23}$$

The integral with respect to $d\mathbf{r}$ in (V22) can be rewritten as

$$\int d\mathbf{r}\, V_{np}(r)\, \phi(\mathbf{r})\, e^{i\mathbf{k}\cdot\mathbf{r}} = -\frac{\hbar^2(\gamma^2 + k^2)}{m} G(k), \tag{V24}$$

with $G(k) \equiv \int d\mathbf{r}\, \phi(\mathbf{r})\, e^{i\mathbf{k}\cdot\mathbf{r}}$. Expanding $v_f$ similarly to (V2),

$$v_f(\xi, \mathbf{r}_n) = \sum_s u_s(\xi)\, F_s(\mathbf{r}_n), \tag{V25}$$

we evaluate the remaining integral (V22) as follows,

$$\begin{aligned}\int d\xi \int_{r_n \geqslant r_0} & d\mathbf{r}_n\, u_i(\xi)\, v_f^*(\xi, \mathbf{r}_n) \exp(i\mathbf{Q}\cdot\mathbf{r}_n) \\ &= \int_{r_n \geqslant r_0} d\mathbf{r}_n\, F_i^*(\mathbf{r}_n) \exp(i\mathbf{Q}\cdot\mathbf{r}_n) \\ &= \left(-\frac{\hbar^2(\kappa^2 + Q^2)}{2m}\right)^{-1} \int_{r_n \geqslant r_0} d\mathbf{r}_n \left(\frac{-\hbar^2}{2m}\right) \\ &\quad \times [\exp(i\mathbf{Q}\cdot\mathbf{r}_n)\, \nabla^2 F_i^*(\mathbf{r}_n) - F_i^*(\mathbf{r}_n)\, \nabla^2 \exp(i\mathbf{Q}\cdot\mathbf{r}_n)]. \end{aligned} \tag{V26}$$

In the last step we have used the fact that the neutron is captured by the nucleus $u_i(\xi)$ with the binding energy of $\hbar^2\kappa^2/(2m)$ [see (V15)], and that the neutron is free of interaction for $r_n > r_0$ [that is, $F_i(\mathbf{r}_n)$ is the free-particle solution with the energy $-\hbar^2\kappa^2/(2m)$; this is a minor assumption]. On using (V24), (V26) and

$$\frac{\hbar^2 k_d^2}{4m} - \frac{\hbar^2\gamma^2}{m} = \frac{\hbar^2 k_p^2}{2m} - \frac{\hbar^2\kappa^2}{2m},$$

we have

$$M^* = \frac{\hbar^2}{2m} G(k) \sum_l i^l \sqrt{4\pi(2l + 1)}\, \delta_{m,0}\, c_{l,0} r_0^2\, W\{j_l(Qr_n), h_l^{(1)}(i\kappa r_n)\}_{r_0}, \tag{V27}$$

where $W$ denotes the Wronskian and $c_{l,0}$ is the expansion coefficient of $F_i$,

$$F_i(\mathbf{r}) = \sum_l c_{l,m} h_l^{(1)}(i\kappa r)\, Y_{lm}(\vartheta, \phi). \tag{V28}$$

If the Hulthén wave function for $\phi(r)$ is used,

$$\phi(r) = \frac{N}{\sqrt{4\pi}}(e^{-\gamma r} - e^{-\zeta r})/r, \quad N = \left(\frac{1}{2\gamma} - \frac{\rho}{2}\right)^{-1/2},$$

where $\rho$ is the effective range of the deuteron [see (E31)] and $\zeta > \gamma$, $G(k)$ is calculated simply to be

$$G(k) = \sqrt{4\pi}N\left(\frac{1}{k^2 + \gamma^2} - \frac{1}{k^2 + \zeta^2}\right). \tag{V29}$$

The assumption (V4) would have led to $G(k)$ of (V29) without the second term.

We will finally include the considerations of the spins and the finite mass of the nucleus. To get the correct expression it can be shown that the definition of $\mathbf{Q}$ must be modified to

$$\mathbf{Q} = \mathbf{k}_d - (M_i/M_f)\,\mathbf{k}_p,$$

where $M_i$ and $M_f$ are the masses of the initial and the final nucleus. $\mathbf{k}_d$, $\mathbf{k}_p$, $\mathbf{r}_n$ and $\kappa$ should be considered to refer to the relative coordinates. To include the spin we need only multiply the cross section by the factor $(2j_f + 1)/(2j_i + 1)$, $j_f$ and $j_i$ being the spins of the final and initial state respectively. There is one important point to be noticed. The final nuclear spin $\mathbf{j}_f$ is the sum of the initial nuclear spin $\mathbf{j}_i$, the orbital angular momentum $\mathbf{l}$ of the captured nucleon, and the spin $\frac{1}{2}$ of the captured neutron.

$$\mathbf{j}_i = \mathbf{j}_f + \mathbf{l} + \tfrac{1}{2}. \tag{V30}$$

Therefore $l$ is restricted by,

$$|\mathbf{j}_i + \mathbf{j}_f + \tfrac{1}{2}|_{\min} \leqslant l \leqslant j_i + j_f + \tfrac{1}{2}. \tag{V31}$$

Furthermore $l$ must have a definite parity, odd or even, depending on whether there is or is not a parity change from the initial to the final nucleus. In many cases the possible value of $l$ is uniquely determined.

The differential cross section is obtained from (V20) by using (V27). It can be shown that, if the spins are properly taken into account, the contributions from different $l$-values do not interfere. The final formula for $(d, p)$ stripping reaction is given by,

$$\frac{d\sigma(\mathbf{k}_d, \mathbf{k}_p)}{d\Omega} = \frac{2m^2}{\pi\hbar^4}\frac{k_p}{k_d}\frac{(2j_f + 1)}{(2j_i + 1)}G^2(k)\sum_l\left(\frac{\hbar^2 r_0^2}{2m}\right)^2 c_{l,0}^2$$
$$\times\left[\frac{\partial h_l^{(1)}(i\kappa r)}{\partial r}j_l(Qr) - h_l^{(1)}(i\kappa r)\frac{\partial j_l(Qr)}{\partial r}\right]^2_{r=r_0}. \tag{V32}$$

The factor $G^2(k)$ diminishes the cross section for large angle. Figs. 1 and 2 show typical examples of stripping cross section. If the independent-particle model (i.e., shell model) is good, only one term is predominant in the expansion (V28). An example is given in Fig. 1. Fig. 2 shows that $l = 2$ is the main orbit, but the orbit $l = 0$ is certainly mixed (configuration mixing) in the shell configuration. Since the effect on the cross section is larger for small

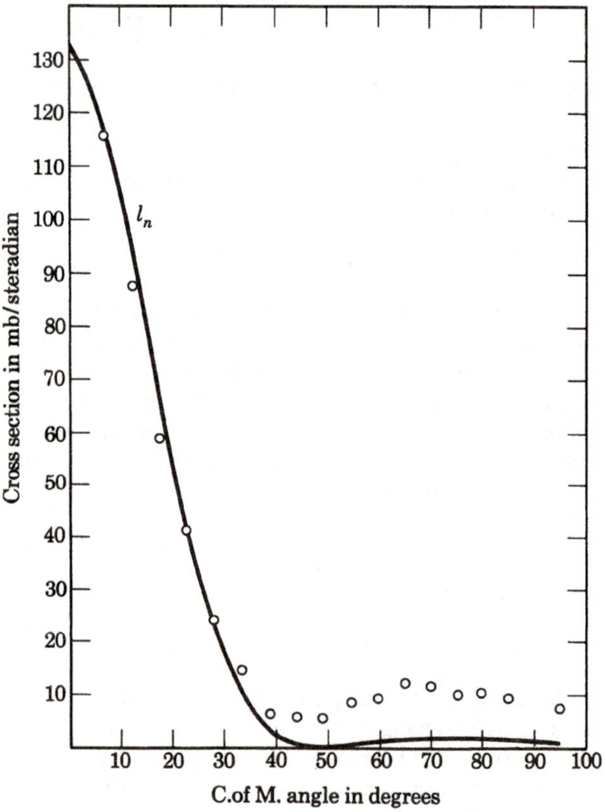


**Figure 1.** Angular distribution of protons from $F^{19}(d,p)$, $F^{20*}$ (3.49-MeV level) reaction [ref. 11, Bromley, D. A., Brunner, J. A., and Fulbright, H. W., *Phys. Rev.* **89**, 396 (1953)].

$l$, we can detect a small mixing of $l = 0$ to $l = 2$. (V32) can be rewritten by using the reduced width of the captured neutron: $\gamma_n^2 = (\hbar^2 r_0^2 c_l^2/m)[h^{(1)}(r_0)]^2$. See (T28). Therefore a comparison with the experimental data supplies us the values of the reduced width and the "nuclear radius" (in a sense) $r_0$. These are not necessarily identical with the nuclear radius and reduced width# determined by other methods, but they provide a qualitative check for the theories.

Although the angular distribution for 3 ~ 20 MeV incident deuteron is

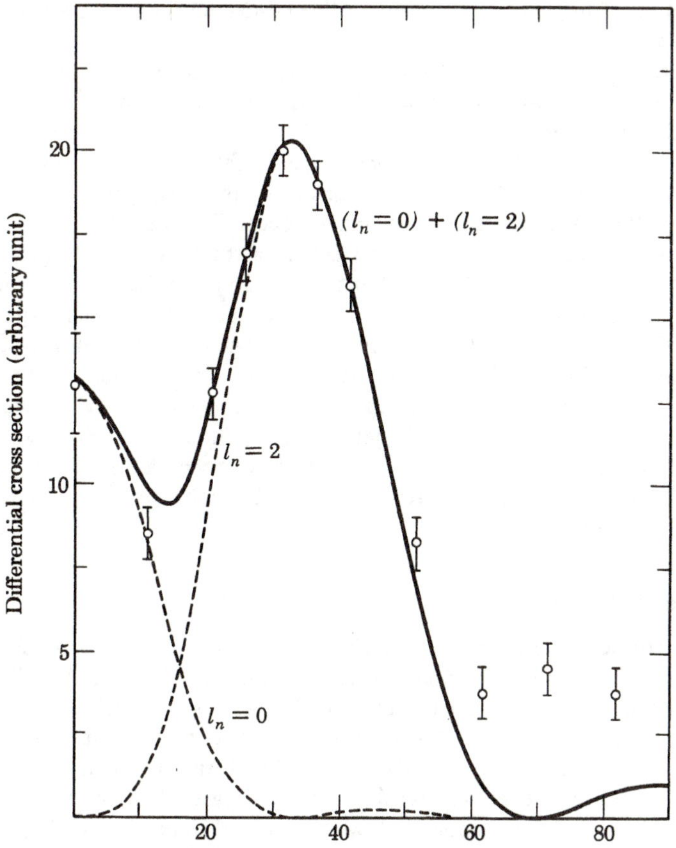


**Figure 2.** The angular distribution of the protons associated with the ground state for the reaction $Cl^{35}(d,p)Cl^{36}$. Incident energy = 6.9 MeV, energy released = 6.3 MeV, $R = 5.5 \times 10^{-13}$ cm [ref. 12, Okai, S., and Sano, M., *Progr. Theor. Phys.* **14**, 399 (1955)].

---

# The reduced width derived from the $(d, p)$ reaction is usually smaller than the width derived from other reactions by a factor 2 ~ 5. This difference may come from several sources. The potential depth $V_0$ in (V7) can be smaller and tend to zero smoothly at $r_n = r_0$. The cross section (V32) will be made smaller by considering the interaction of proton with the final nucleus, and the Coulomb interaction.

explained remarkably well by Butler's theory above, the experimental total cross section is usually larger than the predicted value. The theory fails to explain the angular distribution at large angles for the low energy (especially less than 3 MeV) region. To understand these discrepancies, we shall consider the Coulomb interaction and various indirect (including compound formation) processes.

There are other effects of the compound nucleus formation such as the polarization of the outgoing proton (or neutron), which is not expected on the original Butler theory.

The large cross sections at large angles ($>90°$) in some cases [such as $B^{11}(d, n)C^{12}$] observed have been explained by Madansky and Owen (ref. 15) as due to the "heavy-particle stripping" process. The idea is that a stripping of the target nucleus can occur as well as that of the deuteron. The proton (or neutron) which is "left behind" after the collision may have come from the target nucleus ("at its surface"). This nucleon gives rise to the large cross sections in the backward directions. It is clear that this "heavy-particle stripping reaction" will be automatically taken into account in the usual theory of stripping reactions if the final state nucleon wave function is properly antisymmetrized. Calculations using the Born approximation (Bhatia, Huang, Huby and Newns, ref. 6) have been carried out for many cases (Owen and Madansky, ref. 16) and the results account well for the experimental data.

## 3. DIRECT PROCESSES

Let us consider the inelastic scattering of nucleons (say, proton). For this process both the compound and direct scattering are again possible. When the incident nucleon energy is sufficiently high, various final nuclear states are energetically possible. On a statistical theory based on the compound nucleus model, we can find the probabilities of the final nucleus being left in these various states. The experimental data are roughly explainable on this idea, but the observed cross section for low excited final states is much larger than that expected on this model. This larger cross section has been considered to be due to direct processes.

The mechanism of such direct processes has not yet been so fully clarified as in the stripping (and pick-up) reaction. There may be various types of possible mechanism which do not lead to the formation of a compound nucleus. The cross section may be large if the incident particle collide essentially with one nucleon in the initial nucleus. If we assume that the initial and final states of the nucleus can be written as

$$\psi_i = \psi(\text{core})\,\phi_i(\mathbf{r}), \qquad \psi_f = \psi(\text{core})\,\phi_f(\mathbf{r}),$$

where $\mathbf{r}$ is the coordinate of one of the nucleon $a$ in the nucleus, the simple collision between the incident proton and the nucleon $a$ results in a direct process. If the final state is one with two nucleons excited, the reaction cross section may be smaller. The contribution of the direct process becomes smaller and smaller if the mechanism of excitation becomes more complicated. As to the spatial region where the direct reaction "takes place", there may be the volume interaction (i.e., the interaction in the whole region inside the nucleus) and the surface interaction. But we shall continue to consider a surface interaction as the cause of the direct processes by analogy with $(d, p)$ reaction.

Let us assume that the incident proton interacts only with nucleons in the outermost shell of the initial nucleus. The outermost nucleon (say, neutron) has the least binding energy, and this wave function has a long tail with which the incident proton interacts (i.e., overlaps). The matrix element can be written in a similar form to (V1) and (V3) as,

$$M \cong \int d\mathbf{r}_n\, d\mathbf{r}_p\, d\boldsymbol{\xi}\; v_i(\boldsymbol{\xi}, \mathbf{r}_n)\, V_{np}\, v_f(\boldsymbol{\xi}, \mathbf{r}_n) \exp(i\mathbf{Q}\cdot\mathbf{r}_p), \tag{V33}$$

where $\mathbf{Q} = \mathbf{k}_p - \mathbf{k}_{p'}$, $\mathbf{k}_p$ and $\mathbf{k}_{p'}$ being the incident and outgoing proton's wave vectors respectively, $v_i$ and $v_f$ are the wave functions of the initial and final nucleus respectively. On expanding $v_i$ and $v_f$ in terms of eigenstates $u_s(\boldsymbol{\xi})$ of the nucleus in which the outermost neutron is absent,

$$v_i(\boldsymbol{\xi}, \mathbf{r}_n) = \sum_s F_s^{(i)}(\mathbf{r}_n)\, u_s(\boldsymbol{\xi}),$$

$$v_f(\boldsymbol{\xi}, \mathbf{r}_n) = \sum_t F_t^{(f)}(\mathbf{r}_n)\, u_t(\boldsymbol{\xi}),$$

$M$ becomes

$$M = \sum_s \int d\mathbf{r}_n\, d\mathbf{r}_p\, F_s^{(i)}(\mathbf{r}_n)\, F_s^{(f)}(\mathbf{r}_n)\, V_{np} \exp(i\mathbf{Q}\cdot\mathbf{r}_p). \tag{V34}$$

$F_s^{(i)}$ and $F_s^{(f)}$ can be generally expanded in terms of spherical harmonics, and $l_i$ and $l_f$ are restricted by the following vector addition relation

$$|\mathbf{J}_i + \mathbf{J}_s + \tfrac{1}{2}|_{\min} \leqslant l_i \leqslant J_i + J_s + \tfrac{1}{2}, \tag{V35}$$

$$|\mathbf{J}_f + \mathbf{J}_s + \tfrac{1}{2}|_{\min} \leqslant l_f \leqslant J_f + J_s + \tfrac{1}{2}. \tag{V36}$$

Furthermore $l_i$ and $l_f$ are both odd or even depending on parity consideration. The angular momentum $l$ is constructed from $l_i$ and $l_f$, subject to the condition,

$$|\mathbf{l}_i + \mathbf{l}_f|_{\min} \leqslant l \leqslant l_i + l_f, \tag{V37}$$

$$|\mathbf{J}_i + \mathbf{J}_f|_{\min} \leqslant l \leqslant J_i + J_f, \tag{V38}$$

where in (V38) parity consideration has been used, and the possibility of the spin flip has been discarded. If the main contribution to $M$ comes from a particular value of $s$ in (V34), and the conditions (V35) – (V38) together with the parity rule restrict $l$ to one value, the matrix element $M$ has a simple behavior,

$$M \propto \int d\mathbf{r}_n\, d\mathbf{r}_p\, f(\mathbf{r}_n)\; Y_l^{m}(\vartheta_n, \phi_n)\; V_{np} \exp(i\mathbf{Q}\cdot\mathbf{r}_p) \qquad \text{(V39)}$$
$$\propto \delta_{m,0}\, j_l(Qr_0),$$

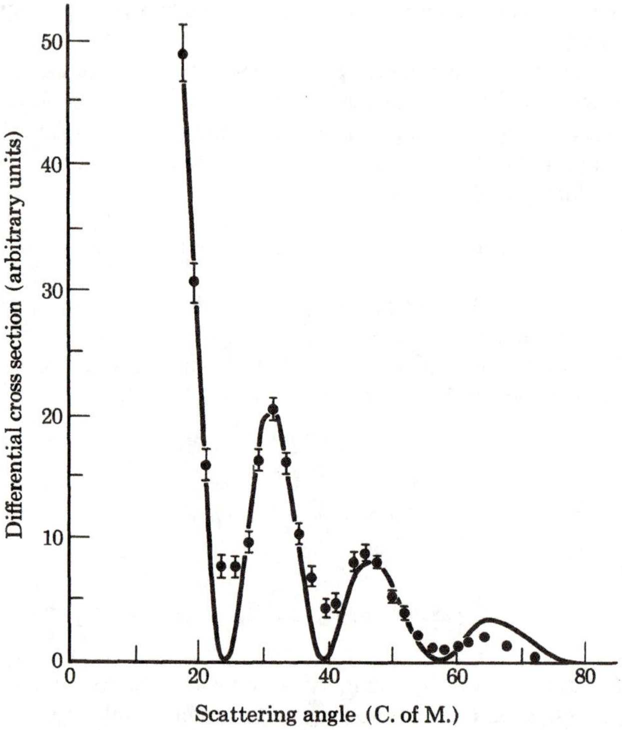


**Figure 3.** Comparison between theoretical and experimental angular distributions for the reaction $Mg^{24}(\alpha, \alpha')\, Mg^{24*}$ proceeding to the 1.37 MeV level. Incident energy = 31.5 MeV (lab.). Experimental data from Watter, H. J., *Phys. Rev.* **103**, 1763 (1956). Theoretical curve: $l = 2$, $r_0 = 5.5 \times 10^{-13}$ cm (ref. 9b).

where (V4) and (V13) have been used. The $j_l^2(Qr_0)$ dependence of angular distribution has been observed experimentally in many cases (see Fig. 3). The information on $l$ derived from the angular data is useful for fixing the parity, spin property, and the individual nucleon orbit of the nucleus. However, the observed angular distribution can not always be fitted to the $j_l^2(Qr_0)$

curve, especially for large angles. This shows that there are other direct processes.

One possible direct process may be the following. Let $\zeta$ be the coordinate which describes a collective motion, such as rotation or surface vibration of the nucleus. Let us consider that the $(p, p')$ reaction changes the nuclear collective state from $u_i(\zeta)$ to $u_f(\zeta)$. The matrix element for this process may be written in the first approximation as

$$M = \int u_i(\zeta)\, u_f^*(\zeta)\, V(\mathbf{r}_p, \zeta) \exp(i\mathbf{Q}\cdot\mathbf{r}_p)\, d\mathbf{r}_p\, d\zeta.$$

This matrix element will not be small. The more complicated the mechanism, the smaller the cross section of the direct process. It must be emphasized that the surface interaction process is nothing but one example of many possible direct reactions.

We have seen that both the compound nuclear process and the direct process are important for describing nuclear reactions. Different reactions are treated by different theories. A more unified theory for various problems of nuclear reactions and nuclear structure is one of the goals of present nuclear studies.

## REFERENCES

The following are recent monographs and review articles on direct reactions (including stripping and pick-up reactions).

**1a.** Butler, S. T. and Hittmair, O. H., *Nuclear Stripping Reactions*, John Wiley and Sons, New York (1956).

**1b.** Breit G., article in Handbuch der Physik, Vol. **41/1**, Springer (1959).

**1c.** Tobocman, W., *Theory of Direct Nuclear Reactions*, Oxford Univ. Press (1961).

**2a.** Levinson, C. A., article on direct reactions in inelastic scattering, in **Nuclear Spectroscopy**, Acad. Press, New York (1960).

**2b.** Banerjee, M. K., article on stripping and pick-up reactions, in **Nuclear Spectroscopy**, Acad. Press, New York (1960).

**2c.** Proc. International Conf. on Nuclear Structure, Toronto University Press (1960). See especially "Direct reactions" reported by N. Austern. Also *Fast Neutron Physics*, Intersc. Publ., New York (1960), Vol. II, Chap. V.

**3.** Butler, S. T., Proc. Roy. Soc. **A208**, 559 (1951). The original paper on the angular distribution of stripping reactions.

**4.** Oppenheimer, J. R. and Phillips, M., Phys. Rev. **48**, 500 (1935).

**5.** Serber, R., Phys. Rev. **72**, 1008 (1947).

**6.** Bhatia, A. B., Huang Kun, Huby, R. and Newns, H. C., Phil. Mag. **43**, 485 (1952); Daitch, P. B. and French, J. A., Phys. Rev. **87**, 900 (1952); Gerjuoy, E., Phys. Rev. **91**, 645 (1953); Newns, H. C., Proc. Phys. Soc. **A66**, 477 (1953). Simplified derivations of Butler's formula by using a modified Born approximation.

**7.** Horowitz, J. and Messiah, A. M. L., J. Phys. Radium, **14**, 695 (1953); Fujimoto, Y., Kikuchi, K. and Yoshida, S., Progr. Theor, Phys. **11**, 264 (1954). Effect of nuclear interaction in *(d, p)*, and derivation of reduced widths.

**8.** Tobocman, W. and Kalos, M. H., Phys. Rev. **97**, 132 (1955); Tobocman, W., Phys. Rev. **115**, 98 (1959). Inclusion of nuclear and Coulomb effect in *(d, p)* reaction.

**9a.** Austern, N., Butler, S. T. and McManus, H., Phys. Rev. **92**, 350 (1953). Surface direct reaction theory.

**9b.** Butler, S. T., Phys. Rev. **106**, 272 (1957). Surface direct reaction theory.

**10.** Hayakawa, S. and Yoshida, S., Proc. Phys. Soc. (London), **A68**, 656 (1955); Prog. Theor. Phys. (Kyoto) **14**, 1 (1955); Brink, D. M., Proc. Phys. Soc. (London) **A68**, 994 (1955). Collective excitation in inelastic scattering.

**11.** Bromley, D. A., Bruner, J. A., Fulbright, H. W., Phys. Rev. **89**, 396 (1953). See Fig. 1 of the text.

**12.** Okai, S. and Sano, M., Progr. Theor. Phys. **14**, 399 (1955). See Fig. 2 of the text.

**13.** Levinson, C. A. and Banerjee, M. K., Annals of Phys. **2**, 471; 499 (1957); **3**, 67 (1958).

**14.** Glendennig, N. K., Phys. Rev. **114**, 1297 (1959).

**15.** Madansky, L. and Owen, G. E., Phys. Rev. **99**, 1608 (1955).

**16.** Owen, G. E. and Madansky, L., Phys. Rev. **105**, 1766 (1957); **113**, 1575 (1959). Also see Hasegawa and Ichikawa, Proc. Theor. Phys. (Kyoto) **21**, 569 (1959).

chapter

6

# Scattering Matrix $S$ and Derivative Matrix $R$

In Chapter 1, (A34)-(A35), we have introduced the $S$ matrix (usually called the scattering matrix, or the collision matrix) for the simple case of the scattering of a particle by a central field. In Chapter 4, we have introduced the $S$ matrix in a general and formal manner. We shall in the present section give the definition of the $S$ matrix for a more general situation than in (A34)-(A35) but in a more explicit form than that in Chapter 4. We shall also bring out some of the important properties of the $S$ matrix.

Another matrix, called the derivative matrix, has been introduced by Wigner and Eisenbud. It is related to the $S$ matrix and is particularly useful in dealing with nuclear reactions. We shall also attempt an elementary account of it in the next section.

section

# W Scattering Matrix S

## 1. DEFINITION

Let us consider, for definiteness but without loss of generality, the collision between two composite particles $A$ and $B$, in a process of the type ($M1$)

$$A + B \to C + D. \tag{W1}$$

Let us specify a system in collision by the "channel" which is specified by the channel index $\Gamma$, the channel spin $S$, and the orbital angular momentum $l$ of the relative motion. The channel index $\Gamma$ denotes the species of the two component parts and the states of these particles (the ground state, for example).

Let $\chi_{\Gamma,S}^{m_s}$ denote the eigenfunction of the separated systems $C$ and $D$ that correspond to the *total* orbital and spin angular momentum $S$ and its $z$ component $m_s$ of the individual $C$ and $D$ (i.e., not including the angular momentum due to the relative motion of $C$ and $D$). $\chi_{\Gamma,S}^{m_s}$ is built up of the "internal" eigenfunctions of the separated $C$ and $D$, including the spins of the particles in $C$ and $D$. $S$ is called the "channel spin". Let $Y_{l,m_l}(\vartheta_\gamma, \phi_\gamma)$ be the angular wave function of the relative motion of $C$ and $D$. Next we construct the eigenfunction $\phi_\gamma^M$ of the channel $\gamma$ corresponding to the total angular momentum $J_\gamma$ and its $z$ component $M_\gamma = \mu$ in a manner analogous to (H19)

$$\phi_\gamma^M = \frac{1}{r_\Gamma(v_\Gamma)^{1/2}} \sum_{m_l} C_{m_l m_s}^{J\mu l S}\, Y_l^{m_l}(\vartheta_\Gamma, \phi_\Gamma)\, \chi_{\Gamma,S}^{m_s}. \tag{W2}$$

Here and in the following, $\gamma$ stands for the set $\Gamma$, $J$, $\mu$, $l$ and $S$, and thus specifies the energy of relative motion and the internal energies of the particles. The summation is taken over $m_l$ such that $m_l + m_s = \mu$. $v_\Gamma$ is the velocity corresponding to the "channel coordinate" $\mathbf{r}_\Gamma$.

The total wave function $\psi$ which describes the collision is expressed by

$$\psi = \sum_\gamma u_\gamma(\mathbf{r}_\Gamma)\, \phi_\gamma^M. \tag{W2a}$$

The function $u_\gamma(\mathbf{r}_\Gamma)$ in the relative motion has ingoing and outgoing waves in general, i.e., asymptotically,

$$u_\gamma(\mathbf{r}_\Gamma) \to A_\gamma \exp\left[-i\left(k_\Gamma r_\Gamma - \frac{l\pi}{2}\right)\right] - B_\gamma \exp\left[i\left(k_\Gamma r_\Gamma - \frac{l\pi}{2}\right)\right]\cdot \tag{W2b}$$

Let the initial state $A + B$ of (W1) be denoted by the channel index $\Gamma$ and the channel spin $S$. The suffix $S$ will be sometimes omitted for simplicity. The $l$ component of the plane wave representing the relative motion of $A$ and $B$ is asymptotically

$$\exp(ik_\Gamma z_\Gamma) \to \frac{2l+1}{2k_\Gamma r_\Gamma}\, i^{l+1}$$
$$\times \{\exp[-i(k_\Gamma r_\Gamma - \tfrac{1}{2}l\pi)] - \exp[i(k_\Gamma r_\Gamma - \tfrac{1}{2}l\pi)]\}\, P_l(\cos\vartheta_\Gamma). \tag{W3}$$

The total flux of the ingoing $l$-wave is, from (W3)#

$$v_\Gamma\left(\frac{2l+1}{2k_\Gamma}\right)^2 2\pi \int_0^\pi [P_l(\cos\vartheta_\Gamma)]^2 \sin\vartheta_\Gamma\, d\vartheta_\Gamma = v_\Gamma\, \frac{(2l+1)\pi}{k_\Gamma^2}. \tag{W4}$$

Let

$$\mathfrak{A}_\Gamma = \frac{1}{k_\Gamma}\sqrt{(2l+1)\,\pi v_\Gamma}\cdot i^{l+1} \tag{W5}$$

be called the probability flux amplitude of the ingoing and outgoing $l$-component of the plane wave. Then (W3) can be written

$$\exp[ik_\Gamma z_\Gamma] \to \frac{1}{r_\Gamma\sqrt{v_\Gamma}}\{\mathfrak{A}_\Gamma \exp[-i(k_\Gamma r_\Gamma - \tfrac{1}{2}l\pi)] - \mathfrak{A}_\Gamma \exp[i(k_\Gamma r_\Gamma - \tfrac{1}{2}l\pi)]\}\, Y_l^0, \tag{W6}$$

where

$$Y_l^0 = \frac{\sqrt{2l+1}}{2\pi}\, P_l.$$

---

# This can also be obtained from the flux of $v_\alpha$ per unit area per second times the area of the annular ring between the impact parameters

$$b_l = \frac{l\hbar}{m_\Gamma v_\Gamma} = \frac{l}{k_\Gamma} \quad \text{and} \quad b_{l+1} = \frac{l+1}{k_\Gamma},$$

namely,

$$v_\Gamma \frac{\pi}{k_\Gamma^2}[(l+1)^2 - l^2] = v_\Gamma\, \frac{(2l+1)\pi}{k_\Gamma^2}.$$

Now we shall decompose the initial state represented by the plane wave into the wave functions in all possible channels. By using the orthogonality relation of the Wigner vector addition (or the Clebsch–Gordon) coefficients, we have from (W2),

$$Y_l^m(\vartheta_\Gamma, \phi_\Gamma)\, \chi_{\Gamma, S}^{m_s} = \sum_{J=|l-S|}^{l+S} C_{0, m_s=\mu}^{J\mu lS} (r_\Gamma^2 v_\Gamma)^{1/2}\, \phi_\gamma^\mu. \tag{W6a}$$

On substituting (W6a) into (W6), we get

$$\exp[ik_\Gamma z_\Gamma]\, \chi_{\Gamma, S}^{m_s} \to \sum_{J=0}^{\infty} \sum_{\mu=-J}^{J} \sum_{l=|J-S|}^{J+S} C_{0,\mu}^{J\mu lS}$$
$$\times \left\{ \mathfrak{A}_\Gamma \exp\left[-i\left(k_\Gamma r_\Gamma - \frac{l\pi}{2}\right)\right] - \mathfrak{A}_\Gamma \exp\left[i\left(k_\Gamma r_\Gamma - \frac{l\pi}{2}\right)\right]\right\} \phi_\gamma^\mu$$
$$\equiv \sum_\beta \left\{ A_\beta \exp\left[-i\left(k_\beta r_\beta - \frac{l\pi}{2}\right)\right] - A_\beta \exp\left[i\left(k_\beta r_\beta - \frac{l\pi}{2}\right)\right]\right\} \phi_\beta^\mu, \tag{W6b}$$

where $\beta$ represents a set of $J$, $\mu$, $l$, $S$ and $\boldsymbol{\beta}$, and $\sum_\beta$ indicates the summation with respect to possible $J$, $\mu$ and $l$. $A_\beta$ is given by

$$A_\beta = \begin{cases} C_{0\mu}^{J\mu lS} \mathfrak{A}_\Gamma & \text{in channels corresponding to } \Gamma, \\ 0 & \text{in other channels.} \end{cases} \tag{W6c}$$

After the collision, the system may go into any of all the channels possible for the initial total energy of the system.

Now let us consider a different initial condition in which the system before collision is in a definite channel, $\alpha = (J, \mu, l, S, \boldsymbol{\alpha})$. The collision initiated from a particular channel $\alpha$ is described by the total wave function $\psi_\alpha$,

$$\psi_\alpha = \sum_\beta u_{\alpha\beta}(\mathbf{r}_\beta)\, \phi_\beta^M. \tag{W7a}$$

If $\beta$ is the same as $\alpha$ (i.e., for the elastic scattering $\alpha \to \alpha$), there are both ingoing and outgoing spherical waves as in (W6b) and (W7a). On account of the interactions, the outgoing amplitude $B_\alpha$ is different from $A_\alpha$, i.e., $B_\alpha = S_{\alpha\alpha} A_\alpha$, where $S_{\alpha\alpha}$ is a constant.

$$u_{\alpha\alpha} \to A_\alpha \left\{ \exp\left[-i\left(k_\alpha r_\alpha - \frac{l\pi}{2}\right)\right] - S_{\alpha\alpha} \exp\left[i\left(k_\alpha r_\alpha - \frac{l\pi}{2}\right)\right]\right\},$$

[similar in form to (A34)]. For $\beta \neq \alpha$, there is no incoming spherical wave but only outgoing scattered wave, i.e.,

$$A_\beta = 0 \quad \text{for} \quad \beta \neq \alpha, \qquad B_\beta \equiv \sum_\alpha S_{\alpha\beta} A_\alpha$$
$$u_{\alpha\beta} \to -S_{\alpha\beta} A_\alpha \exp\left[i\left(k_\alpha r_\alpha - \frac{l\pi}{2}\right)\right], \tag{W7b}$$

where $S_{\alpha\beta}$ is the ratio of the amplitudes of the scattered and the ingoing part of the incident plane wave. The quantities $S_{\alpha\beta}$ are uniquely determined by the Schrödinger equation of the whole system (i.e., in terms of the Hamiltonian of the whole system). The calculations of $S_{\alpha\beta}$ is possible in principle but difficult in general. But formally, the matrix $S$, whose elements are $S_{\alpha\beta}$, contain the information about all the possible collisions from any initial channel $\alpha$ to any channel $\beta$.

On account of the linearity of the Schrödinger equation, the total wave function $\psi$ in (W2a) is expressible by a linear combination of $\psi_\alpha$ in (W7a). The initial (i.e., boundary) condition of $\psi$ has been given by (W6b). We have

$$\psi = \sum A_\alpha \psi_\alpha \tag{W8}$$

where $A_\alpha$ is given by (W6c). The corresponding $B_\beta$ is obtained from (W7b)

$$B_\beta = \sum_\alpha S_{\alpha\beta} A_\alpha. \tag{W7c}$$

The wave function $\psi$ with arbitrary initial condition (i.e., arbitrary $A_\alpha$) can be written in the form of (W8), and the outgoing amplitude $B_\beta$ is given by (W7b) in general. In this sense $\psi_\alpha$ form a complete set. $S$ is called the scattering matrix (or operator) or the collision matrix, or simply the $S$ matrix. It was first introduced by Wheeler, and its importance was brought out by Heisenberg who attempts, in the same spirit in which he initiated the matrix mechanics, to formulate a theory in which the Hamiltonian–Schrödinger equation formalism is replaced by a theory in which one deals only with the directly observable quantities, for example, the $S$ matrix describing the cross sections of collision processes. While attempts to build an entirely new theory on the $S$ matrix alone have not been successful, the $S$ matrix has since played a very important role, both in the ordinary quantum mechanics and in the quantum theory of fields. In the following we shall give the general properties of the $S$ matrix for collision processes in the quantum mechanics of "particles".

The channel $\beta$ covers all elastic, inelastic and rearrangement collisions. The scattering $\alpha \to \alpha$ is an elastic scattering, but the collision $\alpha \to \beta\ (\neq \alpha)$ may also be an elastic scattering. As an example, consider the process $(A + B)$ $l_\alpha = 0 \to (A + B)\ l_\beta = 2$, i.e., the orbital angular momentum changes from 0 to 2 upon a compensating change in the spin angular momenta, but otherwise the two "particles" $A$ and $B$ remain $A$ and $B$. (This may be regarded as an "elastic" collision.) An example of this type is the nucleon–nucleon scattering treated in Chapter 1, H.4. Another example is the exchange scattering which is regarded as an elastic scattering when identical particles are involved, although here $\beta \neq \alpha$.

For the elastic scattering $\alpha \to \alpha$, the total cross section is, as shown in (A37), Chapter 1,

$$\sigma_{\alpha\alpha} = \frac{\pi}{k_\alpha^2}(2l_\alpha + 1)|1 - S_{\alpha\alpha}|^2. \tag{W9}$$

For the process $\alpha \to \beta$ ($\beta \neq \alpha$), the cross section (for the partial wave $l_\alpha$) is equal to the outgoing flux through the channel $\beta$ divided by the incident flux, i.e., from (W5),

$$\begin{aligned}\sigma_{\alpha\beta} &= \frac{1}{v_\alpha}\left|S_{\beta\alpha}\frac{\sqrt{(2l_\alpha + 1)\pi v_\alpha}}{k_\alpha}\right|^2 \\ &= \frac{(2l_\alpha + 1)\pi}{k_\alpha^2}|S_{\alpha\beta}|^2.\end{aligned} \tag{W10}$$

The differential cross sections are also given in terms of the matrix elements $S_{\alpha\alpha}$, $S_{\alpha\beta}$; but one must be careful in considering which waves in various channels are coherent (i.e., interfere with each other). For example the cross section for elastic scattering is given by

$$\sigma_e(\vartheta) = \frac{1}{4k_\alpha^2}\sum_{l_\alpha}|(2l_\alpha + 1)(1 - S_{\alpha\alpha})P_{l\alpha}(\cos\vartheta_\alpha)|^2. \tag{W11}$$

For angular distribution formulas, see subsect. 4 below.

## 2. UNITARITY OF $S$ MATRIX

In Chapter 4, (P23), (Q43), it is shown that $S$ is unitary. The physical meaning of this unitary property can now be seen. From (W10) and (W7b), one gets for the total outgoing flux cross section for scattering into all channels $\beta$ (including $\alpha$)

$$\sigma = \sum_\beta \sigma_{\alpha\beta} = \frac{(2l_\alpha + 1)\pi}{k_\alpha^2}\sum_\beta |S_{\alpha\beta}|^2.$$

But this total cross section must be the area calculated in (W4), namely, $(2l_\alpha + 1)\pi/k_\alpha^2$. Hence one obtains

$$\sum_\beta S_{\alpha\beta}^* S_{\alpha\beta} = 1. \tag{W12}$$

A more general relation than this can be obtained as follows. Consider the system starting in all channels $\alpha$. The total incoming flux is, according to (W4), (W6c), given by $\sum_\alpha |A_\alpha|^2$. The total outgoing flux is $\sum_\beta |B_\beta|^2$. The requirement of *conservation of flux* is, on using (W7b),

$$\sum_\alpha A_\alpha^* A_\alpha = \sum_{\beta\alpha\gamma} S_{\alpha\beta}^* A_\alpha^* S_{\gamma\beta} A_\gamma.$$

To be true for arbitrary $A_\alpha$, we must have

$$\sum_\beta S^*_{\alpha\beta} S_{\gamma\beta} = \delta_{\alpha\gamma},$$

or

$$\sum_\beta S_{\gamma\beta} S^+_{\beta\alpha} = \delta_{\alpha\gamma},$$

or

$$SS^\dagger = S^\dagger S = 1. \tag{W13}$$

Thus the requirement of the conservation of flux or probability leads to the unitary nature of $S$. [See remark at end of Sect. A.]

### 3. SYMMETRY OF $S$ MATRIX

In addition to being unitary, the $S$ matrix can be shown to be symmetric on considerations of the invariance requirement with respect to the reversal of time.

#### *i) Time reversal*

In the Newtonian mechanics, the equation of motion of a planet is invariant under time reversal. Consider the position $\mathbf{x}$ of the planet (regarded as a material point) as a function of time $t$, i.e.,

$$\mathbf{x} = \mathbf{x}(t)$$

is a solution of the equation of motion. Suppose that at an arbitrary instant the momentum of the planet is reversed (leaving the energy unchanged), and let the "time-reversed motion" be denoted by the function $\mathbf{x}^{(r)} = \mathbf{x}^{(r)}(t)$, which is exactly the same orbit as the original motion $\mathbf{x} = \mathbf{x}(t)$ traversed in the reversed time. This relation,

$$\mathbf{x}^{(r)}(t) = \mathbf{x}(-t), \tag{W14}$$

is a consequence of the invariance of the equation of motion with respect to time reversal.

In quantum mechanics, the behavior of a system under time reversal has been studied by Wigner (1932). On changing the sign of time $t$ and taking the complex conjugate of the Schrödinger equation

$$i\hbar \frac{\partial \psi(t)}{\partial t} = H\psi(t), \tag{W15}$$

we have

$$i\hbar \frac{\partial \psi^*(-t)}{\partial t} = H^*\psi^*(-t). \tag{W16}$$

If $H$ is real so that $H^* = H$, and is independent of $t$, then (W16) shows that $\psi^*(-t)$ satisfies the same Schrödinger equation as $\psi(t)$. The meaning of $\psi^*(-t)$ is perhaps made clearer by the following considerations.

Let us define a time reversed state by

$$\psi^{(r)}(t) \equiv W\psi(-t), \tag{W17}$$

where $W$ indicates Wigner's time-reversal operator. To determine $W$, we lay down the following requirement (by analogy with classical mechanics) that the Schrödinger equation be satisfied by both $\psi(t)$ and the time-reversed state $\psi^{(r)}(t)$, i.e.,

$$i\hbar \frac{\partial \psi^{(r)}(t)}{\partial t} = H\psi^{(r)}(t). \tag{W18}$$

By the transformation (W17) from $\psi(-t)$ to $\psi^{(r)}(t)$, a physical observable (operator) $Q$ is transformed into

$$Q^{(r)} = WQW^{-1}. \tag{W19}$$

Operators (containing odd orders of differentiation with respect to time) such as momentum are expected to change their sign

$$P^{(r)} = WPW^{-1} = -P, \text{ etc.,} \tag{W20}$$

while operators (containing even order of differentiation with respect to $t$) such as energy and coordinates remain unchanged

$$H^{(r)} = WHW^{-1} = H. \tag{W21}$$

The time-reversed system has the same physical properties as the original system, namely, if $\psi(t)$ has the momentum expectation value $p$,

$$(\psi(t), P\psi(t)) = p, \tag{W22a}$$

then $\psi^{(r)}(t)$ must have the expectation value $-p$

$$(\psi^{(r)}(t), P\psi^{(r)}(t)) = -p,$$

or, equivalently, by (W20),

$$(\psi^{(r)}(t), P^{(r)}\,\psi^{(r)}(t)) = p. \tag{W22b}$$

For any operator $Q$, we have the relation

$$(\psi^{(r)}, Q^{(r)}\,\psi^{(r)}) = (\psi, Q\psi), \tag{W23}$$

of which (W22a) + (W22b) is a special case.

We shall now use (W18), (W20), (W21) and (W23) as the conditions which the operator $W$ must satisfy. If the Hamiltonian $H$ does not contain spin operators, $H$ is real, i.e., $H^* = H$. In this case, if $W$ is the operator of taking the complex conjugate, all conditions are fulfilled, namely, (W18) is identical with (W16); $\mathbf{p}_i^{(r)} = \mathbf{p}_i^* = (-i\hbar\nabla)^* = i\hbar\nabla = -\mathbf{p}_i$ confirms (W20); and (W21) and (W23) are obviously satisfied. One simple example of $\psi^{(r)}(t) = \psi^*(-t)$ will help the understanding, namely,

$$\psi(t) = \exp\,[i(\mathbf{px} - Et)/\hbar],$$
$$\psi^{(r)}(t) = \exp\,[i(-\mathbf{px} - Et)/\hbar].$$

If the Hamiltonian $H$ is not real, for example, if the system contains a particle of spin $\frac{1}{2}$, $W$ is set#

$$W = UC, \tag{W17a}$$

where $C$ is the operator of taking complex conjugate. From (W23) it is seen that $U$ must be unitary, because

$$\begin{aligned}(\psi, Q\psi) = (\psi^{(r)}, Q^{(r)}\psi^{(r)}) &= (UC\psi, UCQC^{-1}U^{-1}UC\psi)\\ &= (U\psi^*, U(Q\psi)^*)\\ &= (\psi^*, \tilde{U}^*U(Q\psi)^*)\\ &= (\psi, \tilde{U}U^*Q\psi)^*\\ &= (\psi, \tilde{U}U^*Q\psi).\end{aligned}$$

The spin operators $\boldsymbol{\sigma}$ have the same transformation property as the angular momentum $\mathbf{L}$. Therefore we impose the condition analogous to (W20),

$$\sigma_i^{(r)} \equiv W\sigma_i W^{-1} = -\sigma_i.$$

On using the explicit representation of $\sigma$,

$$\sigma_x = \begin{pmatrix}0 & 1\\ 1 & 0\end{pmatrix}, \qquad \sigma_y = \begin{pmatrix}0 & -i\\ i & 0\end{pmatrix}, \qquad \sigma_z = \begin{pmatrix}1 & 0\\ 0 & -1\end{pmatrix},$$

we get

$$\sigma_i U = -U\sigma_i^*,$$

or

$$\sigma_x U = -U\sigma_x, \qquad \sigma_y U = U\sigma_y, \qquad \sigma_z U = -U\sigma_z, \tag{W24}$$

i.e., $U$ must commute with $\sigma_y$ and anticommute with $\sigma_x$, $\sigma_y$, and one may therefore take

$$U = \sigma_y, \tag{W25}$$

which satisfies these conditions. (W25) will guarantee the validity of all the conditions (W18, 20, 21, 23).##

---

# See footnote following (W37a) below.

## A spin-orbit interaction term in $H$ is transformed into

$$\begin{aligned}(\mathbf{L}\cdot\boldsymbol{\sigma})^{(r)} \equiv W(\mathbf{L}\cdot\boldsymbol{\sigma})\,W^{-1} &= \sigma_y[l_x\sigma_x + l_y\sigma_y + l_z\sigma_z]^*\,\sigma_y\\ &= (-l_x\sigma_x + l_y\sigma_y - l_z\sigma_z)^* = \mathbf{L}\cdot\boldsymbol{\sigma}.\end{aligned}$$

If the Hamiltonian contains $n$ particles of spin $\frac{1}{2}$, it is easily seen that the transformation function $U$ is

$$U = \sigma_y(1)\,\sigma_y(2)\,\ldots\,\sigma_y(n). \tag{W25a}$$

For the operator $U$ for any field (for example, electromagnetic field) see Coester (ref. 10) and Schwinger (ref. 11).

Let $Q_i$ be a complete set of commuting operators, and $\psi_0$ be an eigenfunction of $Q_i$,

$$Q_i\psi_0 = q_i\psi_0. \tag{W26}$$

According to the discussion around (W20) and (W21), we have

$$WQ_iW^{-1} = \pm Q_i, \tag{W27}$$

where the minus sign is for velocity-like operators (e.g., momentum, spin, etc.) and plus sign for coordinate-like operators. On using (W26) and (W27) we get

$$\begin{aligned} Q_i\psi^{(r)} &= Q_iW\psi_0 = \pm WQ_i\psi_0 \\ &= \pm q_iW\psi_0 = \pm q_i\psi_0^{(r)}. \end{aligned} \tag{W28}$$

The comparison of (W28) with (W26) shows that $\psi_0^{(r)}$ is the state in which momenta and spins are reversed from the (original) $\psi_0$.

Now consider the time development of $\psi$. Let $\psi$ start from $\psi_0$ at $t = -T\ (T > 0)$,

$$\psi(-T) = \psi_0, \quad \text{at } t = -T < 0. \tag{W29a}$$

$\psi$ will evolve with the time factor $e^{-iHt/\hbar}$, and

$$\psi(0) = \exp(-iHT/\hbar)\,\psi_0, \quad \text{at } t = 0. \tag{W29b}$$

We reverse the direction of momenta and spins at $t = 0$ in just the same way we have done at the beginning of this subsection for the motion of planet. This means that $\psi(0)$ is replaced by $\psi^{(r)}(0)$ at $t = 0$, and the state will evolve in accordance with the Schrödinger equation from the initial condition $\psi = \psi^{(r)}(0)$ at $t = 0$. From (W29b),

$$\begin{aligned} \psi^{(r)}(0) &= W\psi(0) \\ &= W\exp(-iHT/\hbar)\,W^{-1}W\psi_0 \\ &= \exp(iHT/\hbar)\,\psi_0^{(r)}. \end{aligned} \tag{W29c}$$

The state at $t = T\ (T > 0)$ is obtained by multiplying this by the factor $\exp(-iHT/\hbar)$ so that we have

$$\psi^{(r)}(T) = \psi_0^{(r)}. \tag{W29d}$$

Since $\psi_0^{(r)}$ is the corresponding state $\psi_0$ with opposite direction of $\mathbf{p}$ and $\boldsymbol{\sigma}$, on comparing (W29a) and (W29d) we see that the state comes back at $t = T$ to the original $\psi_0$ (at $t = -T$) but with reversed direction of momenta and spins. This shows an analogy to the classical mechanics clearly.

We shall now find the general features of the time reversed state of the $\Psi$ in (W29) for channel $\beta$. By (W23), the time reversed state of $\Psi$ is obtained, apart from $t \to t'$, by taking the complex conjugate and operating on it by the unitary $U$. Referring to (W2), we have

$$Y_l^{*m_l} = i^{2m_l} Y_l^{-m_l}, \tag{W30}$$

where the phases of $Y_l^m(\vartheta, \phi)$ are those defined in Condon and Shortley's book. For the spin wave functions

$$\sigma_y \chi^{m_s} = i^{2m_s} \chi^{-m_s}, \quad m_s = \pm\tfrac{1}{2}, \tag{W31}$$

and

$$\sigma_y \psi_{j_1}^{*m_1} = \sigma_y(Y_{l_1}^{*ml_1} \chi^{m_{s_1}}) = i^{2m_1} \psi_{j_1}^{-m_1}, \quad m_1 = m_{l_1} + m_{s_1}. \tag{W32}$$

For $\psi_j^m$ constructed from $\psi_{j_1}^{m_1}$, $\psi_{j_2}^{m_2}$ according to

$$\psi_j^m = (i)^{j_1+j_2-j} \sum C_{m_1m_2}^{jmj_1j_2} \psi_{j_1}^{m_1} \psi_{j_2}^{m_2}, \tag{W33}$$

where $C_{m_1m_2}^{jmj_1j_2}$ are the Wigner vector addition (or the Clebsch–Gordon) coefficients which are real and which have the symmetry property

$$C_{-m_1-m_2}^{j-mj_1j_2} = (-1)^{j_1+j_2-j} C_{m_1m_2}^{jmj_1j_2},$$

$\psi_j^m$ satisfies

$$\sigma_y(1)\sigma_y(2)(\psi_j^m)^* = i^{2m}\psi_j^{-m}. \tag{W34}$$

The above results (W30)–(W34) can be summarized by saying that the time-reversed wave function is obtained by taking the complex conjugate of the radial part, changing the signs of all the magnetic quantum numbers and multiplying the resulting function by $i^{2M}$ where $M$ is the total magnetic quantum number.

### *ii) Reciprocity relation*

On applying the above rule of making the time-reversed wave functions to (W2b),

$$u_\gamma(r_\Gamma) \to A_\gamma \exp\left[-i\left(k_\Gamma r_\Gamma - \frac{l\pi}{2}\right)\right] - B_\gamma \exp\left[i\left(k_\Gamma r_\Gamma - \frac{l\pi}{2}\right)\right], \tag{W2b}$$

we have

$$u_\gamma^{(r)}(r_\Gamma) \to -(i^{2M})\left\{B_{\gamma'}^* \exp\left[-i\left(k_\Gamma r_\Gamma - \frac{l\pi}{2}\right)\right] - A_{\gamma'}^* \exp\left[i\left(k_\Gamma r_\Gamma - \frac{l\pi}{2}\right)\right]\right\}. \tag{W35}$$

Here $\gamma'$ differs from $\gamma$ only in having the sign of all magnetic quantum numbers reversed. Since the time-reversed state also satisfies the original Schrödinger equation, $A_\alpha$ and $B_\beta$ in (W7b) can be replaced by $B^*_{\alpha'}$ and $A^*_{\beta'}$ respectively.

$$A^*_{\beta'} = \sum_\alpha S_{\alpha\beta} B^*_{\alpha'}. \tag{W36a}$$

If (W7b) is used once again, we get

$$A^*_{\beta'} = \sum_{\alpha\gamma'} S_{\alpha\beta} S^*_{\gamma'\alpha'} A^*_{\gamma'}. \tag{W36b}$$

From (W36b), we get

$$\sum_\alpha S_{\gamma'\alpha'} S^*_{\alpha\beta} = \delta_{\beta'\gamma'}.$$

From these relations, together with the unitarity property $SS^\dagger = 1$ in (W13), it follows that

$$S_{\beta'\alpha'} = S_{\alpha\beta}, \tag{W37}$$

which states that the probability of a scattering process $\alpha \to \beta$ is the same as that of the time reversed process $\beta \to \alpha$. This relation, known as the reciprocity theorem, has its origin in the possibility of finding a time reversed $\psi^{(r)}(t)$ such that the Schrödinger equation for it in the reversed time $t'$ is the same as that for $\psi(t)$ for the "direct" time $t$. This is in fact the same relation as the "microscopic reversibility" in classical physics.

If the process $\alpha \to \beta$ does not depend on the angular momentum $M$, then $S_{\beta'\alpha'} = S_{\beta\alpha}$, and (W37) becomes the simple one:

$$S_{\beta\alpha} = S_{\alpha\beta} \tag{W37a}$$

i.e., $S$ is not only unitary (W13), but is symmetric.#

---

# Remarks concerning the phase of the basic vectors. It has been pointed out by Huby (ref. 8) that there is an inconsistent phasing of the matrix elements (or basic vectors) in the work of Wigner and Eisenbud (ref. 7), Blatt and Biedenharn (ref. 8) and others. The errors so introduced are trivial in most cases, but sometimes lead to incorrect results. We shall only indicate below the modifications of the formulas in sections W and X made necessary by a correct phasing.

We have derived, after (W34), a general rule for obtaining the time-reversed wave function. With the choice of (W17a), (W25) or (W25a), we have to construct $\psi_J^m$ in (W33) with the extra factor $(i)^{j_1+j_2-j}$ to obtain this general rule. This is, however, not the same way as that of constructing the angular momentum functions in (W12) on which our definition of the scattering matrix elements is based.

A more satisfactory manner of obtaining the time-reversed wave function may be as follows. We first note that (W24) allows for a constant factor for $U$. If the factor of $W$ is so chosen that for an angular momentum vector $|j, m\rangle$,

$$W|j_n, m_n\rangle = (i)^{2m_n - 2j_n}|j_n, -m_n\rangle, \quad n=1,2, \tag{1}$$

A general $N \times N$ matrix with complex elements has $2N^2$ real parameters. The unitary and symmetry conditions reduce this number greatly. From (P48), we have the relation between the $K$ and the $S$ matrices

$$S = \frac{1 - \frac{1}{2}iK}{1 + \frac{1}{2}iK}, \quad \text{or} \quad K = \frac{2}{i}\frac{1 - S}{1 + S}.$$

For $S$ both unitary and symmetric, $K$ is the Hermitian and symmetric so that it is real and symmetric. Thus the number of independent real parameters in $K$, and hence in $S$, is $\frac{1}{2}N(N + 1)$.

### *iii) Detailed balance theorem*

In this connection we shall briefly refer to the detailed balance theorem which is often used to derive the transition probability of the inverse of a given process. The transition probability $w_{\alpha\beta}$ from $\Phi_\alpha$ to $\Phi_\beta$ is given in the first order perturbation theory by (O14),

$$w_{\alpha\beta} = \frac{2\pi}{\hbar} |(\Phi_\beta, V\Phi_\alpha)|^2 \rho_\beta, \tag{W38}$$

where $V$ is the perturbing Hamiltonian and $\rho_\beta$ is the number of final states $\beta$ per unit energy interval. From the hermitian property of $V$, we have $(\Phi_\alpha, V\Phi_\beta) = (\Phi_\beta, V\Phi_\alpha)^*$, and from (W38),

$$\frac{w_{\beta\alpha}}{\rho_\alpha} = \frac{w_{\alpha\beta}}{\rho_\beta}. \tag{W39}$$

(W39) may be equivalently written in terms of the cross sections by means of the relation (O19),

$$\frac{1}{v_\alpha \rho_\alpha}\frac{d\sigma_{\beta\alpha}}{d\Omega_\alpha} = \frac{1}{v_\beta \rho_\beta}\frac{d\sigma_{\alpha\beta}}{d\Omega_\beta}, \tag{W39a}$$

---

then the $|j, m>$

$$|j, m> = \sum C^{jmj_1j_2}_{m_1m_2} |j_1, m_1>|j_2, m_2>$$

can be shown to obey the same transformation as (1)

$$W|j, m> = (i)^{2m-2j}|j, -m>. \tag{2}$$

Thus the time-reversed wave function is obtained by taking the complex conjugate of the radial part, changing the sign of all the magnetic quantum numbers and multiplying the resulting function by $(i)^{2M-2J}$ where $M$ is the total magnetic quantum number. However, if $|j, m>$ represents a spherical harmonic, we must take $i^l Y_{lm}$ instead of $Y_{lm}$ in order to be consistent with (1) or (2). Therefore all the $S_{\alpha\beta}$ in sections W and X should be replaced by $(i)^{l_\beta - l_\alpha}S_{\alpha\beta}$. For example (W37a) should be changed to

$$(i)^{l_\alpha - l_\beta}S_{\beta\alpha} = (i)^{l_\beta - l_\alpha}S_{\alpha\beta}.$$

where $v_\alpha$, $v_\beta$ are the initial and the final velocities connected by the relation of energy conservation.

From (P26), $w_{\alpha\beta} = |(\Phi_\alpha, S\Phi_\beta)|^2$, it is seen that the detailed-balance theorem (W39) or (W39a) is essentially based on the equality

$$|S_{\alpha\beta}| = |S_{\beta\alpha}|. \tag{W40}$$

(W40) is not valid in general, for example, when spins are present (see (W37)). However, (W39a) will be still valid even if the perturbation $V$ is not weak, if the spins are not measured. In general, we have, from (O19) and (P28), in terms of the transition operator $T = S - 1$,

$$\frac{d\sigma_{\beta\alpha}}{d\sigma_{\alpha\beta}} = \frac{v_\alpha \rho'_\alpha}{v_\beta \rho'_\beta} \cdot \frac{\sum |T_{\beta\alpha}|^2}{N_\beta} \cdot \frac{N_\alpha}{\sum |T_{\alpha\beta}|^2} = \frac{v_\alpha \rho_\alpha}{v_\beta \rho_\beta}, \tag{W41}$$

where $\rho_\alpha \equiv \rho'_\alpha N_\alpha$, $N_\alpha$ being the number of possible spin states, $\rho'_\alpha$ the state density not considering the spin states. The summation on $|T|^2$ indicates all summations with respect to both initial and final spin states. In the derivation of (W41) we have used the equality: $\sum |T_{\beta\alpha}|^2 = \sum |T_{\alpha\beta}|^2$, which is a consequence of (W37). (W39) or (W39a) does not always hold if the spin directions are measured. For this case we only have the relation (W37).

(W39a) or (W41) can be expressed, by using such an expression as (O17) for the densities of states $\rho_\alpha$, $\rho_\beta$, in the more familiar form

$$v_\alpha^2 \, d\sigma_{\alpha\beta} = v_\beta^2 \, d\sigma_{\beta\alpha}. \tag{W42}$$

In this connection we may remark that in classical statistical mechanics, detailed balance is usually taken to be both the sufficient and necessary condition for equilibrium. It has been shown, however, by Stueckelberg (ref. 12) that the detailed balance is not necessary, and the condition for equilibrium is

$$\sum_k A_{ik}\omega_k = \sum_k A_{ki}\omega_i, \tag{W43}$$

where $A_{ik}$ is the transition probability from the state $k$ at time $t$ to state $i$ is any later time $t'$, and $\omega_i$ are the probabilities of the system being in states $i$ at the time $t$.

### *iv) Parity conservation*

In addition to the "conservation of flux" expressed by the unitarity of $S$ in (W13), and the "time-reversal invariance", the parity invariance (H10A) imposes a further restriction on the matrix elements of $S$. Thus if the interactions $V$ in a system are invariant under the space inversion operation, the matrix elements in (P29)

$$(\Phi_f, V\psi_i)$$

vanish whenever the initial and the final state $\psi_i$, $\Phi_f$ have opposite parities. Since, by (P29), the matrix elements of $S$ (and of the transition operator $T$)

$$(\Phi_f, S\Phi_i), \qquad (\Phi_f, T\Phi_i)$$

are proportional to $(\Phi_f, V\psi_i)$, it follows that the matrix elements of $S$ between channels of different parities vanish.

## 4. THE GENERAL FORMULA FOR DIFFERENTIAL CROSS SECTIONS

We shall derive the formula for differential cross sections. First consider the case in which the initial state is specified by $\gamma$ and the channel spin $S$. The amplitude of the outgoing spherical wave $B_\beta$ in the channel $\gamma'S'l'$ is, according to (W7b), (W6c) and (W5),

$$B^{J\mu}_{\Gamma'S'l'} = k_\alpha^{-1}(\pi v_\Gamma)^{1/2} \sum_{l=|J-S|}^{J+S} C^{J\mu lS}_{om_s}(i)^{l+1}\sqrt{2l+1}\, S^J_{\Gamma'S'l',\Gamma Sl}. \tag{W44}$$

To get the cross section, the total wave function $\psi$ is divided into two parts, the wave function $\psi_{inc}$ describing the initial state which is expressed by a plane wave of the form (W6b), and the wave function $\psi_{col}$ which contains the outgoing waves only. $\psi_{col}$ has the following form for $\gamma'S'$

$$\psi_{col}(\Gamma'S') = \left(\frac{\pi v_\Gamma}{k_\Gamma^2 v_{\Gamma'}}\right)^{1/2} \sum_{J=0}^{\infty} \sum_{\mu=-J}^{J} \sum_{l=|J-S|}^{J+S} \sum_{l'=|J-S'|}^{J+S'}$$
$$C^{J\mu lS}_{om_s}\, i^{l+1}\sqrt{2l+1}\,\exp\left[i\left(k_{\Gamma'}r_{\Gamma'} - \frac{l'\pi}{2}\right)\right](\delta_{\Gamma'\Gamma}\delta_{S'S}\delta_{l'l} - S^J_{\Gamma'S'l',\Gamma Sl})\,\phi^\mu_{Jl'S'}. \tag{W45}$$

If (W45) is written in the form

$$\psi_{col}(\Gamma'S') = \frac{i}{k_\Gamma}\sqrt{\frac{v_\Gamma}{v_{\Gamma'}}}\,\frac{\exp(ik_{\Gamma'}r_{\Gamma'})}{r_{\Gamma'}} \sum_{m_s'=-S'}^{+S'} q_{\Gamma'S'm_s',\Gamma Sm_s}(\vartheta,\varphi)\,\chi^{m_s'}_{S'}, \tag{W46}$$

the differential cross section from $(\Gamma Sm_s)$ into $(\Gamma'S'm_s')$ is given by

$$\frac{d\sigma_{\Gamma'S'm_s',\Gamma Sm_s}}{d\Omega} = \frac{1}{k_\Gamma^2}\,|q_{\Gamma'S'm_s',\Gamma Sm_s}(\vartheta,\varphi)|^2. \tag{W47}$$

The cross section for an unpolarized beam is obtained by averaging over the incident spin directions $m_s$ and summing over the final spin directions $m_s'$.

$$\frac{d\sigma_{\Gamma'S',\Gamma S}}{d\Omega} = \frac{1}{2S+1}\sum_{m_s=-S}^{S}\sum_{m_s'=-S'}^{S'}\frac{d\sigma_{\Gamma'Sm_s',\Gamma Sm_s}}{d\Omega}. \tag{W48}$$

The cross section (W48) does not depend on the angle $\varphi$, which can be seen on simple symmetry considerations. If the experiment is so arranged that the spin values $S$ and $S'$ are not detectable (as in the usual cases), the cross section is obtained by averaging over the possible values of $S$ and summing over the possible values of $S'$

$$\frac{d\sigma_{\Gamma',\Gamma}}{d\Omega} = \sum_{S=|I-i|}^{I+i} \sum_{S'=|I'-i'|}^{I'+i'} \frac{2S+1}{(2I+1)(2i+1)} \frac{d\sigma_{\Gamma'S',\Gamma S}}{d\Omega}, \tag{W49}$$

where $I$ and $i$ are the spins of the target and incident particles, and $I'$ and $i'$ the spins of the residual and outgoing particles. The factor $(2S + 1)/[(2I + 1)(2i + 1)]$ is the probability that the initial state is in the channel spin $S$.

The explicit formula for (W48) was given by Blatt and Biedenham (ref. 8) by making use of the Racah coefficients. $\phi_J^\mu$ in (W45) has one Clebsch–Gordon coefficient [see (W2)], hence $q(\vartheta, \varphi)$ has two coefficients. Accordingly (W47) and (W48) contain four coefficients. On performing all the summation with respect to the magnetic quantum numbers, we finally get the formula,

$$\frac{d\sigma_{\Gamma'S',\Gamma S}}{d\Omega} = \frac{1}{(2S+1)k_\Gamma^2} \sum_{L=0}^{\infty} B_L(\Gamma'S', \Gamma S)\, P_L(\cos\vartheta), \tag{W50a}$$

where

$$B_L(\Gamma'S', \Gamma S) = \frac{(-1)^{S'-S}}{4} \sum_{J_1 J_2 l_1 l_2 l_1' l_2'}^{\infty} Z(l_1 J_1 l_2 J_2, SL)\, Z(l_1' J_1 l_2' J_2, S'L)$$
$$\times \operatorname{Re}\,[(\delta_{\Gamma'\Gamma}\delta_{S'S}\delta_{l_1'l_1} - S^{J_1}_{\Gamma'S'l_1',\Gamma S l_1})^*(\delta_{\Gamma'\Gamma}\delta_{S'S}\delta_{l_2'l_2} - S^{J_2}_{\Gamma'S'l_2',\Gamma S l_2})]. \tag{W50b}$$

Re indicates the "real part", * the "complex conjugate", and $Z$ is related to the Racah coefficient $W$ defined in Racah [1942, Phys. Rev. **62**, 438, Equations (36) and (36')]. For the properties of $Z$ the reader is referred to Biedenharn, Blatt and Rose, Rev. Mod. Phys. **24**, 248 (1952),

$$Z(l_1 J_1 l_2 J_2, SL) = i^{L-l_1+l_2}(2l_1+1)^{1/2}(2l_2+1)^{1/2}(2J_1+1)^{1/2}(2J_2+1)^{1/2}$$
$$\times W(l_1 J_1 l_2 J_2, SL)\, C_{00}^{L0 l_1 l_2}.$$

The summation in (W50b) is simpler than its appearance because of the selection rule for non-vanishing $Z$ coefficients. For a given $J_1$ for example, the summation over the other five $J_2, \ldots l_2'$ quantum numbers is restricted to only a few sets. Thus essentially the summation from 0 to $\infty$ is "one-dimensional". $B_L$ is symmetric in the arguments, i.e., $B_L(\Gamma S, \Gamma'S') = B_L(\Gamma'S', \Gamma S)$. The following conditions must always be fulfilled:

$$l_1 + l_2 - L = \text{even}, \qquad l_1' + l_2' - L = \text{even}$$

$$l_1 + l_1' = \begin{pmatrix}\text{even}\\ \text{odd}\end{pmatrix}, \quad l_2 + l_2' = \begin{pmatrix}\text{even}\\ \text{odd}\end{pmatrix} \quad \text{for} \quad \begin{pmatrix}\text{no in parity change}\\ \text{yes in parity change}\end{pmatrix}.$$

The parity change is no (yes) if the product of the parities of $A$, $B$, $C$ and $D$ in (W1) is plus (minus). For $S = S' = 0$, the $Z$ coefficient reduces to

$$Z(l_1J_1l_2J_2, 0L) = \delta_{l_1J_1}\delta_{l_2J_2}(-i)^{L-l_1+l_2}\,[(2l_1+1)(2l_2+1)]^{1/2}\,C_{00}^{L0l_1l_2}. \quad \text{(W51)}$$

The differential cross section for elastic scattering by a central potential is obtained by substituting (W51) into (W50) and using

$$S^{J}_{\Gamma sl',\Gamma sl} = \delta_{l'l}\exp[2i\delta_l].$$

The cross section is

$$\frac{d\sigma}{d\Omega} = \frac{1}{k^2}\sum_{L=0}^{\infty} B_L P_L(\cos\vartheta),$$

$$B_L = \sum_{l=0}^{\infty}\sum_{l'=|l-L|}^{l+L}(2l+1)(2l'+1)(C_{00}^{L0ll'})^2 \sin\delta_l \sin\delta_{l'}\cos(\delta_l - \delta_{l'}).$$

## 5. COLLISION LIFETIME MATRIX Q AND DELAY TIME MATRIX $\Delta t$

In Sect. A.7, the collision lifetime $Q_l$ in an elastic collision is related to the diagonal matrix element of the $S$ matrix by (A61b)

$$Q_l = -i\hbar\frac{dS_l}{dE}S_l^*, \qquad S_l = \exp(2i\delta_l). \quad \text{(W52)}$$

There $Q_l$ is identical with the delay time $(\Delta t)_l$ in (A61). $Q_l$, $\delta_l$ are functions of the energy $E$ and are properties of the scattering.

The usefulness of the concept of the collision lifetime becomes more obvious when we consider collisions of a more complex kind. For example, in those nuclear reactions which are best interpreted on the compound nucleus theory (Sect. T), the collision lifetime can naturally be associated with the lifetime of the excited state of the compound nucleus. In the case of atomic reactions (in gaseous chemical kinetics), say, the collision of an atom with a diatomic molecule $BC$, leading to the formation of the molecule $AB$,

$$A + BC \rightarrow AB + C,$$

it is permissible to regard the complex $A$, $B$, $C$ as existing in a "compound state" for a time $\tau$ before going into $AB + C$ or back into $A + BC$ (ref. 13). For this picture to be valid for both the nuclear and the atomic reactions, it is necessary for the lifetime of the compound state to be long compared with the characteristic times of the nuclear or atomic system. When this is the case, it is possible to introduce the collision lifetime.

For inelastic collisions (including rearrangement collisions), the scattering is described by the $S$ matrix (W7b). The collision lifetime is represented, instead of (W52), by a matrix $Q$ whose elements $Q_{ij}$ refer to the reaction leading from the (ingoing) channel $l$ to the (outgoing) channel $j$. With a definition similar to (A62), it can be shown that $Q_{lj}$ are related to the $S$ matrix by

$$Q_{lj} = i\hbar \sum_n S_{ln} \frac{d}{dE} S^*_{jn} = i\hbar \sum_n S_{ln} \frac{d}{dE} S^\dagger_{nj} \tag{W53}$$

(ref. 14). Since $SS^\dagger = 1$, $\frac{d}{dE} SS^\dagger = 0$ and we may write

$$Q_{lj} = -i\hbar \sum \frac{dS_{ln}}{dE} S^\dagger_{nj}, \tag{W54}$$

or, in matrix form, and denoting by $t = -i\hbar \frac{\partial}{\partial E}$,

$$Q = -StS^\dagger = S(tS)^\dagger, \tag{W53a}$$

$$Q = (tS)\, S^\dagger. \tag{W54a}$$

From (W53a, b), it follows that $Q^\dagger = Q$, i.e., $Q$ is Hermitian.

From the relation (W53) or (W54), it is possible to obtain $S$ in terms of $Q$. Using $SS^\dagger = 1$ again, and the condition that, as $E \to \infty$, $S \to 1$ and $Q \to 0$, we have

$$S = 1 - \frac{i}{\hbar} \int_E^\infty Q(E')\, S(E')\, dE' \tag{W55}$$

which may be iterated to give $S$ in terms of $Q$. Thus the collision may be described in terms of either the matrix $Q$ or $S$.

For the delay time $(\Delta t)_{lj}$, the relation (A61) $(\Delta t)_l = Q_l$ for elastic collisions no longer holds. Instead, a calculation similar to the wave packet treatment leads to the following expression for the delay time of a particle in the ingoing channel $l$ and the outgoing channel $j$ (ref. 14)

$$\Delta t_{lj} = \mathrm{Re}\, [-i\hbar(S_{lj})^{-1}\, dS_{lj}/dE]. \tag{W56}$$

It is seen from this and (W53) that the delay-time matrix $\Delta t$ is different from the lifetime matrix $Q$. However, the average delay time $\Delta t_l$ for the particle coming from the $l$th channel is

$$\overline{\Delta t_l} = \sum_j S^*_{lj} S_{lj} \Delta t_{lj} = -i\hbar \sum_j S^*_{lj}\, dS_{lj}/dE$$

which, as seen from (W54), is just the $l$th diagonal element of the $Q$ matrix.

The eigenfunction of the collision lifetime matrix $Q$ and the corresponding eigenvalue can be interpreted as the metastable state and its lifetime (not the decay time#), which is formed in the course of the scattering process, if

# The eigenfunctions are symmetrical in their ingoing and outgoing part, and half of the eigenvalue is identified as the decay lifetime.

the lifetime is "long". This statement will be made clearer by the following example. Let there be two possible decay modes of such a metastable state with total energy $E$ (with different decay energies in general). The $S$ matrix will be approximated by neglecting the potential scattering for the (total) energy $E$ which is close to $E_0$.

$$S = \begin{pmatrix} 1 - \dfrac{ig_1^2}{E - E_0 + (i\Gamma/2)}, & -\dfrac{ig_1g_2}{E - E_0 + (i\Gamma/2)} \\ -\dfrac{ig_1g_2}{E - E_0 + (i\Gamma/2)}, & 1 - \dfrac{ig_2^2}{E - E_0 + (i\Gamma/2)} \end{pmatrix},$$

$$g_i = \text{real},$$

$$g_1^2 + g_2^2 = \Gamma.$$

The two eigenvalues of $Q = i\hbar S\,(dS^\dagger/dE)$ are zero and $\hbar\Gamma[(E - E_0)^2 + (\Gamma/2)^2]^{-1}$ with the eigenvectors (in the present representation) respectively.

$$\begin{pmatrix} -g_2/\sqrt{\Gamma} \\ g_1/\sqrt{\Gamma} \end{pmatrix}, \quad \begin{pmatrix} g_1/\sqrt{\Gamma} \\ g_2/\sqrt{\Gamma} \end{pmatrix}.$$

The former corresponds to no-scattering (if the potential scattering is taken into account, this represents the "quickly" scattered wave), and the latter the scattering through the metastable state with the lifetime of $\hbar\Gamma[(E - E_0)^2 + (\Gamma/2)^2]^{-1}$. These two states are also the eigenfunctions of the $S$ matrix with the eigenvalues of 1 (no scattering) and $1 - i\Gamma[E - E_0 + (i\Gamma/2)]^{-1}$ respectively.

### REFERENCES


1. Wheeler, J. A., Phys. Rev. **52**, 1107 (1937);
2. Heisenberg, W., Zeits. f. Physik **120**, 513, 673 (1943);
3. Møller, C., Kgl., Danske Vid. Sels. Mat. Phys. Med. **23**, 1 (1945). These are the classic papers on $S$ matrix.

The general properties on $S$ matrix are treated in many places, for example:

4. Blatt, J. M. and Weisskopf, V. F., *Theoretical Nuclear Physics*, John Wiley & Sons, New York (1952), Chap. X, and
5. Sachs, R. G., *Nuclear Theory*, Addison-Wesley Publ. Co., Cambridge (1953), Chap. 10.

The relations of the unitarity and symmetry of the $S$ matrix with the invariance properties of the Schrödinger equation was first pointed out by

6. Breit, G., Phys. Rev. **58**, 1068 (1940).

The general expression for differential collision cross sections for an unpolarized beam in terms of *S* matrix is given in many papers, for example (W33), (W35), (W48) in

**7.** Wigner, E. P. and Eisenbud, L., Phys. Rev. **72**, 291 (1947).

**8.** Subsection 4 is based on Blatt, J. M. and Biedenharn, L. C., Rev. Mod. Phys. **24**, 258 (1952). Also see the references cited in this paper. For a correction, see Huby, R., Proc. Phys. Soc. (London) A67, 1103 (1954).

The classic paper on time reversal principle is:

**9.** Wigner, E. P., Göttingen Nachr, **31**, 546 (1932).

**10.** Also see Coester, F., Phys. Rev. **84**, 1259 (1951);

**11.** Schwinger, J., Phys. Rev. **82**, 914 (1951).

**12.** Stueckelberg, E. C. G., Helv. Phys-Acta **25**, 577 (1952). The correct proof is given by Pauli, quoted in Stueckelberg's paper, also by M. Inagaki, G. Wanders and C. Piron, Helv. Phys. Acta **27**, 71 (1954).

**13.** Cf. Bauer, E. and Wu, T. Y., J. Chem. Phys. **21**, 726, 2072 (1953), a quantum mechanical calculation of the rates of some chemical reactions.

**14.** Smith, F. T., Phys. Rev. **118**, 349; **119**, 2098 (1960). See refs. 15, 16 of Section A.

section

# X The *R*, or Derivative, Matrix

We have seen in the preceding section that any collision process can be described in terms of the $S$ matrix. The $R$ matrix introduced by Wigner and developed by Wigner and Eisenbud (ref. 2) is also designed for the description of nuclear reactions. It has been extensively applied to a large body of nuclear reactions which are associated with the "compound nucleus" view, and involve the resonance phenomenon in particular. The theory is not concerned with the detailed physical mechanisms "inside the nucleus" in a nuclear reaction; it aims to describe the observed reactions in terms of some parameters such as the radius $a$ of a "nuclear sphere", the "energy levels" and the "reduced widths". No assumptions are made about the wave functions "inside the nuclear sphere"; only their properties at the surface $r = a$ are employed, and these are expressed in terms of the logarithmic derivatives of the wave functions at $r = a$, which form the matrix elements of the $R$ matrix. In a sense the theory is very general, being free of any detailed physical models (except the general view of the compound nucleus theory). It differs from the original hope of the $S$ matrix theory in that it does not attempt to do away with the assumption of the existence of a Hamiltonian.

A very complete treatment of the $R$ matrix theory and its application can be found in the recent excellent monograph of Breit (ref. 1b), and to this the reader is referred for further treatments of this subject and for references to the large literature. In the following, we shall only introduce the basic ideas of the method of the theory.

## 1. INTRODUCTION: ONE-CHANNEL CASE

To illustrate the basic ideas for the introduction of the $R$ matrix, we shall take the simple case of the one-channel process of the collision of a slow neutron with a nucleus (ref. 1b). The interaction $V(r)$ between the neutron and the nucleus is represented by

$$V(r) = \begin{cases} V(r), & 0 < r \leqslant a, \\ 0, & a < r, \end{cases}$$

where $a$ is the radius of the nuclear sphere, for which we shall appeal to other sources for information. The radial wave function for the $s$ wave inside the nucleus ($0 \leqslant r \leqslant a$) is given by

$$\frac{d^2\mathscr{F}}{dr^2} + [k^2 - U(r)]\,\mathscr{F}(r) = 0, \quad U(r) = \frac{2m}{\hbar^2} V(r), \tag{X1}$$

with

$$\mathscr{F}(0) = 0.$$

Let $w_\lambda(r)$ be a complete set of eigenfunctions in the range $0 \leqslant r \leqslant a$ of

$$\left[\frac{d^2}{dr^2} + \varepsilon_\lambda - U(r)\right] w_\lambda(r) = 0, \quad \varepsilon_\lambda = \frac{2m}{\hbar^2} E_\lambda, \tag{X2}$$

subject to the boundary conditions#

$$w_\lambda(0) = 0, \qquad \left.\frac{dw_\lambda}{dr}\right|_{r=a} = 0. \tag{X3}$$

Let the $w_\lambda$ be normalized according to $\int_0^R w_\lambda^2(r)\, dr = 1$. From (X1) and (X2), we obtain

$$\left|\mathscr{F}\frac{dw_\lambda}{dr} - w_\lambda \frac{d\mathscr{F}}{dr}\right|_0^a + (\varepsilon_\lambda - \varepsilon)\int_0^a \mathscr{F} w_\lambda\, dr = 0$$

or

$$\int_0^a \mathscr{F}(r)\, w_\lambda(r)\, dr = \frac{w_\lambda(a)}{\varepsilon_\lambda - \varepsilon}\left(\frac{d\mathscr{F}}{dr}\right)_a, \tag{X4}$$

where $\varepsilon = k^2 = (2m/\hbar^2)\,E$. Thus if we only know the value of $(d\mathscr{F}/dr)_a$, it is possible to expand $\mathscr{F}$ in terms of the $w_\lambda(r)$, namely##

$$\mathscr{F}(kr) = \left(\frac{d\mathscr{F}}{dr}\right)_a \sum_\lambda \frac{w_\lambda(a)}{\varepsilon_\lambda - \varepsilon}\, w_\lambda(r). \tag{X5}$$

---

# Note that the condition at $r = a$ in (X3) is slightly different from the condition (T55) in the Kapur–Peierls formalism.

## Note that the series for $d\mathscr{F}/dr$ obtained from (X5) by differentiating term by term is not convergent at $r = a$, since by (X3), the series vanishes at $r = a$ and will not be consistent with the supposedly given value of $(d\mathscr{F}/dr)_a$.

In other words, the specification of $(d\mathscr{F}/dr)_a$ determines completely the wave function $\mathscr{F}(kr)$ inside the nucleus. From (X5), we define $R$ by

$$R \equiv \frac{1}{\left(\dfrac{1}{\mathscr{F}}\dfrac{d\mathscr{F}}{dr}\right)_a} = \sum_\lambda \frac{w_\lambda^2(a)}{\varepsilon_\lambda - \varepsilon}, \tag{X6}$$

which is a function of the energy $\varepsilon$. $R$ is characterized by the parameters $w_\lambda^2(a)$ and $\varepsilon_\lambda$ which are the properties of the nucleus, as determined by $U(r)$ through (X2).

Now for the field-free region outside the nucleus, $a < r < \infty$, we can write the $s$ waves in the form

$$\begin{aligned} \mathscr{F}(kr) &= \frac{1}{k}\left(\frac{d\mathscr{F}}{dr}\right)_a \sin k(r-a) + \mathscr{F}(ka)\cos k(r-a) \\ &= \left(\frac{d\mathscr{F}}{dr}\right)_a \left\{\frac{1}{k}\sin k(r-a) + R\cos k(r-a)\right\} \\ &= \left(\frac{d\mathscr{F}}{dr}\right)_a \frac{1}{k}\sin (kr+\delta_0), \end{aligned} \tag{X7}$$

where the phase shift is now expressed in terms of $R$

$$\tan (\delta_0 + ka) = kR. \tag{X8}$$

For waves with $l \neq 0$, we may generalize (X7) by making the replacements

$$\frac{1}{k}\sin kr \rightarrow F_l(kr), \qquad \cos kr \rightarrow G_l(kr),$$

where $F_l$, $G_l$ are the regular and irregular solutions (at $r = 0$) of the field-free equation

$$\left[\frac{dr^2}{d^2} + k^2 - \frac{l(l+1)}{r^2}\right] v_l(kr) = 0 \tag{X9}$$

and satisfy the asymptotic conditions

$$\begin{aligned} F_l(kr) &= krj_l(kr) \rightarrow \sin\left(kr - \frac{l}{2}\pi\right), \\ G_l(kr) &= -krn_l(kr) \rightarrow \cos\left(kr - \frac{l}{2}\pi\right). \end{aligned} \tag{X10}$$

In view of these, it follows that the Wronskian is

$$\left(\frac{dF_l}{dr}G_l - \frac{dG_l}{dr}F_l\right) = \text{constant} = k.$$

From this property, it is seen that the functions $\mathscr{S}(kr)$, $\mathscr{C}(kr)$ defined by

$$\begin{aligned}\mathscr{S}(kr) &= \frac{1}{k}\,[G_l(ka)F_l(kr) - F_l(ka)G_l(kr)],\\ \mathscr{C}(kr) &= \left[\frac{dF_l(kr)}{d(kr)}\right]_a G_l(kr) - \left[\frac{dG_l(kr)}{d(kr)}\right]_a F_l(kr)\end{aligned} \tag{X11}$$

satisfy the following conditions

$$\begin{aligned}\mathscr{S}(ka) &= 0, \qquad \left.\frac{d\mathscr{S}}{dr}\right|_{r=a} = 1,\\ \mathscr{C}(ka) &= 1, \qquad \left.\frac{d\mathscr{C}}{dr}\right|_{r=a} = 0.\end{aligned} \tag{X12}$$

Hence (X7) may be replaced by

$$\mathscr{F}_l(kr) = \left(\frac{d\mathscr{F}_l}{dr}\right)_{r=a} [\mathscr{S}_l(kr) + R_l\mathscr{C}_l(kr)]. \tag{X13}$$

It follows from (X10) and (X13) that

$$\mathscr{F}_l(kr) \to \frac{1}{k}\left(\frac{d\mathscr{F}_l}{dr}\right)_{r=a} \sin\left(kr - \frac{l}{2}\pi + \delta_l\right), \tag{X13a}$$

where

$$\tan\delta_l = \frac{kRF'_l - F_l}{G_l - kRG'_l}, \tag{X14}$$

where $G_l \equiv G_l(a)$, $F'_l \equiv dF_l(ka)/d(ka)$, etc. The relation between $R$ and the $S$ matrix is given by (X14) and the usual relation $S_l = \exp(2i\delta_l)$. Thus knowing $R$ and the nucleus radius $a$, one can calculate $S$.

In this simple case of a one-particle, one-channel process, the above result does not bring out the power of the theory. We shall hence take up the many-channel reactions and generalize the above quantity $R$ to a matrix.

## 2. GENERAL FORMULA: MANY-CHANNEL REACTIONS

We shall now consider a collision between two nuclear particles having many possible channels. In Sect. W, we have expressed the total wave function with initial plane-wave conditions by (W8)

$$\psi = \sum A_\alpha \psi_\alpha,$$

where $\psi_\alpha$ are the wave functions describing the collisions initiated from the particular channels $\alpha$, and $A_\alpha$ are the coefficients given by (W6c). For other given initial conditions, we may take the coefficients $A_\alpha$ as arbitrary and determine them by means of the given initial conditions. The reaction leading to the channel $\beta$ is there described by the $S$ matrix by (W7b)

$$B_\beta = \sum_\alpha S_{\alpha\beta} A_\alpha.$$

In the theory of the $R$ matrix, we do not work directly with the $\psi_\alpha$ given in (W7a), and hence not with $A_\alpha$, and hence not with the $S$ matrix, but proceed as follows:

Firstly we generalize the concept of the three-dimensional nuclear sphere of radius $a$ for the one-particle one-channel case to a hypersphere having different radii $a_\beta$ for the channels $\beta$. Inside the sphere, the wave function is very complicated, but we can "select" the basic functions $\psi_\alpha$ (which are different from the previous $\psi_\alpha$) by setting the normal derivatives

$$\psi_\alpha = \sum_\beta \sqrt{k_\beta} \mathscr{F}_\beta \phi_\beta^M, \tag{X15a}$$

$$\left.\frac{d\mathscr{F}_\alpha}{dr_\alpha}\right)_{r_\alpha = a_\alpha} \neq 0, \qquad \left.\frac{d\mathscr{F}_\beta}{dr_\beta}\right)_{r_\beta = a_\beta} = 0, \quad (l = 0), \quad \text{for all } \beta \neq \alpha.$$

By a generalization of (X13) to the many-channel case, the wave functions *outside* the hypersphere will be of the form (X13) for the channel $\alpha$, and of the form

$$\mathscr{F}_l(kr_\beta) = \mathscr{F}_l(ka_\beta)\, \mathscr{C}_l(kr_\beta), \quad \beta \neq \alpha.$$

These $\mathscr{S}(kr)$, $\mathscr{C}(kr)$, through expressions similar to (X11), have asymptotic forms which are combinations of $\sin\,[kr_\beta - (l_\beta\pi/2)]$, $\cos\,[kr_\beta - (l_\beta\pi/2)]$. Such a function outside the channel radius $r_\beta$ may now be written

$$\mathscr{F}_\beta = [\mathscr{S}_\beta(k_\beta r_\beta)\, \delta_{\alpha\beta} + R_{\alpha\beta}\mathscr{C}_\beta(k_\beta r_\beta)], \quad r_\beta > a_\beta \tag{X15}$$

where $\mathscr{S}_\beta(k_\beta r_\beta)$ and $\mathscr{C}_\beta(k_\beta r_\beta)$ are real functions and satisfy the conditions#

$$\begin{aligned} \mathscr{S}_\beta(k_\beta a_\beta) &= 0 \qquad \frac{d}{dr_\beta}\mathscr{S}_\beta(k_\beta r_\beta)|_{r_\beta = a_\beta} = 1, \\ \mathscr{C}_\beta(k_\beta a_\beta) &= 1 \qquad \frac{d}{dr_\beta}\mathscr{C}_\beta(k_\beta r_\beta)|_{r_\beta = a_\beta} = -\frac{l_\beta}{a_\beta}. \end{aligned} \tag{X16}$$

So far, the procedure is general and formal. To calculate reaction cross sections, the procedure is qualitatively as follows. Since the Schrödinger equation is linear, it is possible to form linear combinations of the functions $\psi_\alpha$ in (X15) in such a way as to satisfy the given initial condition of the collision process and to exhibit the asymptotic forms

$$\exp\,[i(k_\beta r_\beta - \tfrac{1}{2}l_\beta\pi)], \qquad \exp\,[-i(k_\beta r_\beta - \tfrac{1}{2}l_\beta\pi)].$$

# If Coulomb interaction exists for $r_\beta > a_\beta$, the definition of $\mathscr{S}$ and $\mathscr{C}$ must be modified.

From the ratio of the coefficients of the outgoing to those of the ingoing wave, we obtain the $S$ matrix and hence the cross sections.

In the above scheme, the $R_{\alpha\beta}$, forming a matrix, are, analogously with (X6), the reciprocal of the logarithmic normal derivatives of the wave function on the hypersphere such that beyond $r_\beta = a_\beta$, there is no interaction between the component parts of the system in the reaction channel $\beta$. The $R_{\alpha\beta}$ are "in principle" obtainable from the Schrödinger equation if the interactions of the system in the many-dimensional configuration space were known. In the application of the theory, however, the $R_{\alpha\beta}$ are regarded as parameters [which in turn are functions of such parameters as $w_\lambda(a_\beta)$, $\varepsilon_{\lambda\beta}$, $a_\beta$ in (X6)]. The advantages of the theory are its generality (since no specific assumptions are made concerning the mechanism of the nuclear reactions) and the success in representing the phenomenon of resonance reactions.

It must be noted that other slightly different definitions of $\mathscr{S}(kr)$ and $\mathscr{C}(kr)$ than the ones given in (X16) are possible. See ref. 1a.

To proceed with the discussion of the $R$ matrix, we recall from the discussion in Sect. W.3, that the time-reversal state $\psi_\alpha^{(r)}$ of $\psi_\alpha$ is also a solution of the Schrödinger equation,

$$\psi_\alpha^{(r)} = i^{2M}[\mathscr{S}_\beta(k_\beta r_{\beta'})\delta_{\alpha\beta} + R^*_{\alpha'\beta'}\mathscr{C}_\beta(k_\beta r_{\beta'})]\sqrt{k_\beta}\,\phi_\beta^{-M}; \quad \text{for} \quad r_{\beta'} \geqq a_\beta. \tag{X17}$$

The channel $\beta'$ differs from $\beta$ only in having the sign of all the magnetic quantum numbers reversed (see Sect. W.3). A comparison of (X15) and (X17) leads to $R^*_{\alpha'\beta'} = R_{\alpha\beta}$. If we assume that the scattering does not depend on $M$, then $R_{\alpha'\beta'} = R_{\alpha\beta}$, and in this case, $R_{\alpha\beta}$ must be real.

The $R$ matrix and the $S$ matrix are related to each other. Let us express the $S$ matrix in terms of the $R$ matrix. Since $\mathscr{S}_\beta$ and $\mathscr{C}_\beta$ are real functions, we can write

$$\mathscr{S}_\beta(k_\beta r_\beta) \to \frac{i}{2\sqrt{k_\beta}}\,[\xi_\beta \exp(-ik_\beta r_\beta) - \xi_\beta^* \exp(ik_\beta r_\beta)], \tag{X18}$$

$$\mathscr{C}_\beta(k_\beta r_\beta) \to \frac{1}{2\sqrt{k_\beta}}\,[\zeta_\beta\omega_\beta^* \exp(-ik_\beta r_\beta) + \zeta_\beta\omega_\beta \exp(ik_\beta r_\beta)]. \tag{X19}$$

$\xi_\beta$, $\zeta_\beta$ and $\omega_\beta$ are functions of energy. In (X19) $\zeta_\beta\omega_\beta^*$ is the corresponding constant of $\xi_\beta$ in (X18), but $\zeta_\beta$ is assumed to be positive and $\omega_\beta$ of modulus 1. From the Wronskian of $\mathscr{S}$ and $\mathscr{C}$, $\mathscr{C}_\beta(d\mathscr{S}_\beta/dr_\beta) - \mathscr{S}_\beta(d\mathscr{C}_\beta/dr_\beta) = 1$, we get

$$\zeta_\beta\omega_\beta\xi_\beta + \zeta_\beta\omega_\beta^*\xi_\beta = 2.$$

We then define a real quantity $\varepsilon_\beta$ by

$$\zeta_\beta\omega_\beta\xi_\beta = 1 - i\varepsilon_\beta, \qquad \zeta_\beta\omega_\beta^*\xi_\beta^* = 1 + i\varepsilon_\beta. \tag{X20}$$

Substituting (X18) and (X19) into (X15) and (X15a) we get

$$\psi_\alpha = \sum_\beta \left\{ \frac{(-i)^{l_\beta}}{2} (R_{\alpha\beta}\zeta_\beta\omega_\beta^* + i\delta_{\alpha\beta}\xi_\beta) \exp\left[-i\left(k_\beta r_\beta - \frac{l_\beta\pi}{2}\right)\right] - \frac{(i)^{l_\beta+2}}{2} (R_{\alpha\beta}\zeta_\beta\omega_\beta - i\delta_{\alpha\beta}\xi_\beta^*) \exp\left[i\left(k_\beta r_\beta - \frac{l_\beta\pi}{2}\right)\right] \right\} \phi_\beta^M. \quad \text{(X21)}$$

If (W2b) and (W7c) are compared with (X21), we have $n^2$ simultaneous linear equations for $S_{\beta\gamma}$ ($n$ = number of channels),

$$(R_{\alpha\beta}\zeta_\beta\omega_\beta - i\delta_{\alpha\beta}\xi_\beta^*) = \sum_\gamma S_{\beta\gamma}(-i)^{l_\beta+l_\gamma+2}(R_{\alpha\gamma}\zeta_\gamma\omega_\gamma^* + i\delta_{\alpha\gamma}\xi_\gamma). \quad \text{(X22)}$$

We shall write the solution of Equation (X22) in a matrix form. We define the diagonal matrices $\zeta$, $\omega$ and $\xi$, of which the elements are $\zeta_\beta$, $\omega_\beta$ and $\xi_\beta$ respectively. Equation (X22) can then be written as

$$(R\zeta\omega - i\xi^*)_{\alpha\beta} = \sum_\gamma S_{\beta\gamma}(-i)^{l_\beta+l_\gamma+2}(R\zeta\omega^* + i\xi)_{\alpha\gamma}.$$

Multiplying by $(R\zeta\omega^* + i\xi)^{-1}_{\sigma\alpha}$ and summing over $\alpha$, we have

$$S_{\beta\sigma} = (+i)^{l_\beta+l_\sigma+2}\left(\frac{R\zeta\omega - i\xi^*}{R\zeta\omega^* + i\xi}\right)_{\sigma\beta}. \quad \text{(X23)}$$

The formula (X23) can be slightly modified by using (X6) and noting that $\zeta$, $\omega$ and $\varepsilon$, being diagonal matrices, are mutually commutable and that the inverse of a product operator $(AB)^{-1}$ equals $B^{-1}A^{-1}$. Thus

$$S_{\beta\sigma} = (i)^{l_\beta+l_\sigma}\left(\omega\,\frac{1 + i\zeta R\zeta + i\varepsilon}{1 - i\zeta R\zeta - i\varepsilon}\,\omega\right)_{\sigma\beta}. \quad \text{(X23a)}$$

The matrix elements of $\zeta$ and $\varepsilon$ are real, and those of $\omega$ are of modulus 1, therefore $R$ should be a real and symmetric matrix if $S$ is unitary and symmetric. Once $R$ is obtained, $S$ can be constructed by (X23) or (X23a) and the collision is thus described by the $R$ matrix.

The matrix elements of $\zeta$, $\omega$ and $\varepsilon$ are given in Table 1 for $l_\alpha \leqq 2$, when there is no external Coulomb potential, as assumed in the present discussion.

TABLE 1. *The diagonal matrix elements in $\zeta$, $\omega$ and $\epsilon$. (Taken from reference 2, Wigner and Eisenbud)*

| $l_\alpha$ | $\zeta_\alpha$ | $\omega_\alpha$ | $\varepsilon_\alpha$ |
|---|---|---|---|
| 0 | $\sqrt{k_\alpha}$ | $\exp(-ik_\alpha a_\alpha)$ | 0 |
| 1 | $\sqrt{k_\alpha}$ | $\exp(-ik_\alpha a_\alpha)$ | $-1/(k_\alpha a_\alpha)$ |
| 2 | $\frac{1}{k_\alpha a_\alpha}\sqrt{k_\alpha(1 + k_\alpha^2 a_\alpha^2)}$ | $-i\exp[-i(k_\alpha a_\alpha - \tan^{-1}k_\alpha a_\alpha)]$ | $-[3 + 2(k_\alpha a_\alpha)^2]/(k_\alpha a_\alpha)^3$ |

The formula (X23a) is particularly simple if the number of open channels is only one. All matrices reduce to constants. Using Table 1 we easily have

$$S_{\alpha\alpha} = \exp(2i\delta_0), \qquad \delta_0 = \tan^{-1}(k_\alpha R_{\alpha\alpha}) - k_\alpha a_\alpha; \quad l_\alpha = 0, \tag{X23b}$$

$$\begin{aligned} S_{\alpha\alpha} = \exp(2i\delta_1), \qquad \delta_1 &= \tan^{-1}[k_\alpha R_{\alpha\alpha} - (k_\alpha a_\alpha)^{-1}] - k_\alpha a_\alpha - (\pi/2), \\ &= k_\alpha^3(a_\alpha^2 R_{\alpha\alpha} - a_\alpha^3/3) + O(k^5); \quad l_\alpha = 1. \end{aligned} \tag{X23c}$$

The next step is to define the "compound states", $\phi_s$, and to express $\psi_\alpha$ as linear combinations of $\phi_s$ inside the radius $a_\beta$. The $R$ matrix will be simply expressed by the coefficient of $\phi_s$ in $\psi_s$. The definition of $\phi_s$ is

$$\frac{\partial(r_\beta\phi_s)}{\partial\pi_\beta}|(r_\beta\phi_s)|_{r_\beta = a_\beta} = -\frac{l_\beta}{a_\beta}, \quad \text{for all } \beta, \tag{X24}$$

where $\phi_s$ is a solution of the Schrödinger equation. (X24) is similar to the condition for $\mathscr{C}$ in (X16). In other words, $\psi_\alpha$ given by (X15) would be $\phi_s$ if the first term containing $\mathscr{S}_\alpha$ does not exist. The condition (X24) would be satisfied only for discrete values of the energy $E_s$. $\phi_s$ is normalized inside the region $r_\beta \leqq a_\beta$ (for all $\beta$).

$$H\phi_s = E_s\phi_s, \qquad \int |\phi_s|^2\, d\tau = 1, \quad (r_\beta \leqq a_\beta).$$

The eigenstates $\phi_s$ form an orthonormal (and complete) set for $r_\beta \leqq a_\beta$. It should be noted that as the number of channels increase with the energy, the condition (X24) must be applied as soon as a channel is newly opened. $\psi_\alpha$ can be expanded in terms of $\phi_s$ for $r_\beta \leqq a_\beta$,

$$\psi_\alpha = \sum c_{\alpha s}\phi_s, \qquad c_{\alpha s} = \int \psi_\alpha\phi_s^*\, d\tau, \qquad r_\beta \leqslant a_\beta.$$

By virtue of the wave equation satisfied by $\psi_\alpha$ and $\phi_s$, we have

$$(E - E_s)\int \psi_\alpha\phi_s^*\, d\tau = \int (\phi_s^* H\psi_\alpha - \psi_\alpha H\phi_s^*)\, d\tau.$$

If Green's theorem is applied to $\phi_s^*$ and $\psi_\alpha$, we have

$$(E - E_s)\int \zeta_\alpha\phi_s^*\, d\tau = -\sum \frac{\hbar^2}{2M_\beta}\int \left(\phi_s^*\frac{\partial\psi_\alpha}{\partial r_\beta} - \psi_\alpha\frac{\partial\phi_s^*}{\partial r_\beta}\right) d\sigma_\beta, \quad (r_\beta = a_\beta)$$

where $d\sigma_\beta$ is the "surface" element (i.e., $d\tau$ without the factor $dr_\beta$). The surface integral is much simplified by the boundary conditions (X16) and (X24). If we write the values of $\phi_s$ at the "nuclear surface" (i.e., $r_\beta = a_\beta$) as

$$\phi_s = v_{s\beta}\phi_\beta^M \quad \text{at} \quad r_\beta = a_\beta \quad \text{for channel } \beta,$$

we get

$$(E - E_s)\int \psi_\alpha \phi_s^* \, d\tau = -\frac{\hbar^2}{2M_\alpha}\int \phi_s^* \frac{\partial \mathscr{S}_\alpha(k_\alpha r_\alpha)}{\partial r_\alpha}\sqrt{k_\alpha}\,\phi_\alpha^M \, d\sigma_\alpha$$
$$= -\sqrt{\frac{\hbar}{2}}\,\gamma_{s\alpha},$$

where#

$$\gamma_{s\alpha} = \frac{\hbar \nu_{s\alpha}}{\sqrt{2M_\alpha v_\alpha}}. \tag{X25}$$

The coefficients $c_{\alpha s}$ are now given by

$$\psi_\alpha = \sum_s c_{\alpha s}\phi_s, \qquad c_{\alpha s} = \sqrt{\frac{\hbar}{2}}\frac{\gamma_{s\alpha}}{(E_s - E)}. \tag{X26}$$

The expansion (X26) shows the energy dependence of $\psi_\alpha$, and (X26) is valid inside the "nuclear surface" $r_\beta = a_\beta$. Although $\gamma_{s\beta}$ can be determined if the value $\nu_{s\beta}$ of $\phi_s$ at $r_\beta = a_\beta$ is known, $\gamma_{s\beta}$ are often considered as (adjustable) parameters in an actual application of the theory to nuclear reaction phenomena.

Let us evaluate both sides of the first equation of (X26) at $r_\beta = a_\beta$ for a particular (but any) channel $\beta$. By Equation (X15) the left-hand side is equal to $R_{\alpha\beta}\sqrt{M_\beta v_\beta}\,\phi_\beta^M/\sqrt{\hbar}$ (at $r_\beta = a_\beta$), and the right-hand side becomes $\sum_s (\hbar/2)^{1/2}\gamma_{s\alpha}\nu_{s\beta}\phi_\beta^M/(E_s - E)$. Using (X25), we finally have

$$R_{\alpha\beta} = \sum_s \frac{\gamma_{s\alpha}\gamma_{s\beta}}{E_s - E}. \tag{X27}$$

It is clear from (X27) that the $R$ matrix elements are symmetric and real. The energy dependence of $R_{\alpha\beta}$ is also clearly seen from (X27). All the elements $R_{\alpha\beta}$ have poles at the same energies $E_s$. The meaning of the poles of $R_{\alpha\beta}$ for the $S$ matrix is seen from (X23b) and (X23c). Apart from the term $k_\alpha a_\alpha$, etc., the

---

# $\gamma_{\beta s}$ is a real quantity. The reason is shown as follows. The time reversed solution $\phi_M^{(r)}$ is constructed from $\phi$ as

$$\phi_M^{(r)} = \sum_t \sigma_{y_t}\,\phi_M^*.$$

(see Sect. W.3). Then it follows that, for any $\phi_M$ and $\phi'_M$,

$$\int \phi_M^* \phi'_M \, d\tau = \int \phi_M^{\prime(r)*}\,\phi_M^{(r)}\, d\tau.$$

On the other hand,

$$\int \phi_M^{\prime(r)*}\,\phi_M^{(r)}\, d\tau = \int \phi'_{-M}\,\phi^*_{-M}\, d\tau = \int \phi'_M \phi_M^* \, d\tau.$$

Since $\int \phi_M^* \phi'_M \, d\tau$ is thus real, $\gamma_{\beta s}$ should be real.

phase shift varies rapidly for small $k_\alpha$ when the energy $E$ passes the characteristic energy $E_s$, and changes its value by about $\pi$ when $E$ increases from below to just above $E_s$ The scattering cross section will correspondingly have a sharp maximum at $E \sim E_s$. This is also true for any reactions (i.e., $S_{\alpha\beta}$, $\alpha \neq \beta$). We shall derive the (generalized) one-level formula from (X23a).

We have seen that, if the energy $E$ is close to $E_s$, the variation of the phase shift is mainly due to the one term in (X27) corresponding to $E \sim E_s$. Therefore the following approximation can be made when $E$ is close to $E_\lambda$,

$$R_{\alpha\beta} \doteqdot \frac{\gamma_{\lambda\alpha}\gamma_{\lambda\beta}}{E_\lambda - E} + R'_{\alpha\beta}. \tag{X27a}$$

$R'_{\alpha\beta}$ represents all the terms in (X27) except $\gamma_{\lambda\alpha}\gamma_{\lambda\beta}/(E_\lambda - E)$, and may be considered a constant matrix which is independent of the energy $E$. It is often a good approximation to put $R'_{\alpha\beta} = 0$. Now we define $\varepsilon'$, $\alpha_\lambda$, $\Gamma_\lambda$ and $\Delta_\lambda$ as follows,

$$\varepsilon' \equiv \varepsilon + \zeta R' \zeta. \tag{X27b}$$

($\varepsilon$ is given in (X20)). Let $\gamma_\lambda$ be a vector with components $\gamma_{\lambda\beta}$, and define a new vector $\alpha_\lambda$ by

$$\begin{aligned} \alpha_\lambda &\equiv (1 - i\varepsilon')\,\zeta\gamma_\lambda, \\ \Gamma_\lambda &\equiv \sum_\beta \Gamma_{\lambda\beta}, \qquad \Gamma_{\lambda\beta} \equiv 2|(\alpha_\lambda)_\beta|^2, \\ \Delta_\lambda &\equiv \sum_\beta \Delta_{\lambda\beta}, \qquad \Delta_{\lambda\beta} \equiv (\alpha_\lambda)^*_\beta(\varepsilon'\alpha_\lambda)_\beta. \end{aligned} \tag{X27c}$$

With the help of these quantities it can be shown that

$$(-i)^{l_\beta + l_\sigma} S_{\beta\sigma} = \left[\omega\left(\frac{1 + i\varepsilon'}{1 - i\varepsilon'}\right)\omega\right]_{\sigma\beta} + \frac{2i(\omega\alpha_\lambda)_\sigma(\omega\alpha_\lambda)_\beta}{E_\lambda + \Delta_\lambda - E - \frac{i}{2}\Gamma_\lambda}. \tag{X28}$$

According to (X28), the scattering matrix element is expressed as a sum of two terms. The first term is a slowly varying function of the energy and gives the so-called "potential scattering". The second term represents the resonance part of the collision. A more detailed discussion in connection with the interpretation of the formulas will be given in the next subsection. If $\varepsilon'$ is diagonal, the total cross section corresponding to $\alpha \to \beta$ ($\beta \neq \alpha$) is, from (W10), given by

$$\sigma_{\alpha\beta} = (2l + 1)\frac{\pi}{k_\alpha^2}\frac{\Gamma_{\lambda\alpha}\Gamma_{\lambda\beta}}{(E_\lambda + \Delta_\lambda - E)^2 + (\Gamma_\lambda^2/4)}. \tag{X28a}$$

The formula (X28a) can be interpreted by the compound nucleus formation and the subsequent decay into the final state. [Also see the discussion around (T50) and (T51).] The decay branching ratio $\Gamma_\beta/\Gamma$ is independent of the incident channel, provided that a single term $E_s$ in (X27) is retained to derive (X28) and (X28a), (i.e., $\varepsilon' - \varepsilon = 0$). This means that the compound state formed does not remember the history of how the state is formed. We know that the nuclear reaction is caused partly by the compound formation and partly by the direct reaction. These two are extreme concepts. The actual process may lie between the extreme cases. However, if the cross section fits the one-level formula (X28a) and (X28) with $\varepsilon' - \varepsilon = 0$, we may picture the compound state as really being formed for this resonance reaction (including elastic scattering).#

It is also possible to give an explicit expression for the cross section if we take up two terms in (X27) which are adjacent to each other [E. Wigner, Phys. Rev. **70**, 606 (1946)]. If we deal with all the terms in (X27), the matrix algebra involved in (X23a) is too difficult to handle except for the cases where the number of open channels is one or two. The explicit formulas for $\Gamma_{\lambda\beta}$ and $\Delta_{\lambda\beta}$ in terms of $\gamma_{\lambda\beta}$, $k_\beta$ and $a_\beta$ are given in Table 2 on the assumption that $R'_{\alpha\beta} = 0$.

TABLE 2. $\Gamma_{\lambda\beta}$ *and* $\Delta_{\lambda\beta}$ *for* $R'_{\alpha\beta} = 0$

| $l_\beta$ | $\Gamma_{\lambda\beta}$ | $\Delta_{\lambda\beta}$ |
|---|---|---|
| 0 | $k_\beta \gamma_{\lambda\beta}^2$ | 0 |
| 1 | $\dfrac{k_\beta^3 a_\beta^2 \gamma_{\lambda\beta}^2}{1 + (k_\beta a_\beta)^2}$ | $-\dfrac{k_\beta^2 a_\beta \gamma_{\lambda\beta}^2}{1 + (k_\beta a_\beta)^2}$ |
| 2 | $\dfrac{k_\beta^5 a_\beta^4 \gamma_{\lambda s}^2 [1 + (k_\beta a_\beta)^2]}{[3 + 2(k_\beta a_\beta)^2]^2 + (k_\beta a_\beta)^6}$ | $\dfrac{-k_\beta^2 a_\beta \gamma_{\lambda\beta}^2 [1 + (k_\beta a_\beta)^2][3 + 2(k_\beta a_\beta)^2]}{[3 + 2(k_\beta a_\beta)^2]^2 + (k_\beta a_\beta)^6}$ |

### 3. INTERPRETATION OF THE FORMULA

We have derived the general formula for the $R$ matrix elements (X27), and also for the $S$ matrix elements (X28). The $R$ matrix depends on the values of $a_\beta$. If we modify the boundary conditions (X16) for $\mathscr{S}$ and $\mathscr{C}$ and also (X24) for $\phi_s$, $\gamma_{s\beta}$ and $E_s$ appearing in the expression of the $R$ matrix will change their values. The value of $a_\beta$ can be arbitrary, and there is no *a priori* reason why the choice of the boundary conditions must be exactly (X16) and (X24). Therefore the theory has an ambiguity in the arbitrariness of the choice of $a_\beta$,

# If the effect of $\varepsilon' - \varepsilon$ is not small, the (effective) partial widths derived from the experimental data are no longer independent of the incident channel. The experimental test of this independence is not certain. The available reactions are limited (one of the best examples is to consider the compound nucleus of $N^{15}$) because the different initial channels do not correspond to the same compound state for slow collisions.

and the values of $\Gamma_{\lambda\beta}$ and $\Delta_{\lambda\beta}$ can be changed according to a corresponding change of the boundary conditions (X16) and (X24). Although the $S$ matrix (X28) is free from this ambiguity, we must introduce a physical argument in order to remove the ambiguity when we apply the general $R$ matrix theory to a physical situation.

It is clear from (X28a) that the $\Gamma_\lambda$ defined in (X27c) should be interpreted as the "total width" at the "compound state" $\phi_\lambda$. $\phi_\lambda$ has been mathematically defined by (X24). The condition (X24) is equivalent to requiring that the radial part of $\phi_s$ in the channel $\beta$ satisfy the same condition which $\mathscr{C}_\beta$ satisfies, namely (X16). $a_\beta$ should be equal to the nuclear radius outside of which the outgoing particle $C$ (into the channel $\beta$) does not interact with the residual nucleus $D$ (on account of the short range of the interactions between $C$ and $D$).

It has been shown for $l_\beta = 0$ that the amplitude of the function inside the nuclear surface may be large compared with that outside the surface if the derivative of the wave function at the surface vanishes [see (T9)]. This suggests the existence of the well-defined compound state when (X16) is satisfied for $l_\beta = 0$. For $l_\beta \geqq 1$, the same will probably also be true, because for very small $k_\beta$, the condition (X16) will select the "irregular solution" of the Schrödinger equation as $\mathscr{C}_\beta(kr)$. The eigenvalue $E_s$ corresponding to $\phi_s$ is not exactly the same as $E_{res}$ [see (X28a) and (X29)]. This fact is not necessarily due to the inadequacy of the condition (X24) for defining the compound state. According to (X27b) and (X27c), $\Delta_\lambda$ has two different origins, namely, the contributions from the far-away levels, and those from $\varepsilon$. If the centrifugal barrier is higher than the incident (or outgoing) energy, the low energy particles will find it much more difficult to come in (or go out) than particles of higher energies. Owing to this factor the center of the resonance reaction $E_{res}$ does not necessarily coincide with the energy of the compound nucleus. $\varepsilon$ is usually considered as representing the shift of $E_{res}$ from $E_\lambda$ due to this penetrability factor. This shifting is often observed in experiments. Now let us look at the $\Gamma_{\lambda\beta}$ in Table 2. We see that $\Gamma_{\lambda\beta}$ can be factored as follows [see (T21)–(T22), (T22a)],

$$\Gamma_{\lambda\beta} \cong QP_\beta \frac{K\gamma_{\lambda\beta}^2}{4}.$$

Thus $\Gamma_{\lambda\beta}$ has the correct behavior as the decay rate of the compound nucleus. The separable probability $S$ (see Sect. T.6) can be made equal to $K\gamma_{\lambda\beta}^2/4$. This probability can be calculated on adequate models for the compound nucleus, and it is worthwhile to compare the calculated value with $K\gamma_{\lambda\beta}^2/4$ deduced from the experimental data in order to understand the reaction mechanism. In this connection the "strength function" $\gamma_{\lambda\beta}^2$/(level spacing of the compound nucleus) is often used [see (T26) and Fig. 5 in Sect. U].

We have confined our discussion to reactions. Let us consider the elastic scattering. The first term, the potential scattering, of (X27) also contributes to the elastic scattering ($\sigma = \beta$). If we neglect $\varepsilon'$ for $l_\beta = 0$, the first term is $\exp(-2ik_\alpha a_\alpha)$ which is just the same as the scattering amplitude due to an impenetrable sphere (i.e., hard core) of radius $a_\alpha$. Then the total cross section for $\alpha \to \alpha$ can be written

$$\sigma = \frac{\pi}{k_\alpha^2}|1 - S_{\alpha\alpha}|^2 = \frac{\pi}{k_\alpha^2}\left|1 - \exp(-2ik_\alpha a_\alpha)\left(1 + \frac{i\Gamma_{\lambda\alpha}}{E_\lambda - E - \frac{i}{2}\Gamma_\lambda}\right)\right|^2.$$

The potential term is unimportant for $l_\beta \geqq 1$ in the low energy region because of the strong centrifugal barrier.

The $R$ matrix is also used when one tries to derive the "gross structure" of the nuclear reactions (see ref. 4 of Sect. U). Historically, the derivation of the many-level formula under a general condition has been attempted by many authors. The first successful attempt was made by Kapur and Peierls (1938) (Sect. T.9). The main idea for the formulation of the problem used by Wigner already appears in their work and there are many common features between the two papers. Kapur and Peierls adopt a boundary condition depending on energy for the definition of the compound states. Because of this energy dependence, many constants in their treatment are actually energy dependent. The $R$-matrix formalism of Wigner and his collaborators makes the energy dependence of all expressions as explicit as possible, and overcomes this shortcoming of the Kapur–Peierls formalism. However the Kapur and Peierls theory is also often used because of the simple expression of the scattering matrix element in terms of many-level parameters.

## 4. ANGULAR DISTRIBUTION IN THE ONE-LEVEL-APPROXIMATION

The scattering matrix is given by (X28). The crudest, but usually adopted, approximation is made by writing (for notations, see Sect. W)

$$S^{J_0}_{\Gamma's'l',\Gamma sl} =$$

$$(i)^{l+l'}\exp(i\xi_{\Gamma l})\left[\delta_{\Gamma'\Gamma}\delta_{s's}\delta_{l'l} + i\frac{g_{\Gamma sl}g_{\Gamma's'l'}}{E_{res} - E - \dfrac{i\Gamma\lambda}{2}}\right]\exp(i\xi_{\Gamma'l'}), \quad \text{(X29)}$$

where $E_{res} \equiv E_\lambda + \Delta_\lambda$, $g_{\Gamma sl} = \pm(2\alpha_{\lambda\Gamma sl})^{1/2} = \pm(\Gamma_{\Gamma sl})^{1/2}$ and $\xi_{\Gamma l}$ is a real number. $\xi_{\Gamma l}$ is of the order of magnitude of the potential scattering phase for neutrons, hence $\xi_{\Gamma 0} \cong k_\Gamma a_\Gamma$ and $\xi_{\Gamma l} \cong 0$ (for $l \geqq 1$). The sign of $g_{\Gamma sl}$ can be fixed by measuring the angular distribution [see (X31)]. For charged particles, $\xi_{\Gamma l}$ is the sum of the "potential scattering phase" and $-\eta_\Gamma \log(2k_\Gamma r_\Gamma)$ where $\eta_\Gamma = (ZZ'e^2/\hbar v_\Gamma)$, and $r_\Gamma$ is the screening radius for the Coulomb field. $r_\Gamma$ does not appear in the final expression. We assume that the resonance level occurs for the particular value of $J = J_0$ and parity (even or odd) only, namely,

$$S^J_{\Gamma' S' l', \Gamma S l} = \delta_{\Gamma'\Gamma}\delta_{S'S}\delta_{l'l} \exp(2i\xi_{\Gamma l}) \quad \text{for } J \neq J_0$$
$$\text{or } (J = J_0 \text{ but different parity}). \qquad \text{(X30)}$$

We consider the reaction cross section $(\Gamma, S) \to (\Gamma', S')$ for which either $\Gamma \neq \Gamma'$, or $S' \neq S$, or both. On substituting (X29) and (X30) into (W50b) we have

$$B_L(\Gamma'S', \Gamma S) = \frac{(-1)^{S'-S}}{4\left[(E - E_{res})^2 + \left(\frac{\Gamma}{2}\right)^2\right]} \sum_{l_1=|J_0-S|}^{J_0+S} \sum_{l_2=|J_0-S|}^{J_0+S} \sum_{l_1'=|J_0-S'|}^{J_0+S'} \sum_{l_2'=J_0-S'|}^{J_0+S'}$$
$$\times Z(l_1J_0l_2J_0, SL)\, Z(l_1'J_0l_2'J_0, S'L)\, g_{\Gamma S l_1}\, g_{\Gamma S l_2}\, g_{\Gamma' S' l_1'}\, g_{\Gamma' S' l_2'}$$
$$\times \cos[\xi_{\Gamma l_1} - \xi_{\Gamma l_2} + \xi_{\Gamma' l_1'} - \xi_{\Gamma' l_2'}]. \qquad \text{(X31)}$$

The summation on $l_1, l_2, l_1', l_2'$ is also restricted by the parity consideration. If the "channel parity" of $\Gamma$ is equal (opposite) to the compound nucleus parity, $l_1$ and $l_2$ must be even (odd). The same argument applies to $l_1', l_2'$ and $\Gamma'$. By virtue of the property of the $Z$ coefficients, $L$ is restricted by

$$L \leqslant 2J_0, \qquad L = \text{even}. \qquad \text{(X32)}$$

If the measured angular distribution is not symmetric (i.e., $L$ = odd is necessary), the one-level approximation is not a good one, and it indicates the effect from the other levels. If the higher partial waves of incident and outgoing particles are so suppressed by the penetration barrier that $g_{\Gamma S l} = 0$ for $l > l_{max}$ and $g_{\Gamma' S' l'} = 0$ for $l' > l'_{max}$, we have

$$L \leqq 2l_{max}, \qquad L \leqq 2l'_{max}.$$

The elastic scattering cross section formula in the one-level approximation has the hard sphere scattering term, the interference term between the resonance scattering and the hard sphere scattering as well as the term like (X31). We shall not go into further details and the reader is referred to the paper of Blatt and Biedenharn (ref. 5).

## REFERENCES

Comprehensive treatments of the $R$ matrix theory are given in the following articles:

**1a.** Lane, A. M. and Thomas, R. G., Rev. Modern Phys. **30**, 257 (1958).

**1b.** Breit, G., article in Handbuch der Physik, Vol. **41/1**, Springer (1959).

The original paper on the $R$ matrix formalism is that of

**2.** Wigner, E. P. and Eisenbud, L., Phys. Rev. **72**, 29 (1947).

This paper follows the development in two previous papers:

3. Wigner, E. P., Phys. Rev. **70**, 15; 606 (1946).

Subsequent developments of the *R* matrix theory are:

4. Teichman, T. and Wigner, E. P., Phys. Rev. **87**, 123 (1952); Teichman, T., Phys. Rev. 77, 506 (1950); Ehrman, J. B., Phys. Rev. **81**, 412 (1951); Thomas, R. G., Phys. Rev. **97**, 224 (1955); **100**, 25 (1955).

For nuclear resonance reactions, see the references given in Section T.

5. Blatt, J. M. and Biedenharn, L. C., Phys. Rev. **86**, 399 (1952).

chapter

7

# Dispersion Relations

In recent years an important development in the quantum field theory and in the study of collisions of elementary particles (between a $\pi$ meson and a nucleon, for example) is the theory of dispersion relations. The term "dispersion relation" comes, historically, from the observation of Kronig and Kramers that in the theory of dispersion of light by (gaseous) atoms (or molecules), a relation exists between the real and the imaginary parts of the complex index of refraction. It has been found that an extension of the formulation of such general relations to the scattering matrix $S$ is possible both in ordinary quantum mechanics and in the quantum field theory. These relations will be between the real and the imaginary parts of the scattering amplitude, or the matrix elements of the $S$ matrix. It has been established that such relations in the field theory can be obtained on a very general principle, namely, that of "causality". Briefly, this principle of causality simply means that at any point $P(\mathbf{r})$, there will be no scattered wave until an incident wave falling on and being scattered by a scatterer has reached $P(\mathbf{r})$. Such relations have been called "dispersion relations" after that found by Kramers for the ordinary dispersion law in electromagnetic theory. In the last few years, a considerable literature has grown up on various aspects of this subject, such as the foundation of the dispersion relations (causality), the application to various systems (classical electromagnetic waves, nonrelativistic quantum mechanics of collision processes, field theory, etc.). In the following we shall confine ourselves to an elementary account of the origin of the dispersion relations and their application to the scattering of particles in ordinary quantum mechanics (i.e., non-field theoretic).

section Y

# Dispersion Relation and Causality in Optics: Observations of Kronig and Kramers

## 1. SOME MATHEMATICAL PRELIMINARIES

Before presenting the dispersion relations, we shall briefly state some mathematical theorems. It is thought that even if they are given without details, it may be helpful to separate the mathematical matter and the physical principles and results (such as the causality principle and the dispersion relations).

### *i) Cauchy's integral theorem*

If $f(z)$, a function of a complex variable, is analytic in a region $|z| \leqslant R$, then the integral over a contour $\Gamma$ inside $R$ is given by

$$\frac{1}{2\pi i}\oint \frac{f(z)\,dz}{z - z_0} = \begin{cases} f(z_0), & \text{if } z_0 \text{ lies inside } \Gamma, \\ 0, & \text{if } z_0 \text{ lies outside } \Gamma, \end{cases} \tag{Y1}$$

and

$$\frac{1}{\pi i} P \oint \frac{f(z)\,dz}{z - z_0} = f(z_0), \qquad \text{if } z_0 \text{ lies on } \Gamma. \tag{Y2}$$

Here $P$ stands for "taking the principal value".

If $f(z)$ is analytic in the upper half of the complex plane and on the real axis and has the asymptotic property

$$|f(z)| \leqslant \frac{1}{|z|}, \quad \text{as} \quad |z| \to \infty \tag{Y3}$$

and if $z_0$ lies on the real axis, i.e., $z_0 = \kappa_0$, then we can choose for $\Gamma$ in (Y2) the real axis and an infinite semi-circle on the upper half of the $z$-plane. By (Y3), the contribution from the semi-circle vanishes and we have from (Y2)

$$\frac{1}{\pi i} P \int_{-\infty}^{\infty} \frac{f(x)\,dx}{x - x_0} = f(x_0), \tag{Y4}$$

since the path of integration is the real axis on which $f(z) \to f(x)$.

### *ii) Analytic continuation*

Consider an integral of a complex function of a real variable $x$

$$\int_{-\infty}^{\infty} \frac{f(x)\,dx}{x - x_0}, \tag{Y5}$$

where $f(x)$ is analytic on the real axis. If, by a continuation of $f(x)$ into the upper half of the complex plane $z = x + iy$, the function $f(z)$ is analytic in the upper half of the $z$-plane and satisfies the asymptotic condition (Y3) above, then one can evaluate the integral (Y5) by a contour integral of $f(z)$ along the real axis and an infinite semi-circle in the upper half of the $z$-plane. By applying the Cauchy integral (Y2) one then has

$$f(x_0) = \frac{1}{\pi i} P \int_{-\infty}^{\infty} \frac{f(x)\,dx}{x - x_0}. \tag{Y6}$$

From (Y6), one obtains

$$f(x_0) - f(0) = \frac{x_0}{\pi i} P \int_{-\infty}^{\infty} \frac{f(x)\,dx}{x(x - x_0)},$$

and if $f(x)$ is differentiable at $x = 0$,

$$f(x_0) - f(0) - x_0 f'(0) = \frac{x_0^2}{\pi i} P \int_{-\infty}^{\infty} \frac{f(x)\,dx}{x^2(x - x_0)}.$$

To repeat, the application of (Y6) to (Y5) depends on the *analytic* behavior of the continuation of $f(x)$ into the complex $z = x + iy$ plane *and* the asymptotic behavior such as (Y3) of $f(z)$.

### *iii) Hilbert transforms and "dispersion relations"*

On taking the real part of both sides of (Y6), one obtains

$$\operatorname{Re} f(x_0) = \frac{1}{\pi} P \int_{-\infty}^{\infty} \frac{\operatorname{Im} f(x)\,dx}{x - x_0}. \tag{Y7}$$

On taking the imaginary parts of both sides of (Y6), one obtains

$$\operatorname{Im} f(x_0) = -\frac{1}{\pi} P \int_{-\infty}^{\infty} \frac{\operatorname{Re} f(x)\,dx}{x - x_0}. \tag{Y8}$$

In general, if two functions $U(x)$, $V(x)$ are such that

$$U(x_0) = \frac{1}{\pi} P \int_{-\infty}^{\infty} \frac{V(x)\, dx}{x - x_0},$$
$$V(x_0) = -\frac{1}{\pi} P \int_{-\infty}^{\infty} \frac{U(x)\, dx}{x - x_0}, \tag{Y9}$$

the two functions are said to be the Hilbert transform of each other. (Y7), (Y8) show that the real and the imaginary part of $f(x)$ satisfying (Y6) are the Hilbert transform of each other.

If $f(x)$ has the symmetry property

$$f(-x) = f^*(x), \tag{Y10}$$

where * indicates the complex conjugate, then (Y7) can be shown to be

$$\operatorname{Re} f(x_0) = \frac{2}{\pi} P \int_0^{\infty} \frac{\operatorname{Im} f(x) x\, dx}{x^2 - x_0^2}. \tag{Y11}$$

If $f(x)$ is such that

$$f(-x) = -f^*(x), \tag{Y10a}$$

then (Y7) becomes

$$\operatorname{Re} f(x_0) = \frac{2x_0}{\pi} P \int_0^{\infty} \frac{\operatorname{Im} f(x)\, dx}{x^2 - x_0^2}. \tag{Y11a}$$

From (Y11), one obtains, by taking the difference $f(x_1) - f(x_0)$,

$$\operatorname{Re}\,[f(x_1) - f(x_0)] = \frac{2}{\pi}(x_1^2 - x_0^2)\, P \int_0^{\infty} \frac{\operatorname{Im} f(x) x\, dx}{(x^2 - x_1^2)(x^2 - x_0^2)}. \tag{Y12}$$

If $x_0 = 0$, this becomes

$$\operatorname{Re}\,[f(x_1) - f(0)] = \frac{2x_1^2}{\pi} P \int_0^{\infty} \frac{\operatorname{Im} f(x)\, dx}{x(x^2 - x_1^2)}, \tag{Y12a}$$

which, on account of the factor $\frac{1}{x^2}$ in the integrand relative to (Y7), is more rapidly convergent. One can obtain a still more convergent integral by taking a second difference.

Equation (Y7), or its equivalents (Y8), (Y11), (Y12), (Y12a) is the mathematical form of the "dispersion relations" to be taken up in the following sections.

### *iv) Fourier transforms, causal transform*

Let $f(\omega)$ be a complex function of the real variable $\omega$,

$$f(\omega) = \operatorname{Re} f(\omega) + i \operatorname{Im} f(\omega),$$

such that it is square integrable

$$\int_{-\infty}^{\infty} |f(\omega)|^2\, d\omega = \text{finite}.$$

Let us assume that by the continuation into the upper half of the complex plane $\Omega = \omega + i\kappa$, the function $f(\Omega)$ is analytic in the upper half of the $\Omega$-plane, and is quadratically integrable along any line parallel to the real axis in the upper half of the $\Omega$-plane. Such a function $f(\omega)$ is called a "causal transform."

Let $F(t), f(\omega)$ be the Fourier transform of each other

$$F(t) = \frac{1}{\sqrt{2\pi}} \int_{-\infty}^{\infty} d\omega\, f(\omega)\, e^{i\omega t}, \tag{Y13}$$

$$f(\omega) = \frac{1}{\sqrt{2\pi}} \int_{-\infty}^{\infty} dt\, F(t)\, e^{i\omega t}. \tag{Y13a}$$

Theorem: The necessary and sufficient condition for a quadratically integrable $F(t)$ to vanish for $t < 0$ is that its Fourier transform $f(\omega)$ be a causal transform. (Y14)

Theorem: The necessary and sufficient condition for $f(\omega)$ to be a causal transform is that $f(\omega)$ satisfy (Y9), i.e.

$\operatorname{Re} f(\omega)$ and $\operatorname{Im} f(\omega)$ are the Hilbert transform of each other. (Y14a)

Without attempting a proof (for which reference may be made to ref. 1), let us indicate that the condition $F(t) = 0$ for $t < 0$ is sufficient condition for $f(\omega)$ to satisfy (Y6). By hypothesis,

$$f(\omega) = \frac{1}{\sqrt{2\pi}} \int_{0}^{\infty} dt\, F(t)\, e^{i\omega t}. \tag{Y15}$$

Now for $t > 0$, $e^{i\omega t}$ becomes $e^{i\omega t}e^{-\kappa t}$ when the analytic continuation is made into the upper half of the complex plane $\Omega = \omega + i\kappa$, $\kappa > 0$. Hence, for $t > 0$, we have by (Y2)–(Y4)

$$\frac{1}{\pi i} P \int_{-\infty}^{\infty} d\omega' \cdot \frac{e^{i\omega' t}}{\omega' - \omega} = e^{i\omega t}.$$

On putting this expression for $e^{i\omega t}$ into (Y15), we have

$$f(\omega) = \frac{1}{\pi i} \int_{0}^{\infty} dt\, F(t)\, P \int_{-\infty}^{\infty} d\omega' \frac{e^{i\omega' t}}{\omega' - \omega},$$

and on assuming $f(\omega)$, and hence $F(t)$, to be such (uniform convergence) as to permit the interchange of the order of integration, we have, on using (Y15),

$$f(\omega) = \frac{1}{\pi i} P \int_{-\infty}^{\infty} \frac{f(\omega')\, d\omega'}{\omega' - \omega}, \tag{Y16}$$

which is (Y6), and from which the pair of dispersion relations follows.

### v) "Analytic Poisson formula", causal factor

Let $f(\Omega)$ be any function which is analytic and has bounded $|f(\Omega)|$ in the upper half of the complex $\Omega = \omega + i\kappa$ plane. Then for $\kappa > 0$, it can be shown that $f(\Omega)$ is expressible by the "Analytic Poisson formula" (ref. 3),

$$f(\Omega) = \frac{1}{\pi i}\int_{-\infty}^{\infty} \frac{1 + \omega'\Omega}{1 + \omega'^2} \operatorname{Re} f(\omega') \frac{d\omega'}{\omega' - \Omega} + if_0, \tag{Y17}$$

where $f_0$ is some real constant.

On taking the limit $\kappa \to 0^+$ (i.e., from above), this formula becomes

$$\begin{aligned} f(\omega) &= \operatorname{Re} f(\omega) + i \operatorname{Im} f(\omega) \\ &= \frac{1}{\pi i} \lim_{\kappa \to 0^+} \int_{-\infty}^{\infty} \frac{1 + \omega'(\omega + i\kappa)}{1 + \omega'^2} \operatorname{Re} f(\omega') \frac{d\omega'}{\omega' - (\omega + i\kappa)} + if_0. \end{aligned} \tag{Y17a}$$

Similarly, by considering the function $g(\Omega) = if(\Omega)$, one obtains

$$f(\omega) = \frac{1}{\pi} \lim_{\kappa \to 0^+} \int_{-\infty}^{\infty} \frac{i + \omega'(\omega + i\kappa)}{1 + \omega'^2} \operatorname{Im} f(\omega') \frac{d\omega'}{\omega' - (\omega + i\kappa)} + f_0', \tag{Y17b}$$

$f_0'$ being a real constant.

A function $f(\omega)$ such that $f(\omega)\,\phi(\omega)$, where $\phi(\omega)$ is any causal transform, is a causal transform, is called a causal factor. A causal factor satisfies (Y17).

### vi) Dispersion relations for a function $f(\omega)$ that may diverge as $\omega \to \infty$, and is not itself a causal transform

If $f(\omega)$ satisfies the following conditions ($\omega$ real)

i) $f(\omega)$ is integrable over any finite interval in $\omega$,

ii) $\dfrac{f(\omega)}{\omega^{j-1}}$ is bounded as $\omega \to \infty$ for some positive integer $j$,

iii) $\dfrac{f(\omega)}{(\omega + \mu)^j}$, where $\mu = \beta + i\gamma$, $\gamma > 0$, is a causal transform,

iv) $f(\omega)$ is $j$-times differentiable at some point $\omega = \omega_0$,

it is possible to construct a function $B(\omega)$ from $f(\omega)$ such that $B(\omega)$ is a causal transform, i.e., $B(\omega)$ satisfies the relations (Y7) and (Y8).

Define $B(\omega)$ by

$$B(\omega) = f(\omega) - \sum_{p=0}^{j-1} \frac{1}{p!} \left(\frac{d^p f(\omega)}{d\omega^p}\right)_{\omega = \omega_0} (\omega - \omega_0)^p \tag{Y18}$$

so that $B(\omega_0) = 0$ up to order $(\omega - \omega_0)^j$. The functions $\dfrac{f(\omega)}{(\omega + \mu)^j}$, $\dfrac{(\omega - \omega_0)^p}{(\omega + \mu)^j}$,

$p \leqslant j - 1$, are all causal transforms. $\dfrac{B(\omega)}{(\omega + \mu)^j}$ being a sum of causal transforms, is hence a causal transform. It is bounded at $\omega_0$ and square-integrable. It can be proved that $\dfrac{B(\omega)}{(\omega - \omega_0)^j}$ is a causal transform.

To prove this, define

$$C_m(\omega) = \frac{B(\omega)}{(\omega - \omega_0)^{m-1}(\omega + \mu)^{j-m}}, \quad m = 1, 2, \ldots, j,$$

so that $C_m(\omega_0) = 0$. Each $C_m(\omega)$, $\dfrac{C_m(\omega)}{\omega - \omega_0}$ is square-integrable. Suppose $\dfrac{C_m(\omega)}{\omega + \mu}$ is a causal transform, i.e., by (Y14a) and (Y9),

$$\frac{C_m(\omega)}{\omega + \mu} = \frac{1}{\pi i} P \int_{-\infty}^{\infty} \frac{C_m(\nu)\, d\nu}{(\nu + \mu)(\mu - \omega)}.$$

By means of the identity

$$\frac{\omega + \mu}{(\nu + \mu)(\nu - \omega)} \equiv \frac{\omega - \omega_0}{(\nu - \omega_0)(\nu - \omega)} + \frac{(\mu + \omega_0)}{(\nu + \mu)(\nu - \omega_0)},$$

one has

$$C_m(\omega) = (\omega - \omega_0) \frac{1}{\pi i} P \int_{-\infty}^{\infty} \frac{C_m(\nu)\, d\nu}{(\nu - \omega_0)(\nu - \omega)} + (\mu + \omega_0) \frac{1}{\pi i} P \int_{-\infty}^{\infty} \frac{C_m(\nu)\, d\nu}{(\nu + \mu)(\nu - \omega_0)}.$$

The second term is seen to be $C_m(\omega_0)$ which vanishes. Hence

$$\frac{C_m(\omega)}{\omega - \omega_0} = \frac{1}{\pi i} P \int_{-\infty}^{\infty} \frac{C_m(\nu)\, d\nu}{(\nu - \omega_0)(\nu - \omega)}.$$

By (Y14a) and (Y9), it follows that $\dfrac{C_m(\omega)}{\omega - \omega_0}$ is a causal transform. Hence, if $\dfrac{C_m(\omega)}{\omega + \mu}$ is a causal transform, then $\dfrac{C_{m+1}(\omega)}{\omega + \mu} = \dfrac{C_m(\omega)}{\omega - \omega_0}$ is a causal transform. Now $\dfrac{C_1(\omega)}{\omega + \mu} = \dfrac{B(\omega)}{(\omega + \mu)^j}$ and the latter is a causal transform. Hence $\dfrac{C_2(\omega)}{\omega + \mu}$ and $\dfrac{C_1(\omega)}{\omega - \omega_0}$ are causal transforms. By repeating the same procedure we can prove finally that $\dfrac{B(\omega)}{(\omega - \omega_0)^j} = \left(\dfrac{C_j}{\omega - \omega_0}\right)$ is a causal transform, i.e.,

$$B(\omega) = \frac{(\omega - \omega_0)^j}{\pi i} P \int_{-\infty}^{\infty} \frac{B(\nu)\, d\nu}{(\nu - \omega_0)^j(\nu - \omega_0)}, \tag{Y19}$$

where $B(\omega)$ is defined by (Y18). This dispersion relation for $B(\omega)$ gives a relation between the real and the imaginary parts of $f(\omega)$, together with the $j$ (complex) numbers $f(\omega_0)$ and $\left.\dfrac{d^p f(\omega)}{d\omega^p}\right|_{\omega_0}$, $p = 1, \ldots, j - 1$.

For example, if $f(\omega)$ is twice differentiable at $\omega_0$, (Y19) can be written in the explicit form

$$f(\omega) - f(\omega_0) - (\omega - \omega_0)f'(\omega_0) = \frac{(\omega - \omega_0)^2}{\pi i} P \int_{-\infty}^{\infty} \frac{f(\nu) - f(\omega_0) - (\nu - \omega_0)f'(\omega_0)}{(\nu - \omega_0)^2(\nu - \omega)}\, d\nu, \quad \text{(Y20)}$$

and, in particular, for $\omega_0 = 0$,

$$f(\omega) - f(0) = \omega f'(0) = \frac{\omega^2}{\pi i} P \int_{-\infty}^{\infty} \frac{f(\nu) - f(0) - \nu f'(0)}{\nu^2(\nu - \omega)}\, d\nu, \quad \text{(Y20a)}$$

which is a "dispersion relation" for $f(\omega)$, which, while not itself a causal transform, satisfies the conditions i)–iv) stated at the beginning of this subsection.

The function $f(\omega)$ is said to be a causal amplitude if its product with any causal transform is a causal transform. Thus if $f(\omega)$ is a causal amplitude, then $\dfrac{1}{(\omega + \beta + i\gamma)^2} f(\omega)$, $\gamma$ real and $>0$, is a causal transform.

If $f(-\omega) = f^*(\omega)$, then, from (Y20a),

$$\operatorname{Re} f(\omega) = \frac{2\omega^2}{\pi} P \int_0^{\infty} \frac{\operatorname{Im} f(\nu)\, d\nu}{\nu(\nu^2 - \omega^2)} + \operatorname{Re} f(0). \quad \text{(Y20b)}$$

## 2. DISPERSION RELATION FOR INDEX OF REFRACTION

Before showing the relation first obtained by Kronig—a relation which has become the forerunner of a general class of relations called the dispersion relations—it is perhaps worthwhile to make a brief review of the theory of dispersion of light in a gas.

In the classical electron theory, a harmonically bound electron of natural angular frequency $\omega$ under the action of a periodic electric field is described by

$$\ddot{x} + \omega^2 x = \frac{eE_0}{m} e^{i\nu t}. \quad \text{(Y21)}$$

The solution is

$$x = e^{i\omega t} + \left(\frac{eE_0}{m}\right) \frac{e^{i\nu t}}{\omega^2 - \nu^2}. \quad \text{(Y22)}$$

If there are $f$ independent charged particles in each atom, the induced electric dipole moment per atom is

$$\mathfrak{M}(t) = E_0 \frac{fe^2}{m} \cdot \frac{e^{i\nu t}}{\omega^2 - \nu^2}. \quad \text{(Y23)}$$

If there are $N$ atoms per unit volume, the electric polarization $P$ is $N\mathfrak{M}$, and from the relation $D = E + 4\pi P = KE$ between the electric displacement $D$, the dielectric constant $K$, and the relation $K = \mathfrak{N}^2$ between $K$ and the index of refraction $\mathfrak{N}$ in the electromagnetic theory of light, one obtains

$$K(\nu) = \mathfrak{N}^2(\nu) = 1 + \frac{4\pi N f e^2}{m(\omega^2 - \nu^2)}. \tag{Y24}$$

If in (Y21) a term representing the radiation damping is added, i.e.,

$$\ddot{x} + \omega^2 x = \frac{eE_0}{m} e^{i\nu t} + \frac{2e^2}{3mc^3} \dddot{x}, \tag{Y25}$$

which may be approximated by

$$\ddot{x} + \omega^2 x = \frac{eE_0}{m} e^{i\nu t} - \gamma \dot{x}, \tag{Y26}$$

$$\gamma = \frac{2e^2}{3mc^3} \omega^2,$$

the solution $x$ is

$$\frac{eE_0}{m} \frac{e^{i\nu t}}{\omega^2 - \nu^2 + i\gamma\nu}. \tag{Y27}$$

The $\mathfrak{M}(t)$ and $\mathfrak{N}(\nu)$ corresponding to (Y27) are now complex, i.e.,

$$\mathfrak{N}^2(\nu) \equiv n^2(1 - i\kappa)^2 = 1 + \frac{4\pi N f e^2}{m} \frac{1}{\omega^2 - \nu^2 + i\gamma\nu}. \tag{Y28}$$

The real and the imaginary parts of the complex index of refraction $\mathfrak{N}(\nu)$ are

$$\mathrm{Re}\,(\mathfrak{N}^2) = n^2(1 - \kappa^2) = 1 + \frac{4\pi N f e^2}{m} \frac{\omega^2 - \nu^2}{(\omega^2 - \nu^2)^2 + (\gamma\nu)^2}, \tag{Y29}$$

$$\mathrm{Im}\,(\mathfrak{N}^2) = -2n^2\kappa = -\frac{4\pi N f e^2}{m} \frac{\gamma\nu}{(\omega^2 - \nu^2)^2 + (\gamma\nu)^2}, \tag{Y30}$$

where $n(\nu)$, $\kappa(\nu)$ are functions of $\nu$. The term $(\gamma\nu)^2$ in (Y29) and (Y30) give the "width" of the "line" at $\nu = \omega$, due to the (radiation) damping of the oscillators. Let us consider the integral (remembering $n \simeq 1$)

$$\frac{1}{\pi}\int_0^\infty 2n\kappa\nu\, d\nu = \frac{1}{\pi}\int_0^\infty \frac{4\pi N f e^2}{m} \gamma \frac{\nu^2\, d\nu}{(\omega^2 - \nu^2)^2 + \gamma^2\nu^2} = \frac{2\pi N f e^2}{m}, \tag{Y31}$$

which is independent of the damping $\gamma$. By (Y31), we see that

$$\frac{\omega^2 - \nu^2}{(\omega^2 - \nu^2)^2 + (\gamma\nu)^2}\cdot\frac{1}{\pi}\int_0^\infty 2n(\nu')\kappa(\nu')\nu'\, d\nu' = \frac{1}{2}\cdot\frac{4\pi N f e^2}{m} \frac{(\omega^2 - \nu^2)}{(\omega^2 - \nu^2)^2 + (\gamma\nu)^2}, \tag{Y32}$$

and by (Y30), the left-hand side, and by (Y29) the right-hand side are, respectively (remembering $n \simeq 1$),

$$-\frac{\omega^2 - \nu^2}{(\omega^2 - \nu^2)^2 + (\gamma\nu)^2}\frac{1}{\pi}\int_0^\infty \text{Im}\,\mathfrak{N}^2(\nu')\,\nu'\,d\nu' = \tfrac{1}{2}\,\text{Re}\,(\mathfrak{N}^2 - 1), \quad \text{(Y33)}$$

$$\simeq n - 1. \quad \text{(Y33a)}$$

This establishes a relation between the real part of the complex $\mathfrak{N}^2(\nu)$ and the imaginary part.

The above results are for the classical electron theory for a gas [dilute, so that the effect of the neighboring atoms can be neglected in the equation of motion (Y25)]. In the quantum theory, the oscillator frequency $\omega$ is to be replaced by the transition frequencies $\omega_j$ (from one state, say 0, to another $j$), $f$ is to be replaced by the oscillator strength $f_j$ for that transition $\omega_j$, and the expression for $\mathfrak{N}^2(\nu)$ in (Y28) becomes

$$K(\nu) = n^2(\nu)[1 - i\kappa(\nu)]^2 = 1 + \sum_j \frac{4\pi N f_j e^2}{m}\frac{1}{(\omega_j^2 - \nu^2) + i\gamma_j\nu}. \quad \text{(Y34)}$$

The oscillator strength $f_j$ is related to the Einstein absorption coefficient by

$$f_j = \frac{m\hbar\omega_j}{\pi e^2}B_0^j, \quad \text{(Y35)}$$

and the absorption coefficient $\alpha(\omega)$ is related to $B_0^j$ by

$$B_0^j\hbar\omega_j\rho(\omega_j) = \alpha(\omega_j)\,c\rho(\omega_j), \quad \text{(Y36)}$$

where $c$ is the velocity of light and $\rho(\omega_j)$ the energy density per unit frequency range. From (Y35) and (Y36), one obtains

$$\pi e^2 f_j = mc\alpha(\omega_j). \quad \text{(Y37)}$$

(Y24) can now be written

$$n^2 - 1 = \sum_j \frac{4\pi N f_j e^2}{m(\omega_j^2 - \nu^2)} = \sum_j \frac{4Nc\alpha(\omega_j)}{\omega_j^2 - \nu^2}. \quad \text{(Y38)}$$

In the preceding relations (Y34)–(Y38), the summation is taken over all transitions. In the region of continuous absorption, such as beyond the limit of the $K$ series in the case of X-rays, the relation (Y37) is replaced by

$$\pi e^2 f(\omega)\,d\omega = mc\alpha(\omega)\,d\omega, \quad \text{(Y39)}$$

and (Y38) by

$$n^2(\nu) - 1 = \int_0^\infty \frac{4Nc\alpha(\omega)\,d\omega}{\omega^2 - \nu^2}. \quad \text{(Y40)}$$

On integrating (Y39),

$$\int \alpha(\omega)\,d\omega = \frac{\pi e^2 f}{mc}, \quad f = \int f(\omega)\,d\omega, \quad \text{(Y41)}$$

and comparing this with (Y31), one obtains the following relation between the absorption coefficient $\alpha(\omega)$ and $\kappa(\nu)$#

$$\pi N c \alpha(\omega) = n\omega\kappa(\omega). \tag{Y42}$$

Hence (Y40) can be written in terms of $\kappa(\omega)$

$$n^2(\nu) - 1 = \frac{2}{\pi} P \int_0^\infty \frac{2n\kappa(\omega)\omega \, d\omega}{\omega^2 - \nu^2}, \tag{Y43}$$

or by (Y28), in the form

$$\text{Re}\,(\mathfrak{N}^2 - 1) = \frac{2}{\pi} P \int_0^\infty \frac{\text{Im}\,\mathfrak{N}^2(\omega)\omega \, d\omega}{\omega^2 - \nu^2}, \tag{Y44}$$

which is of the form (Y11), and hence, by (Y13), can also be put in the form

$$\text{Re}\,[\mathfrak{N}^2(\nu) - \mathfrak{N}^2(0)] = \frac{2\nu^2}{\pi} P \int_0^\infty \frac{\text{Im}\,\mathfrak{N}^2(\omega)\, d\omega}{\omega(\omega^2 - \nu)}. \tag{Y45}$$

This is perhaps the first instance of a relation between the real and the imaginary part of a function. Other relations of this type have now been called "dispersion relations", after the dispersion relation (Y44) of Kronig and Kramers. For the relation between the dispersion relation (Y44) and the causality condition, see the remark at the end of the following section.

### 3. DISPERSION RELATION FOR SCATTERING OF LIGHT: CAUSALITY

Had the relation (Y44) been an isolated special situation peculiar to the index of refraction of a gas, it would have been only of some interest but of no great importance. It was found, however, that relations of the form (Y6), or (Y7), are of a general nature, that they owe their origin to some general principles (namely, that of causality) and not to any particular models. This was recognized early by Kronig who showed that a relation of the type (Y7) for the (coherent) scattered amplitude (in the forward direction) results from the principle of causality alone.

Consider a wave packet incident along the direction of the $x$ axis on a scatterer at the origin. Suppose the wave packet arrives at $x = 0$ at $t = 0$. The

---

# The electric field propagates according to $E_0 \exp\{i\nu[t-(\mathfrak{N}/c)x]\}$, and on using the complex $\mathfrak{N}$ in (Y28), $E_0 \exp[-(n\kappa\nu/c)x] \exp\{i\nu[t - (n/c)x]\}$. The absorption coefficient (for energy) is $\dfrac{2n\kappa\nu}{c}$ for the frequency $\nu$. The "total" value integrated over $\nu$ is, from (Y31) $\dfrac{1}{c}\displaystyle\int 2n\kappa\nu \, d\nu = \dfrac{2\pi^2 N f e^2}{mc}$. But this is related to the absorption coefficient $\alpha = \displaystyle\int \alpha(\nu)\, d\nu$ in the expression (Y41). Hence (Y42).

causality requirement means that at a point $x$ (in the forward direction), there should not be any scattered wave until a time $t \geqslant x/c$. The condition on the incident wave can be expressed by writing the amplitude of the wave packet—a superposition of waves of different frequencies

$$A\left(t - \frac{x}{c}\right) = \int_{-\infty}^{\infty} d\omega a(\omega) \exp\left[-i\omega\left(t - \frac{x}{c}\right)\right], \tag{Y46}$$

where at $x = 0$, $A(t) = 0$ for $t < 0$. The situation is seen to be that represented by (Y15). From the physical meaning of $A(t)$, we require $A(t)$ to be quadratically integrable and also real. From the reality of $A(t)$, it follows that

$$a(-\omega) = a^*(\omega) \tag{Y47}$$

which is the condition (Y10). From (Y15)–(Y20), it follows that $a(\omega)$ satisfies (Y20). Hence $a(\omega)$ is, according to the theorems (Y14), (Y14a), a causal transform. Let us choose the particular causal transform

$$a(\omega) = \sum_n \frac{a_n}{\omega - \beta_n + i\gamma}, \tag{Y48}$$

where

$$\beta_n, \gamma \text{ are real and } > 0.$$

With this $a(\omega)$, a continuation of $\omega$ into the upper half of the complex $\Omega = \omega + i\kappa$ plane brings about a factor $e^{\kappa t}$ which vanishes for $t < 0$ and $\kappa \to \infty$ so that the integral (Y46) can be replaced by a contour integral over the real axis $\omega$ and an infinite semi-circle in the upper half of the $\Omega$-plane. But since the poles of $a(\omega)$ are $(\beta_n - i\gamma)$, the contour integral vanishes according to the Cauchy theorem (Y1). Hence,

$$A(t) = 0 \quad \text{for} \quad t < 0. \tag{Y49}$$

This expresses the initial condition that the incident wave arrives at $x = 0$ at $t = 0$.

According to (Y15)–(Y20), this condition (Y49) leads to the dispersion relations for $a(\omega)$.

For the scattered wave in the (forward) direction $\mathbf{x}$, let us write the Fourier transform

$$F\left(t - \frac{x}{c}\right) = \int_{-\infty}^{\infty} d\omega f(\omega) a(\omega) \exp\left[-i\omega\left(t - \frac{x}{c}\right)\right], \tag{Y50}$$

where $f(\omega)$ is the scattering amplitude. The causality condition requires that there be no scattered waves at $x$ for $t < x/c$, i.e.,

$$F\left(t - \frac{x}{c}\right) = 0, \quad \text{for} \quad t - \frac{x}{c} < 0. \tag{Y51}$$

The amplitude $F[t - (x/c)]$, for $0 < t - (x/c)$, must be quadratically integrable. From these and the theorems (Y14), (Y14a), it follows that $f(\omega)a(\omega)$ must be a causal transform, i.e., $g(\omega) = f(\omega)a(\omega)$ satisfies the dispersion relation (Y11) or (Y12a).

$$\operatorname{Re} g(\omega) = \frac{2}{\pi} P \int_0^\infty \frac{\operatorname{Im} g(\omega')\omega' \, d\omega'}{\omega'^2 - \omega^2}. \tag{Y52}$$

$$\operatorname{Re} [g(\omega) - g(0)] = \frac{2\omega^2}{\pi} P \int_0^\infty \frac{\operatorname{Im} g(\omega') \, d\omega'}{\omega'(\omega'^2 - \omega^2)}. \tag{Y52a}$$

Thus from the causality condition (Y51), it follows that $a(\omega)$ and $f(\omega)a(\omega)$ are causal transforms. From this, it can be shown that $f(\omega)$ has the following properties:

(i) $f(\omega)a(\omega)$ is a causal transform for every causal transform $a(\omega)$;

(ii) $f(\omega)$ can be analytically continued into the upper half of the complex $\Omega = \omega + i\kappa$ plane. This analytical property is clear from the analyticity of $f(\Omega)a(\Omega)$, the latter following from $f(\omega)a(\omega)$ being a causal transform. $f(\omega)$, having the property (i), is thus a causal factor [see definition following (Y17b)].

(iii) $|f(\Omega)| \geqslant 1$ in the upper $\Omega$-plane. Proof: From the analyticity of $f(\Omega)a(\Omega)$ and Cauchy's theorem, one has

$$f(\Omega)a(\Omega) = \frac{1}{2\pi i} \oint \frac{f(\Omega')a(\Omega') \, d\Omega'}{\Omega' - \Omega},$$

for any causal transform $a(\Omega)$. If one takes

$$a(\Omega') = \frac{1}{\Omega' - \Omega^*} = \frac{1}{\Omega' - \omega + i\kappa}, \quad a(\Omega) = \frac{1}{2i\kappa}, \tag{Y53}$$

one has

$$f(\Omega)a(\Omega) = \frac{1}{2\pi i} \oint \frac{f(\Omega')}{(\Omega' - \omega)^2 + \kappa^2} \, d\Omega',$$

or

$$f(\Omega) \frac{1}{2i\kappa} = \frac{1}{2\pi i} \int_{-\infty}^\infty \frac{f(\omega') \, d\omega'}{(\omega' - \omega)^2 + \kappa^2}. \tag{Y54}$$

Since $|f(\omega)| \leqslant 1$, one has

$$|f(\Omega)| \leqslant \frac{\kappa}{\pi} \int_{-\infty}^\infty \frac{d\omega'}{(\omega' - \omega)^2 + \kappa^2} = 1. \tag{Y55}$$

(iv) By (iii), (Y55), and the formula (Y19), the $f(\omega)$ satisfies the "generalized dispersion relations" (Y20) and (Y20a).

If the symmetry (Y10) is assumed, i.e.,

$$f(-\omega) = f^*(\omega), \tag{Y56}$$

(Y20), (Y20a) become, respectively,

$$f(\omega) = \frac{2}{\pi i} \lim_{\kappa \to 0+} (\omega + i\kappa) \int_0^\infty \frac{\operatorname{Re} f(\omega')\, d\omega'}{\omega'^2 - (\omega + i\kappa)^2} + if_0, \tag{Y57}$$

$$f(\omega) = \frac{2(1 + \omega^2)}{\pi} \lim_{\kappa \to 0+} \int_0^\infty \frac{\operatorname{Im} f(\omega')\omega'\, d\omega'}{(\omega'^2 + 1)[\omega'^2 - (\omega + i\kappa)^2]} + f_0'. \tag{Y57a}$$

We have indicated briefly how the causality condition (Y51) has led to the dispersion relation (Y52) for $f(\omega)a(\omega)$ and the generalized dispersion relation (Y57) or (Y57a) for $f(\omega)$. The important point to note is that no detailed knowledge about the scattering process enters in the establishment of these relations.#

Let us next consider, instead of a point scatterer and the scattering of light (plane waves) in the forward direction represented by (Y50), the scattering by a sphere of radius $a$, and in a direction making an angle $\vartheta$ with the incident beam ($x$-axis). (Y50) now becomes

$$F\left(t - \frac{r}{c}\right) = \int_{-\infty}^{\infty} d\omega f(\omega, \vartheta) a(\omega) \exp\left[-i\omega\left(t - \frac{r}{c}\right)\right]. \tag{Y60}$$

The shortest path of the ray scattered from the surface of the sphere through an angle $\vartheta$ is shorter by $2a \sin(\vartheta/2)$ than that scattered by the center in the same direction. Hence the causality requirement is that

$$F\left(t - \frac{r}{c}\right) = 0 \quad \text{for} \quad \left[t - \frac{1}{c}\left(r - 2a \sin\frac{\vartheta}{2}\right)\right] < 0. \tag{Y61}$$

We may now rewrite (Y60) in the form

$$F\left(t - \frac{s}{c}\right) = \int_{-\infty}^{\infty} d\omega f(\omega, \vartheta) \exp\left[\frac{2i\omega a \sin(\vartheta/2)}{c}\right] a(\omega) \exp\left[-i\omega\left(t - \frac{s}{c}\right)\right], \tag{Y62}$$

---

# In deriving (Y52), (Y57), we have assumed the symmetry relation (Y47), (Y56). For the antisymmetric relation $f(-x) = -f^*(x)$ instead of (Y10), the corresponding dispersion relation is (Y11a). Had we assumed

$$a(-\omega) = -a^*(\omega), \qquad f(-\omega) = -f^*(\omega), \tag{Y58}$$

(Y52) would have read

$$\operatorname{Re} g(\omega) = \frac{2\omega}{\pi} P \int_0^\infty \frac{\operatorname{Im} g(\omega')\, d\omega'}{\omega'^2 - \omega^2},$$

and (Y57), (Y57a) become, respectively,

$$f(\omega) = \frac{2}{\pi i} \lim_{\kappa \to 0+} (1 + \Omega^2) \int_0^\infty \frac{\omega' \operatorname{Re} f(\omega')\, d\omega'}{(1 + \omega'^2)(\omega'^2 - \Omega^2)}, \tag{Y59}$$

$$f(\omega) = \frac{2}{\pi} \lim_{\kappa \to 0+} \int_0^\infty \frac{\operatorname{Im} f(\omega')\, d\omega'}{\omega'^2 - \Omega^2}, \tag{Y59a}$$

where $\Omega = \omega + i\kappa$.

where $s = r - 2a \sin(\vartheta/2)$, and we impose the causality condition

$$F = 0 \quad \text{for} \quad t - \frac{s}{c} < 0.$$

By the theorems (Y14)–(Y14a), $f(\omega, \vartheta) \exp\{[2i\omega a \sin(\vartheta/2)]/c\}\, a(\omega)$ is a causal transform, and $f(\omega, \vartheta)\, e^{[2i\omega a \sin(\vartheta/2)]/c}$ is a causal amplitude. As in (Y57), this satisfies a generalized dispersion relation.

Let us now come back to the dispersion relation (Y44) for the index of refraction. It was shown by Kramers that in a medium with a complex refractive index $\mathfrak{N}$ *satisfying* the *dispersion relation* (Y44), no signal can travel with a speed greater than $c$, that of light. Historically, Sommerfeld and Brillouin had found much earlier that with a complex $\mathfrak{N}$, no signal can travel faster than the velocity $c$, although the dispersion relation (Y44) was apparently not noted until Kronig in 1926. The connection between the dispersion relation and causality was suggested and proved by Kronig in 1942.

## REFERENCES

The mathematical theorems quoted here are treated in

1. Titchmarsh, E. C., *Theory of Fourier Integrals*. Clarendon Press, Oxford, 1948.

2. Wiener, N., *The Fourier Integral and Certain of Its Applications*. Cambridge University Press, 1933.

3. Toll, J. S., Phys. Rev. **104**, 1760 (1956).

4. Born, M. and Wolf, E., *Principles of Optics*. Pergamon Press, Chapter X, 2, Hilbert Transforms.

The relation (Y40) probably was first given by

5. Kronig, R. de L., Jour. Opt. Soc. Am. **12**, 547 (1926).

The treatment given in Sect. 2 here is essentially that given by

6. Kramers, H. A., Estratto dagli Atti de Congresso Internazionale de Fisici Como (Bologna) (1927).

The relation between causality and the dispersion relations was probably first proved by

7. Kronig, R. de L., Ned. Tijdschr. Natuurk. **9**, 402 (1942).

8. van Kampen, N. G., Phys. Rev. **89**, 1072 (1953); **91**, 1267 (1953). Ref. 8 treats the equivalence of causality and the dispersion relations for electromagnetic field and for non-relativistic particles.

The application of the causality principle to the formulation of the dispersion relations in the quantum theory was first made in the following works

9. Gell-Mann, M., Goldberger, L. M. and Thirring, W. E., Phys. Rev. **95**, 1612 (1954).

10. Goldberger, M. L., Phys. Rev. **97**, 508 (1955); **99**, 979 (1955).

For a selected bibliography and the contributions of Bogoliubov, see

11. Bogoliubov, N. N. and Shirkov, D. V., *Introduction to the Theory of Quantized Fields*. English trans. Interscience Publ. New York (1959). Chap. IX.

An extensive bibliography on the dispersion relations and related subjects has been given by

12. Taylor, J. G., *Lectures on Dispersion Relations in Quantum Field Theory and Related Topics*. Department of Physics, University of Maryland, New issue (1959).

section

# Z Dispersion Relations: Scattering by a Potential

Since the scattering of a particle by a potential is completely described by the phase shifts, or the $S$ matrix, one expects that the dispersion relations, which result from the causality conditions, will furnish relations for the phase shifts or the $S$ matrix. We shall give an elementary account of the theory of dispersion relations for the scattering in ordinary quantum mechanics (i.e., non-field theory) by means of examples.

Before proceeding to the study of dispersion relations, let us make a few remarks of a general nature.

i) Notations for spherical Bessel functions. The Hankel functions $H^{(1)}_{l+\frac{1}{2}}(x)$, $h^{(1)}_l(x)$ have been given in (A11). In the following, we shall for brevity write $h_l(x)$ for $h^{(1)}_l(x)$. From (A11), it is seen that

$$ih_l(x) = (-1)^l j_{-l}(x) + ij_l(x), \tag{Z1}$$

$$j_l(x) = \tfrac{1}{2}[h_l(x) + h_l^*(x)]. \tag{Z2}$$

$h_l(x)$ has the further properties

$$h_l(-x) = (-1)^l h_l^*(x), \tag{Z3}$$

and

$$xh_l(x) = (-i)^{l+1}\, e^{ix} q_l(x), \tag{Z4}$$

where

$$q_l(x) = \sum_{n=0}^{l} \frac{(l+n)!}{n!(l-n)!}\left(\frac{i}{2x}\right)^n. \tag{Z5}$$

**ii)** The $S$ matrix is introduced in (A34). Let us recall its meaning in this connection. Consider a superposition of plane waves (including light),

$$\psi(\mathbf{r}, t) = \int_{-\infty}^{\infty} d\omega \varphi(\omega) \exp [i(kz - \omega t)].$$

This can be analyzed into spherical waves according to (A8),

$$\psi(\mathbf{r}, t) = \int_{-\infty}^{\infty} d\omega \varphi(\omega) \sum_l (2l + 1)\, i^l j_l(kr)\, P_l (\cos \vartheta)\, e^{-i\omega t},$$

and by (Z2) and (Z3),

$$\psi(\mathbf{r}, t) = \int_{-\infty}^{\infty} d\omega \varphi(\omega) \sum_l \left(\frac{2l + 1}{2}\right) i^l [h_l(kr) + (-1)^l h_l(-kr)]\, P_l (\cos \vartheta)\, e^{-i\omega t}. \tag{Z6}$$

The $h_l(-kr)$, $h_l(kr)$ correspond, respectively, to the ingoing and outgoing spherical waves. If there is a scattering potential at $r = 0$, the outgoing spherical wave will be represented by

$$\int_{-\infty}^{\infty} d\omega \varphi(\omega) \sum_l \left(\frac{2l + 1}{2}\right) i^l S_l(k)\, h_l(kr)\, P_l (\cos \vartheta)\, e^{-i\omega t}. \tag{Z6a}$$

The *scattered* wave is given by

$$\begin{aligned}\psi_{sc}(\mathbf{r}, t) &= \int_{-\infty}^{\infty} d\omega \varphi(\omega) \sum_{l=0} \left(\frac{2l + 1}{2}\right) i^l [S_l(k) - 1]\, h_l(kr)\, P_l (\cos \vartheta)\, e^{-i\omega t} \\ &\overset{(X5)}{=} \int_{-\infty}^{\infty} d\omega \varphi(\omega) \frac{\exp [i(kr - \omega t)]}{2ikr} \sum_l (2l + 1)[S_l(k) - 1] q_l(kr) \\ &\qquad \times P_l (\cos \vartheta). \end{aligned} \tag{Z7a}$$

In the expressions above $k$ and $\omega$ are related by the relations: $\omega = kc$ for light, $\omega = k^2\hbar/m$ for non-relativistic particles, and $\hbar\omega = c\sqrt{\hbar^2 k^2 + m^2 c^2}$ for relativistic particles.

**iii)** Causality condition and dispersion relations

The ingoing part of the wave in (Z6) can be written, by the use of (Z4), for each individual $l$, and in the forward direction,

$$\psi_{\text{in}}\left(t + \frac{1}{c} r\right) = \int_{-\infty}^{\infty} d\omega \varphi(\omega)(-1)^l \frac{1}{ikr} q_l(-kr) \exp \left[-i\omega\left(t + \frac{1}{c} r\right)\right]. \tag{Z8}$$

The causality condition requires that

$$\psi_{\text{in}}\left(t + \frac{1}{c} r\right) = 0 \quad \text{for} \quad t + \frac{1}{c} r < 0. \tag{Z9}$$

It is assumed, on physical ground, that $\psi_{\text{in}}$ is square-integrable in $t$. By the theorems (Y14)–(Y14a), the function

$$\varphi(\omega)\frac{1}{kr}q_l(-kr) \tag{Z10}$$

must be a causal transform. Similarly, the outgoing wave $\psi_{\text{out}}$

$$\begin{aligned}\psi_{\text{out}}\left(t-\frac{1}{c}r\right) &= \int_{-\infty}^{\infty} d\omega\varphi(\omega)S_l(\omega)\frac{1}{ikr}q_l(kr)\exp\left[-i\omega\left(t-\frac{1}{c}r\right)\right],\\ &= 0 \quad \text{for} \quad t-\frac{1}{c}r<0,\end{aligned} \tag{Z11}$$

and the assumption of square-integrability of $\psi_{\text{out}}$ in time and the theorems (Y14)–(Y14a) require that

$$\varphi(\omega)S_l(\omega)\frac{1}{kr}q_l(kr) \tag{Z12}$$

be a causal transform. From these, one obtains the dispersion relations for $S_l(\omega)$.

Since $S_l(\omega)$ is related to the phase shifts by (A35),

$$S_l(k) = \exp\left[2i\delta_l(k)\right], \tag{Z13}$$

one obtains the dispersion relations for $\delta_l(k)$ [or $\delta_l(\omega)$] from those for $S_l(k)$ [or $S_l(\omega)$].

## I. SCATTERING OF A SCALAR PARTICLE OF ZERO REST MASS

As a simple example illustrating the dispersion relations for the scattering matrix, let us take the scattering of a scalar wave of zero rest mass so that the energy momentum relation and wave equation are

$$\frac{\hbar\omega}{c} = \hbar k, \tag{Z14}$$

and

$$\left(\nabla^2 - \frac{1}{c^2}\frac{\partial^2}{\partial t^2}\right)\psi = 0. \tag{Z15}$$

An incident plane wave packet is analyzed into ingoing and outgoing spherical waves as in (Z6), and the causality condition for the ingoing wave is given by (Z8) and (Z9). The scattered wave is given by (Z7a), and in the forward direction,

$$\begin{aligned}\psi_{\text{sc}}\left(t-\frac{r}{c}\right) = \int_{-\infty}^{\infty} d\omega \exp\left[-i\omega\left(t-\frac{r}{c}\right)\right]\varphi(\omega)\\ \times\left\{\frac{1}{2ikr}\sum_{l=0}^{\infty}(2l+1)[S_l(\omega)-1]q_l(kr)\right\}.\end{aligned} \tag{Z16}$$

The causality condition is

$$\psi_{sc}\left(t - \frac{r}{c}\right) = 0 \quad \text{for} \quad t - \frac{r}{c} < 0. \tag{Z17}$$

This, together with the square-integrability (assumed on physical grounds) of $\psi_{sc}[t - (r/c)]$ in time requires that $\varphi(\omega)\{\ldots\}$ in (Z16) be a causal transform. Let us investigate the behavior of the series $\sum_{l=0}^{\infty}$ in (Z16). We shall first consider the case where there are *no bound states* in the field. From the behavior of the phase shifts $\delta_l(\omega)$ (which are small for large $l$), it can be shown that, for $l \gg (ka)$, $(ka)^2$, 1, where $a$ is the range of the scattering potential,

$$|S_l(\omega) - 1| \lesssim l\left[\frac{l!}{(2l+1)!}\right]^2 (2ka)^{2l+1} \cong \frac{1}{2e}\left(\frac{eka}{2l}\right)^{2l+1}. \tag{Z18}$$

The behavior of $q_l(kr)$ is given by (Z5). It is seen that the series $\sum_l$ in (Z16) is absolutely convergent. Thus, on using (Z5), one has, for the curly bracket expression in (Z16),

$$\begin{aligned} F(\omega, r) &= \frac{1}{2ikr}\sum_{l=0}^{\infty}(2l+1)[S_l(\omega) - 1]\sum_{n=0}^{l}\frac{(l+n)!}{n!(l-n)!}\left(\frac{i}{2kr}\right)^n, \\ rF(\omega, r) &= \sum_{n=0}^{\infty}\left(\frac{ik}{2r}\right)^n \frac{1}{n!} f_n(\omega), \end{aligned} \tag{Z19}$$

where

$$f_n(\omega) = \frac{1}{ik^{2n+1}}\sum_{l=n}^{\infty} a_{nl}[S_l(\omega) - 1], \tag{Z20}$$

and

$$a_{nl} = \begin{cases} \frac{1}{2}(2l+1)\dfrac{(l+n)!}{(l-n)!}, & l \geqslant n, \\ 0, & l < n. \end{cases} \tag{Z21}$$

Next, to investigate the behavior of $F(\omega, r)$ for large $\omega$, one assumes $|S_l(\omega)|$ to be bounded (see below) and replaces the summation over $l$ in (Z20) by an integration to the limit $l \simeq ka$, as contributions from $l > ka$ are unimportant [see (A27)]. It can be seen that $f_n(\omega)$ behaves as $k$ as $k \to \infty$.

For small $\omega$ (i.e., $k$), and $l$, $n \gg ka$, $(ka)^2$, 1, the behavior of $f_n(\omega)$ is obtained from (Z18) and (Z21), by using Stirling's formula for the factorials,

$$\begin{aligned} |f_n(\omega)| &\simeq \sqrt{\frac{\pi}{2e}}\,(a)^{2n+1}\sum_{\alpha=0}^{\infty}\frac{1}{\alpha!}\left(\frac{e(ka)^2}{2n}\right)^{\alpha} \\ &\simeq \sqrt{\frac{\pi}{2e}}\,a^{2n+1}\exp\frac{e(ka)^2}{2n} \simeq \sqrt{\frac{\pi}{2e}}\,a^{2n+1} \end{aligned} \tag{Z22}$$

so that $|f_n(\omega)|$ is absolutely convergent.

One can solve (Z20) for $S_l(\omega) - 1$ in terms of the $f_n(\omega)$

$$S_l(\omega) - 1 = i \sum_{n=l}^{\infty} b_{ln} k^{2n+1} f_n(\omega), \tag{Z23}$$

where

$$b_{ln} = \frac{2(-1)^{n-l}}{(n-l)!(n+l+1)!}. \tag{Z24}$$

$a_{nm}$, $b_{ln}$ being reciprocal matrices, satisfy the relation

$$\sum_{n=l}^{\infty} b_{ln} a_{nm} = \delta_{lm}. \tag{Z25}$$

Now, for large $r$ (the distance from the scattering center to the point of observation), only the first term $n = 0$ is important in (Z19), and one obtains from (Z19)–(Z21)

$$rF(\omega, r) \rightarrow f_0(\omega) = \frac{1}{2ik} \sum_{l=0}^{\infty} (2l+1)[S_l(\omega) - 1], \tag{Z26}$$

which is the familiar formula (A36) for the scattered amplitude in the forward direction.

From (Z19), for $r$ large, one has

$$2r[rF(\omega, r) - f_0(\omega)] \rightarrow ikf_1(\omega), \tag{Z27}$$

and

$$-(2r)^2\left[rF(\omega, r) - f_0(\omega) - \frac{ik}{2r} f_1(\omega)\right] \rightarrow k^2 f_2(\omega), \text{ etc.} \tag{Z28}$$

Each $f_n(\omega)$ has the properties (for large $l$, large and small $\omega$) as shown above.

Let us come back to (Z16)–(Z17) concerning the requirement that

$$\varphi(\omega)F(\omega, r) = \varphi(\omega)\frac{1}{2ikr}\sum (2l+1)[S_l(\omega) - 1]\, q_l(kr) \tag{Z29}$$

$$= \varphi(\omega)\frac{1}{r}\sum_{n=0}^{\infty}\left(\frac{ik}{2r}\right)^n \frac{1}{n!} f_n(\omega) \tag{Z29a}$$

be a causal transform. That $\varphi(\omega)rF(\omega, r)$, in the limit of large $r$, must be a causal transform means that $\varphi(\omega)f_0(\omega)$ must be a causal transform. If $\varphi(\omega)$ is chosen to be such a causal transform that $\psi_{sc}\left(t - \frac{r}{c}\right)$ in (Z16) is square-integrable, then $f_0(\omega)$ must be a causal factor. But since $f_n(\omega)$ diverges only as $\omega$ for large $\omega$, it follows that $f_0(\omega)$ is a causal amplitude and hence satisfies the dispersion relation (Y20a, b). From (Z27), since $f_1(\omega)$ is the difference between two causal amplitudes, it follows that $f_1(\omega)$ is one. Similarly for all $f_n(\omega)$.

If the $f_n(\omega)$ are twice differentiable at $\omega = 0$, one has on using the relation (Y20a) for $f_n(\omega)$,

$$f_n(\omega) - f_n(0) - \omega f_n'(0) = \frac{\omega^2}{\pi i} P \int_{-\infty}^{\infty} \frac{f_n(\nu) - f_n(0) - \nu f_n'(0)}{\nu^2(\nu - \omega)} d\nu, \quad \text{(Z30)}$$

and using (Z20) for $f_n(\omega)$, one has, from (Z23), (Z30), $\left(k = \frac{\omega}{c}, k(\nu) = \frac{\nu}{c}\right)$

$$S_l(\omega) - 1 = \sum_{n=l}^{\infty} b_{ln} k^{2n+1} \left[\frac{\omega^2}{\pi i} P \int_{-\infty}^{\infty} \frac{d\nu}{\nu^2(\nu - \omega)} \left\{ \sum_{m=n}^{\infty} a_{nm} [S_m(\nu) - 1] \frac{1}{k^{2n+1}(\nu)} - if_n(0) - i\nu f_n'(0) \right\} + if_n(0) + i\omega f_n'(0) \right], \quad \text{(Z31)}$$

where

$$f_n(0) = \text{Re} f_n(0) = \tfrac{1}{2}(2n+1)! \left[\frac{\text{Im } S_n(\omega)}{k^{2n+1}(\omega)}\right]_{\omega=0},$$

$$f_n'(0) = i \,\text{Im} f_n'(0) = \frac{i}{2}(2n+1)! \left[\frac{d}{d\omega}\left(\frac{1 - \text{Re } S_n(\omega)}{k^{2n+1}(\omega)}\right)\right]_{\omega=0}. \quad \text{(Z32)}$$

by (Z18) and the symmetry relation $f_n(-\omega) = f_n^*(\varepsilon)$.

On taking the imaginary part of (Z31), and using (Y20b), one obtains $\left(k = \frac{\omega}{c}\right)$,

$$\text{Im } S_l(\omega) = \sum_{n=l}^{\infty} b_{ln} k^{2n+1} \left[\frac{2\omega^2}{\pi} P \int_0^{\infty} \frac{d\nu}{\nu(\nu^2 - \omega^2)} \sum_{m=n}^{\infty} \frac{a_{nm}}{k^{2n+1}} \text{Re} \times \text{Re}\{1 - S_m(\nu)\} + f_n(0)\right]. \quad \text{(Z33)}$$

From this relation and (Z20), it follows that if one knows the imaginary part of the scattering amplitude $f(\omega, \vartheta)$ [i.e., the real part of $1 - S_m(\omega)$] and the low energy limit $f_n(0)$, then the real part of $f(\omega, \vartheta)$ is determined.

If we express the $S$ matrix in terms of the phase shifts (real) by (Z13), we have

$$\sin 2\delta_l(\omega) = \sum_{n=l}^{\infty} b_{ln} \omega^{2n+1} \times \left[\frac{4\omega^2}{\pi} P \int_0^{\infty} \frac{d\nu}{\nu(\nu^2 - \omega^2)} \sum_{m=n}^{\infty} \frac{a_{nm}}{\nu^{2n+1}} \sin^2 \delta_m(\nu) + f_n(0)\right]. \quad \text{(Z34)}$$

The meaning of this dispersion relation is as follows. From the low-energy limit $f_n(0)$, one calculates $\sin 2\delta_l$, first neglecting $\sin^2 \delta_n(\nu)$ which is of the second order in the scattering potential. The $f_n(0)$ and $\sin 2\delta_l(\omega)$ are then obtained to the second order by iteration. Or one may start with the phase shifts $\delta_0(\omega)$ for the $s$ waves to the lowest order, and obtain $f_n(0)$, and then $\sin 2\delta_l$ to the same order by (Z34). Again higher order calculations are carried out by iteration. This iteration procedure is valid, however, only if the perturbation expansion (in powers of the scattering potential, or "coupling constant" in the case of fields) is convergent.

As mentioned before [preceding (Z18)], we have assumed that there are no bound states of the particle in the scattering field. As shown in Chapter 5, the analytic property of $S_l(\omega)$ is closely connected with the existence or non-existence of bound states. If there are no bound states for the angular momentum $l$, then $S_l(\omega)$ is bounded for real $\omega$. All the results above (Z18)–(Z34) are valid for that case.

If there is one bound state for $l = 0$ say, at $\omega = \omega_B$, and if $\omega_B$ is of the nature of a simple pole, i.e., $S_l(\omega) = g/(\omega - \omega_B)$ in the neighborhood of $\omega_B$, one can insure all the necessary properties for $\varphi(\omega)\, F(\omega, r)$ in (Z29) to be a causal transform by choosing $\varphi(\omega)$ to contain a factor $\omega - \omega_B$

$$\varphi(\omega) = \frac{\omega - \omega_B}{\omega - i\gamma} \times \text{causal transform, } \operatorname{Re}\gamma > 0. \tag{Z35}$$

The $f_0(\omega)$ however, has a singularity on the real axis at $\omega_B$. The dispersion relation (Z30) can still be used for $f_n(\omega)$, and $S_n(\omega)$, if the integration is carried out along the real axis with a detour along a small semicircle about $\omega_B$ in the upper $\Omega = \omega + i\kappa$ plane. The integral in (Z31) will then consist of the principal value of (Z31), plus the contribution from the semi-circle, namely, $\pi i \times$ residue of $f_0(\omega)$ at $\omega_B$. From (Z21), (Z24), $a_{00} = \frac{1}{2}$, $b_{00} = 2$, one has, from (Z31)

$$S_0(\omega) - 1 = i\frac{k}{k_B}\left(\frac{\omega}{\omega_B}\right)^2 \frac{g}{\omega - \omega_B} + \sum_{n=0}^{\infty} b_{0n} k^{2n+1} \left\{ \frac{\omega^2}{\pi i} P \int_{-\infty}^{\infty} \frac{d\nu}{\nu^2(\nu - \omega)} \right.$$

$$\left. \times \left[ \frac{1}{k^{2n+1}(\nu)} \sum_{l=n}^{\infty} a_{nl}\{S_l(\omega) - 1\} - if_n(0) - i\nu f_n'(0) \right] + if_n(0) + i\omega f_n'(0) \right\}. \tag{Z36}$$

If $\omega = \omega_B$ for $l = 0$ is the only bound state, then for all the other $S_l(\omega)$, $l \neq 0$, (Z31) is valid.

On taking the imaginary part of both sides of (Z36) as in (Z33), it is seen that from the real part of $S_l(\omega)$, the low energy limits $f_n(0)$, *and* the constants of the bound state ($\omega_B$, $g$), the imaginary part of $S_l(\omega)$ is determined.

If there are bound states for many $l$, one can extend the method above by properly choosing the $\varphi(\omega)$ in (Z35) and replacing the term in $g$ in (Z36) by a sum over the contributions from all the bound states.

Entirely analogous calculations for the scattering of a (unquantized) Klein–Gordon field have been carried out. It differs from the case treated above only in having, instead of (Z14), (Z15), the energy-momentum relation

$$k^2 = \left(\frac{\omega}{c}\right)^2 - \left(\frac{mc}{\hbar}\right)^2, \tag{Z37}$$

and

$$\left[\nabla^2 - \frac{1}{c^2}\frac{\partial^2}{\partial t^2} - \left(\frac{mc}{\hbar}\right)^2\right]\psi = 0. \tag{Z38}$$

Here one has to distinguish between the physical region

$$k = \sqrt{\left(\frac{\omega}{c}\right)^2 - \left(\frac{mc}{\hbar}\right)^2} \quad \text{for} \quad \begin{matrix} \hbar\omega > mc^2 \\ \hbar\omega < -mc^2 \end{matrix} \tag{Z39a}$$

and the unphysical region

$$k = i\sqrt{\left|\left(\frac{\omega}{c}\right)^2 - \left(\frac{mc}{\hbar}\right)^2\right|} \quad \text{for} \quad |\omega| < mc^2/\hbar. \tag{Z39b}$$

Except for this, most of the calculations given above are the same.

A similar treatment of a Dirac field has also been given.

## 2. DISPERSION RELATIONS IN NON-RELATIVISTIC THEORY

In the preceding section, we have essentially treated the scattering amplitude of "light" in the forward direction and derived the dispersion relation for the $S$ matrix from the "strict causality" condition (Z17). It is of importance to see what one can obtain in the way of dispersion relations if one deals with the scattering of a particle by a potential in the non-relativistic Schrödinger quantum mechanics in which the definition of causality in terms of wave packets of infinitely sharp edges and maximum velocity $c$ has to be modified. Also it is of importance to see if dispersion relations exist for the scattering amplitude corresponding to a given momentum transfer. Such studies have been made by many authors. The following is an account of the work of Khuri based on the ordinary (i.e., non-field theoretic) quantum mechanics above.

Let the Schrödinger equation of the particle in the potential field $V(r)$ (assumed to be central for simplicity, but extension to a non-central field is possible) be

$$[\nabla^2 + k^2 - \lambda V(r)]\,\psi(\mathbf{x}) = 0, \tag{Z40}$$

which can be transformed into the integral equation (B23)

$$\psi(\mathbf{x}) = e^{i\mathbf{k}\cdot\mathbf{x}} + \lambda \int K(\mathbf{x}, \mathbf{y})\,\psi(\mathbf{y})\,d\mathbf{y}, \tag{Z41}$$

with

$$K(\mathbf{x}, \mathbf{y}) = -\frac{1}{4\pi}\frac{1}{|\mathbf{x}-\mathbf{y}|}\exp(ik|\mathbf{x}-\mathbf{y}|)\,V(y). \tag{Z42}$$

As in Sect. B, (B25), (B45), (B46), for the scattering amplitude $f(k, \tau)$ for the momentum transfer $\boldsymbol{\tau}$

$$\boldsymbol{\tau} = \mathbf{k} - \mathbf{k}', \qquad |\boldsymbol{\tau}| = 2k \sin\frac{\vartheta}{2}, \quad \vartheta = \text{scattering angle}, \tag{Z43}$$

one has

$$f(k, \tau) = f_B(k, \tau) + g(k, \tau), \tag{Z44}$$

$$g(k, \tau) = G_2(k, \tau) + G_3(k, \tau) + \frac{G_4(k, \tau)}{\Delta(\lambda^2, k)} + \frac{G_5(k, \tau)}{\Delta(\lambda^2, k)}, \tag{Z45}$$

and

$$f_B(k, \tau) = -\frac{\lambda}{4\pi} \int e^{i\boldsymbol{\tau}\cdot\mathbf{y}} V(\mathbf{y})\, d\mathbf{y}, \tag{Z46}$$

the $G_j(k, \tau)$ being given in (B43), and $f_B(k, \tau)$ is the scattering amplitude in the Born approximation.

For the purpose of obtaining the dispersion relations for the scattering amplitude $f(k, \tau)$, it is necessary to know the analytic properties of the $G_j(k, \tau)$ and $\Delta(\lambda^2, k)$ as functions of $k$ and $\tau$. As shown in Sect. B (also see Sect. F) the zeroes of $\Delta(\lambda^2, k)$ in (Z45) are connected with the bound states of the system (Z40). It is hence necessary to investigate the behaviors of $G_j(k, \tau)$, $\Delta(\lambda^2, k)$ for complex as well as real values of $k$ ($\mathbf{k} = k_r + i\kappa$). The requisite information has been found by Jost, Pais and Khuri and summarized in Sect. B. We shall refer to that section for these properties of $G_j(k, \tau)$, $\Delta(\lambda^2, k)$, and consequently of $g(k, \tau)$ and $f(k, \tau)$.

According to (ii), (vi) of Sect. B, for $\tau$ real and $\frac{1}{2}\tau \leqslant \alpha$, $g(k, \tau)$ is analytic in $k$; is regular in the upper half of the complex $k$-plane ($\kappa > 0$); is continuous and uniformly bounded for $\kappa \geqslant 0$, except for a finite [on account of the condition (B49b) on $V(r)$] number $N$ of poles at the zeroes of $\Delta_1(\lambda^2, k)$ which for real $\lambda$ all lie on the positive imaginary axis of the $k$-plane; $g(k, \tau) \to 0$ as $|\mathbf{k}| \to \infty$, and $g(k, \tau)$ has branch points at $k = k_r = \pm\frac{1}{2}\tau$. (See ref. 4, however.)

From these properties, it is possible to obtain the dispersion relation for $f(k, \tau)$ in (Z44). It is convenient, however, to effect a transformation from the momentum $k$ to the energy $E = k^2$. Let

$$\phi(E, \tau) \equiv g(k, \tau), \tag{Z47}$$

$$M(E, \tau) \equiv f(k, \tau) = f_B(k, \tau) + g(k, \tau). \tag{Z48}$$

The transformation $E = k^2$ maps the positive imaginary axis of the $k$-plane onto the negative real axis of the complex $E$-plane, and hence the zeroes of $\Delta_1(\lambda^2, k)$ on the positive imaginary axis of the $k$-plane (corresponding to the bound states $E_j = -\kappa_j^2 < 0$) onto the negative real axis of $E$. $\phi(E, \tau)$ is analytic everywhere on the $E$-plane, except for the finite number of poles at $E_j$, $j = 1, \ldots, N$, and a branch cut on the positive real axis (on account of the branch points $k = \pm\frac{1}{2}\tau$ or $E = \frac{1}{4}\tau^2$).

To obtain the value of $\phi(E, \tau)$ for the energy $E$ and the momentum transfer $\tau$, we shall apply the Cauchy theorem (Y1), (Y2) and carry out a contour integration of $\frac{\phi(E', \tau)}{E' - E} dE'$ over a path as shown in Figure 1. The large circle is to enclose all the poles $E_1 \ldots E_n$ and is eventually extended to an infinite radius. These poles contribute the value

$$2\pi i \sum_{j=1}^{N} \frac{R_j(\tau)}{E_j - E}, \tag{Z49}$$

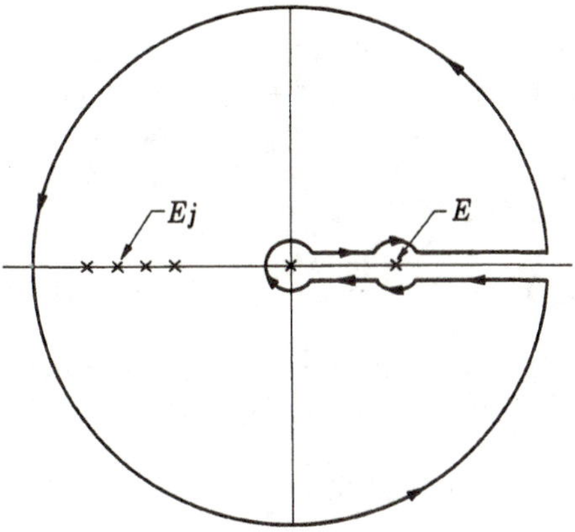


**Figure 1.** Path of integration.

where $R_j(\tau)$ are the residues of $\phi(E, \tau)$ at $E = E_j$. The contribution from the pole $E = E_0 = 0$ is denoted by

$$2\pi i \frac{R_0(\tau)}{E_0 - E}. \tag{Z49a}$$

The $\phi(E, \tau)$ has the property $\phi(E + i\varepsilon, \tau) = \phi^*(E - i\varepsilon, \tau)$, which follows from $g(k, \tau) = g^*(-k, \tau)$, which in turn follows from (B63) and $\Delta(\lambda^2, k) = \Delta^*(\lambda^2, -k)$. The value of $\phi(E, \tau)$ at $E$ (on the positive real axis) is given by Cauchy's theorem#

$$\phi(E, \tau) = \frac{1}{\pi i} P \int_0^\infty \frac{\phi(E', \tau)}{E' - E} dE' + \sum_{j=0}^{N} \frac{R_j(\tau)}{E_j - E} \tag{Z50}$$

# The region $0 < k < \frac{\tau}{2}$ is "unphysical" since according to (Z43), $\tau = 2k \sin(\vartheta/2)$, where $\vartheta$ is the scattering angle. This region is mapped into $0 < E < \frac{1}{4}\tau^2$ on the positive real axis of the $E$-plane. In the following integrals (where the path is eventually made to approach the real axis in the limit), the integration includes this unphysical region $0 < E' < \frac{1}{4}\tau^2$, and the question arises as to the behavior of $\phi(E, \tau)$ in this region. It can be shown, however, that $M(E, \tau)$ is actually convergent in this region if $\tau < \alpha$.

which is a dispersion relation. From (viii), Sect. B, it follows that all the residues $R_j$ are real. On taking the real part of (Z50), one obtains

$$\operatorname{Re}\phi(E,\tau) = \frac{1}{\pi} P \int_0^\infty \frac{\operatorname{Im}\phi(E',\tau)}{E'-E}\,dE' + \sum_{j=0}^{N} \frac{R_j(\tau)}{E_j'-E}, \tag{Z51}$$

and by (Z48), for the scattered amplitude,#

$$\operatorname{Re} M(E,\tau) = \frac{1}{\pi} P \int_0^\infty \frac{\operatorname{Im} M(E',\tau)}{E'-E}\,dE' + \sum_{j=0}^{N} \frac{R_j(\tau)}{E_j-E} + M_B(E,\tau). \tag{Z52}$$

This dispersion relation (Z52), when it is valid, gives the relation between the real and the imaginary part of the scattering amplitudes for a given momentum transfer $\tau = 2k \sin \vartheta/2$. The effect of the bound states is brought out explicitly. If there are no bound states in an attractive field, or if one is dealing with a repulsive field, (Z52) reduces to

$$\operatorname{Re} M(E,\tau) = \frac{1}{\pi} P \int_0^\infty \frac{\operatorname{Im} M(E,\tau)}{E'-E}\,dE' + M_B(E,\tau). \tag{Z53}$$

A particularly simple relation results for the forward scattering $\tau = 0$. Since we know, from the "optical theorem" (A21), (N19)

$$\operatorname{Im} M(E,0) = \frac{k}{4\pi}\sigma(E) \tag{Z54}$$

connecting the total cross section (elastic + inelastic) at the energy $E$ and $\operatorname{Im} M(E, 0)$, we have

$$\operatorname{Re} M(E,0) = \frac{1}{4\pi^2} P \int_0^\infty \frac{k\sigma(E')\,dE'}{E'-E} + M_B(E,0) \tag{Z55}$$

which gives the real part of $M(E, 0)$ from an integral of $\sigma(E)$ over the energy spectrum, which is, according to (A17) and (A19),

$$\operatorname{Re} M(E,0) = \frac{1}{2k}\sum_l (2l+1)\sin 2\delta_l(E), \tag{Z56}$$

$$= \sqrt{\sigma(E,0)}\cos\eta(0). \tag{Z56a}$$

Here $M_B$ is, by (Z46), given by

$$4\pi M_B(E,0) = -\lambda \int V(y)\, y^2\, dy$$

---

# (Z52) is valid for $\frac{1}{2}\tau \leqslant \alpha$. This restricts the usefulness of the dispersion relation to potentials which fall off sufficiently fast with $r$, say, the Gaussian law. This restriction of course arises from the condition for the convergence of the series expansion of the Fredholm method.

For forward scattering, $\tau = 0$ (Z43), and then (Z52) will be valid for $V(r)$ falling off as $1/r^4$ or faster.

which is finite according to the condition (B48b) for the whole method to be valid.

It is to be noted that the dispersion relations (Z53)–(Z55) have been obtained only with the outgoing wave condition, without the explicit use of the "strict causality" condition of the preceding section for light. [See Nussenzveig (ref. 4), however.]

### 3. SCATTERING OF AN ELECTRON BY A HYDROGEN ATOM

As an example of the application of the dispersion relations of the preceding section, let us consider the scattering of an electron by a hydrogen atom. This problem has been treated in detail in Sect. L, Chapter 3 (general theory; Born approximation; variational method). For convenience we shall collect the pertinent formulas (L10), (L24), (L20a), (L20b), (L27), (L28), (L30) here.

For definiteness, let an electron (coordinate $\mathbf{r}_1$) of initial momentum $\mathbf{k}$ and energy $E = k^2$ be incident on an atom in the ground state $\xi_0(\mathbf{r}_2)$, of energy $W_0$ ($= -1$ rydberg). The total energy of the system is denoted by $E_0$ so that for various inelastic scatterings (momentum of scattered electron $= k_m$, atom in state $\xi_m$),

$$E_0 = k^2 + W_0 = E + W_0 = k_m^2 + W_m. \tag{Z57}$$

Now

$$H = H_0' + V', \qquad V'(\mathbf{r}_1, \mathbf{r}_2) = -\frac{2}{r_1} + \frac{2}{|\mathbf{r}_1 - \mathbf{r}_2|},$$

$$H_0' = -\nabla_1^2 - \nabla_2^2 - \frac{2}{r_2}, \tag{L10}$$

$$(H_0' - E_0) \exp(i\mathbf{k}_m \cdot \mathbf{r}_1)\, \xi_m(\mathbf{r}_2) = 0,$$

$$\psi^{(+)}(\mathbf{r}_1, \mathbf{r}_2) = \exp(i\mathbf{k} \cdot \mathbf{r}_1)\, \xi_0(\mathbf{r}_2) + \frac{1}{E_0 - H_0' + i\varepsilon} V'(\mathbf{r}_1, \mathbf{r}_2)\, \psi^{(+)}, \tag{L20a}$$

$$f_m(E, \vartheta) = -\frac{1}{4\pi} \int\!\!\int \exp(-i\mathbf{k}_m \cdot \mathbf{r}_1)\, \xi_m^*(\mathbf{r}_2)\, V'(\mathbf{r}_1, \mathbf{r}_2)\, \psi^{(+)}\, d\mathbf{r}_1\, d\mathbf{r}_2. \tag{L24}$$

$$H = H_0'' + V'', \qquad V''(\mathbf{r}_1, \mathbf{r}_2) = -\frac{2}{r_2} + \frac{2}{|\mathbf{r}_1 - \mathbf{r}_2|}, \tag{L27}$$

$$H_0'' = -\nabla_2^2 - \nabla_1^2 - \frac{2}{r_1},$$

$$(H_0'' - E_0) \exp(i\mathbf{k}_m \cdot \mathbf{r}_2)\, \xi_m(\mathbf{r}_1) = 0,$$

$$g_m(E, \vartheta) = -\frac{1}{4\pi} \int\!\!\int \exp(-i\mathbf{k}_m \cdot \mathbf{r}_2)\, \xi_m^*(\mathbf{r}_1)\, V''(\mathbf{r}_1, \mathbf{r}_2)\, \psi^{(+)}\, d\mathbf{r}_1\, d\mathbf{r}_2, \tag{L30}$$

where the $\psi^{(+)}$ is the same as the $\psi^{(+)}$ in (L24). [See paragraph following (L33), and (M9′)–(M14′)].

The dispersion relations (Z52) applied to the elastically scattered $f_0(E, \vartheta)$ and $g_0\,(E, \vartheta)$ in the forward direction ($\vartheta = 0$) are

$$\begin{aligned} \operatorname{Re} f_0(E, 0) &= f_0(E, 0)_{\text{Born}} + \frac{1}{\pi} P \int_0^\infty \frac{\operatorname{Im} f_0(E'\ 0)\, dE'}{E' - E} + \frac{R_f}{E_{(-)} - E}, \\ \operatorname{Re} g_0(E, 0) &= g_0(E, 0)_{\text{Born}} + \frac{1}{\pi} P \int_0^\infty \frac{\operatorname{Im} g_0(E', 0)\, dE'}{E' - E} + \frac{R_g}{E_{(-)} - E}, \end{aligned} \tag{Z58}$$

where $R_f$, $R_g$ are the residues of $f_0(E, 0)$, $g_0(E, 0)$ respectively at their pole due to the bound state $1s^2\,{}^1S$ of the negative ion $H^-$—the only bound state known experimentally and theoretically. In (X58), $E_{(-)} = -0.05545$ rydberg, i.e., the energy of $1s^2\,{}^1S$ of $H^-$ is $W_0 + E_{(-)} = -1.05545$ rydberg.

To obtain the residues $R_f$, $R_g$, it is more convenient to work with the scattering amplitudes for the singlet and the triplet state scatterings

$$\begin{aligned} {}^1F_0(E, \vartheta) &= f_0(E, \vartheta) + g_0(E, \vartheta), \\ {}^3F_0(E, \vartheta) &= f_0(E, \vartheta) - g_0(E, \vartheta). \end{aligned} \tag{Z59}$$

On referring to (L35), and the relation

$$V''(\mathbf{r}_2, \mathbf{r}_1) = V'(\mathbf{r}_1, \mathbf{r}_2), \tag{Z60}$$

we have

$$\left.\begin{matrix} {}^1F_0(E, \vartheta) \\ {}^3F_0(E, \vartheta) \end{matrix}\right\} = -\frac{1}{4\pi} \iint \exp(-i\mathbf{k}'\cdot\mathbf{r}_1)\, \xi_m^*(\mathbf{r}_2)\, V'(\mathbf{r}_1, \mathbf{r}_2) \left\{\begin{matrix} \Psi_s'^{(+)} \\ \Psi_a'^{(+)} \end{matrix}\right\} d\mathbf{r}_1\, d\mathbf{r}_2 \tag{Z61}$$

where, on using (L20b),

$$\begin{aligned} \left.\begin{matrix} \Psi_s'^{(+)} \\ \Psi_a'^{(+)} \end{matrix}\right\} &= \exp(i\mathbf{k}\cdot\mathbf{r}_1)\, \xi_0^*(\mathbf{r}_2) \pm \exp(i\mathbf{k}\cdot\mathbf{r}_2)\, \xi_0^*(\mathbf{r}_1) \\ &+ \frac{1}{E_0 - H + i\varepsilon} \{V'(\mathbf{r}_1, \mathbf{r}_2) \exp(i\mathbf{k}\cdot\mathbf{r}_1)\, \xi_0^*(\mathbf{r}_2) \pm V'(\mathbf{r}_2, \mathbf{r}_1) \exp(i\mathbf{k}\cdot\mathbf{r}_2)\, \xi_0^*(\mathbf{r}_1)\}. \end{aligned} \tag{Z62}$$

On applying (B22) to the operator $1/(E_0 - H + i\varepsilon)$, i.e., expanding the expression $\{\ldots\}$ in (Z62) in the complete set of eigenfunctions of $E_0 - H$ for the singlet and triplet states respectively, since we assume $1s^2\,{}^1S$ to be the only bound state of the negative hydrogen ion $H^-$, we have no poles for the triplet states amplitudes, i.e.,

$${}^3R = 0, \tag{Z63}$$

and for the singlet states, on remembering $E = k^2$,

$$\begin{aligned} {}^1R = &\iint d\mathbf{r}_1\, d\mathbf{r}_2 \exp(-\sqrt{|E_{(-)}|}\,\mathbf{n}'\cdot\mathbf{r}_1)\, \xi^*(\mathbf{r}_2)\, V'(\mathbf{r}_1, \mathbf{r}_2)\, \Phi_{(-)}(\mathbf{r}_1, \mathbf{r}_2) \\ &\times 2 \iint d\mathbf{r}_1'\, d\mathbf{r}_2'\, \Phi_{(-)}^*(\mathbf{r}_1', \mathbf{r}_2')\, V'(\mathbf{r}_1', \mathbf{r}_2') \exp(-\sqrt{|E_{(-)}|}\,\mathbf{n}\cdot\mathbf{r}_1')\, \xi_0(\mathbf{r}_2'). \end{aligned} \tag{Z64}$$

Here $\Phi_{(-)}(\mathbf{r}_1, \mathbf{r}_2)$ is the eigenfunction of $1s^2\ {}^1S$ of $H^-$ which is symmetric with respect to the interchange of $\mathbf{r}_1$ and $\mathbf{r}_2$

$$[H - (W_0 + E_{(-)})]\, \Phi_{(-)}(\mathbf{r}_1, \mathbf{r}_2) = 0, \tag{Z65}$$

$$\Phi_{(-)}(\mathbf{r}_2, \mathbf{r}_1) = \Phi_{(-)}(\mathbf{r}_1, \mathbf{r}_2). \tag{Z66}$$

$\mathbf{n}$, $\mathbf{n}'$ are unit vectors in the direction of $\mathbf{k}$ and $\mathbf{k}'$ respectively. The factor 2 in (Z64) comes from the fact that the two terms in $\{\ldots\}$ of (Z62) give the same contribution to ${}^1R$ on account of (Z66).

On applying the optical theorem (N19) relating the forward scattering amplitude to the total cross section (elastic + all inelastic)

$$\begin{aligned} \operatorname{Im} {}^1F_0(E, 0) &= \frac{k}{4\pi}\, {}^1\sigma(E), \\ \operatorname{Im} {}^3F_0(E, 0) &= \frac{k}{4\pi}\, {}^3\sigma(E), \end{aligned} \tag{Z67}$$

the dispersion relations (Z58) become

$$\operatorname{Re} {}^1F_0(E, 0) = {}^1F_0(E, 0)_{\text{Born}} + \frac{1}{4\pi^2} P \int_0^\infty \frac{\sqrt{E'}\, {}^1\sigma(E')\, dE'}{E' - E} + \frac{{}^1R}{E_{(-)} - E}, \tag{Z68}$$

$$\operatorname{Re} {}^3F_0(E, 0) = {}^3F_0(E, 0)_{\text{Born}} + \frac{1}{4\pi^2} P \int_0^\infty \frac{\sqrt{E'}\, {}^3\sigma(E')\, dE'}{E' - E}. \tag{Z68a}$$

In terms of the total cross section $\sigma_m(E)$ for an unpolarized beam of incident electrons,

$$\sigma(E) = \tfrac{1}{4}\, {}^1\sigma(E) + \tfrac{3}{4}\, {}^3\sigma(E), \tag{Z69}$$

one has

$$\begin{aligned} \operatorname{Re} (\tfrac{1}{4}\, {}^1F_0 + \tfrac{3}{4}\, {}^3F_0) &= (\tfrac{1}{4}\, {}^1F_0 + \tfrac{3}{4}\, {}^3F_0)_{\text{Born}} \\ &\quad + \frac{1}{4\pi^2} P \int_0^\infty \frac{\sqrt{E'}\, \sigma(E')\, dE'}{E' - E} + \frac{{}^1R}{4(E_{(-)} - E)}, \end{aligned} \tag{Z70}$$

or

$$\operatorname{Re} (f_0 - \tfrac{1}{2} g_0) = (f_0 - \tfrac{1}{2} g_0)_{\text{Born}} + \frac{1}{4\pi^2} P \int_0^\infty \frac{\sqrt{E'}\, \sigma(E')\, dE'}{E' - E} + \frac{{}^1R}{4(E_{(-)} - E)}, \tag{Z70a}$$

where, for the forward direction, from (Z64),

$${}^1R = 2 \left| \iint d\mathbf{r}_1\, d\mathbf{r}_2 \exp(-\sqrt{|E_{(-)}|}\, \mathbf{n}\cdot\mathbf{r}_1)\, \xi_0^*(\mathbf{r}_2)\, V'(\mathbf{r}_1, \mathbf{r}_2)\, \Phi_{(-)}(\mathbf{r}_1, \mathbf{r}_2) \right|^2, \tag{Z71}$$

which can be readily evaluated for any variational wave function $\Phi_{(-)}$ of the Hylleraas type for the $1s^2\ {}^1S$ state. The Born values for $f_0(E, 0)$, $g_0(E, 0)$ are known from (L45). Hence (Z70a) furnishes a relation between the exact $\operatorname{Re} [f_0(E, 0) - \tfrac{1}{2} g_0(E, 0)]$ at a given energy $E$, and the spectrum of the total (elastic + inelastic) cross section $\sigma(E)$ for all $E$.

Theoretically, the $s$-wave contributions to $f_0(E, 0)$, $g_0(E, 0)$ at "zero energies" have been calculated by a number of authors by the effective range method (see Chapter 3, Sect. L). For energies $E$ below 1 rydberg, the phase shifts for both the singlet and triplet state have been calculated by the variational method of Hulthén and Kohn. Unfortunately, no reliable results for the $p$ waves ($l = 1$) are available for these energies.

Experimentally $\sigma(E)$ has been measured for energies from $\sim 1$ to $\sim 10$ eV by two groups of investigators, with rather different results. Bederson has found a rather sharp and pronounced maximum in $\sigma(E)$ at about $E \simeq 3.2$ eV, while Fite *et al.* have found a monotonic decrease of $\sigma(E)$ with $E$ (see Sect. L.6). (Z70a) may now be used as a test of the two sets of observed $\sigma(E)$. Recently, Gerjuoy and Krall (ref. 5) have applied (Z70a) to such a test. The energy $E$ is chosen to be near zero, and the observed $\sigma(E)$ is extrapolated beyond 10 eV to the Born approximation values for high energies. It is found that the results of Fite *et al.* are in much better agreement with this relation. It seems that a more sensitive test would be to apply (Z70a) to $E$ near 3.2 eV where Bederson's $\sigma(E)$ has a high maximum, thereby accentuating the difference in the integral in (Z70a) between the two sets of observed $\sigma(E)$. But this calls for a more reliable $p$ wave contribution to $f_0(E, 0)$, $g_0(E, 0)$ at this energy (see Sect. L.6).

## 4. MANDELSTAM REPRESENTATION FOR POTENTIAL SCATTERING#

In the previous subsection 2, in the case of the nonrelativistic scattering of a particle by a potential, we have discussed the work of Khuri, who has shown that a dispersion relation can be written down in the energy variable for values of the squared momentum transfer $\tau^2$ less than or equal to $4m^2$. Let us denote by $t$ the value of $\tau^2$ in this section. Two questions come naturally to our mind.

1) Is it possible that the scattering amplitude is actually analytic in the momentum transfer variable $t$ in a region larger than that mentioned in the work of Khuri?
2) If this is so, can the analyticity of the scattering amplitude in both the energy and the momentum transfer variable be expressed in a single two-dimensional representation?

In connection with the second question above, in field theory, an intuitive conjecture was made by Mandelstam (ref. 6). From certain considerations, Mandelstam made the conjecture that the scattering amplitude is an analytic

---

# The authors wish to thank Dr. M. K. Sundaresan for his help in preparing this subsection.

function of both the energy and momentum transfer variables, so that by applying Cauchy's theorem, he could get a simple integral representation for the scattering amplitude. Such a representation is now given the name of Mandelstam representation.

In this section we follow rather closely the establishment of Mandelstam representation for potential scattering presented in the work of Blankenbecler, Goldberger, Khuri and Treiman (ref. 6). The answer to the first question has been given by these workers, by an adaptation of a method in field theory due to Lehmann, to show that the scattering amplitude, apart from the first Born approximation, *is actually analytic in $t$ and regular inside an ellipse in the $t$-plane which intersects the real axis at values of $t$ not less than $4m^2$.* In fact these authors have shown further, using the Fredholm method mentioned before, that the scattering amplitude (excluding the Born term) is analytic in the entire $t$-plane if a cut is introduced in the plane from $-\infty$ to $-4m^2$. After having established the analyticity of the scattering amplitude in terms of $t$ and knowing already the dependence on the energy variable, a two dimensional (Mandelstam) representation for the amplitude is established as an answer to the second question above. Thus the conjecture due to Mandelstam, while of intuitive appeal in field theory, is actually established for potential scattering in the work of the above authors. Below we shall give a brief outline of the steps in setting up the Mandelstam representation for the scattering amplitude.

### *i) Analyticity in the t-plane excluding the cut.*

The work of Blankenbecler, Goldberger, Khuri and Treiman is limited to static central potentials. The assumptions that have already been made by Khuri concerning the potential are retained here also, namely,

$$|V(r)| < \frac{M}{r^2},$$
$$\int_{M'}^{\infty} r^2\, dr|V(r)| < \infty, \tag{Z72}$$
$$\int_0^{M''} dr\, r|V(r)| < \infty,$$

where $M$, $M'$ and $M''$ are positive numbers. These restrictions on the potential guarantee the existence of its three dimensional Fourier transform. The Fredholm series solution developed by Khuri will thus be applicable here. A further assumption besides (Z72) has to be made on the potential in the present work, namely that the potential has the representation

$$rV(r) = \int_0^{\infty} d\mu\, e^{-i\mu r}\, \sigma(\mu). \tag{Z73}$$

This simply means that the potential can be represented as a linear combination of Yukawa potentials with different ranges. If the range of $V(r)$ is $m^{-1}$, that is if

$$\int_0^\infty dr\, r|V(r)|e^\alpha < \infty \quad \text{for} \quad 0 \leqslant \alpha \leqslant m,$$

then $\sigma(\mu)$ is zero for $\mu < m$. The possibility of representing $rV(r)$ as in (Z73) (assuming $\sigma(\mu)$ is bounded almost everywhere except perhaps for singularities) is guaranteed if: (i) $rV(r)$ has derivatives of all orders $(0 < r < \infty)$ and (ii) $d^k(rV)/dr^k < (\text{const. } k!)/r^{k+1}$ for $k = 0, 1, 2, \ldots$ $(0 < r < \infty)$. The second and third inequality in (Z72) imply that $\sigma(\mu)/\mu \to 0$ as $\mu \to 0$ and $\sigma(\mu) \to 0$ as $\mu \to \infty$ respectively.

In the following discussion of the analyticity properties of the scattering amplitude we shall find it convenient to separate out the first Born approximation to the scattering amplitude and discuss it separately. The first Born approximation is given by

$$f_B(t) = -\frac{M}{2\pi\hbar^2}\int d\mathbf{r}\, e^{-i\boldsymbol{\tau}\cdot\mathbf{r}}V(r) = -\frac{2M}{\hbar^2}\int_0^\infty d\mu\, \frac{\sigma(\mu)}{\mu^2 + t}, \tag{Z74}$$

where $\boldsymbol{\tau} = \mathbf{k}_f - \mathbf{k}_i$, $t = \tau^2$, and $\mathbf{k}_i$, $\mathbf{k}_f$ are the initial and final wave number vectors respectively. $M$ is the reduced mass of the system. $f_B(t)$ considered as a function of $t$ is obviously analytic in the $t$-plane cut from $-\infty$ to $-m^2$, the effective lower limit on the integral in (Z74) being $m \geqslant 0$. It is possible that $f_B(t)$ may have isolated poles on the negative real axis rather than a branch cut if $\sigma(\mu)$ is a linear combination of $\delta$ functions and its derivatives. As $\sigma(\mu) \to 0$ when $\mu \to \infty$, $f_B(t) \to 0$ as $t \to \infty$.

We shall now indicate the essential steps of the procedure to show that the scattering amplitude, for fixed value of the energy variable, is analytic in the $t$-plane (apart from the cut from $-\infty$ to $-4m^2$ on the real axis). Let us choose the system of units such that $\hbar^2/2M = 1$, so the energy becomes simply $k^2$. This we denote by $s$ below. Then the scattering amplitude $f$ is a function of two variables $s$ and $t$, $f(s, t)$, and we shall try to prove the statement made above for $f'(s, t) \equiv f(s, t) - f_B(t)$. The steps can be divided into two parts. The first part uses the Lehmann procedure to show that $f'(s, t)$ for fixed $s > 0$, is analytic in an ellipse in the $t$-plane which includes the physical region. The second part, a modification of the Lehmann method, is concerned with the establishment of the fact that the domain of analyticity in the $t$-plane is in fact larger than the ellipse that has been mentioned above. This is effected by studying the Fredholm series solution of the integral equation for the scattering amplitude, showing that each term is analytic in $t$ and regular in the $t$-plane with a cut in it extending from $-\infty$ to $-4m^2$, and finally showing also that the series converges uniformly in any closed finite region of the $t$-plane that does not include a point on the cut.

We now present the Lehmann procedure in the non-relativistic theory of scattering of a particle by a potential. The scattering amplitude satisfies the well known integral equation (ref. 7)

$$f(\mathbf{k}_f, \mathbf{k}_i) = f_B(\mathbf{k}_f - \mathbf{k}_i) - \int \frac{dk'}{2(\pi)^3} \frac{\tilde{V}(\mathbf{k}_f - \mathbf{k}')}{k'^2 - k^2 - i\varepsilon} f(\mathbf{k}', \mathbf{k}_i), \tag{Z75}$$

where $\tilde{V}$ is the Fourier transform of the potential, so that $f_B(t) = -\frac{1}{4\pi}\tilde{V}(t)$, [see (B44)]. A formal solution to (Z75) is

$$f(\mathbf{k}_f, \mathbf{k}_i) = f_B(\mathbf{k}_f, \mathbf{k}_i) + \int d\mathbf{q}_1 \int d\mathbf{q}_2\, \tilde{V}(\mathbf{k}_f - \mathbf{q}_1) G(\mathbf{q}_1, \mathbf{q}_2; k^2) \tilde{V}(q_2 - k_i), \tag{Z76}$$

where $G(\mathbf{q}_1, \mathbf{q}_2; k^2)$ is the Fourier transform of the full Green's function for the problem:

$$G(\mathbf{q}_1, \mathbf{q}_2; k^2) = -\frac{1}{4\pi} \int \frac{d\mathbf{r}_1}{2\pi^3} \int \frac{d\mathbf{r}_2}{(2\pi)^3} \exp(-i\mathbf{q}_1 \cdot \mathbf{r}_1)$$
$$\times \left\langle \mathbf{r}_1 \left| \frac{1}{k^2 + i\varepsilon + \nabla^2 - V} \right| \mathbf{r}_2 \right\rangle \exp(i\mathbf{q}_2 \cdot \mathbf{r}_2). \tag{Z77}$$

Writing the potential in the form (Z73), we have from (Z76)

$$f'(\mathbf{k}_f, \mathbf{k}_i) \equiv f(\mathbf{k}_f, \mathbf{k}_i) - f_B(\mathbf{k}_f, \mathbf{k}_i)$$
$$= \int d\mu_1 \int d\mu_2 \sigma(\mu_1)\, \sigma(\mu_2) \int d\mathbf{q}_1 \int d\mathbf{q}_2$$
$$\times \frac{1}{[\mu_1^2 + (\mathbf{k}_f - \mathbf{q}_1)^2]} G(\mathbf{q}_1, \mathbf{q}_2; k^2) \frac{1}{[\mu_2^2 + (\mathbf{q}_2 - \mathbf{k}_i)^2]}. \tag{Z78}$$

A special coordinate system is now chosen to effect some of the integrations in (Z78):

$$\begin{aligned} \mathbf{k}_i &= k(1, 0, 0), \\ \mathbf{k}_f &= k(\cos\vartheta, \sin\vartheta, 0), \\ \mathbf{q}_j &= q_j(\sin\beta_j \cos\alpha_j, \sin\beta_j \sin\alpha_j, \cos\beta_j) \quad j = 1, 2. \end{aligned} \tag{Z79}$$

$\vartheta$ is obviously the scattering angle as it is the angle between $\mathbf{k}_i$ and $\mathbf{k}_j$. Now on introducing two new variables

$$\begin{aligned} \chi &= \alpha_1 - \alpha_2, \\ \lambda_j &= \frac{\mu_j^2 + k^2 + q_j^2}{2kq_j \sin\beta_j} \quad (j = 1, 2), \end{aligned} \tag{Z80}$$

if $m^{-1}$ is the "range" of the potential, it is easy to show that for fixed $k$, there exists a minimum value $\lambda_0$ for $\lambda_j$ when $\mu_j$, $q_j$ and $\beta_j$ vary over their respective ranges. This value is

$$\lambda_0 = \left(1 + \frac{m^2}{k^2}\right)^{1/2}. \tag{Z81}$$

With these new variables (Z78) can be transformed to

$$f'(k, \cos\vartheta) = \int_{\lambda_0}^{\infty} d\lambda_1 \int_{\lambda_0}^{\infty} d\lambda_2 \int_0^{2\pi} d\chi \int_0^{2\pi} d\alpha_1 \times \frac{w(\lambda_1, \lambda_2, k, \chi)}{[\lambda_1 - \cos(\vartheta - \phi_1)][\lambda_2 - \cos(\alpha_1 - \chi)]}, \qquad \text{(Z82)}$$

where $w$ is some weight function which involves $\sigma(\mu)$'s and the $G$. If the integration over $\alpha_1$ is performed in (Z82), we obtain

$$f'(k, \cos\vartheta) = 2\pi \int_{\lambda_0}^{\infty} d\lambda_1 \int_{\lambda_0}^{\infty} d\lambda_2 \int_0^{2\pi} d\chi\, w(\lambda_1, \lambda_2, k, \chi) \times \frac{(\lambda_1/\sqrt{\lambda_1^2 - 1}) + (\lambda_2/\sqrt{\lambda_2^2 - 1})}{[\lambda_1\lambda_2 + (\lambda_1^2 - 1)^{1/2}(\lambda_2^2 - 1)^{1/2} - \cos(\vartheta - \chi)]}. \qquad \text{(Z83)}$$

Finally we introduce $y = [\lambda_1\lambda_2 + (\lambda_1^2 - 1)^{1/2}(\lambda_2^2 - 1)^{1/2}]$ as a new variable, and we get:

$$f'(k, \cos\vartheta) = \int_{y_0}^{\infty} dy \int_0^{2\pi} \frac{\bar{w}(y, k, \chi)}{[y - \cos(\vartheta - \chi)]},$$

where $\bar{w}$ is a new weight function and $y_0$ is the minimum value of $y$ for fixed $k$ and is given by

$$y_0 = 2\lambda_0^2 - 1 = 1 + \frac{2m^2}{k^2}.$$

By an exactly similar procedure we can establish an identical representation for the complex conjugate of $f'$. Thus both Re $f'$ and Im $f'$ can be represented in a similar manner. Lehmann's arguments lead to the conclusion that *both* Re $f'$ *and* Im $f'$ *are for given real* $k^2 \geq 0$, *analytic functions of* $\cos\vartheta$, *regular inside an ellipse in the* $\cos\vartheta$ *plane centered at the origin with semi-major axis* $y_0 = 1 + (2m^2/k^2)$ *and semi-minor axis* $(y_0^2 - 1)^{1/2}$. Since $t = 2k^2(1 - \cos\vartheta)$, in the $t$-plane, we will have analyticity inside an ellipse which intersects the real axis at $t = -4m^2$ and $t = 4m^2 + 4k^2$.

Now we shall proceed to establish that in fact the region of analyticity of the scattering amplitude is much larger than that obtained above. To do this, let us consider the Fredholm solution to the integral equation for the scattering amplitude (ref. 7)

$$f(\mathbf{k}_f, \mathbf{k}_i) = f_B(\mathbf{k}_f, \mathbf{k}_i) + \int \frac{d\mathbf{p}}{(2\pi)^3} \frac{N(\mathbf{k}_f, \mathbf{p}; k)}{D(k)} f_B(\mathbf{p} - \mathbf{k}_i), \qquad \text{(Z84)}$$

where $N$ and $D$ are given by the well known uniformly convergent series expansions and in which $D$ does not depend upon the relative directions of

$\mathbf{k}_i$ and $\mathbf{k}_f$. Thus in our discussions of the analyticity of $f$ in $t$, $D(k)$ need not be considered. $N(\mathbf{k}_f, \mathbf{p}; k)$ is given by

$$N(\mathbf{k}_f, \mathbf{p}; k) = \frac{1}{(2\pi)^3}\frac{\tilde{V}(\mathbf{k}_f, \mathbf{p})}{p^2 - k^2 - i\varepsilon} + \sum_{n=1}^{\infty}(-1)^n \frac{(2\pi)^{-3n}}{n!}\frac{1}{p^2 - k^2 - i\varepsilon}$$
$$\times \int \prod_{i=1}^{n} \frac{d\mathbf{p}_i}{p_i^2 - k^2 - i\varepsilon} \begin{vmatrix} \tilde{V}(\mathbf{k}_f, \mathbf{p}) & \tilde{V}(\mathbf{k}_f, \mathbf{p}_1) & \cdots & \cdots & \tilde{V}(\mathbf{k}_f, \mathbf{p}_n) \\ \tilde{V}(\mathbf{p}_1, \mathbf{p}) & 0 & \tilde{V}(\mathbf{p}_1, \mathbf{p}_2) & \cdots & \tilde{V}(\mathbf{p}_1, \mathbf{p}_n) \\ \cdot & \cdot & \cdot & \cdot & \cdot \\ \tilde{V}(\mathbf{p}_n, \mathbf{p}) & \tilde{V}(\mathbf{p}_n, \mathbf{p}_1) & \cdots & \cdots & 0 \end{vmatrix}, \tag{Z85}$$

where $\tilde{V}(\mathbf{p}_i, \mathbf{p}_j) \equiv \tilde{V}(\mathbf{p}_i - \mathbf{p}_j)$.

It can be seen that this series converges uniformly for physical values of $\mathbf{k}_f$, $\mathbf{p}$ and $k$. If we consider the $n$th term in the above series, and expand the determinant, it will have the form:

$$N_l^{(n)}(\mathbf{k}_f, \mathbf{p}; k) = \int \prod_{j=1}^{l} \frac{d\mathbf{p}_j}{p_j^2 - k^2 - i\varepsilon}\frac{F_l^{(n)}(k)}{p^2 - k^2 - i\varepsilon}\tilde{V}(\mathbf{k}_f, \mathbf{p}_1)\tilde{V}(\mathbf{p}_1, \mathbf{p}_2)\cdots$$
$$\times \tilde{V}(\mathbf{p}_{l-1}, \mathbf{p}_l)\tilde{V}(\mathbf{p}_l, \mathbf{p}), \tag{Z86}$$

where $l \leqslant n$ and $F_l^{(n)}(k)$ is obtained after the integrations over the other variables, namely, $\mathbf{p}_{l+1}, \mathbf{p}_{l+2}, \ldots, \mathbf{p}_n$ have been carried out.

Substituting the values of $\tilde{V}$ from (Z74), we obtain:

$$f' = \frac{1}{(2\pi)^3}\frac{F_l^{(n)}(k)}{D(k)}\int \prod_{i=1}^{l+1} \frac{\sigma(\mu_i)\, d\mu_i\, dp_i}{2[p_i^2 - k^2 - i\varepsilon]}\frac{\sigma(\mu_{l+2})\, d\mu_{l+2}}{2k^2}$$
$$\times \int d\Omega_1 \ldots \int d\Omega_{l+1}[\lambda_1 - \hat{\mathbf{k}}_f\cdot\hat{\mathbf{p}}_1]^{-1}[\lambda_2 - \hat{\mathbf{p}}_1\cdot\hat{\mathbf{p}}_2]^{-1}\ldots[\lambda_{l+1} - \hat{\mathbf{p}}_{l+1}\cdot\hat{\mathbf{k}}_i]^{-1}, \tag{Z87}$$

where $\hat{\mathbf{k}}_f$, $\hat{\mathbf{k}}_i$ and $\hat{\mathbf{p}}_j$ are unit vectors and the $\lambda_j$ are defined by:

$$\lambda_j = \frac{\mu_j^2 + p_{j-1}^2 + p_j^2}{2p_{j-1}p_j}, \quad 1 < j < l + 2,$$
$$\lambda_1 = \frac{\mu_1^2 + k^2 + p_1^2}{2kp_1},$$
$$\lambda_{l+2} = \frac{\mu_{l+2}^2 + p_{l+1}^2 + k^2}{2p_{l+1}k}. \tag{Z88}$$

We can carry out the angular integrations in (Z87) by following a method similar to the one in Lehmann's method. For the basic integral of the form

$$I = \int d\Omega_{\mathbf{p}} \frac{1}{[\tau_1 - \hat{\mathbf{p}}_1\cdot\hat{\mathbf{p}}][\tau_2 - \hat{\mathbf{p}}_2\cdot\hat{\mathbf{p}}]} \quad (\tau_1, \tau_2 \geqslant 1),$$

the result of integration over $d\Omega_{\hat{p}}$ may be written in the form:

$$I = 4\pi \int_{\eta_0}^{\infty} d\eta \frac{1}{(\eta - \hat{\mathbf{p}}_1 \cdot \hat{\mathbf{p}}_2)} \frac{1}{K(\eta)},$$

where

$$\eta_0 = \tau_1\tau_2 + [(\tau_1^2 - 1)(\tau_2^2 - 1)]^{1/2},$$

and

$$K(\eta) = [(\tau_1\tau_2 - \eta)^2 - (\tau_1^2 - 1)(\tau_2^2 - 1)]^{1/2}. \qquad \text{(Z89)}$$

By repeated integrations $f_l'^{(n)}(k, \cos\vartheta)$ can be written in the form:

$$f_l'^{(n)}(k, \cos\vartheta) = \int_{\eta_{0l}^{(n)}} d\eta \frac{\phi_l^{(n)}(\eta, k)}{\eta - \cos\vartheta}, \qquad \text{(Z90)}$$

where for a potential of "range" $m^{-1}$,

$$\eta_{0l}^{(n)} \geqslant 1 + \frac{2m^2}{k^2}. \qquad \text{(Z91)}$$

A representation identical to (Z90) can be established for the complex conjugate of $f_l'^{(n)}$. Thus both the real and imaginary parts of $f_l'^{(n)}$ have a representation of the form (Z90). From this result, we come to the conclusion that for fixed value of $k$, *every term* in the Fredholm expansion of $f'(k, \cos\vartheta)$ *is analytic in* $\cos\vartheta$ *regular in the* $\cos\vartheta$ *plane except for a cut on the positive real axis from* $1 + (2m^2/k^2)$ *to* $\infty$.

Finally, it can be shown that the Fredholm series converges uniformly for any finite region in the $\cos\vartheta$ plane not including the cut. Thus we come to the conclusion that the whole $f'(k, \cos\vartheta)$ is analytic in the $\cos\vartheta$ plane excluding the cut. In the $t$-plane $f'(s, t)$, for fixed $s$, is analytic in the $t$-plane except for the cut from $t = -\infty$ to $-4m^2$.

### *ii) Mandelstam representation*

Now we have all the material we need to write down the Mandelstam representation for the potential scattering amplitude. The dispersion relation which we have so far studied [in terms of the variables $s = k^2$, $t = 2k^2(1 - \cos\vartheta)$] is

$$f(s, t) = f_B(s, t) + \sum_{i=0}^{r} \frac{\Gamma_i(t)}{s + s_i} + \frac{1}{\pi}\int_0^{\infty} ds' \frac{\operatorname{Im} f(s', t)}{s' - s}, \quad (\operatorname{Im} s \neq 0). \qquad \text{(Z92)}$$

The $s_i$ here are the negative energies of the bound states in our system of units and $\Gamma_i(t)$ are polynomials in $t$, of degree $l_i$ where $l_i$ is the angular momentum of the $i$th bound state.

It has been shown that $\operatorname{Im} f(s', t)$, for $s' \geqslant 0$ is analytic in the $t$-plane cut along the negative real axis from $-\infty$ to $-4m^2$. The residues $\Gamma_i$ at the poles $s_i$ are simple polynomials in $t$. Thus the last two terms on the right-hand side of (Z92) can be extended into the cut $t$-plane to define $f'(s, t)$ which will be an analytic function of two complex variables $s$ and $t$. This function will be regular in the product of the two cut planes (except at the poles $s_i$ in the $s$-plane). It can be easily established that the function $f'(s, t)$ obtained by analytic continuation of the right-hand side of (Z92) is identical with the actual $f'(s, t)$.

Finally, in order to obtain an integral representation for $f(s, t)$, we simply have to find a suitable representation for $\operatorname{Im} f(s', t)$ which manifestly exhibits the analyticity of this function as a function of $t$. If we take the simplest assumption that $\operatorname{Im} f(s', t) \to 0$ as $|t| \to \infty$ (although this assumption is not consistent with the unitarity condition if bound states are present), we can write for $\operatorname{Im} f(s', t)$:

$$\operatorname{Im} f(s', t) = (-1)^n \frac{t^{n+1}}{\pi} \int_0^\infty dt' \frac{\rho(s', t')}{t'^{n+1}(t' + t)} + \sum_{j=0}^{n} \frac{t^j}{j!} g_j(s'), \tag{Z93}$$

where the degree $n$ of the polynomial in (Z93) is not specified at the moment. It is clear that $g_j(s')$ is the $j$th derivative with respect to $t$ of $\operatorname{Im} f(s', t)$ evaluated at $t = 0$. The lower limit of integration in (Z93) is formally written as zero although in general it depends on $s'$ and is never less than $4m^2$. The region over which the function $\rho$ is non-vanishing in the $(s, t)$ plane is to be determined from the unitarity condition as will be indicated below. Substituting for $\operatorname{Im} f(s', t)$ from (Z93) into (Z92), we get the (Mandelstam) representation for $f$ in the form:

$$\begin{aligned} f(s, t) &= f_B(t) \\ &+ \sum_{i=1}^{r} \frac{\Gamma_i(t)}{s + s_i} + (-1)^n t^{n+1} \int_0^\infty \frac{ds'}{\pi} \int \frac{dt'}{\pi} \frac{\rho(s', t)}{t'^{(n+1)}(t' + t)} \frac{1}{s' - s - i\varepsilon} \\ &+ \sum_{j=0}^{n} t^j \int_0^\infty ds' \frac{g_j(s')}{s' - s - i\varepsilon}. \end{aligned} \tag{Z94}$$

### *iii) Determination of the function $\rho$ from the unitarity condition*

We consider here the simplest case of no bound states and shall show how the unitarity principle can be used to set up an equation to determine $\rho$. Let us assume that

$$\operatorname{Im} f(s, t) \to 0 \quad \text{as} \quad |t| \to \infty,$$

a condition which is consistent with unitarity when no bound states are present. Here the Mandelstam representation becomes

$$f(s, t) = f_B(t) + \int_0^\infty \frac{ds'}{\pi} \int_0^\infty \frac{dt'}{\pi} \frac{\rho(s', t')}{(t' + t)(s' - s - i\varepsilon)}, \tag{Z95}$$

where

$$f_B(t) = -\int_0^\infty d\mu \frac{\sigma(\mu)}{\mu^2 + t}.$$

The unitarity condition can be written as

$$\operatorname{Im} f(s, t) = \frac{\sqrt{s}}{4\pi} \int d\Omega' f^*[s, (\mathbf{k}_f - \mathbf{k}')^2] f[s, (\mathbf{k}' - \mathbf{k}_i)^2], \tag{Z96}$$

where $t = (\mathbf{k}_f - \mathbf{k}_i)^2$, $k_f^2 = k_i^2 = k'^2 = s$. Equation (Z96) is valid only for the physical region $t \leqslant 4s$. For larger $t$, it can be given a meaning by analytic continuation. If we substitute the form (Z95) for $f(s, t)$ in (Z96), it is easily seen that the integral

$$I = 4s^2 \int d\Omega' \frac{1}{[t_1 + (\mathbf{k}_f - \mathbf{k}')^2][t_2 + (\mathbf{k}' - \mathbf{k}_i)^2]} \tag{Z97}$$

occurs in every term. This integral can be written in the form

$$I = \int d\Omega' \frac{1}{(\tau_1 - \hat{\mathbf{k}}_f \cdot \hat{\mathbf{k}}')(\tau_2 - \hat{\mathbf{k}}' \cdot \hat{\mathbf{k}}_i)}, \tag{Z98}$$

where

$$\tau_1 = 1 + \frac{t_1}{2s}, \qquad \tau_2 = 1 + \frac{t_2}{2s}, \qquad \hat{\mathbf{k}}_i = \frac{\mathbf{k}_i}{k_i} \text{ etc.}$$

This integral is the same as that already encountered just below (Z88), and the result of evaluating it can be written as

$$I = 4\pi \int_{t_0}^{\infty} \frac{dt'}{t' + t} \frac{1}{K(1 + t'/2s)} \tag{Z99}$$

with $t_0 = 2s\{\tau_1\tau_2 - 1 + [(\tau_1^2 - 1)(\tau_1^2 - 1)]^{1/2}\}$, and the function $K$ being defined in (Z89). From (Z95) we have

$$\operatorname{Im} f(s, t) = \frac{1}{\pi} \int dt' \frac{\rho(s, t')}{t' + t}. \tag{Z100}$$

Using (Z99) in (Z96), we find the equation of $\rho(s, t)$ in the form

$$\begin{aligned} \rho(s, t) = &\int d\mu_1 \sigma(\mu_1) \int d\mu_2 \sigma(\mu_2)\, K(s, t; \mu_1^2, \mu_2^2) \\ &- 2P \int_0^{\infty} \frac{ds_1}{\pi} \int_0^{\infty} \frac{dt_1}{\pi} \frac{\rho(s_1, t_1)}{(s_1 - s)} \int d\mu_2 \sigma(\mu_2)\, K(s, t; t_1, \mu_2^2) \\ &+ \int_0^{\infty} \frac{ds_1}{\pi} \int_0^{\infty} \frac{dt_1}{\pi} \int_0^{\infty} \frac{ds_2}{\pi} \int_0^{\infty} \frac{dt_2}{\pi} \frac{\rho(s_1, t_1)\rho(s_2, t_2)}{(s_1 - s + i\varepsilon)(s_2 - s - i\varepsilon)} K(s, t; t_1, t_2), \end{aligned} \tag{Z101}$$

with

$$K(s, t; t_1, t_2) = \frac{\pi}{2} \frac{\vartheta\left\{t - t_1 - t_2 - \dfrac{t_1 t_2}{2s} - \dfrac{(t_1 t_2)^{1/2}}{2s}[16s^2 + 4s(t_1 + t_2) + t_1 t_2]^{1/2}\right\}}{\{s[t - (t_1^{1/2} + t_2^{1/2})^2][t - (t_1^{1/2} - t_2^{1/2})^2] - t t_1 t_2\}^{1/2}}.$$

Equation (Z101) has to be solved to find the function $\rho(s, t)$. It is easily seen that the function $K(s, t; t_1, t_2)$ considered as a function of $s$ and $t$ is non-vanishing in the region bounded by the curve $t_1 t_2/s$ for small $s$ and the line $t = (t_1^{1/2} + t_2^{1/2})^2$ for large $s$. In the case when the "range" of the potential is finite, that is when $m \neq 0$, a well defined procedure can be evolved to construct $\rho$ in a sequence of steps. It turns out that there is a finite region in the $(s, t)$ plane, where $\rho(s, t)$ is given *exactly* by the second Born approximation [the first term on the right-hand side in (Z101)]. Knowing $\rho$ in this region, one can construct it in a larger region by substituting this value of $\rho$ into the second term on the right-hand side of (Z101), etc. Ultimately, one arrives at an expression for $\rho$ which is a polynomial of finite degree in terms of a coupling strength parameter characterizing the number of times the potential enters into each term. Using this value of $\rho$, one arrives at an expression for $f(s, t)$ which is the limit of a sequence of polynomials. This sequence converges if our assumption about $\text{Im} f(s, t) \to 0$, as $|t| \to \infty$, holds. It is worthwhile noting that even if the power series representation for the function $f$ diverges it can be represented by a convergent sequence of polynomials.

This completes our discussion of the Mandelstam representation for potential scattering in the case when no bound states are present. This simple case has been presented above with a view to giving an idea of the procedure involved in the new method for treating potential scattering based simply on causality and unitarity conditions. Generalizations of the above procedure to cases involving exchange potentials, systems having bound states, etc., are available, but for a study of these recourse must be made to the original paper of Blankenbecler *et al.*

## REFERENCES

The present section is based on the following works


1. Wong, D. Y. and Toll, J. S., Annals of Phys. **1**, 91 (1957). Causality and dispersion relation for $S$ matrix for scalar and vector field.

2. Knight, J. M. and Toll, J. S., Annals of Phys. **3**, 49 (1958). Extension of above work to unquantized Klein-Gordon and Dirac fields.

3. Khuri, N. N., Phys. Rev. **107**, 1148 (1957). Gives the dispersion relation for the scattered amplitude $f(k, \tau)$ for the momentum transfer in the scattering of a (Schrödinger) particle by a potential, on the outgoing wave condition without using explicitly the "strict causality" requirement. Fredholm's method of solving the integral equation, as given by Jost and Pais (see Sect. B) is used.

4. Nussenzveig, H. M., Physica **26**, 209 (1960). The problem of finding the dispersion relations for fixed momentum transfer is attacked by the method of van Kampen (ref. 8, Sect. Y), in contrast to the approach of Khuri (ref. 3

above, and refs. 7, 8 below) in which the use of the formal (Fredholm's) solution of the Schrödinger equation implies assumptions equivalent to the "causality condition". The physical assumptions that have to be introduced to give the dispersion relations are discussed in detail. The branch points of $f(k, \tau)$ at $k = \pm\frac{1}{2}\tau$ of Khuri [see Sect. B, (B45); sentence preceding (B50); paragraph preceding (Z47)], do not exist, according to Nussenzveig.

5. Gerjuoy, E. and Krall, N. A., Phys. Rev. **119,** 705; **120,** 143 (1960). Dispersion relation applied to the electron-hydrogen atom scattering.

6. Mandelstam, S., Phys. Rev. **112,** 1344 (1958); **115,** 1741, 1752 (1959); Blankenbecler, R., Goldberger, M. L., Khuri, N. N. and Treiman, S. B., Annals of Phys. **10**, 62 (1960). See also Regge, T., Nuo. Cim. **14,** 951 (1959); Klein, A., J. Math. Phys. **1,** 41 (1960); Bowesck, Nuo. Cim. **14,** 516 (1959).

7. Jost, R. and Pais, A., Phys. Rev. **82,** 840 (1951).

Other papers on dispersion relations for the scattering of a particle by a potential (non-field theoretic) are:

8. Wong, D. Y., Phys. Rev. **107,** 302 (1957). Treats the scattering of a particle by a potential, assuming the convergence of the iteration solution of the Schrödinger integral equation.

9. Khuri, N. N. and Treiman, S. B., Phys. Rev. **109**, 198 (1958). Treats the dispersion relations for the scattering of a Dirac particle by a potential. The method used is an extension of that of ref. 3.

10. Klein, A. and Zemach, C., Annals of Phys. **7**, 440 (1959). Treats the dispersion of the scattering of a non-relativistic particle by a central potential on the basis of the Hermitian character of the Hamiltonian and the properties of the Born series for the scattering amplitude.

# Index

A CATALOG OF SELECTED

# DOVER BOOKS

IN SCIENCE AND MATHEMATICS

# Engineering

DE RE METALLICA, Georgius Agricola. The famous Hoover translation of greatest treatise on technological chemistry, engineering, geology, mining of early modern times (1556). All 289 original woodcuts. 638pp. 6¾ x 11. 0-486-60006-8

FUNDAMENTALS OF ASTRODYNAMICS, Roger Bate et al. Modern approach developed by U.S. Air Force Academy. Designed as a first course. Problems, exercises. Numerous illustrations. 455pp. 5⅜ x 8½. 0-486-60061-0

DYNAMICS OF FLUIDS IN POROUS MEDIA, Jacob Bear. For advanced students of ground water hydrology, soil mechanics and physics, drainage and irrigation engineering and more. 335 illustrations. Exercises, with answers. 784pp. 6⅛ x 9¼. 0-486-65675-6

THEORY OF VISCOELASTICITY (SECOND EDITION), Richard M. Christensen. Complete consistent description of the linear theory of the viscoelastic behavior of materials. Problem-solving techniques discussed. 1982 edition. 29 figures. xiv+364pp. 6⅛ x 9¼. 0-486-42880-X

MECHANICS, J. P. Den Hartog. A classic introductory text or refresher. Hundreds of applications and design problems illuminate fundamentals of trusses, loaded beams and cables, etc. 334 answered problems. 462pp. 5⅜ x 8½. 0-486-60754-2

MECHANICAL VIBRATIONS, J. P. Den Hartog. Classic textbook offers lucid explanations and illustrative models, applying theories of vibrations to a variety of practical industrial engineering problems. Numerous figures. 233 problems, solutions. Appendix. Index. Preface. 436pp. 5⅜ x 8½. 0-486-64785-4

STRENGTH OF MATERIALS, J. P. Den Hartog. Full, clear treatment of basic material (tension, torsion, bending, etc.) plus advanced material on engineering methods, applications. 350 answered problems. 323pp. 5⅜ x 8½. 0-486-60755-0

A HISTORY OF MECHANICS, René Dugas. Monumental study of mechanical principles from antiquity to quantum mechanics. Contributions of ancient Greeks, Galileo, Leonardo, Kepler, Lagrange, many others. 671pp. 5⅜ x 8½. 0-486-65632-2

STABILITY THEORY AND ITS APPLICATIONS TO STRUCTURAL MECHANICS, Clive L. Dym. Self-contained text focuses on Koiter postbuckling analyses, with mathematical notions of stability of motion. Basing minimum energy principles for static stability upon dynamic concepts of stability of motion, it develops asymptotic buckling and postbuckling analyses from potential energy considerations, with applications to columns, plates, and arches. 1974 ed. 208pp. 5⅜ x 8½. 0-486-42541-X

BASIC ELECTRICITY, U.S. Bureau of Naval Personnel. Originally a training course; best nontechnical coverage. Topics include batteries, circuits, conductors, AC and DC, inductance and capacitance, generators, motors, transformers, amplifiers, etc. Many questions with answers. 349 illustrations. 1969 edition. 448pp. 6½ x 9¼. 0-486-20973-3

ROCKETS, Robert Goddard. Two of the most significant publications in the history of rocketry and jet propulsion: "A Method of Reaching Extreme Altitudes" (1919) and "Liquid Propellant Rocket Development" (1936). 128pp. 5⅜ x 8½. 0-486-42537-1

STATISTICAL MECHANICS: PRINCIPLES AND APPLICATIONS, Terrell L. Hill. Standard text covers fundamentals of statistical mechanics, applications to fluctuation theory, imperfect gases, distribution functions, more. 448pp. 5⅜ x 8½. 0-486-65390-0

ENGINEERING AND TECHNOLOGY 1650–1750: ILLUSTRATIONS AND TEXTS FROM ORIGINAL SOURCES, Martin Jensen. Highly readable text with more than 200 contemporary drawings and detailed engravings of engineering projects dealing with surveying, leveling, materials, hand tools, lifting equipment, transport and erection, piling, bailing, water supply, hydraulic engineering, and more. Among the specific projects outlined-transporting a 50-ton stone to the Louvre, erecting an obelisk, building timber locks, and dredging canals. 207pp. 8⅜ x 11¼. 0-486-42232-1

THE VARIATIONAL PRINCIPLES OF MECHANICS, Cornelius Lanczos. Graduate level coverage of calculus of variations, equations of motion, relativistic mechanics, more. First inexpensive paperbound edition of classic treatise. Index. Bibliography. 418pp. 5⅜ x 8½. 0-486-65067-7

PROTECTION OF ELECTRONIC CIRCUITS FROM OVERVOLTAGES, Ronald B. Standler. Five-part treatment presents practical rules and strategies for circuits designed to protect electronic systems from damage by transient overvoltages. 1989 ed. xxiv+434pp. 6⅛ x 9¼. 0-486-42552-5

ROTARY WING AERODYNAMICS, W. Z. Stepniewski. Clear, concise text covers aerodynamic phenomena of the rotor and offers guidelines for helicopter performance evaluation. Originally prepared for NASA. 537 figures. 640pp. 6⅛ x 9¼. 0-486-64647-5

INTRODUCTION TO SPACE DYNAMICS, William Tyrrell Thomson. Comprehensive, classic introduction to space-flight engineering for advanced undergraduate and graduate students. Includes vector algebra, kinematics, transformation of coordinates. Bibliography. Index. 352pp. 5⅜ x 8½. 0-486-65113-4

HISTORY OF STRENGTH OF MATERIALS, Stephen P. Timoshenko. Excellent historical survey of the strength of materials with many references to the theories of elasticity and structure. 245 figures. 452pp. 5⅜ x 8½. 0-486-61187-6

ANALYTICAL FRACTURE MECHANICS, David J. Unger. Self-contained text supplements standard fracture mechanics texts by focusing on analytical methods for determining crack-tip stress and strain fields. 336pp. 6⅛ x 9¼. 0-486-41737-9

STATISTICAL MECHANICS OF ELASTICITY, J. H. Weiner. Advanced, self-contained treatment illustrates general principles and elastic behavior of solids. Part 1, based on classical mechanics, studies thermoelastic behavior of crystalline and polymeric solids. Part 2, based on quantum mechanics, focuses on interatomic force laws, behavior of solids, and thermally activated processes. For students of physics and chemistry and for polymer physicists. 1983 ed. 96 figures. 496pp. 5⅜ x 8½. 0-486-42260-7